ACS SYMPOSIUM SERIES 393

Biogenic Sulfur in the Environment

Eric S. Saltzman, EDITOR
University of Miami

William J. Cooper, EDITOR
Florida International University

Developed from a symposium sponsored
by the Division of Environmental Chemistry
at the 194th Meeting
of the American Chemical Society,
New Orleans, Louisiana
August 30–September 4, 1987

American Chemical Society, Washington, DC 1989

Library of Congress Cataloging-in-Publication Data

Biogenic sulfur in the environment.
  Eric S. Saltzman, editor. William J. Cooper, editor

  Developed from a symposium sponsored by the Division
of Environmental Chemistry at the 194th Meeting of the
American Chemical Society, New Orleans, Louisiana,
August 30–September 4, 1987.

  p.   cm.—(ACS Symposium Series, 0097–6156; 393).

  Includes index.

  ISBN 0–8412–1612–6
  1. Sulfur—Environmental aspects—Congresses.
2. Sulfur cycle—Congresses.

  I. Saltzman, Eric S. 1955– . II. Cooper, William J.
III. American Chemical Society. Division of
Environmental Chemistry. IV. American Chemical
Society. Meeting (194th: 1987: New Orleans, La.).
V. Series

TD196.S95B56   1989
574.5'.222—dc19   89–6566
                                          CIP

ACS Symposium Series

M. Joan Comstock, *Series Editor*

1989 ACS Books Advisory Board

Foreword

The ACS SYMPOSIUM SERIES was founded in 1974 to provide a medium for publishing symposia quickly in book form. The format of the Series parallels that of the continuing ADVANCES IN CHEMISTRY SERIES except that, in order to save time, the papers are not typeset but are reproduced as they are submitted by the authors in camera-ready form. Papers are reviewed under the supervision of the Editors with the assistance of the Series Advisory Board and are selected to maintain the integrity of the symposia; however, verbatim reproductions of previously published papers are not accepted. Both reviews and reports of research are acceptable, because symposia may embrace both types of presentation.

Contents

Preface

THE CHEMISTRY OF SULFUR COMPOUNDS in the environment has taken on a new significance in recent years due to its involvement in the formation of atmospheric aerosols and its impact on acid precipitation, human health, and the radiation balance of the atmosphere. A major focus of research on the sulfur problem is the natural cycling of biologically produced sulfur compounds through the environment. An understanding of the processes involved in this cycle is critical, both to assess the background levels of sulfur upon which anthropogenic emissions are superimposed and to develop mechanistic models for the chemical conversions that biogenic and anthropogenic sulfur undergo after release into the environment.

The content of the symposium on which this book is based was unusual in that it involved an extremely wide range of disciplines, from gas kinetics to biochemistry. The purpose behind this was to help broaden the perspective of investigators beyond that of their own disciplines; to encourage interaction and collaboration on multi-disciplinary aspects of the sulfur cycle; and to promote the transfer of new analytical techniques across disciplinary boundaries. Normally, the research papers reported at the symposium would be published in a diverse array of specialized scientific journals. We felt that publishing the papers in one volume would give scientists who are interested in sulfur cycling ready access to current ideas in fields that are outside their specialties.

Biogenic Sulfur in the Environment is divided into eight sections. The first two sections, which deal with terrestrial and freshwater systems, address the emission of volatile reduced sulfur gases into the atmosphere by soils, plants, and wetlands. This subject has received a lot of attention recently because of the possibility that, at least locally, the emissions of natural sulfur compounds may be substantial even compared to anthropogenic sources of sulfur dioxide. Efforts to quantify fluxes accurately and to inventory these sources on a regional scale have required advances in both analytical techniques and the design of flux experiments. Developing predictive or mechanistic models of such processes is currently in an early stage because it requires an understanding of the diverse factors controlling the natural variability of complex environments. Two chapters address the question of emissions of sulfur gases from higher plants, a source about which little is currently known.

Sulfur cycling in fresh water is also important to understanding the impact of acid precipitation in these environments. The presence of hydrogen sulfide results not only from dissimilatory sulfate reduction, as commonly assumed, but also from putrefaction. Chapter 8 addresses an entirely different question: the potential reactivity of sulfur species toward nucleophilic reactions with halogenated organic compounds of anthropogenic origin. These reactions may be an interesting link between natural sulfur cycling and pollutant transport in subsurface environments.

The third, fourth, and fifth sections investigate the distribution and biological and chemical transformations of reduced sulfur compounds in the oceans. The third section focuses primarily on dimethyl sulfide, which is the predominant form of volatile sulfur in the ocean. Research in the past has concentrated on documenting the distribution of DMS in various oceanic environments. The factors controlling this distribution are not well understood. These chapters examine laboratory and field investigations relating DMS production to productivity and speciation.

The fourth section focuses on marine sediments, stressing the biological transformations of DMS and other organosulfur compounds. The fate of sulfur in sediments is interesting because of its role as a microbial matabolite and because of its ultimate fate as an impurity in fossil fuels.

Hydrogen sulfide formation through dissimilatory sulfur reduction has for years been known as a source of environmental sulfur. This compound has invited recent study because of its possible effect on the redox chemistry of sea water. Both the lifetime and the oceanic concentrations of this reactive and highly toxic compound are the focus of the fifth section.

The sixth, seventh, and eighth sections of this volume deal with the atmospheric cycling of biogenic sulfur compounds. This aspect of the sulfur cycle has received a great deal of attention in recent years because of its obvious relationship to the acid rain problem and the discovery that natural marine sources constitute a major portion of the total global atmospheric sulfur burden. The chapters in these sections focus on three aspects of this cycle: field measurements and techniques used to establish the distributions and fluxes, experimental studies of reaction mechanisms and rates, and numerical simulations of the atmospheric sulfur cycle. Two chapters address the chemical processes involving cloud droplets and aerosols. Such multiphase processes are an important aspect of the atmospheric sulfur cycle, although one which is particularly difficult to study experimentally.

Acknowledgments

Support for the symposium that served as the catalyst for this volume was obtained from the Office of Naval Research and the Division of Environmental Chemistry of the American Chemical Society. This support provided many speakers with travel allowances to present papers at the symposium. The participation of authors, many of whom would not normally attend an ACS meeting, resulted in lively discussions, a more comprehensive coverage of the subject, and new collaborative research efforts after the symposium.

The session chairmen, Patrick L. Brezonik, Richard A. Carrigan, Edward J. Green, John M. C. Plane, and Barrie F. Taylor, gave considerable time during the symposium.

All of the chapters in this volume were reviewed by at least one reviewer. We thank them for their efforts and the timeliness of the reviews.

We appreciate the helpful assistance of the acquisitions editors of the ACS Books Department. Our initial efforts were guided by Susan Robinson, whose good-natured persistence was invaluable. The finishing touches were skillfully handled by Robin Giroux.

We wish to thank Maria Fernandez-Reynardus for her energetic administrative assistance throughout the review process. All of the final word processing of the chapters was accomplished under the watchful guidance of Marva Loi. Without her, the production of this book would not have been possible. She was assisted by Sonia Ling, who provided invaluable support in word processing. We would also like to thank Joseph Prospero, chairman of the Division of Marine and Atmospheric Chemistry at the University of Miami, for generously allowing us the use of divisional facilities to produce *Biogenic Sulfur in the Environment*.

ERIC S. SALTZMAN
University of Miami
Miami, FL 33149

WILLIAM J. COOPER
Florida International University
Miami, FL 33199

December 13, 1988

TERRESTRIAL AND FRESHWATER SYSTEMS: EMISSIONS

Chapter 1

Biogenic Sulfur Emissions

A Review

Viney P. Aneja[1] and William J. Cooper[2]

[1]Department of Marine, Earth, and Atmospheric Sciences,
North Carolina State University, Raleigh, NC 27695
[2]Drinking Water Research Center,
Florida International University, Miami, FL 33199

Biogenic sulfur emission rates are reviewed for important components of the sulfur cycle. A summary of emission estimates is provided for vegetation, coastal and wetland ecosystems, inland soils, and oceanic environments. One area which is briefly reviewed, emissions from plants, may play a significant role in global sulfur cycling and very little work has been reported covering this subject. An important trend is that estimates of biogenic emissions are being lowered for terrestrial and wetlands areas. This, coupled with decreased wetland acreage, may significantly decrease local estimates of biogenic sulfur to acid precipitation.

Sulfur in the atmosphere originates either from natural processes or anthropogenic activity. The natural biogenic sources are thought to constitute a large fraction (estimates as high as 50 % have been reported) of the atmospheric sulfur burden (1-8). As such, these natural sources may have a substantial impact on global sulfur cycling.

Biogenic sulfur, that is sulfur compounds which result from biological processes, are only one component of the natural sulfur cycle. The first measurements of biogenic sulfur fluxes were those of Aneja and co-workers (9,10). Numerous studies have been published in the last decade which add to our understanding of the natural sulfur emissions (11-29). This research has provided experimental data, which is helpful in refining the biogenic emission estimates necessary to refine the global sulfur cycle. This data base, although sparse for detailed quantitative estimates, is helpful for estimating emissions from both terrestrial and oceanic environments.

In the initial attempts at developing global sulfur budgets, biogenic emissions were usually obtained from the amount of sulfur necessary to balance the cycle. This resulted in considerable scatter in the biogenic estimates, from 34 Tg S yr^{-1} (4) to 267 Tg S yr^{-1} (7), where Tg = 10^{12} g. It is possible with the existing data to begin to make estimates of biogenic emissions based on direct measurements. However, additional data are necessary to assess biogenic sulfur emissions independent of other portions of the global sulfur cycle.

0097–6156/89/0393–0002$06.00/0

The objectives of this overview chapter are:
1. To review the extant data base of biogenic sulfur emissions for terrestrial and oceanic environments and to summarize direct estimates of emissions where possible.
2. To integrate the other chapters of this section into the discussion.

Biogenic Sulfur Emissions from Vegetation

It has been recognized for some time that sulfur is essential for plant growth. It is used in amino acids, and many other biochemicals. The biological transformation of sulfur compounds in natural ecosystems is closely coupled to the formation of living biomass and to the subsequent decomposition and remineralization of the biomass. Plants contain an average sulfur content of 0.25 % (dry weight basis) (30). Hence sulfur may be released directly from vegetation or during the process(es) of decomposition of the organic matter. Data on sulfur released by vegetation is scanty and the mechanism(s) of release are not known in any detail. A summary of available data is provided below and a more detailed discussion follows in the next chapter (31).

Sulfur compounds are known to be volatilized from living plant leaves (32), and from decaying leaves (33). It has been estimated that sulfur emission rates from decaying leaves are about 10 to 100 times higher than those from living leaves of the same species (33). Many fungi and bacteria release sulfur compounds (34) during plant decomposition. Some plants are known to emit H_2S (11,29,31,35,36). Emission rates of H_2S from several lawns and from a pine forest on aerobic soils, in France, ranged from 0.006 to 0.25 g S m^{-2} yr^{-1} (35). However, in the Ivory Coast, West Africa, emission rates of H_2S from humid forests ranged between 0.24 and 2.4 g S m^{-2} yr^{-1} (36).

Some plants are also known to emit dimethyl sulfide, DMS, (13,37), carbonyl sulfide, COS, (29), and carbon disulfide, CS_2, (13,37-41), and possibly ethyl mercaptan (40,41). A study conducted in a tropical rain forest which focused on *Stryphnodendron excelsum* is trteated in more detail in a following chapter (41). It is quite possible that additional studies as described in Chapter 5 (41) will lead to the discovery of other terrestrial "hot spots" which may be important in biogenic sulfur cycling.

Emission rates of sulfur from crops including corn, soybeans, oats, alfalfa, and miscellaneous vegetables have been measured (27-29). The flux from crops range from 0.008 to 0.3 g S m^{-2} yr^{-1}. H_2S and DMS are the two primary sulfur species being emitted by crops. Emission rates of sulfur from a variety of plants are summarized in Table I.

It is also possible that plants emit volatile sulfur containing compounds which are not easily analyzed by current gas chromatographic methods. Thus, the use of other analytical methods may reveal compounds as yet unidentified which serve as a source of volatile biogenic sulfur compounds.

Biogenic Sulfur Emissions from Wetlands

The tidal flats of marine environments are areas of extreme complexity and biological activity. They serve as both sources and sinks of a wide variety of compounds and materials. They are in a constant state of mass, energy and momentum flux with the surrounding environment. In these areas sulfur plays a major role in biological processes, principally because of the relatively high concentration of sulfate ion in marine waters.

Sulfate ion is the major electron acceptor for respiration in anoxic marine sediments and may account for 25 % of the total sediment respiration in near shore sediments, 0.3 to 3 g C m^{-2} day^{-1}. In salt marsh sediments, where total

Table I. Biogenic Emissions of Sulfur From Vegetation

Plant	Mean Sample Temperature (°C)	Primary Sulfur Species	Emission Rate g S m^{-2} yr^{-1}	Ref.
Spartina				
alterniflora, N.C.	30	DMS	0.66	[12]
S. alterniflora, N.C.	25	CS_2	0.20	[13]
Stryphnodendron				
excelsum, Costa Rica	NR*			[41]
4 meters from tree trunk		CS_2	1.5	
		H_2S	0.022	
16 meters from tree trunk		CS_2	0.15	
		H_2S	not detected	
Lawn, France	22	H_2S	0.24	[35]
Pine Forest, France	10	H_2S	0.023	[35]
Humid Forest,				
Ivory Coast	25	H_2S	0.88	[36]
Crops				
Oats (with soil), IA	35.4	DMS, H_2S	0.016	[29]
Corn, IA, OH	28.9	DMS, H_2S	0.032	[29]
Soybeans, IA	32.8	DMS, H_2S	0.066	[29]
Alfalfa, WA	22.4	DMS, H_2S	0.056	[29]
Trees				
Deciduous, IA, OH, NC	29.5	DMS, H_2S, COS	0.007	[29]
Coniferous, NC	29.2	COS	0.005	[29]
Crops				
Soybeans, IA	25.5		0.0018	[27]
Carrots, OH	22		0.113	[27]
Onions, OH	22		0.104	[27]
Grass, IA	25.5		0.018	[27]
Crops				
Soybeans,	30	DMS	0.037	[28]
Oats,	30	DMS	0.023	[28]
Orchard Grass,	30	DMS	0.008	[28]
Purple Clover,	30	DMS	0.007	[28]
Corn,	30	DMS	0.273	[28]

*NR = not reported

respiration rates may be 2.5 to 5 g C m^{-2} day^{-1}, sulfate ion respiration may account for up to 90 % of the total (e.g. 19,23). Detailed discussions of the processes and variables affecting H_2S emissions may be found in other papers (e.g. 19,23,26,42-46), and in other chapters of this volume (47-50).

There have been a number of studies of biogenic emissions of sulfur gases other than H_2S reported in the literature. Most of these have been concerned with high productivity sources, such as salt marshes and tidal areas and are summarized in Table II.

There are a number of factors which affect the emission rates of biogenic sulfur from wetlands. In a recent study these have been investigated for wetlands in Florida, USA, (57-59) and are summarized in a chapter in this volume (60). These factors are divided into spatial, seasonal, diel and tidal components. In addition, other variables which affect emissions are temperature, insolation, and soil inundation. When these factors are taken into account in estimating emissions, and emission rates are obtained by integrating over the appropriate cycle, the emission estimates are up to two orders of magnitude lower than earlier estimates. However, using these methods results in large uncertainties in the emission estimates, and considerable additional data are required to better refine and extend emission estimates to other environments.

Biogenic Emissions From Land

Sulfur flux measurements from various locations are summarized in Table III. In three inland soils in France the H_2S flux data ranged from 0.19 to 0.24 g S m^{-2} yr^{-1} (35,61). In a broad and diverse inland study area in the U.S. 27 soils were examined, and total sulfur flux reported from 0.013 to 0.33 g S m^{-2} yr^{-1}. The primary sulfur species was H_2S and the flux ranged from undetectable to 0.16 g S m^{-2} yr^{-1}. However, several other gases, DMS, CS_2 and COS were observed in some locales.

Several recent studies have reported additional data for the emission rates of biogenic sulfur species. Lamb and co-workers (29) measured emissions rates in several regions of the U.S. and observed H_2S, COS and DMS, during the summer of 1985. The total flux of the sulfur species can be summarized for two soils, mollisol and histol, averaging 0.008 and 0.114 g S m^{-2} yr^{-1}, respectively.

Goldan and co-workers (28) measured sulfur fluxes from bare soils at two mid-continent sites, also during the summer of 1985. The principal sulfur species were COS, H_2S, DMS, and CS_2, all of which were strongly correlated with air temperature. The emission rate of the sulfur species ranged from 0.003 to 0.008 g S m^{-2} yr^{-1}.

An important trend is becoming evident as additional data is being published. The trend is that recent terrestrial emission estimates appear to be greater than twenty times lower than reported in earlier studies (15,16,37,63). This trend is discussed in more detail in Chapter 2 (64), as are some of the possible explanations for the differences.

Lower terrestrial and coastal emission estimates combined with the increasing loss in wetlands (65-68), although it may not significantly impact the global sulfur cycle, may be an important consideration in local contributions of natural emissions to acid precipitation.

Biogenic Emissions for Oceanic Environments

This topic is included in order to complete a discussion of biogenic sulfur emissions. The reader is referred to more comprehensive reviews in this book and other sources for more details (e.g. 69-73).

Table II. Biogenic Emissions of Sulfur Compounds
from Coastal Ecosystems

Source	Month of	Emission Rate (g S m^{-2} yr^{-1})						Ref.
	Year	H$_2$S	DMS	DMDS	CH$_3$SH	CS$_2$	COS	
Salt Marsh NY	10/11	0.55	0.15	0.018	0.064	-	-	(9,10)
Swamps and tidal flats, Denmark		0.044	-	-	-	-	-	(17)
Coastal area Denmark		~19	-	-	-	-	-	(51)
Salt marsh N. Carolina	7/8/9	0.5	0.66	-	-	0.2	0.03	(12,13,52)
Salt Marshes N. Carolina	5/7/10	0.033	0.538	0.0005	0.00026	0.035	0.012	(15,16)
Delaware	8	0.096	0.48	0.00053	-	0.07	0.012	
Massachusetts	8	-	0.06	0.006	-	0.028	0.004	
Virginia		-	1.87	0.04	0.22	1.38	0.03	
Salt marsh Virginia	8	9.5	-	-	-	-	-	(53)
Salt marsh Virginia	8/9	0.0013	-	-	-	-	-	(54)
Salt marsh Massachusetts	1 yr.	2.05	2.88	0.42	-	0.16	0.3	(55)
Salt marsh N. Carolina	8	0.5	-	-	-	-	-	(26)
Salt marsh N. Carolina	8	0.33	0.083	0.00064	-	0.0017	0.052	(29)
Salt marsh N.Carolina		0.05	0.1	-	0.0037	0.0032	0.0042	(28)

Table III. Biogenic Emissions of Sulfur Compounds from Inland Soils

Source	Month of	Emission Rate (g S m^{-2} yr^{-1})						Ref.
	Year	H$_2$S	DMS	DMDS	CH$_3$SH	CS$_2$	COS	
Equatorial forest Ivory Coast	1/10 11/12	0.07	-	-	-	-	-	(35)
Lawn, France	4/5/6/11/12	0.044	-	-	-	-	-	(35)
Dry inland soil N. Carolina	8/9/10	<0.01	<0.01	<0.05	<0.05	<0.05	<0.01	(62)
Mollisol, Iowa	7	0.15	0.0032	-	-	0.017	0.017	(15,16)
Histosol, Ohio	7	0.047	0.0032	-	-	0.006	0.012	(15,16)
Inceptisols, Ohio	7	0.04	0.002	0.0007	0.001	0.01	0.003	(15,16)
Mollisol, Iowa	7	0.0003	0.0005	-	-	0.0003	0.0015	(28)
Histisol, Ohio	7	0.0024	0.0003	-	-	0.0007	0.0036	(28)
Mollisol, Indiana	7	0.0002	0.0005	-	-	0.00023	0.0029	(29)
Histisol, Ohio	7	0.009	0.0004	-	-	0.00004	0.015	(29)

Oceans and the marine environment are the major source of biogenic sulfur. The reasons for this are the generally abundant phytoplankton in surface oceans and the areal extent of these waters. A summary of the emission rates of DMS, the major biogenic sulfur species in marine environments, is presented in Table IV. Estimates are included which were determined directly and from model calculations.

The first report of DMS in the ocean appeared in 1972 ([33]). The authors suggested that DMS might be more important than H_2S as a biogenic sulfur source for balancing the global sulfur budgets. Preliminary estimates of DMS sea-to-air flux based on the limited data were made by Liss and Slater ([74]).

DMS is produced in oceanic waters by benthic and to a greater extent by planktonic marine algae ([78]), suggesting that it is ubiquitous in the surface ocean ([20,69-71,75,79,80]). Its distribution has been characterized by "hot spots" with high DMS concentrations superimposed on more or less constant level of approximately 1-3 nM. These high concentrations, hot spots, may be the result of blooms of e.g. *Phaeocystis poucheti*, which are known to produce DMS. The most comprehensive survey relating DMS and its precursor DMSP (dimethyl-propionosulfonate) to marine phytoplankton appears in another section of this book ([81]), with over 120 phytoplankton clones having been surveyed.

Estimates of DMS flux from oceanic environments have resulted from direct measurement of DMS in surface waters ([22]) and from model calculations ([76,77]). At this time it is not possible to determine which of the numbers, if any, are correct, however, there is little doubt that the marine environment is one of the major sources of biogenic sulfur.

There is considerable debate regarding the marine environment and the possibility of it being a major source of H_2S. In a coastal area, integrated over tidal and diel cycles, a rate of H_2S of 0.10 g S m^{-2} yr^{-1}, has been reported ([26]).

Table IV. Biogenic Emissions of DMS from Open Oceans
and the Marine Environment

Environment	Emission	Reference
	(g S m^{-2} yr^{-1})	
Open Ocean*	0.02	([74])
Ocean	0.075	([75])
Ocean	0.106	([22])
Ocean*	0.022	([76])
Gulf of Mexico*	0.037	([77])

*Based on model calculations

Emission Flux Measurement Methods

To complete the overview on sulfur emissions, a brief discussion of methods used to estimate emissions is appropriate. There are primarily two methods that may be used to measure earth-atmosphere flux of gases. In the dynamic chamber method, an open-bottom chamber is placed over a surface of interest such as mud, soil, or water, with or without vegetation, to capture the gases emanating from the surface. A carrier gas is introduced into the chamber and mixed with the natural gases. The carrier gas is usually, but not necessarily free of the species being detected. The effluent gas from the chamber is sampled and analyzed for the compounds of interest and the flux is estimated by mass balance. While this technique is easy to use, there are indications that care should be taken to minimize the changes which the chamber itself may exert upon the emitting surface (82).

The second method is the micrometerological method (vertical gradient). The concentration of the gas of interest is measured at various altitudes above the source along with the wind speed and direction. To determine the vertical concentration profile, samples obtained simultaneously at various elevations on a tower must be analyzed. This requires the ability to determine very precisely small differences in concentrations (at low ambient concentrations) among the vertical samples. Estimates of flux are made by applying turbulent diffusion theory to the concentration profile data. This method, although reasonably simple in concept, is very difficult in practice and requires considerable supporting micrometerological data.

Summary

The goal of developing estimates of biogenic sulfur emissions, to provide data for use in estimating regional and global fluxes of biogenic sulfur to the atmosphere, based upon direct measurements is far from being accomplished. With improved analytical methods and a better understanding of the factors affecting biogenic emissions improved estimates are being obtained. Biogenic emissions remain a major area of interest where information is required on anthropogenic impacts on ecosystems and for refining models of global sulfur cycling.

Future studies of biogenic emissions should be designed to include sufficient data for uncertainty analyses of flux estimates. It is also important to conduct intercomparisons of different sampling and measurement methods, as well as the methods used for estimating emissions, i.e. the dynamic chamber vs the micrometerological methods. Additional data are required to confirm emission estimates for those environments which have been characterized, and to extend the emission estimates to environments which have not been studied. Studies in ecosystems such as tropical rain forests should include surveys designed to identify potential terrestrial "hot spots" of volatile sulfur emissions.

Literature Cited

1. Junge, C. E. Air Chemistry and Radioactivity; Academic Press: New York, 1963; p 382.

2. Robinson, E.; Robinson, R. C. Sources, Abundance, and Fate of Gaseous Atmospheric Pollutants; SRI Report PR-6755; American Petroleum Institute: New York, 1968; p 123.

3. Conway, E. J. Proc. R. Irish Acad. 1943, A48, 119-59.

4. Eriksson, E. J. Geophys. Res. 1963, 68, 4001-8.

5. Kellogg, W. W.; Cadle, R. D.; Allen, E. R.; Lazrus, A. L.; Martell, E. A. Science 1972, 175, 587-96.

6. Friend, J. P. In Chemistry of the Lower Atmosphere; Rasool, S. I., Ed.; Plenum Press: New York, 1973; pp 177-201.

7. Granat, L.; Rodhe, H.; Halberg, R. O. Ecol. Bull., SCOPE Report 7 1976, 22, 89-134.

8. Ryaboshapko, A. G. In The Global Biogeochemical Sulfur Cycle; Ivanov, M. V.; Freney, J. R., Eds.; John Wiley and Sons: New York, 1983; pp 203-78.

9. Aneja, V. P. M. S. Thesis, North Carolina State University, 1975.

10. Hill, F. B.; Aneja, V. P.; Felder, R. M. Environ. Sci. Health 1978, 13, 199-225.

11. Wilson, L. G.; Bressan, R. A.; Filner, P. Plant Physiol. 1978, 61, 184-9.

12. Aneja, V. P.; Overton, J. H., Jr.; Cupitt, L. T.; Durham, J. L.; Wilson, W. E. Tellus 1979, 31, 174-8.

13. Aneja, V. P.; Overton, J. H., Jr.; Cupitt, L. T.; Durham, J. L.; Wilson, W. E. Nature 1979, 282, 493-6.

14. Adams, D. F.; Farwell, S. O.; Pack, M. R.; Bamesberger, W. L. J. Air Pollut. Control Assoc. 1979, 29, 380-3.

15. Adams, D. F.; Farwell, S. O.; Pack, M. R.; Robinson, E. J. Air Pollut. Control Assoc. 1981, 31, 1083-9.

16. Adams, D. F.; Farwell, S. O.; Robinson, E.; Pack, M. R.; Bamesberger, W. L. Environ. Sci. Technol. 1981, 15, 1493-8.

17. Jaeschke, W.; Claude, H.; Herrmann, J. J. Geophys. Res. 1980, 85, 5639-44.

18. Goldberg, A. B.; Maroulis, P. J.; Wilner, L. A.; Bandy, A. R. Atmos. Environ. 1981, 15, 11-18.

19. Ingvorsen, K.; Jorgensen, B. B. Atmos. Environ. 1982, 16, 855-65.

20. Barnard, W. R.; Andreae, M. O.; Watkins, W. E.; Bingemer, H.; Georgii, H. W. J. Geophys. Res. 1982, 87, 8787-93.

21. Aneja, V. P.; Aneja, A. P.; Adams, D. F. J. Air Pollut. Control Assoc. 1982, 32, 803-7.

22. Andreae, M. O.; Raemdonk, H. Science 1983, 221, 744-7.

23. Howarth, R. W.; Gilbin, A. Limnol. Oceangr. 1983, 24, 999-1013.

24. Aneja, V. P. Environmental Impact of Natural Emissions; Air Pollution Control Association: Pittsburg, Pa., 1984; 430 pp.

25. Steudler, P. A.; Peterson, B. J. Atmos. Environ. 1985, 19, 1411-6.

26. Aneja, V. P. Tellus 1986, 388, 81-6.

27. MacTaggart, D. L.; Adams, D. F.; Farwell, S. O. J. Atmos. Chem. 1987, 5, 417-37.

28. Goldan, P. D.; Kuster, W. C.; Albritton, D. L.; Fehsenfeld, F. C. J. Atmos. Chem. 1987, 5, 439-67.

29. Lamb, B.; Westburg, H.; Allwine, G.; Bamesberger, L.; Guenther, A. J. Atmos. Chem. 1987, 5, 469-91.

30. Maynard, D. G.; Stewart, J. W. B.; Bettamy, J. R. Biogeochemistry 1986, 1, 97-111.

31. Rennenberg, H. In Biogenic Sulfur in the Environment; Saltzman, E. S.; Cooper, W. J., Eds.; ACS Symposium Series 393, Chapter 4; American Chemical Society: Washington, DC, 1988.

32. Spaleny, J. Plant Soil 1977, 48, 557-63.

33. Lovelock, J. L.; Maggs, R. J.; Rasmussen, R. A. Nature 1972, 237, 452-3.

34. Clarke, P. H. J. Gen. Microbiol. 1953, 8, 397.

35. Delmas, R.; Baudet, J.; Servant, J.; Baziard, Y. J. Geophys. Res. 1980, 85, 4468-74.

36. Delmas, R.; Servant, J. Tellus 1983, 35B, 110-20.

37. Adams, D. F.; Farwell, S. O. Biogenic Sulfur Emissions in the SURE and Extended SURE Regions; Final Report, EPRI Proj 856-1 and 856-2; Electric Power Research Institute: Palo Alto, CA, 1980; F- p 13.

38. Baily, S. D.; Bazinet, M. L.; Driscoll, J. L.; McCarthy, A. J. J. Food Sci. 1961, 26, 163-70.

39. Westburg, H.; Lamb, B. In Environmental Impact of Natural Emissions; Aneja, V. P., Ed.; Air Pollution Control Association: Pittsburg, PA, 1984; pp 41-53.

40. Haines, B.; Black, M.; Fail, J., Jr.; McHargue, L. A.; Howell, G. In Effects of Acidic Deposition on Forests, Wetlands, and Agricultural Ecosystems; Hutichson, T. C.; Meema, K. M., Eds.; Springer-Verlag, 1988; pp 599-610.

41. Haines, B.; Black, M.; Bayer, C. In Biogenic Sulfur in the Environment; Saltzman, E. S.; Cooper, W. J., Eds.; ACS Symposium Series 393, Chapter 5; American Chemical Society: Washington, DC, 1988.

42. Hansen, M. H.; Ingvorsen, K.; Jorgensen, B. B. Limnol. Oceanogr. 1978, 23, 68-76.

43. Howarth, R. W.; Teal, J. M. Limnol. Oceanogr. 1979, 24, 999-1013.

44. Jorgensen, B. B. Nature 1982, 296, 644-5.

45. Jorgensen, B. B. Philos. Trans. Royal Soc. Lond. B 1982, 298, 543-56.

46. Ingvorsen, K.; Jorgensen, B. B. Atmos. Environ. 1982, 16, 855-65.

47. Taylor, B. F.; Kiene, R. P. In Biogenic Sulfur in the Environment; Saltzman, E. S.; Cooper, W. J., Eds.; ACS Symposium Series 393, Chapter 13; American Chemical Society: Washington, DC, 1988.

48. Fischer, U. In Biogenic Sulfur in the Environment; Saltzman, E. S.; Cooper, W. J., Eds.; ACS Symposium Series 393, Chapter 17; American Chemical Society: Washington, DC, 1988.

49. Vairavamurthy, A.; Mopper, K. In Biogenic Sulfur in the Environment; Saltzman, E. S.; Cooper, W. J., Eds.; ACS Symposium Series 393, Chapter 15; American Chemical Society: Washington, DC, 1988.

50. Kiene, R. P.; Taylor, B. F. In Biogenic Sulfur in the Environment; Saltzman, E. S.; Cooper, W. J., Eds.; ACS Symposium Series 393, Chapter 14; American Chemical Society: Washington, DC, 1988.

51. Hansen, M. H.; Ingvorsen, K.; Jorgensen, B. B. Limnol. Oceangr. 1978, 23, 68-76.

52. Aneja, V. P.; Corse, E. W.; Cupitt, L. T.; King, J.C.; Overton, J. H.; Rader, R. E.; Richards, M. H.; Sher, M. J.; Whitkus, R. J. Biogenic Sulfur Sources Strength Field Study; Northrop Services, Inc., Report No. ESC-TR-79-22; Research Triangle Park: NC, 1979; p 189.

53. Goldberg, A.; Maroulis, P.; Wilner, L.; Bandy, A. Atmos. Environ. 1981, 15, 11-18.

54. Carroll, M. A. Ph.D. Dissertation, Massachusetts Institute of Technology, Cambridge, MA, 1983.

55. Steudler, P. A.; Petersen, B. J. Nature 1984, 311, 455-7.

56. Jorgensen, B. B.; Okholm-Hansen, B. Atmos. Environ. 1985, 11, 1737-49.

57. Cooper, D. J.; de Mello, W. Z.; Cooper, W. J.; Zika, R. G.; Saltzman, E. S.; Prospero, J. M.; Savoie, D. L. Atmos. Environ. 1987, 21, 7-12.

58. de Mello., W. Z.; Cooper, D. J.; Cooper, W. J.; Saltzman, E. S.; Zika, R. G.; Savoie, D. L.; Prospero, J. M. Atmos. Environ. 1987, 21, 987-90.

59. Cooper, W. J.; Cooper, D. J.; Saltzman, E. S.; de Mello, W. Z.; Savoie, D. L.; Zika, R. G.; Prospero, J. M. Atmos. Environ. 1987, 21, 1491-5.

60. Cooper, D. J.; Cooper, W. J.; de Mello, W. Z.; Saltzman, E. S.; Zika, R. G.; In Biogenic Sulfur in the Environment; Saltzman, E. S.; Cooper, W. J., Eds.; ACS Symposium Series 393, Chapter 3; American Chemical Society: Washington, DC, 1988.

61. Delmas, R.; Servant, J. Tellus 1983, 35B, 110-20.

62. Aneja, V. P.; Overton, J. M.; Aneja, A. P. J. Air Poll. Con. Assoc. 1981, 31, 256-8.

63. Adams, D. F.; Farwell, S. O. In Air Pollution; Stern, A. C., Ed.; Academic Press: New York, 1986; Vol. 3, pp 53-7.

64. Guenther, A.; Lamb, B.; Westberg, H. In Biogenic Sulfur in the Environment; Saltzman, E. S.; Cooper, W. J., Eds.; ACS Symposium Series 393, Chapter 2; American Chemical Society: Washington, DC, 1988.

65. Tiner, R. W., Jr. Wetlands of the United States: Current Status and Recent Trends; U.S. Department of the Interior, Fish and Wildlife Service, 1984; p 59.

66. Abbruzzese, B.; Allen, A. B.; Henderson, S.; Kentula, M. E. In Proceedings of Symposium 87-Wetlands/Peatlands; Edmunton: Alberta, Canada, 1987; p 19.

67. Tiner, R. W., Jr.; Finn, J. T. Status and Recent Trends of Wetlands in Five Mid-Atlantic States: Delaware, Maryland, Pennsylvania, Virginia and West Virginia; U.S. Fish and Wildlife Service, Region 5; National Wetlands Inventory Project: Newton Corner, MA; and U.S. Environmental Protection Agency, Region III; Cooperative Publication: Phil., PA, 1986; p 40.

68. Peters, D. D.; Bohn, J. A. In Coastal Zone 87; Magoon, O. T.; Converse, H.; Miner, D.; Tobin, L. T.; Clark, D.; Domural, G., Eds.; American Society of Civil Engineers: 1987; Vol. 1, pp 1171-82.

69. Cooper, D. J.; Saltzman, E. S. In Biogenic Sulfur in the Environment; Saltzman, E. S.; Cooper, W. J., Eds.; ACS Symposium Series 393, Chapter 20; American Chemical Society: Washington, DC, 1988.

70. Turner, S. M; Liss, P. S. In Biogenic Sulfur in the Environment; Saltzman, E. S.; Cooper, W. J., Eds.; ACS Symposium Series 393, Chapter 12; American Chemical Society: Washington, DC, 1988.

71. Berresheim, H.; Andreae, M. O.; Ayers, G. P.; Gillet, R. W. In Biogenic Sulfur in the Environment; Saltzman, E. S.; Cooper, W. J., Eds.; ACS Symposium Series 393, Chapter 21; American Chemical Society: Washington, DC, 1988.

72. Bates, T. S.; Cline, J. D.; Gammon, R. H.; Kelly-Hansen, S. R. J. Geophys. Res. 1987, 92, 2930-8.

73. Andreae, M. O. In Role of Air-Sea Exchange in Geochemical Cycling; Baut-Menard, P.; Liss, P. S.; Merlivat, L., Eds.; Reidel: Dordrecht, 1986; pp 331-62.

74. Liss, P. S.; Slater, P. G. Nature 1974, 247, 181-4.
75. Nguyen, B. C.; Gaudy, A.; Bonsang, B.; Lambert, G. Nature 1978, 275, 637-9.
76. Luria, M.; Meagher, J. F. Proceedings of the 7th International Clean Air Congress, 1986, pp 295-302.
77. Luria, M; VanValin, C. C.; Wellman, D. L.; Pueschel, R. F. Environ. Sci. Technol. 1986, 20, 91-5.
78. Challenger, F. Adv. Enzymology 1951, 8, 1269-81.
79. Andrea, M. O.; Barnard, W. R. Mar. Chem. 1984, 14, 267-79.
80. Cline, J.; Bates, T. S. Geophys. Res. Lett. 1983, 10, 949-52.
81. Keller, M. D.; Bellows, W. K.; Guillard, R. R. L. In Biogenic Sulfur in the Environment; Saltzman, E. S.; Cooper, W. J., Eds.; ACS Symposium Series 393, Chapter 11; American Chemical Society: Washington, DC, 1988.
82. Kuster, W. C.; Goldan, P. D. Environ. Sci. Technol. 1987, 21, 810-5.

RECEIVED November 5, 1988

Chapter 2

U.S. National Biogenic Sulfur Emissions Inventory

A. Guenther, B. Lamb, and H. Westberg

Laboratory for Atmospheric Research, College of Engineering,
Washington State University, Pullman, WA 99164–2730

A U. S. national biogenic sulfur emissions inventory with county
spatial and monthly temporal scales has been developed using
temperature dependent emission algorithms and available
biomass, land use and climatic data. Emissions of dimethyl
sulfide (DMS), carbonyl sulfide (COS), hydrogen sulfide (H_2S),
carbon disulfide (CS_2), and dimethyl disulfide (DMDS) were
estimated for natural sources which include water and soil
surfaces, deciduous and coniferous leaf biomass, and agricultural
crops. The best estimate of 16100 MT of sulfur per year was
predicted with emission algorithms developed from emission rate
data reported by Lamb et al. ([1]) and is a factor of 22 lower than
an upper bound estimate based on data reported by Adams et al.
([2]). The predicted 16100 MT represents 0.13% of the 13 million
MT annual emission of anthropogenic sulfur reported by
Toothman et al. ([3]).

Recent concern for the role of sulfate in acidic deposition has intensified the
need for a more accurate estimation of natural sulfur emissions in the United
States. The magnitude and the spatial and temporal distribution of natural
sulfur emissions must be quantified in order to be useful in efforts to predict the
effectiveness of various control strategies for acid deposition. The inventory
estimates described in this paper predict monthly sulfur emissions from each
county in the contiguous United States. Improvements in the methodology used
to calculate this natural emissions inventory may be a useful guide in the
calculation of the national emissions inventories of other naturally emitted
compounds which have the potential to make significant contributions to
regional air quality.
   One of the major difficulties in estimating a national inventory is the high
degree of variability in local natural sulfur emissions. This can be illustrated by
a comparison of H_2S emission rates reported for tidal areas and salt marshes.
Micrometeorological techniques have indicated an H_2S flux of 80 ng · m⁻² ·
min⁻¹ over a tidal flat ([4]). Goldberg et al. ([5]) used a similar method to
estimate emissions from a salt marsh that ranged from a minimum of 800 ng S ·
m⁻² · min⁻¹ in December to 3 x 10⁴ ng S · m⁻² · min⁻¹ in July. Other investigators
have used dynamic enclosure methods and have reported a wide range of H_2S

0097–6156/89/0393–0014$06.00/0

emissions. Seasonal variations in H_2S flux in marine intertidal areas of Denmark have been estimated to range over several orders of magnitude (6) up to peak emissions of 6700 ng S · m^{-2} · min^{-1} (7) and 1700 ng S · m^{-2} · min^{-1} (8) which were attributed to a tidal pumping action. Aneja et al. (9-11), and Adams et al. (2,12) have measured natural H_2S emissions at tidal and salt marsh sites along the U. S. Atlantic coast that varied from 10 to 1 x 10^5 ng S · m^{-2} · min^{-1}. Carroll (13) has reported lower rates of H_2S emission (0.6 to 20 ng S · m^{-2} · min^{-1}) from a marsh. Annual variations in the monthly H_2S flux, measured by Steudler and Peterson (14) in a salt water marsh, varied from removing 3 x 10^3 ng S · m^{-2} · min^{-1} to emitting 1 x 10^4 ng S · m^{-2} · min^{-1} into the atmosphere. Hourly fluctuations in H_2S ground to atmosphere flux rates for one day in October varied from an uptake of 2 x 10^4 ng S · m^{-1} · min^{-1} to a release of 3 x 10^4 ng S · m^{-2} · min^{-1} to he atmosphere. H_2S emissions variability of over five orders of magnitude have been observed in Florida salt water marshes where reported rates range from nearly 1 ng S · m^{-2} · min^{-1} to 10^5 ng S · m^{-2} · min^{-1} (15-17). The large range of H_2S emission rate estimates represented in these various studies demonstrates the variability of only one compound, H_2S, in one general source category, salt marshes and tidal zones. At the present time, it is not known whether the variability in reported marshland H_2S emissions is due to environmental conditions or measurement problems, or a combination of both.

The quantity of natural sulfur emitted to the atmosphere is dependent upon the availability of sulfur, the level of natural sulfur-reducing activity, and the environment into which the gases are released. At present, there is a lack of information on specific mechanisms of biological sulfur release. As a result, algorithms designed to predict natural sulfur emissions must be empirically based on analyses of correlations between observed natural sulfur emissions and environmental parameters. In order to extrapolate the available emission rate data, emission functions must be based upon parameters which are measurable and available on an appropriate scale of temporal and spatial resolution.

The development of a natural sulfur emissions inventory for the contiguous U. S. is a necessary step toward understanding the natural component of acid deposition on a regional and national scale. A county length scale and a monthly time scale constitute appropriate levels of resolution for regional modeling. Emission estimates are based on the emission rate data described by Lamb et al. (1) which are limited to measurements of natural sulfur emissions in Ohio, Iowa, North Carolina, Washington and Idaho. The uncertainty associated with this inventory is carefully considered by analyses of model sensitivity.

Methodology

Over 300 field measurements of natural sulfur emission rates were collected for use in this inventory. Most of these data were obtained from sites in Iowa, Ohio, and North Carolina during 1985 and have been described by Lamb et al. (1). Additional unpublished emissions data collected in Washington and Idaho have also been used. Bare Mollisol, Histosol, and marshland soils were sampled in addition to surface areas with row crops (celery, carrots and onions) or natural vegetation. Above ground emissions from agricultural crops (soybean, corn, and alfalfa) and forest canopies (ash, oak, maple, hickory, and pine) were also measured. These data were collected using a dynamic enclosure technique with capillary gas chromatography similar to that described by Farwell et al. (1979). Sulfur-free sweep air was blown through a polycarbonate or Teflon enclosure and a portion of the effluent air was cryo-trapped at -183°C. Individual sulfur compounds were then separated using fused silica capillary gas chromatography and detected with a sulfur specific flame photometric detector. The presence of water vapor in this system greatly reduces its ability to detect H_2S and MeSH. In response to this, H_2S measurements were made using silver nitrate ($AgNO_3$) impregnated filters and fluorescence quenching as described by Natusch et al. (19) and Jaeschke and Herrmann (20).

Emission Rate Algorithms. In order to compile a natural emissions inventory, emission rate functions must be determined for the sources included in the inventory. The emission rate for a specific source will vary depending upon certain environmental conditions. Analyses of sulfur emission measurements collected by Adams et al. (2) and later studies (21,22) suggest that temperature plays an important role in determining sulfur flux. While the mechanisms controlling the release of natural sulfur emissions are not well understood, field observations have demonstrated characteristic trends in temperature-flux patterns. Sulfur emissions tend to increase logarithmically with increasing temperature for normal ambient temperatures (10^oC to 35^oC).

As temperatures fall below 0^oC, sulfur emissions fall below the lower detection limit of the method which is approximately 1 ng S · m^{-2} · min^{-1} for surface fluxes and 4 x 10^{-2} ng S · kg^{-1} · hour^{-1} for vegetative emissions. The low emission rate below 0^oC is not unexpected considering the minimal biological enzymatic activity which occurs at those temperatures (23). Although biological enzymatic activity normally reaches a plateau at high temperatures, this saturation point was not observed in our studies which included numerous measurements taken at temperatures above 35^oC. This is higher than any regional monthly average temperature reported for the contiguous U. S. during 1980 (24).

The emission characteristics observed in measurement studies suggest that the following form of the Michaelis-Menten equation could be used to mathematically represent the relationship between ambient temperatures and natural sulfur emissions:

$$\log(F) = \frac{k_a}{1 + \dfrac{k_b}{T}} + k_c \qquad (1)$$

where F is the sulfur emission rate (ng S · m^{-2} · min^{-1}). T is ambient temperature (^oC), k_a and k_b are rate constants determined by a non-linear least squares fit to emission rate data, and k_c is a rate constant set at the lower detection limit. Equation 1 is used to develop natural sulfur emission algorithms that predict very low emission near 0^oC, increase logarithmically at intermediate temperatures, and approach a maximum emission rate at some saturation temperature which appears to be much greater than average ambient temperatures observed in temperate regions.

Extrapolation to the Contiguous U. S. Rate constants for temperature dependent emissions functions for COS, CS_2, DMDS, DMS and H_2S have been developed for five surface categories which include wetland soils, organic soils, other soils, water, and agricultural crops (other than corn) and three vegetation categories which are deciduous and coniferous forest canopies and corn biomass. Additional natural sulfur compounds (e.g. mercaptans) are typically released at rates which are small (<1%) relative to the total sulfur flux. The three soil categories correspond to the three sites visited by Lamb et al. (1), Goldan et al. (21), and MacTaggert and Farwell (22) in 1985 and Adams et al.

(2) during 1977-78. Soil analyses indicate that these sites can be distinguished on the basis of the soil chemistry characteristics listed in Table I. Emission rate data collected over wetlands were used to develop algorithms for predicting emissions from wetland soils which are defined as those soil map units which have a pH of less than 6, a cation exchange coefficient that is less than 25 meq · 100 g^{-1} and are designated as wet in the USDI National Atlas (25). Organic soils are those soil map units which have more than 8% organic matter and a cation exchange coefficient that is greater than 35 meq · 100 g^{-1}. Emissions measured over Histosol soils provide the data for the organic soil emission rate algorithms and Mollisol emissions data were used to generate functions to predict emissions from all other soil map units. Using information available in the Geoecology Data Base (26) and the USDI National Atlas (25), surface areas for each soil category are determined for all counties in the contiguous United States. Water and agricultural cropland (other than corn) surface areas are also determined from the Geoecology Data Base. A lack of extensive measurements of sulfur emissions from water resulted in the emissions function for that category being developed from a combination of water and wetland measurements data. Although corn plants were found to significantly increase emission rates relative to bare soil or natural vegetation surface emissions, Lamb et al. (1) and Goldan et al. (21) found that other crops such as soybeans, alfalfa, oats and row crops produced only a slight increase over bare or naturally vegetated areas. The amount of additional sulfur resulting from these plants is estimated by subtracting out emissions predicted for nearby bare and naturally vegetated soils from emissions predicted for the croplands.

Sulfur emissions from biomass are calculated only for the period between the last spring frost and the first fall frost. Data reported in the Geoecology Data Base and in the USDI National Atlas were used to make estimates of mean frost dates. The Geoecology Data Base also contains potential natural vegetation surface areas which are adjusted for current land-use practices occurring within counties. Deciduous and coniferous forest surface area estimates compiled in this data base were converted to leaf biomass estimates using the relationships listed in Table II. Leaf biomass densities tend to be uniform throughout a vegetation association and are relatively insensitive to site-specific variables (27). Corn is assumed to grow or increase biomass linearly through the growing season with periodic harvests. Estimates of corn biomass are based on the agricultural yield estimates reported on a county basis in the Geoecology Data Base which are then converted to total plant dry weight using relationships available in the scientific literature (28-30). The amount of biomass present during each month increases monthly by a fraction of the total annual yield. For example, if corn requiring four months to mature is growing in a county with a seven month growing season from April to October, then monthly fractions of the total annual yield are assigned as follows:

$$\text{Nov to Mar } 0; \text{ Apr } \frac{1}{7} ; \text{ May } \frac{2}{7} ; \text{ Jun } \frac{3}{7} ; \text{ Jul } \frac{4}{7} ; \text{ Aug } \frac{1}{7} ; \text{ Sep } \frac{2}{7} ; \text{ Oct } \frac{3}{7}$$

In the above example, the growing season is longer than the maturity period which results in simulated harvests occurring in July and in October. The sum of the fractions assigned to each harvest ($\frac{4}{7} + \frac{3}{7}$) is equal to 1 which represents the total annual crop yield.

Monthly emissions of COS, CS_2, DMDS, DMS and H_2S are calculated for each of the 3071 contiguous U. S. counties. In addition to the source factors described above, mean monthly temperatures compiled in the Geoecology Data Base provide the inputs required for the estimation of natural sulfur emissions.

Table I. Soil Characteristics

Soil Category	Site[a]	CEC[b] meq 100 g	pH	Organic Matter %
1 (Wetlands)	Cedar Island, NC	2.2-15.3	5.2-6.9	1-6.5
2 (Organic)	Celleryville, OH	110	5.4	>8
3 (Other)	Ames, IA	21-25	7.1-7.3	3.1-3.6

[a]See Lamb et al. ([1]) for detailed site descriptions.
[b]Cation Exchange Coefficient.

Table II. Conversion Factors: Leaf Biomass Density

Vegetation Type	Average Leaf Biomass Density (kg · ha^{-1})
Coniferous Forest	6500
Deciduous Forest	4500
Sclerophyll Scrub	3000
Grasslands	2500
Tundra, Alpine fields	1800
Desert	1000

Results and Discussion

Inventory Estimates. The inventory estimates described in detail in this section are based on the emission rate data reported by Lamb et al. (1). In order to obtain an uncertainty estimate, these inventory results are compared with estimates based upon data reported by Goldan et al. (21) and Adams et al. (2). The SURE emissions data (2) referred to in the following text have been corrected using the results of recovery efficiency studies conducted by S. Farwell (personal communication, 1985) using a GC system identical to that used during the SURE program.

For discussion purposes, the annual national flux of 16100 metric tons of sulfur (based on the data reported by Lamb et al. (1)) is broken down by source, season, compound and region in Tables III and IV. Seasons are defined as: **spring** - March, April, May; **summer** - June, July, August; **autumn** - September, October, November; and **winter** - December, January, February. Regional emissions are grouped according to the federal region designation scheme while sources are combined into eight categories: wetlands, organic soils, other soils, cropland (increase in emissions resulting from bare or naturally vegetated soils), and water surface areas, as well as corn, deciduous and coniferous biomass. Area and biomass estimates for the eight sources are listed in Table V.

Above ground biomass is estimated to generate slightly more than one half of the national annual total flux. Coniferous canopies are the major biomass emitter with 30% of the total, while soil category 3 (other) is the largest surface source with 22% of all natural sulfur emissions. Although organic and wetland soils have higher per area emission rates, they each contribute only 7% of the total due to their smaller share of the total U. S. land area. Water and non-corn crops each contribute less than three percent of the total emissions on an annual basis. H_2S is the dominant emission from wetlands, organic soils and water. Corn emits predominately DMS while forest canopies and non-corn crops release more COS. A significant portion of the emitted COS may be rapidly recycled into the biosphere by vegetative uptake (21).

The total summer-time flux is predicted to account for 55% of the annual total while winter-time emissions are estimated to contribute only 4%. This is not unexpected considering the strong temperature dependence of the emission functions. In between these two extremes are autumn emissions (24%) and emissions during spring (18%). The higher autumn emissions are due both to increased agricultural biomass and higher temperatures. The autumn flux from agricultural crops is predicted to be more than twice the spring-time flux. The relatively large seasonal temperature variations in the northern regions (1, 2, 3, 5, 7 and 8) result in about 70% of their annual total being emitted in the summer and less than 0.5% in the winter while southern regions (4, 6 and 9) are predicted to have more uniform seasonal emission rates.

The relative contribution that each reduced sulfur compound makes to the total sulfur flux is often of interest because the various compounds behave differently once they enter the atmosphere. Terrestrial biogenic sulfur emissions are dominated by COS (38%), DMS (35%) and H_2S (21%). Emissions of CS_2 and DMDS together represent about 6% of the total. DMS emissions dominate during the summer season with 41% of the total.

On a regional basis, two-thirds of the total national flux originates in regions 4, 5, 6 and 7 which are in the southeastern and midwest portions of the United States. This encompasses about one-half of the total land area. The warmer southern regions (4, 6 and 9) generate about 90% of the inter-time emissions. By comparison, they produce about one-half of the sulfur flux on an annual basis. The regional average fluxes given in Table VI range from about 2 ng S $\cdot$ m^{-2} $\cdot$ min^{-1} in region 8 to about 6 ng S $\cdot$ m^{-1} $\cdot$ min^{-1} in region 4, in

Table III. Source Summary: Surface and Vegetative Emissions Inventory

| Source | Season | Biogenic Flux in Metric Tons of Sulfur | | | | | |
		COS	CS$_2$	DMDS	DMS	H$_2$S	Total Sulfur
Soil Category 1	Spring	60	9	3	42	63	180
Wetlands	Summer	260	23	6	170	280	740
	Autumn	80	11	4	55	85	240
	Winter	10	3	2	8	10	33
	Annual	**410**	**46**	**15**	**270**	**440**	**1200**
Soil Category 2	Spring	8	2	1	4	190	200
Organic	Summer	81	6	4	29	560	680
	Autumn	13	2	2	7	260	290
	Winter	0.1	0.1	0.1	0.1	0.5	1
	Annual	**100**	**10**	**7**	**40**	**1000**	**1200**
Soil Category 3	Spring	150	52	43	260	110	610
Other	Summer	530	110	77	930	300	2000
	Autumn	190	60	48	340	130	770
	Winter	31	18	17	48	28	140
	Annual	**890**	**240**	**190**	**1600**	**570**	**3500**
Water	Spring	17	3	1	12	18	50
	Summer	79	8	2	51	84	220
	Autumn	24	3	1	16	25	70
	Winter	3	1	1	2	2	9
	Annual	**120**	**15**	**5**	**81**	**130**	**350**
Crop Biomass	Spring	13	2	1	3	2	20
(non corn)	Summer	53	11	4	18	16	100
	Autumn	39	5	3	7	7	60
	Winter	1	0.1	0.1	0.1	0.1	1
	Annual	**100**	**18**	**8**	**28**	**25**	**180**
Corn Biomass	Spring	1	2	1	200	27	230
	Summer	4	9	5	1800	140	1900
	Autumn	3	5	3	490	68	570
	Winter	.01	.01	.01	0.6	0.1	1
	Annual	**8**	**16**	**10**	**2500**	**240**	**2700**
Deciduous	Spring	300	13	10	58	110	500
Canopy	Summer	620	27	19	200	260	1100
	Autumn	330	15	11	72	130	560
	Winter	20	1	0.6	3	7	31
	Annual	**1300**	**56**	**41**	**340**	**510**	**2200**
Coniferous	Spring	750	56	34	140	120	1100
Canopy	Summer	1300	78	45	480	180	2100
	Autumn	850	61	37	180	130	1300
	Winter	250	24	16	34	48	370
	Annual	**3200**	**220**	**130**	**840**	**480**	**4800**
All	Spring	1300	140	96	700	640	2900
Sources	Summer	2900	270	160	3600	1800	8800
	Autumn	1500	160	110	1200	840	3800
	Winter	310	48	36	96	96	590
	Annual	**6100**	**620**	**400**	**5600**	**3400**	**16100**

Table IV. Regional Summary: Surface and Vegetative Emissions Inventory

Region	Season	Biogenic Flux in Metric Tons of Sulfur					
		COS	CS2	DMDS	DMS	H2S	Total Sulfur
1,2&3	Spring	89	10	7	46	130	280
	Summer	260	22	14	300	430	1000
	Autumn	120	13	9	93	90	430
	Winter	4	1	1	2	2	11
	Annual	**480**	**45**	**31**	**440**	**760**	**1800**
4	Spring	310	28	17	180	130	660
	Summer	540	44	24	520	300	1400
	Autumn	330	29	18	210	150	740
	Winter	99	13	9	33	33	190
	Annual	**1300**	**110**	**68**	**940**	**620**	**3000**
5	Spring	70	10	7	99	69	260
	Summer	240	28	17	850	270	1400
	Autumn	100	15	10	240	110	480
	Winter	1	0.5	0.5	0.5	1	3
	Annual	**410**	**53**	**35**	**1200**	**450**	**2100**
6	Spring	330	33	22	180	120	680
	Summer	640	56	32	550	300	1600
	Autumn	360	36	23	210	130	760
	Winter	110	16	12	34	30	200
	Annual	**1400**	**140**	**88**	**970**	**580**	**3200**
7	Spring	61	9	7	88	84	250
	Summer	190	23	14	700	220	1200
	Autumn	79	12	9	190	110	400
	Winter	4	2	1	2	2	11
	Annual	**340**	**46**	**31**	**990**	**420**	**1800**
8	Spring	130	17	13	44	35	230
	Summer	420	41	27	350	130	970
	Autumn	160	21	15	71	45	320
	Winter	1	0.2	0.2	0.3	0.3	2
	Annual	**710**	**79**	**55**	**470**	**210**	**1500**
9	Spring	200	20	14	72	51	360
	Summer	390	33	21	270	110	810
	Autumn	240	23	15	110	61	440
	Winter	78	11	8	20	20	140
	Annual	**900**	**87**	**59**	**470**	**240**	**1800**
10	Spring	110	12	9	24	24	180
	Summer	260	22	14	110	58	460
	Autumn	140	15	10	39	31	230
	Winter	22	4	3	4	6	39
	Annual	**520**	**52**	**36**	**170**	**120**	**900**
National	Spring	1300	140	96	700	640	2900
	Summer	2900	270	160	3600	1800	8800
	Autumn	1500	160	110	1200	840	3800
	Winter	310	48	36	96	96	590
	Annual	**6100**	**620**	**400**	**5600**	**3400**	**16100**

Table V. Emission Sources

Surface Area	10^3 km^2	Leaf Biomass	10^{12} g
Soil Class 1 (Wetlands)	352	Deciduous Canopy	576
Soil Class 2 (Organic)	270	Coniferous Canopy	1250
Soil Class 3	7055	Corn	496
Water	154		
Total	7831		
Cropland (Not Corn)	1281		

Based on data compiled in the Geoecology Data Base (26).

Table VI. Regional Summary: Area Flux From All Natural Sulfur Sources

Federal Region	Emission Rate ng S · m^{-2} · min^{-1}
1,2 and 3	5.4
4	5.9
5	4.9
6	4.4
7	4.7
8	1.9
9	3.5
10	2.8
National	4.0

Based on emissions data reported by Lamb et al. (1).

comparison to the national average of 4 ng S · m^{-1} · min^{-1}. The spatial distribution of county averaged flux rates is shown in Figure 1. In general, the highest flux rates are predicted to be in the agriculturally intensive midwestern corn belt region, in the New England deciduous forests, and over the wetlands and organic soils along the Atlantic and Gulf Coasts and the Mississippi river valley. The lowest flux rates are predicted for the Rocky Mountain region. Annual county average emission rates range over three orders of magnitude with a predicted minimum flux of 0.52 ng S · m^{-2} · min^{-1} in Denver county, Colorado and a maximum flux of 20 ng S · m^{-2} · min^{-1} in Monroe county, Florida.

Anthropogenic vs. Biogenic Emission Inventories. The NAPAP 1980 anthropogenic emission inventory has estimated an annual SO$_2$ flux of 13 million MT or 3200 ng S · m^{-2} · min^{-1} (3). In comparison, terrestrial natural emissions are predicted to be 16100 MT which corresponds to 0.13% of the combined natural and anthropogenic total. Table VII contains estimates of seasonal natural emissions and their contribution to total sulfur emissions, reported for the base year of 1980, in each state. Anthropogenic SO$_2$ emissions are relatively constant throughout the year while biogenic sulfur emissions decrease substantially during the winter months. As a result, the national average contribution of natural emissions ranges from 0.03% in winter to 0.28% in summer. Natural emissions are estimated to contribute less than 0.05% of the total annual emissions in Delaware, Indiana, Kentucky, New Jersey, Ohio and Pennsylvania, while over 1% of the total annual emissions are estimated to be from natural sources in Nebraska, Oregon, South Dakota and Vermont. Summer-time natural emissions contribute between 3% and 7% in Nebraska, Oregon, Idaho, South Dakota and Vermont. Natural sulfur emissions are from very diffuse sources spread over large areas while anthropogenic sources are point or concentrated area sources. The diffuse natural emissions are expected to have a negligible impact on urban sites but they should be included in regional budgets and may make significant contributions to sulfur concentrations in remote areas.

Assessment of Uncertainty

Direct evaluation of the accuracy of the emission rate estimates compiled in this natural sulfur emissions inventory is difficult. Our limited understanding of natural sulfur release mechanisms and the wide variety of possible environmental conditions to which the observed data must be extrapolated require a simplified approach to this complex process. A sensitivity analysis of the important components of the modeling procedure can be used indirectly to evaluate the uncertainty which should be associated with the model. The major components affecting the estimation of natural emissions in this inventory are source factors, temperature estimates, emission prediction algorithms and emission rate data.

Source factor estimates are expected to make a relatively minor contribution to the overall uncertainty (less than 10%). Surface area estimates of the various land use categories (e.g. water, marsh, urban, agriculture, and natural vegetation) compiled in the Geoecology Data Base were derived from several different data sources (26). The data are reported for different years so that it is possible to have some inconsistencies due to recent changes in land use (i.e. urbanization). In counties where the sum of the land use categories was greater than the reported county area, the amount of overlap was subtracted from the natural vegetation surface area.

Figure 1. County averaged annual natural sulfur emission rate estimates for all compounds from all sources. Empty = 0.5 to 3, ng S · m^{-2} · min^{-1}; hatched = 3 to 6, ng S · m^{-2}· min^{-1}; solid = 6 to 21, ng S · m^{2}· min^{-1}.

Table VII. Comparison of Anthropogenic and Biogenic Sulfur Emissions

	Biogenic Flux in Metric Tons of Sulfur									
State	Spring	%[a]	Summer	%[a]	Autumn	%[a]	Winter	%[a]	Annual	%[a]
Alabama	87	0.11	170	0.17	93	0.10	22	0.03	370	0.09
Arizona	130	0.12	290	0.40	150	0.18	46	0.04	610	0.15
Arkansas	77	0.84	200	1.6	86	0.91	15	0.15	380	0.92
California	170	0.29	340	0.56	210	0.33	82	0.14	800	0.33
Colorado	47	0.36	220	1.5	59	0.39	0.1	n	320	0.53
Connecticut	11	0.15	36	0.55	17	0.22	n	n	64	0.21
Delaware	2	0.01	10	0.07	4	0.03	0.2	n	16	0.03
Florida	150	0.11	270	0.17	180	0.14	74	0.06	670	0.12
Georgia	120	0.13	230	0.20	120	0.12	28	0.02	500	0.12
Idaho	44	0.60	150	4.6	66	0.97	0.4	n	260	0.99
Illinois	61	0.04	370	0.22	120	0.07	1	n	540	0.08
Indiana	33	0.02	180	0.09	64	0.03	0.7	n	280	0.03
Iowa	44	0.11	350	0.70	110	0.27	n	n	500	0.28
Kansas	78	0.33	260	0.87	100	0.40	5	0.02	450	0.43
Kentucky	41	0.04	120	0.09	53	0.05	5	n	220	0.04
Louisiana	100	0.16	270	0.55	120	0.21	27	0.03	510	0.22
Maine	50	0.30	210	1.7	80	0.57	n	n	340	0.55
Maryland	12	0.03	48	0.15	18	0.06	1	n	79	0.05
Massachusetts	18	0.04	55	0.13	26	0.09	0.1	n	99	0.06
Michigan	42	0.04	200	0.21	78	0.09	n	n	320	0.08
Minnesota	48	0.16	280	1.1	87	0.32	n	n	410	0.30
Mississippi	82	0.26	190	0.53	89	0.25	20	0.05	380	0.28
Missouri	68	0.06	230	0.14	88	0.07	6	n	390	0.07
Montana	63	0.24	220	1.8	81	0.66	n	n	370	0.49
Nebraska	61	0.63	320	3.3	110	1.7	n	n	480	1.3
Nevada	64	0.36	180	1.0	81	0.50	10	0.06	340	0.50
New Hampshire	19	0.15	69	0.66	28	0.23	n	n	120	0.23
New Jersey	5	0.01	20	0.06	8	0.02	0.4	n	34	0.02
New Mexico	95	0.35	230	0.75	100	0.35	21	0.07	450	0.39
New York	42	0.05	180	0.20	72	0.08	0.1	n	300	0.07
North Carolina	80	0.12	200	0.26	93	0.13	17	0.03	390	0.13
North Dakota	12	0.11	73	0.63	20	0.15	n	n	110	0.21
Ohio	27	0.01	140	0.04	52	0.02	0.7	n	220	0.02
Oklahoma	75	0.64	180	1.9	86	0.67	20	0.17	360	0.78
Oregon	80	1.3	180	3.0	100	1.3	25	0.31	380	1.5
Pennsylvania	33	0.01	137	0.06	55	0.02	0.3	n	230	0.02
Rhode Island	3	0.20	9	0.47	5	0.34	n	n	17	0.26
South Carolina	57	0.16	130	0.30	62	0.18	12	0.02	260	0.17
South Dakota	24	0.39	140	5.8	44	0.87	n	n	210	1.16
Tennessee	47	0.04	120	0.09	54	0.05	8	n	230	0.05
Texas	330	0.22	710	0.46	370	0.23	120	0.07	1500	0.25
Utah	51	0.36	160	1.3	59	0.43	2	n	270	0.50
Vermont	19	1.3	70	7.2	28	3.0	n	n	120	2.5
Virginia	46	0.12	130	0.30	49	0.13	7	0.02	240	0.14
Washington	54	0.15	130	0.39	64	0.17	13	0.06	260	0.17
West Virginia	22	0.02	56	0.17	27	0.07	2	0.01	110	0.07
Wisconsin	46	0.05	230	0.29	79	0.12	n	n	360	0.12
Wyoming	37	0.13	150	0.55	52	0.18	n	n	2400	0.12
National	2900	0.10	8800	0.28	3800	0.13	590	0.03	16100	0.13

[a]biogenic contribution to 1980 total sulfur emissions (Toothman 3).
nrepresents negligible emission or contribution.

Vegetative emission source factors also depend on biomass factors to convert surface area estimates to biomass estimates. The leaf biomass factors used in this inventory (Table II) are those suggested by Zimmerman (31). A comparison of leaf biomass factors reported in the literature provides a simple illustration of the variability associated with the conversion factors used in this inventory. Monk et al. (32) reported a biomass factor of 8400 kg · ha^{-1} for trees with 14 cm diameter trunks in an oak hickory forest. Biomass factors of 5700 kg · ha^{-1} (30) and 5800 kg · ha^{-1} (33) have been estimated for mixed deciduous forests. Arnts et al. (34) reported a range of biomass factors between 4400 kg · ha^{-1} and 6340 kg · ha^{-1} for a loblolly pine forest. The variability in these biomass estimates is less than ±25%.

The Geoecology Data Base reports arithmetic mean monthly temperatures for climatic divisions in the United States. This is not an ideal input to the inventory modeling procedure which assumes that sulfur emissions increase logarithmically with temperature. Extreme temperatures, which can generate a disproportionate amount of total emissions, are not well represented by the mean. The impact of this effect can be demonstrated by comparing a flux estimate calculated using a mean temperature to an estimate based on a range of temperatures. A mean temperature flux is based on the average monthly temperature. An example of a weighted temperature flux is given in Equation 2 where a proportion of five fluxes predicted by five different temperatures are summed to determine the emission rate:

$$F_{wt} = 0.1\,F_1 + 0.2\,F_2 + 0.4\,F_3 + 0.2\,F_4 + 0.1\,F_5 \tag{2}$$

where F_1 is the flux resulting from the minimum temperature, F_2 results from the average of minimum and mean temperatures, F_3 is based on mean temperature, F_4 is based on the average of mean and maximum temperatures, and F_5 is based on a maximum temperature.

Equation two has been used to estimate monthly and daily weighted emissions from monthly and daily weighted maximum, minimum and mean temperatures for a specific county. The daily temperatures were obtained from the U. S. Environmental Data Service (24). The total January and July monthly weighted flux estimates were within 4% of estimates predicted with only the mean monthly temperature. Daily weighted temperatures predict a July flux that is within 3% of July mean monthly estimates but the January flux is 25% greater than that predicted for January by monthly temperature statistics. The percent difference depends upon the flatness of the flux vs. temperature curve over the temperature range of the month. The analysis indicates that the use of monthly average temperatures in the calculation of this natural emissions inventory is not likely to overestimate emissions but may underestimate individual emissions estimates by up to 25%.

A comparison of emissions inventories which use the same emission rate data base but different modeling procedures can illustrate the sensitivity of the model to the flux prediction algorithms. Table VIII lists surface emission rates predicted by three different emission rate algorithms which have been fit to the corrected SURE data (2) and extrapolated to cover the original SURE area in the northeastern U. S. The Michaelis-Menten function (Equation 1) generates estimates that result in an annual average emission rate of 29 ng S · m^{-2} · min^{-1}. A segmented line emission algorithm (SLEA) has been used to predict an annual average emission rate of 70 ng S · m^{-2} · min^{-1} (35) which is a factor of 2.4 greater than that predicted by the Michaelis-Menten function.

The SLEA method fits a log$_{10}$ least squares line to the emission rate and temperature data within the range of temperatures sampled in he study. At lower temperatures, emissions are assumed to increase linearly from the lower detection limit at 0°C to the least squares line. At high temperatures the emission rate is constant at the level predicted for the highest measured temperature. The annual average estimate is based upon thirty-five SLEA emission functions which were used to predict the monthly flux of five major sulfur compounds (H_2S, DMS, COS, CS_2 and DMDS) from seven major sources (water, Histosols, Mollisols, Alfisols, Inceptisols, Ultisols and marshlands) on a county scale of resolution.

Table VIII. Model Sensitivity to Emission Algorithm: Annual Average
Natural Sulfur Emission Rate (ng $S \cdot m^{-2} \cdot min^{-1}$) for
Soil Surfaces in the Northeastern U. S.

Emission function	Emission rate ng $S \cdot m^{-2} \cdot min^{-1}$
MM[b]	29
GMFIA[b]	15
SLEA[c]	70

[a]Michaelis-Menten functions based on SURE surface emissions data ([2]).
[b]Geometric mean - forced intercept algorithms based on SURE surface emissions data ([2]).
[c]Segmented line emission algorithms ([35]) based on SURE surface emissions data ([2]).

The third emission function assumes a logarithmic increase in sulfur emissions from the lower detection limit at 0°C to the geometric mean measured flux at the arithmetic mean sampling temperature. These geometric mean - forced intercept (GMFI) functions were developed from the SURE data and used to predict monthly total sulfur emissions from nine soil groups in U. S. counties. The total annual average flux rate of 15 ng $S \cdot m^{-2} \cdot min^{-1}$ predicted for the SURE region with this method is a factor of 2 lower than the flux predicted by the Michaelis-Menten functions. This comparison of these three emission functions suggests that the choice of temperature-flux function can cause a factor of two variation in emission rate estimates.

In addition to the natural variability in sulfur emissions observed by individual investigators at single sites, there is also variability in emission rates observed at single sites by different investigators at different times. This variability could be due to differences in analytical techniques or changes in environmental conditions over a long period of time or combination of both. A comparison of the emission rate data recorded by Lamb et al. ([1]) and Goldan et al. ([21]) at one site demonstrates that the corrected SURE H_2S emission rates ([2]) were approximately two orders of magnitude higher than comparable data recorded by Lamb et al. ([1]) and Goldan et al. ([21]). The emission rates of other natural sulfur compounds compiled in the corrected SURE data base are about an order of magnitude greater than the emissions reported by Lamb et al. ([1]). Independent data collected simultaneously by Lamb et al. ([1]), Mactaggart and Farwell ([22]), and Goldan et al. ([21]) are within a factor of two. Segmented-line emission rate functions were developed from soil emission rates compiled in the Lamb et al. ([1]), Goldan et al. ([21]) and corrected Adams et al. ([2]) data bases and extrapolated to the national level using the inventory procedures described in this paper. This analysis of model sensitivity to emission rate data is summarized in Table IX and demonstrates that the corrected 1978 SURE data results in emission estimates which are a factor of 22 higher than estimates derived from the 1985 data bases.

Table IX. Model Sensitivity to Emission Data: National Annual Natural
Sulfur Emission Rates (MT S) Estimated for all Soil Categories

Data Source	Annual Total (MT)	Average Flux $(ng\ S \cdot m^{-2} \cdot min^{-1})$
SURE[a]	97200	25
NOAA[b]	5040	1.3
WSU[c]	4210	1.1

[a]Corrected emissions data based on Adams et al. (2).
[b]Emissions data reported by Golden et al. (21).
[b]Emissions data reported by Lamb et al. (1).

Conclusions

Temperature dependent natural sulfur emission algorithms, along with biomass
density data from the literature and land use and climatic data from the
Geoecology Data Base, provide the basis of this national inventory of natural
sulfur emission rate estimates. The annual national total of 16100 MT of
naturally-emitted sulfur comprises 0.13% of the combined anthropogenic and
natural sulfur flux. The biogenic contribution is as high as 7% in individual
states during summer. COS is estimated to be the major component of natural
sulfur emissions followed by DMS, H_2S, CS_2, and DMDS. Vegetative emissions
are estimated to be slightly higher than those from soil surfaces. About 55% of
the annual emission is predicted to occur during the high temperatures and
biomass densities of summer.

Three areas of uncertainty in this present inventory of natural sulfur
emissions which need further work include natural variability in complicated
wetland regions, differences in emission rates in the corrected SURE data and
those reported by Lamb et al. (1) and Goldan et al. (21) for inland soil sites, and
biomass emissions for which only a very limited data base exists. The current
difficulty in determining the sources of variability emphasizes the need to better
understand natural sulfur release mechanisms. At present, it may be useful to
consider the emission rates based on the corrected SURE data as an upper
bound to natural emissions and use the emission rates based on data described
by Lamb et al. (1) as a more conservative estimate of natural sulfur emissions.
However, this still leaves a factor of 22 difference between the suggested upper
bound and our best current estimate.

The temperature dependent algorithms used to predict natural sulfur
emissions do not account for all of the variation in observed emissions. Other
important environmental parameters may include, but are not limited to, tidal
flushing, availability of sulfur, soil moisture, soil pH, mineral composition,
ground cover, and solar radiation. A more accurate estimation of the national
sulfur inventory will require a better understanding of the factors which
influence natural emissions and the means to extrapolate any additional
parameters which are determined to be important.

Acknowledgments

This work was supported by the National Oceanic and Atmospheric
Administration (Contracts NA82RAC00151, NA82RAC00152, and
NA85RAC05105) as part of the National Acid Precipitation Assessment
Program.

Literature Cited

1. Lamb, B. K.; Westberg, H. H.; Bamesberger, W. L.; Allwine, G.; Guenther, A. B. J. Atmos. Chem. 1987, 5, 469-91.
2. Adams, D.; Farwell, S.; Robinson, E.; Pack, M.; Bamesberger, W. Environ. Sci. Technol. 1981, 15, 1493-8.
3. Toothman, D. A.; Yates, J. C.; Sabo, E. J. Status report on the development of the NAPAP emission inventory for the 1980 base year and summary of preliminary data. EPA-600/7-84-091, EPA Office of Research and Development, Industrial Environmental Research Laboratory, Research Triangle Park: NC, 1984.
4. Jaeschke, W. Atmos. Environ. 1978, 12, 715-22.
5. Goldberg, A. B.; Maroulis, P. J.; Wilner, L. A.; Bandy, A. R. Atmos. Environ. 1981, 15, 11-18.
6. Ingvorsen, K.; Jorgensen, B. B. Atmos. Environ. 1982, 16, 855-65.
7. Hansen, M. H.; Ingvorsen, K.; Jorgensen, B. B. Limnol. Oceanogr. 1978, 23, 68-76.
8. Jorgensen, B. B.; Okholm-Hansen, B. Atmos. Environ. 1985, 19, 1737-49.
9. Aneja V.; Overton, J.; Cupitt, L.; Durham, J.; Wilson W. Tellus 1979, 31, 174-8.
10. Aneja, V. P.; Overton, J. H.; Aneja, A. J. Air Pollut. Control Assoc. 1981, 31, 256-8.
11. Aneja, V. P. In Environmental Impact of Natural Emissions; Aneja, V. P., Ed.; Air Pollution Control Association: Pittsburg, PA., 1984.
12. Adams, D.; Farwell, S.; Robinson, E.; Pack, M. Biogenic Sulfur Emissions in the SURE Region, EPRI Report EA-1516, Palo Alto: CA, 1980.
13. Carroll, M. A. J. Atmos. Chem. 1986, 4, 375-82.
14. Steudler, P. A.; Peterson B. J. Atmos. Environ. 1985, 19, 1411-16.
15. Cooper, D. J.; de Mello, W. Z.; Cooper, W. J.; Zika, R. G.; Saltzman, E. S.; Prospero, J. M.; Savoie, D. L. Atmos. Environ. 1987a, 21, 7-12.
16. Cooper, W. J.; Cooper, D. J.; Saltzman, E. S.; de Mello, W. Z.; Savoie, D. L.; Zika, R. G.; Prospero, J. M. Atmos. Environ. 1987b, 21, 1491-5.
17. de Mello, W. Z.; Cooper, D. J.; Cooper, W. J.; Saltzman, E. S.; Zika, R. G.; Savoie, D. L.; Prospero, J. M. Atmos. Environ. 1987, 21, 987-90.
18. Farwell, S. O.; Gluck, S. J.; Bamesberger, W. L.; Schutte, T. M.; Adams, D. F. Anal. Chem. 1979, 51, 609-15.
19. Natusch, D.; Klonis, H. B.; Axelrod, H. D.; Teck, R. J.; Lodge J. P. Jr. Anal. Chem. 1972, 44, 2067-70.

20. Jaeschke, W.; Herrmann, J. Int. J. Environ. Anal. Chem. 1981, 10, 107-20.

21. Goldan, P. D.; Kuster, W. C.; Albritton, D. L.; Fehsenfeld, F. C. J. Atmos. Chem. 1987, 5, 439-67.

22. MacTaggart, D.; Farwell, S. J. Atmos. Chem. 1987, 5, 417-38.

23. Tauber, H. The Chemistry and Technology of Enzymes; Wiley: New York, 1949.

24. Climatological data; North Carolina, 1975, National Oceanic and Atmospheric Agency, Environmental Data Service: Asheville, NC, 1980.

25. The National Atlas of the United States of America; U. S. Department of Interior Geological Survey: Washington, DC, 1970.

26. Olson, R. J. Geoecology: A County-level Environmental Data Base for the Coterminous United States. Environmental Sciences Division, Oak Ridge National Laboratory, Publication No. 1537; Oak Ridge, TN, 1980.

27. Satoo, T. In Primary Productivity and Mineral Cycling in Terrestrial Ecosystems; Symposium, 13th annual meeting of the Ecological Society of America; American Association for the Advancement of Science: New York, NY, 1967.

28. Hoffman, G. J. Trans. ASAE 1973, 16, 164-7.

29. Cooper J. P. Photosynthesis and Productivity in Different Environments; Cambridge University Press: Cambridge, 1975.

30. Whittaker, R. H.; Borman, F. H.; Likens, G. E.; Siccama, T. G. Ecol. Monogr. 1974, 44, 233-52.

31. Zimmerman, P. R. Testing for hydrocarbon emissions from vegetation leaf litter and aquatic surfaces, and development of a methodology for compiling biogenic emission inventories. PA Report 450/4-4-79-004, Environmental Protection Agency, 1979.

32. Monk, D. D.; Child, G. I.; Nicholson, S. A. Oikos 1970, 21, 138-41.

33. Lamb, B.; Westberg, H.; Allwine, G. J. Geophys. Res. 1985, 90, 2380-90.

34. Arnts, R. R.; Peterson, W. B.; Seila, R. L.; Gay, B. W. Jr. Atmos. Environ. 1981, 16, 2127-37.

35. Guenther, A. B. M. S. Thesis, Washington State University, Pullman, Washington, 1986.

RECEIVED July 6, 1988

Chapter 3

Variability in Biogenic Sulfur Emissions from Florida Wetlands

David J. Cooper[1], William J. Cooper[2], William Z. de Mello[1],
Eric S. Saltzman[1], and Rod G. Zika[1]

[1]Rosenstiel School of Marine and Atmospheric Science,
University of Miami, Miami, FL 33149–1098
[1]Drinking Water Research Center,
Florida International University, Miami, FL 33199

Emissions of biogenic hydrogen sulfide (H_2S), dimethylsulfide
(DMS), carbon disulfide (CS_2) and dimethyldisulfide (DMDS)
were measured from several wetland ecosystems in Florida. The
factors governing their variability can be divided into spatial,
seasonal, diel and tidal components. These components, in turn,
relate to the effects of temperature, insolation and soil inundation
on both the physicochemical factors controlling sulfur emissions
and the metabolism of bacteria and higher plants. This paper
illustrates these effects and their inter-relationships using data
obtained over several diel periods, repeated several times during
the course of a single seasonal cycle. Ecosystems studied include
a Spartina alterniflora coastal fringe, a brackish Disticlis spicata
marsh, a freshwater Cladium jamaicense swamp, a Juncus
roemerianus marsh, an Avicennia germinans (Black mangrove)
fringe, coastal seawater surfaces, and intertidal mudflats at
several locations.

Models of the global sulfur cycle include biological sources of volatile sulfur
gases as a significant input to the atmosphere. The magnitude of this input,
however, remains a major uncertainty involved in such models. Biological
sources of reduced sulfur include both marine and terrestrial components.
Most recent studies suggest a value of 40 $\pm$ 20 Tg S/yr (1) for the former
component, due to the release of dimethylsulfide (DMS) from the surface
ocean. Estimates of the latter component are highly variable, however,
reflecting the variability and the sparsity of previous direct measurements. We
are hampered by our poor understanding of the processes controlling the
release of reduced sulfur gases from the different ecosystems studied, their
spatial and temporal importance, and doubts concerning earlier analytical
methodology (2).

Many of the previous direct flux measurements have focused on two distinct
ecosystems, intertidal mudflats and Spartina alterniflora salt marshes. These
coastal systems have the potential for large emissions of volatile reduced sulfur
gases due to the availability of sulfate and organic matter. Intertidal mudflats
(3,4) have a tendency towards anoxia, with concomitant production of H_2S via
sulfate reduction. S. alterniflora marshes (4,5) release DMS through the

0097–6156/89/0393–0031$06.00/0

cleavage of dimethylsulphonioproprionate (DMSP), an osmolite found at high concentration within the plant (6). Not surprisingly, flux measurements made from more representative soil types show significantly lower emission rates (7).

This paper summarizes the results of emission measurements made from several wetlands in Florida. The two environments mentioned above were sampled, in addition to sites with predominant plant communities of Cladium jamaicense, Juncus roemerianus, Disticlis spicata, Avicennia germinans, Batis maritima, and coastal seawater surfaces. Complete descriptions of the sites are given in refs. (8-11). These sites were chosen to represent the major wetland plant communities of Florida, and their geographic locations are shown in Figure 1.

Sampling and Analytical Methods

Emissions were measured using a dynamic flow chamber technique (8). H_2S was collected using a silver nitrate impregnated filter method, and analysed by the fluorescence quenching of fluorescein mercuric acetate (8,9,12). DMS, CS_2 and DMDS were collected by cryogenic trapping in a packed Teflon loop, using liquid oxygen, and analysed by gas chromatography with flame photometric detection (9,10). A 12 foot Chromosil 330 column was used with nitrogen as the carrier gas at 30 cm^3 min^{-1}. The GC oven was temperature programmed from 25-100°C at 25°C min^{-1} with an initial hold of two minutes and a final hold of eight minutes. The GC method is quantitative for these three compounds. Low molecular weight thiols and H_2S, however, showed losses in our chromatographic system, and carbonyl sulfide (OCS) could not be quantified due to co-elution with CO_2 and/or hydrocarbons. A sample chromatogram is shown in Figure 2. The $AgNO_3$ filter method is quantitative for H_2S (8). Measurements were made over several diel periods at most of the sampling sites, repeated several times during the course of a seasonal cycle.

Deposition of reduced sulfur compounds to enclosed surfaces could also be measured by adding low loss (5 ng/min, GC Industries) permeation tubes to the inlet of the chambers. This permeation rate adds an additional sulfur load which varied typically between 2 and 5 times that already within the enclosure. These studies were limited to surfaces where large temporal changes in natural emissions were not occurring, and ultimately became limited by the reliability of the permeation devices.

Results and Discussion

Emissions of biogenic sulfur compounds to the atmosphere result from an imbalance between metabolic formation processes and biological or physicochemical consumption processes, determined on the spatial scale of the available methods for measuring emission fluxes. Variability in emissions patterns reflects the complexity of the factors contributing to these processes. Our results show that this variability can be described as the net result of the combined effects of vegetation coverage, temperature, insolation, soil inundation and seasonal flooding. This gives resultant emissions patterns showing spatial, seasonal, diel and tidal effects.

Within the S. alterniflora fringe, spatial variability in both the magnitude and speciation of biogenic sulfur emissions was found to reflect the inhomogeneity of vegetation coverage. This is vividly demonstrated in the data shown in Table I, a summary of measurements made within the Spartina fringe. DMS and H_2S measurements were made from the same chambers, and all sites were within 5 meters of each other. The variations can be explained by the relative contributions to the total emission from soil bacteria versus plant

Figure 1. Sampling site locations: 1. Merritt Island (D. spicata); 2. Rookery Bay (A. germinans, sea surface, mudflat); 3. Old Ingraham highway (C. jamaicense); 4. Flamingo (B. maritima); 5. St. Marks (S. alterniflora, J. roemerianus, sea surface, mudflat)

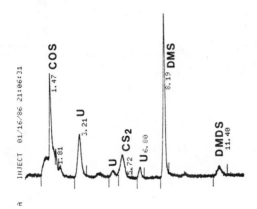

Figure 2. Sample chromatogram. Chamber over C. jamaicense/water. CS_2, DMS and DMDS quantified, OCS not quantified due to co-elution with H_2S, CO_2, and/or HCs. DMS emission rate = 0.11 μg $S/m^2/hr$. Unidentified peaks are marked "U".

Table I. Variation in Biogenic DMS and H_2S Emissions from a
Spartina alterniflora Fringe, St. Marks, Florida
Units of μg S m^{-2} hr^{-1}

Date Soil Temperature (°C)	May, 1985 22-37	Oct, 1985 19-36	Jan, 1986 9-19
Wet sand site - DMS	n.d.	1.3 - 17	0.5 - 1.7
Wet sand site - H_2S	3 - 14866	n.d.	0.09 - 8151
Wet Spartina site - DMS	19 - 59	30 - 70	17 - 96
Wet Spartina site - H_2S	1.5 - 118	n.d. 16 - 69*	0.17 - 9.6
Dry Spartina site - DMS	97 - 556	66 - 147*	10 - 188
Dry Spartina site - H_2S	1.3 - 5.5	n.d.	0.07 - 1.9

*Two adjacent chambers with different vegetation coverage (see text).

metabolic processes. The release of DMS appears to be related to the metabolism of S. alterniflora, with higher fluxes occurring from vegetated sites, peaking in the early afternoon and at times of osmotic stress (10).

Within the vegetated sites the magnitude of the DMS emission measured also reflects the biomass enclosed. Measurements made using two chambers at apparently identical dry sites, in a region only inundated by extreme high tides (Table I), showed emission rates to be approximately a factor of two to four different throughout the day. Later examination of the chamber locations revealed that the root biomass below the ground was a factor of 2-3 greater at the higher emitting site.

The release of H_2S from the Spartina fringe, on the other hand, is consistent with sulfate reduction in the underlying anoxic sediment, which occurs at a shallower depth in the unvegetated areas. H_2S emissions occurred in intense pulses as the rising tide reached the sampling sites, forcing greater than 90% of the integrated emission in just 3% of the tidal cycle (9). This tidal pumping effect was found to be the major driving force in the release of H_2S at all intertidal mudflat sites sampled; at Rookery Bay, Flamingo and St. Marks, and also at two estuarine intertidal areas at St. Marks. This observation is in agreement with studies been made by other workers at various coastal locations (3-5).

Emissions of H_2S from the surface of shallow water bodies with underlying anoxic sediments also appear to be controlled by the movement of the tides. Figure 3 shows the effect of water depth on the emission rate of H_2S from the sea surface at Rookery Bay, measured over the course of several tidal cycles in October 1985 and January 1986. It is likely that oxidation of H_2S in the water column acts as a cap to the emission of H_2S as it diffuses from the sediment below. The data of Figure 3 suggest that any emission of H_2S from the surface of water deeper than about one meter is unlikely. No such effect was found in the emission of DMS from the coastal water surfaces, which is consistent with formation of this gas by algae in the water column, and not by bacteria in the sediment.

The emissions from all soil sites other than the S. alterniflora marsh followed a pattern that was consistent with release of microbial metabolites from the soil surface. H_2S and DMS were the predominant species found, with emission rates 10-100 times greater than those of CS_2 and DMDS. Peak emissions of all compounds were measured in the early afternoon, corresponding to the maximum soil temperature. No effects were found due to the extent of vegetation coverage. The difference between these sites and the S. alterniflora sites provides indirect evidence for the role of DMSP cleavage in the release of DMS from the tissues of the S. alterniflora. Elevated concentrations of DMSP have been reported to occur in the leaves of S. alterniflora, when compared to other American salt-marsh species (6).

The effect of temperature on soil microbial activity is demonstrated in emission measurements made over two days at the J. roemerianus site, shown in Figure 4. Data were obtained during the passage of a cold front on January 1986, clearly showing a diel cycle in the release of DMS and H_2S, which is correlated with soil temperature rather than insolation. Figure 5 shows a plot of DMS and H_2S emissions against temperature, combining the data set of Figure 4 with measurements made on two other sampling trips. The different slopes of the DMS and H_2S data sets may reflect different bacterial processes leading to the formation of these gases. The emission of DMS at this site is more strongly temperature dependent than the emission of H_2S.

The brackish Disticlis spicata and Avicennia germinans sites and the freshwater Cladium jamaicense site experienced seasonal inundation during this study. During the dry season, temperature-related diurnal variations in both

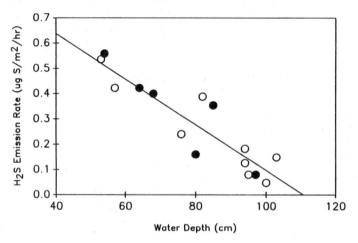

Figure 3. Effect of water column depth on the emission of H$_2$S from a coastal seawater surface, Rookery Bay, Florida, October 1985 and January 1986.

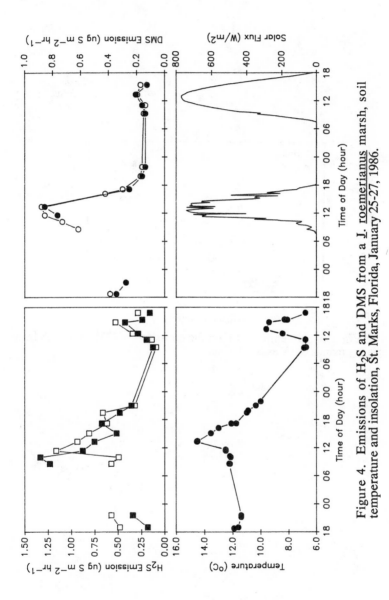

Figure 4. Emissions of H_2S and DMS from a *J. roemerianus* marsh, soil temperature and insolation, St. Marks, Florida, January 25-27, 1986.

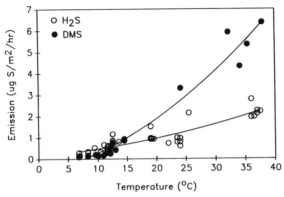

Figure 5. Effect of soil temperature on the emissions of DMS and H_2S from a $\underline{J}$. roemerianus marsh, St. Marks, Florida.

DMS and H_2S emissions from the exposed soils were found. When inundated during the wet season, the emission of DMS from the water surfaces exhibited a similar diurnal cycle, but with a considerably lower emission rate. In contrast, the emission of H_2S from the water surfaces showed a reversed diurnal cycle, with peak emissions occurring before dawn (Figure 6). A precipitous decrease of almost three orders of magnitude in the emission of H_2S after sunrise suggests that thermal stratification and/or light-related removal processes in the water column may be controlling its release during the day.

Although the diel patterns of H_2S emission from the water surface and the soil surface over the Disticlis spicata are reversed (Figure 6), the overall (integrated) emission rate is about the same (60 and 56 μg S m^{-2} hr^{-1}, respectively). Integrated emission rates of DMS were also similar from the two different surfaces, 6.8 and 6.2 μg S m^{-2} hr^{-1}, respectively. Greater than 90% of the DMS emission from the D. Spicata/soil surface occurred between the hours of 09:00 and 20:00.

Possible effects of plant metabolic effects were investigated in a later experiment over the D. spicata/water surface. Two adjacent chambers were suspended on the water surface over a highly vegetated site and a sparsely vegetated site. Again, the H_2S emissions peaked before dawn, and DMS emissions peaked in the early afternoon. The ranges of emission rates of DMS and H_2S measured from these sites are shown in Table II. H_2S emissions are higher from the unvegetated site than the vegetated site, whereas DMS emissions are higher from the vegetated site and almost undetectable from the unvegetated site. This effect suggests that release of DMS may be a result of metabolic processes in the D. spicata, while the release of H_2S is related to sediment processes, and is inhibited by the oxic metabolism of the plant community, in a similar way to S. alterniflora. However, it is interesting that DMSP, the precursor of DMS in Spartina species, has not been found in D. spicata (6). It is possible that other similar compounds may function in the same role. Further study is necessary.

The general variability found within these study sites, and the different patterns found in emissions data at the individual sampling locations, clearly demonstrate the complexity of processes governing the release of biogenic sulfur gases. It is evident that with our current understanding of these processes and their spatial extent, it is extremely difficult to estimate the contribution of biogenic sulfur to the atmosphere. It is clear from the data presented here, however, that the emissions patterns observed in wetlands can be explained by considering the effects of tide, insolation and temperature on the metabolisms of both soil bacteria and macrophytes.

Even in the simplest cases, comprehensive emission measurements are necessary if a meaningful flux calculation is to be made. This can only be achieved by integrating the measurements over the relevant cycle, i.e. tidal (in the case of H_2S from coastal environments), diel (in the case of DMS, CS_2 and DMDS from all locations and H_2S from non-tidal locations), and seasonal (temperature effects and water coverage effects). In addition, the effect of spatial variability, i.e. the effects of changing vegetation coverage and/or soil inundation need also be considered within some ecosystems. For these reasons, we do not attempt to attempt to average our emissions data, and for the purpose of flux extrapolation, use only the sites that have sufficient emissions data (8,9).

Most of the emissions data available in the literature was obtained within S. alterniflora marshes, and hence reflect the complexity of this ecosystem. Our data set extends the available emissions data to less complex systems, and at the same time lowers our estimation of the biogenic contribution to atmospheric sulfur loading. Even our highest estimate of biogenic emissions, that of

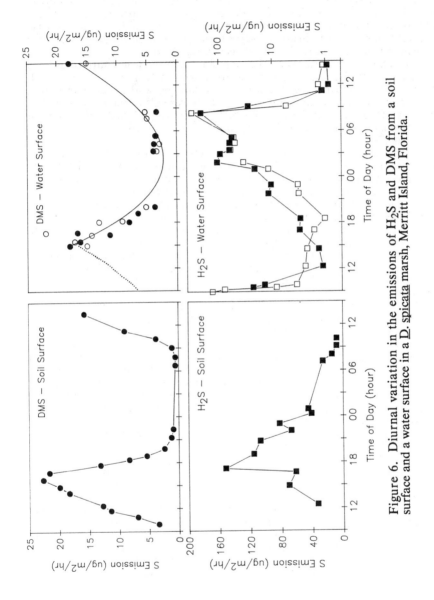

Figure 6. Diurnal variation in the emissions of H_2S and DMS from a soil surface and a water surface in a D. spicata marsh, Merritt Island, Florida.

reference 9, from sand within $\underline{S}$. alterniflora (total 96 μg S m^{-2} hr^{-1} gives a flux of only 5 x 10^9 g S yr^{-1} when extrapolated to Florida's saline wetland coverage (6 x 10^9 m^2, ref. 13), which is most likely an overestimate due to the much lower emission rates measured from the other ecosystems (11). This flux is two orders of magnitude lower than the anthropogenic SO_2 emissions from Florida (14). Extrapolating further to the total area of salt marshes in the world (3.8 x 10^{11} m^2) gives a annual flux of only 0.32 Tg S to the atmosphere, two orders of magnitude less than the flux of DMS calculated from the surface oceans (1).

The results of deposition studies using H_2S are summarized in Table III. The sparsity of the data reflects the problems of conducting this type of experiment. The emission rates shown are the values measured immediately before and after making deposition measurements, and are used to calculate the deposition rate. It is clear in the case of the seawater surface that changes in natural emissions complicate the study to such an extent that interpretation is impossible. Furthermore, it should be noted that the nature of enclosure designs provides elevated concentrations of emitted gaseous species, and it is likely that the measured emission rates already include a depositional component. The concentration of sulfur species added to the chamber in these studies is comparable to that already present, and the actual deposition rates would probably be higher at ambient reduced sulfur levels.

Table II. Variation in Biogenic DMS and H_2S Emissions from a
Disticlis spicata Marsh, Merritt Island, Florida, Jan 30 - Feb 1, 1986
Units of μg S m^{-2} hr^{-1}

	H_2S Emission	DMS Emission
Highly vegetated site	0.39 - 21	2.1 - 5.1
Sparsely vegetated site	1.2 - 125	<0.01 - 0.19

Table III. Deposition of H_2S to Various Substrates.
The Number of Measurements is Shown in Parentheses

Site	Emission Rate (μg S m^{-2} hr^{-1})	Loss to Surface (%)
Dry Inland Lawn	0.15 (2)	100 (2)
Pool in $\underline{C}$. jamaicense	1.32 (2)	75 (1)
Soil in $\underline{C}$. jamaicense	0.70 (2)	66 (2)
Sea Surface (Flamingo)	1.20 (2)	28, 45 & -16
Mudflat (Rookery Bay)	1.66 (2)	56 (1)
Soil in $\underline{A}$. germinans	1.80 (2)	58 (2)
Soil in $\underline{J}$. roemerianus[*]	0.87 (4)	58 (4)
Soil in $\underline{J}$. roemerianus[**]	0.93 (3)	59 (4)

* April, 1985
** May, 1985

All of the surfaces where this study was possible showed uptake of H_2S. Complete loss of added H_2S was found at an inland sandy lawn, and greater than 50% loss was measured at all wetland sites. Complete recovery of both H_2S and DMS was obtained when the experiment was repeated on a Teflon sheet, ruling out losses to the chamber walls. H_2S deposition velocities calculated from this data range from 0.004 to greater than 0.021 cm s^{-1} (8), in agreement with earlier laboratory studies (15). Similar results were found for DMS deposition, although greater natural variability at all sites makes the measurements largely unquantifiable. These results suggest that wetland soils, while emitting relatively low levels of reduced sulfur compounds, also have the potential to remove the same reduced sulfur compounds from the atmosphere at times of elevated concentration.

Summary

Patterns of biogenic sulfur emissions show components due to tidal, diel, spatial, and seasonal effects. Emissions of DMS and H_2S show different temporal patterns, reflecting these various components, and also reflecting the different metabolic effects leading to their formation. DMS and H_2S emissions are 2-3 orders of magnitude greater than those of CS_2 and DMDS.

Production of H_2S appears to occur in anoxic sediments, with the majority of the emission relating to the effects of tidal coverage on either the flushing of porewater or the depth of coverage. Production of DMS appears to be related to the metabolism of either macrophytes or planktonic algae. Emissions of DMS are related to the diel nature of the production processes in the former case.

Calculation of a time-averaged flux clearly requires comprehensive emission measurements over an extended time period, allowing integration over the relevant (tidal, diel or annual) cycles.

The low emission rates measured in this study indicate that salt marshes are a minor source of sulfur to the global atmosphere. Furthermore, our measurement of deposition rates show the potential for losses to the same soil surfaces at times of elevated atmospheric concentration.

Literature Cited

1. Andreae, M. O. In The Biogeochemical Cycling of Sulfur and Nitrogen in the Remote Atmosphere; Galloway, J. N.; Charlson, R. J.; Andreae, M. O.; Rodhe, H., Eds.; NATO ASI Series; Reidel, 1985.

2. Adams, D. F.; Farwell, S. O. In Environmental Impact of Natural Emissions; Aneja, V. P., Ed.; APCA, 1984.

3. Jorgensen, B. B.; Okholm-Hansen, B. Atmos. Environ. 1985, 19, 1737-49.

4. Aneja, V. P. In Environmental Impact of Natural Emissions; Aneja, V. P., Ed.; APCA, 1984.

5. Steudler, P. A.; Peterson, B. J. Atmos. Environ. 1985, 19, 1411-16.

6. Dacey, J. W. H.; King, G. M.; Wakeham, S. G. Nature 1987, 330, 643-45.

7. Adams, D. F.; Farwell, S. O.; Robinson, E.; Pack, M. R. EPRI Report EA-1516, Palo Alto, CA, 1980.

8. Cooper, D. J. M.S. Thesis, University of Miami, Florida, 1986.

9. Cooper, D. J.; de Mello, W. Z.; Cooper, W. J.; Zika, R. G.; Saltzman, E. S.; Prospero, J. M.; Savoie, D. L. Atmos. Environ. 1987, 21, 7-12.

10. de Mello, W. Z.; Cooper, D. J.; Cooper, W. J.; Saltzman, E. S.; Zika, R. G.; Savoie, D. L.; Prospero, J. M. Atmos. Environ. 1987, 21, 987-90.

11. Cooper, W. J.; Cooper, D. J.; Saltzman, E. S.; de Mello, W. Z.; Savoie, D. L.; Zika, R. G.; Prospero, J. M. Atmos. Environ. 1987, 21, 1491-95.

12. Natusch, D. F. S.; Klonis, H. B.; Axelrod, H. D.; Teck, R. J.; Lodge, J. P., Jr. Anal. Chem. 1972, 44, 2067-70.

13. U. S. Fish and Wildlife Service National Wetlands Inventory, St. Petersburg, Florida, 1984.

14. Edgerton, E. S. M.S. Thesis, University of Florida, Gainesville, 1981.

15. Judeikis, H. S.; Wren, A. G. Atmos. Environ. 1977, 11, 1221-24.

RECEIVED August 11, 1988

Chapter 4

Synthesis and Emission of Hydrogen Sulfide by Higher Plants

Heinz Rennenberg

Fraunhofer Institute for Atmospheric Environmental Research,
Kreuzeckbahnstrasse, 19, D–8100 Garmisch–Partenkirchen,
Federal Republic of Germany

Hydrogen sulfide is emitted by higher plants in response to the uptake of sulfate, sulfur dioxide, L- and D-cysteine. For each of these sulfur sources a different pathway of synthesis of hydrogen sulfide has been established. Apparently, these pathways are part of an intracellular sulfur cycle that may operate to maintain the cysteine concentration in plant cells at the observed low level when sulfur is present in excess. Hydrogen sulfide emission by higher plants is not a laboratory artefact but is also observed in the field. Though a proper quantification of the volatile sulfur emitted on a global scale can not be obtained from the data presently available, hydrogen sulfide emission by higher plants seems to be a significant contribution to the biogeochemical cycle of sulfur.

Hydrogen sulfide emission by higher plants was first observed by DeCormis in 1968 (1-2). He showed that leaves fumigated with injurious dosages of sulfur dioxide release hydrogen sulfide into the atmosphere. Emission of hydrogen sulfide was found to be a light-dependent process with the sulfur originating from the sulfur dioxide used for the fumigation of the plants. Since these observations of DeCormis, numerous reports from laboratory experiments have been published showing that plant cells are capable of emitting hydrogen sulfide when they are supplied with an excess of sulfur in the form of sulfur dioxide, sulfate or cysteine. Recently, these laboratory studies have partially been verified by field experiments. The aim of the present report is to summarize the state of the art on the synthesis and emission of hydrogen sulfide by higher plants and to introduce a hypothesis on the physiological role of this process in the regulation of sulfur nutrition.

Substrates for Hydrogen Sulfide Emission by Plants

Many naturally occurring sulfur compounds have been analyzed as potential substrates for the emission of volatile sulfur by plant cells (Table I). It has previously been claimed that volatile sulfur is only released by plant cells in response to sulfur dioxide, sulfate, or the L-stereoisomere of cysteine. The volatile sulfur compound emitted in response to these substrates was found to be exclusively hydrogen sulfide (cf. 3,4). Recent investigations have shown that

0097–6156/89/0393–0044$06.00/0

Table I. Potential Substrates for Volatile Sulfur Emissions
from Leaves of Cucurbit Plants

Substrate	Sulfur Compound Emitted	
	Hydrogen Sulfide	Methyl Mercaptan
Sulfur Dioxide	+	-
Sulfate	+	-
L-cysteine	+	-
D-cysteine	+	-
L-cystine	-	-
D-cystine	-	-
O-methyl-L-cysteine	-	-
Cysteamine	-	-
3-mercaptoproprionic acid	-	-
L-homocysteine	-	-
D/L-cysteic acid	-	-
Taurine	-	-
L-methionine	-	+
D-methionine	-	+
S-methyl-L-cysteine	-	+

hydrogen sulfide is also emitted in response to D-cysteine (5,6). In addition it has been observed that methyl mercaptan is released by plant cells or cell homogenates in response to L-methionine, D-methionine, or S-methyl-L-cysteine (7). Carbonylsulfide is metabolized in plants and, therefore, is a potential substrate for hydrogen sulfide emission as well (8). However, only hydrogen sulfide emission in response to sulfur dioxide, sulfate, or cysteine has been studied in detail.

Sulfur Dioxide. Hydrogen sulfide is emitted in response to sulfur dioxide or sulfite in a light dependent process at rates varying between 0.8 and 8 nmol h^{-1} cm^{-2} leaf area (9-12). The actual amount of hydrogen sulfide emitted from leaf cells exposed to this sulfur source is positively correlated with the rate of influx of sulfur dioxide into the leaves; for the induction of measurable hydrogen sulfide emissions a threshold dosage of 0.08 pmol sulfur dioxide per cell over 15 min has been found (11). Mass-balance calculations of sulfur dioxide fluxes reveal that reduction to, and emission of hydrogen sulfide represents 7 to 15 % of the sulfur dioxide taken up (11). Apparently, the bulk of the sulfur dioxide absorbed is not reduced to hydrogen sulfide, but oxidized to sulfate inside the leaves. In cucurbit plants, young leaves were found to be much more resistant to sulfur dioxide than mature leaves (13). The absorption of sulfur dioxide was, however, higher in young than in mature leaves and both, young and mature leaves, oxidize about 60% of the absorbed sulfur dioxide to sulfate (10). Therefore, neither differences in oxidation nor in absorption of sulfur dioxide can account for the differences in sulfur dioxide sensitivity of cucurbit leaves. The only striking difference between the fates of sulfur dioxide in the leaf types is an up to 100-fold higher rate of emission of hydrogen sulfide in young, sulfur-dioxide resistant leaves. Young, but not mature leaves convert about 10% of the sulfur dioxide absorbed to the reduced product, hydrogen sulfide (10). From these observations it may be assumed that hydrogen sulfide emission is a biochemical mechanism of developmentally regulated resistance in cucurbit plants. One may argue that this explanation is unlikely as it implies that the reduction of 10% of the sulfur dioxide absorbed accomplishes resistance, whereas the oxidation of more than 40% of it does not. A possible solution for this contradiction is that a cellular compartment is the predominant site of injury in which reduction to hydrogen sulfide rather than oxidation to sulfate takes place. For several reasons the chloroplast is likely to be this compartment: It is frequently thought that inhibition of photosynthesis is one of the initial metabolic effects of sulfur dioxide on plants; consequently, the chloroplast is regarded to be the first site of injury caused by sulfur dioxide or its products (14-16). When isolated spinach chloroplasts were exposed to radioactively labeled sulfur dioxide, labeled sulfide was found inside the chloroplast (17); apparently, reduction of sulfur dioxide to sulfide can take place inside the chloroplast.

Sulfate. Hydrogen sulfide emission has also been observed, when whole plants, detached leaves, or leaf discs are fed sulfate (3,12,18-24). The rate of hydrogen sulfide emission in response to this sulfur source was found to be in the same order of magnitude as the emission from sulfur dioxide/sulfite. As also observed with sulfur dioxide/sulfite, the actual amount of hydrogen sulfide emitted from sulfate is highly dependent on the sulfate concentration fed and on light intensity (12). In experiments with leaf discs or detached leaves, emission of hydrogen sulfide was usually much higher than if sulfate was fed to the roots of whole plants (3,12,18-20,22). However, when the roots of the plants were injured, attached leaves emitted hydrogen sulfide in amounts comparable to the emission of detached leaves (12). Apparently, the uptake of sulfate via the root system is a barrier preventing the immediate emission of hydrogen sulfide from sulfate in the soil. The extent to which hydrogen sulfide is emitted from the leaves of plants with intact roots upon prolonged exposure to high concentrations of sulfate in the soil remains to be elucidated.

Hydrogen sulfide emission in response to sulfate is under developmental control (20). In this context one has to distinguish between the development of an individual organ and the development of the entire plant: Higher plants develop new organs, e.g. new leaves, during almost all stages of development. At the same time the entire plant is continually developing. Therefore, the actual developmental stage of a leaf can be determined by both the development of the leaf itself and the development of the plant. The potential to emit hydrogen sulfide in response to sulfate is low in leaves of young cucurbit plants, but high in leaves of 3-4 weeks old cucurbit plants (20). This difference was found to be determined by the age of the plant and to be independent of the age, i.e. the developmental stage, of the leaves.

Cysteine. Emission of hydrogen sulfide by plant cells can also occur in response to L-cysteine (3,6,20,21,25-28). Recently, also the D-stereoisomere of cysteine was found to cause hydrogen sulfide emission from leaf discs of cucurbit plants or cultured tobacco cells (6). The amount of hydrogen sulfide emitted in response to D-cysteine was considerably smaller than the amount of hydrogen sulfide emitted from L-cysteine. Apparently, this difference in the emission of hydrogen sulfide can be explained by differences in the rate of influx between the two stereoisomeres of cysteine (6). Both, hydrogen sulfide emission from L-cysteine (28) and from D-cysteine (Rennenberg, unpublished results) are light-independent processes. The hydrogen sulfide emitted in response to L-cysteine is derived directly from the L-cysteine added (28). As observed with the hydrogen sulfide emission from sulfate, the potential for hydrogen sulfide emission in response to L-cysteine is under developmental control (20). Leaves from young cucurbit plants possess a higher potential for the emission of hydrogen sulfide from L-cysteine than leaves from plants 3-4 weeks old. Independent of the developmental stage of the plant, the potential for hydrogen sulfide emission in response to L-cysteine decreased with increasing age of the individual leaf. Therefore, the potential for hydrogen sulfide emission in response to L-cysteine is determined by the development of both the individual leaf and the entire plant.

Paths of Hydrogen Sulfide Synthesis in Higher Plants

For each of the three precursors of hydrogen sulfide, i.e. sulfur dioxide/sulfite, sulfate, and cysteine, a different biosynthetic pathway has been established. Figure 1 gives an overall view of these three pathways; a suggestion for a path of synthesis of hydrogen sulfide from carbonyl sulfide is included.

Sulfur Dioxide/Sulfite. At least three possible pathways for the synthesis of hydrogen sulfide from sulfur dioxide/sulfite have to be considered (Figure 1). Two observations suggest that light-dependent reduction of sulfate to sulfide may be involved in this pathway: First, the major part of the sulfur dioxide/sulfite absorbed by leaf cells is oxidized to sulfate (10); second, hydrogen sulfide emission in response to sulfur dioxide/sulfite is a light dependent process (12). The reduction of sulfate may be followed by the release of sulfide from carrier-bound sulfide; alternatively, the sulfide moiety of carrier-bound sulfide may be incorporated into cysteine that is subsequently degraded to ammonium, pyruvate, and hydrogen sulfide. However, experiments with ^{35}S-sulfur dioxide and ^{35}S-sulfate show that conversion of sulfite to sulfate

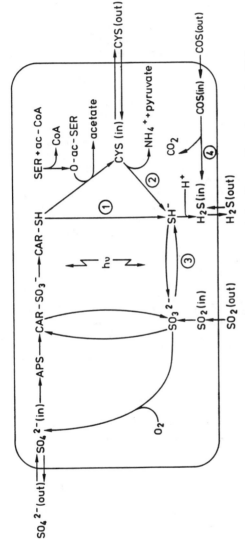

Figure 1. Pathways of hydrogen sulfide synthesis in plants. 1) release of sulfide from carrier-bound sulfide; 2) desulfhydration of cysteine; 3) direct reduction of sulfite; 4) hydrolysis of carbonyl sulfide. CAR, carrier; APS, adenosine-5'-phosphosulfate; SER, serine; O-ac-SER, O-acetyl-serine; CoA, coenzyme A; ac-CoA, acetyl-coenzyme A; CYS, cysteine.

is not involved in the synthesis of hydrogen sulfide from sulfur dioxide/sulfite (10): When cucurbit plants were exposed to unlabeled sulfate and fumigated with ^{35}S-sulfur dioxide, the specific radioactivity of the hydrogen sulfide emitted was comparable to the specific radioactivity of the sulfur dioxide used for the fumigation of the leaves; exposure of cucurbit plants to ^{35}S-sulfate and fumigation with unlabeled sulfur dioxide leads to the emission of almost unlabeled hydrogen sulfide. Therefore, another pathway of hydrogen sulfide production from sulfur dioxide/sulfite is much more plausible, i.e. the direct, light-dependent reduction of sulfite to sulfide (see Figure 1, pathway 3).

Sulfate. As for the production of hydrogen sulfide from sulfur dioxide/sulfite at least three possible pathways for the light-dependent synthesis of hydrogen sulfide in response to sulfate can be assumed, i.e. first the light-dependent reduction of sulfate to carrier-bound sulfide followed by a release of the sulfide moiety from its carrier; second the light-dependent reduction of sulfate to carrier-bound sulfide followed by an incorporation of the sulfide moiety into cysteine and subsequent degradation of cysteine; third the release of sulfite from carrier-bound sulfite followed by reduction of free sulfite to sulfide (see Figure 1).

Experiments on the hydrogen sulfide emission in response to cysteine suggest that free sulfite is not an intermediate in the path of hydrogen sulfide production from sulfate (28-29). Upon feeding of leaf tissue with L-cysteine a sulfite pool sufficiently large to cause injury to the cells is generated (12,22,29). If the pathway of hydrogen sulfide production from sulfate was to proceed via free sulfite, a sulfite pool should be developed causing injury to the leaf tissue in a way reminiscent to the injury caused by L-cysteine. The lack of symptoms of injury when hydrogen sulfide is emitted in response to sulfate (12) makes it unlikely that free sulfite is an intermediate in the synthesis of hydrogen sulfide from sulfate.

Apparently, L-cysteine also is not an intermediate in the conversion of sulfate to hydrogen sulfide. Cysteine synthase and cysteine desulfhydrase are thought to be pyridoxal phosphate-dependent enzymes (30,31) and may therefore be inhibited by aminooxy acetic acid (32). Light-dependent hydrogen sulfide emission in response to sulfate is, however, stimulated rather than inhibited by this compound (3). Inhibition of glutathione synthesis, a major path of sulfate assimilation into organic compounds (22), keeps sulfate reduction unaffected and reduces the incorporation of labeled sulfur into cysteine; emission of hydrogen sulfide in response to sulfate is, however, stimulated by up to 80% (19). These observations indicate that the excess sulfate reduced in the presence of inhibitors of glutathione or cysteine synthesis may be released into the atmosphere in the form of hydrogen sulfide without being incorporated into cysteine.

When cucurbit cells are fed O-acetylserine or its metabolic precursors the rate of hydrogen sulfide emission in response to sulfate declines, and the incorporation of labeled sulfur from 35S-sulfate into cysteine increases (18). Inhibition of the synthesis of the O-acetylserine precursor acetyl coenzyme A by 3-fluoropyruvate (33) enhances hydrogen sulfide emission, but inhibits cysteine synthesis (18). These observations indicate that the availability of O-acetylserine is the rate limiting factor in cysteine synthesis. Hydrogen sulfide may be emitted to the extent the amount of sulfate reduced exceeds the synthesis of O-acetylserine. Therefore, direct release of sulfide from carrier-bound sulfide appears to be responsible for the emission of hydrogen sulfide in response to sulfate (Figure 1, pathway 1).

L-/D-cysteine. Hydrogen sulfide is produced from L-cysteine in a light-independent process that can be inhibited in vivo and in vitro by aminooxy acetic acid, an inhibitor of pyridoxal phosphate-dependent enzymes; the hydrogen sulfide emitted in response to L-cysteine is directly derived from the L-cysteine fed (6,28). Therefore, hydrogen sulfide appears to be produced from L-cysteine by a pyridoxal phosphate-dependent, L-cysteine specific cysteine desulfhydrase. This conclusion is supported by the finding that in cucurbit plants the developmental changes in the potential for hydrogen sulfide emission from L-cysteine, as observed in experiments with leaf discs in vivo, are paralleled by changes in the extractable L-cysteine desulfhydrase activity in vitro (20).

The recent finding of hydrogen sulfide emission in response to D-cysteine reveals, however, that degradation of cysteine in plant cells may be much more complicated (6). Whereas emission of hydrogen sulfide in response to D-cysteine by leaf discs of cucurbit plants or cultured tobacco cells was considerably smaller than in response to L-cysteine, D-cysteine desulfhydrase activity as measured in crude homogenates in vitro was more than one order of magnitude higher than L-cysteine desulfhydrase activity. From this observation one might assume that desulfhydration of D-cysteine is an activity of a stereo-unspecific cysteine desulfhydrase without any physiological significance for the intact cell. This idea is supported by the partial purification of a cystine lyase from cabbage which degraded cystine, cysteine, O-acetylserine, and several S-substituted cysteines by an alpha, beta elimination reaction (34). Several pieces of evidence indicate, however, that desulfhydration of L- and D-cysteine is catalyzed by different enzymes localized in different compartments of the cell: L- and D-cysteine desulfhydrase activity in spinach leaves can partially be purified and separated from each other by column chromatography (35). Both, L- and D-cysteine desulfhydrase activity in homogenates of cucurbit and tobacco cells are inhibited by the reaction products pyruvate and ammonium, but the patterns of inhibition of the activities are entirely different (6). Differential centrifugation of tobacco cell homogenates reveals that L-cysteine desulfhydrase is a soluble enzyme predominantly localized in chloroplasts and mitochondria, whereas D-cysteine desulfhydrase is a soluble enzyme predominantly localized in the cytoplasm (6). Investigations with purified, isolated organells have proven that L-cysteine desulfhydrase is present in chloroplasts of tobacco cells, but also showed that L-cysteine desulfhydrase activity in the mitochondrial fraction is the consequence of an unspecific adsorption of the enzyme to the organell during the homogenization procedure (Rennenberg, unpublished results). These data suggest that in cucurbit, tobacco, and spinach cells desulfhydration of L- and D-cysteine is catalyzed by different enzymes. As C-S lyases from other Brassica species exhibited a higher substrate specificity and did not degrade cysteine, the presence of an unspecific C-S lyase may be restricted to cabbage (36). The physiological role of the cytoplasmic D-cysteine desulfhydrase in plants remains obscure, as the origin of its substrate, D-cysteine, is unknown. The D-enantiomer of cysteine may, however, be produced by a cysteine racemase in a similar way D-alanine is synthesized from L-alanine in pea seedlings (37). Such a racemase may be an additional regulatory device to control the fluxes of L-cysteine into biosynthetic pathways, i.e. protein-, glutathione-, or methionine-synthesis, or into its degradative path, when L-cysteine is present in the cytoplasm in excess.

L-Cysteine desulfhydrase in leaves of cucurbit plants is a constitutive enzyme whose activity can be enhanced by preincubation of leaf discs with L-cysteine, D-cysteine, or structural analogs of L-cysteine at millimolar concentrations; preincubation with cystine does not affect the activity of the enzyme (20,26). Although the stimulation of L-cysteine desulfhydrase activity is already half-maximal within 15 min of preincubation with L- or D-cysteine, the enhanced activity is not the consequence of an activator formed in the leaf discs during preincubation or an inhibitor present in cells not preincubated (26). Stimulation of L-cysteine desulfhydrase in leaf discs is by itself light-independent; the preexposure of the entire plant with light or dark, however, determines the extent to which stimulation can occur. Darkness reduces, and exposure to light after a period of dark restores the potential for stimulation of L-cysteine desulfhydrase activity (26). From these observations it can be concluded that desulfhydration of L-cysteine is controlled by a compound not synthesized, but metabolized in the leaves. The chemical nature of this compound remains to be elucidated.

Carbonyl Sulfide. Recent investigations suggest that vegetation may be the major sink for biogenic carbonyl sulfide emitted from soils (38-42). When stomata are open, mesophyll- rather than stomatal resistance seems to determine the deposition velocity of this compound in plants (39). This finding suggests that carbonyl sulfide is metabolized in plants. Experiments with ^{35}S-carbonyl sulfide showed that, as a consequence of absorption of this compound by the leaves, its sulfur is incorporated into sulfate, methionine, cysteine and glutathione inside the leaves; in addition, a significant part of the radioactivity absorbed by the leaves is translocated to the roots (8). Though the chemical nature of the sulfur translocated has not been analyzed in the experiments, the finding of reduced glutathione (GSH) being the predominant long-distance transport form of reduced sulfur in higher plants (43,44) suggests that radioactively labeled GSH produced from the ^{35}S-carbonyl sulfide in the leaves is translocated to the roots. When ryegrass (Lolium perenne) is used in the experiments, about 17% of the radioactivity absorbed is lost from the plants within 24 hr (8); as exudation by the roots does not appear to be responsible for this loss, it might be assumed that part of the carbonyl sulfide absorbed is re-emitted by the plants. These observations can be explained by the hydrolysis of carbonyl sulfide to hydrogen sulfide and carbon dioxide, and the partial incorporation of hydrogen sulfide into endogenous, organic sulfur compounds, its partial oxidation to sulfate, and its partial emission into the atmosphere. This conclusion is consistent with our present knowledge on sulfur metabolism in plants as summarized in Figure 1. A carbonic anhydrase catalyzing the hydrolysis of carbonyl sulfide to hydrogen sulfide and carbon dioxide remains to be demonstrated in plant cells. Experiments showing that hydrogen sulfide is emitted by plants upon deposition of carbonyl sulfide are needed to test these assumptions.

The Physiological Role(s) of Hydrogen Sulfide Emission

Why is hydrogen sulfide produced and emitted by higher plants? What are the benefits for a plant cell of being able to release hydrogen sulfide into the atmosphere? A first indication for the physiological role(s) of the emission of hydrogen sulfide comes from pulse-chase experiments with L-cysteine (29). In these experiments the emission of radioactively labeled and unlabeled cysteine as well as the labelling pattern of intracellular sulfur compounds was analyzed in leaf discs of cucurbit plants. Upon feeding of ^{35}S-L-cysteine, radioactivity enters the sulfide pool of the cells thereby causing the emission of a first wave of radioactively labeled hydrogen sulfide. However, part of labeled sulfide originating from the desulfhydration of L-cysteine passes through the sulfite and sulfate pool of the cells, is then reduced, and incorporated into the cells' sulfide pool again, thereby giving rise to the emission of labeled hydrogen sulfide in a second wave. Two observations indicate that the second wave of emission of labeled hydrogen sulfide is the consequence of a reaction different from desulfhydration, i.e. the process responsible for the first wave of emission of labeled hydrogen sulfide. First, the radioactivity is not reincorporated into cysteine. Second, the second wave of emission of labeled hydrogen sulfide can be prevented by inhibitors of photosynthetic electron transport or by turning off the light (29). These data suggest the existence of an intracellular sulfur cycle in leaf cells of cucurbit plants (Figure 1). Apparently, the cysteine fed is degraded to sulfide, but only part of it is emitted into the atmosphere in the form of hydrogen sulfide. The balance seems to be oxidized to sulfite, and subsequently to sulfate. The sulfate synthetized this way can then be reduced by the process of light-dependent sulfate assimilation. In the presence of excess exogenous cysteine further synthesis of cysteine may be blocked by the expansion of the intracellular cysteine pool. Under these conditions sulfide may be split off carrier-bound sulfide and may be emitted as hydrogen sulfide (Figure 1).

A possible function of this intracellular sulfur cycle is to buffer, i.e. to homeostatically regulate, the cysteine concentration of the cells. Irrespective of whether sulfate, cysteine, or sulfur dioxide is available as sulfur source, the intracellular sulfur cycle would allow a plant cell to use as much of these compounds as necessary for growth and development. At the same time, it would give a plant cell the possibility to maintain the cysteine pool at an appropriate concentration by emitting excess sulfur into the atmosphere. Thus, emission of hydrogen sulfide may take place when the influx of sulfur in the form of sulfate, cysteine, or sulfur dioxide exceeds the conversion of these sulfur sources into protein, glutathione, methionine, and other sulfur-containing components of the cell.

Maintaining cysteine at the observed low concentration in plant cells (30,45,46) may not only be necessary in the presence of excess sulfur. Changes in light intensity, e.g., will cause tremendous changes in the rate of sulfate reduction. Similar fluctuations in the thiol content of plant cells can not be tolerated, as essential metabolic processes such as protein synthesis (47), glycolysis (cf. 48), etc. are extremely sensitive to even minor changes in the thiol/disulfide status. Obviously, hydrogen sulfide emission out of an intracellular sulfur cycle can prevent such fluctuations. Apparently, also qualitative changes in sulfur nutrition can result in hydrogen sulfide emission. Cultured tobacco cells emit hydrogen sulfide when they are transferred from one sulfur source to another or when a sulfur source is added after a period of sulfur starvation (21). Hydrogen sulfide is, however, only emitted during the first 12 to 16 h after addition of sulfur. From these experiments it can be concluded that a transient hydrogen sulfide emission occurs when the sulfur metabolism of plant cells is adjusted to a new sulfur source or to the availability of sulfur after a period of sulfur starvation. Hydrogen sulfide emission may be necessary under these conditions to prevent the accumulation of toxic amounts of sulfur, e.g. of cysteine, when the uptake of sulfur exceeds its metabolism inside the cells.

It is obvious that getting rid of excess sulfur taken up upon fumigation of plants with sulfur dioxide may be beneficial for plants. However, in rural or remote environments the sulfate available to a plant in the soil is its predominant source of sulfur. One may therefore assume that regulatory processes at the level of the uptake of sulfate by the root system will be sufficient to control the influx of sulfate into plants. Our present knowledge on the regulation of sulfate uptake by plants (cf. 4) suggests that there is still a need for an additional regulatory device whose requirements can be fulfilled by hydrogen sulfide emission from the leaves. When excess sulfate becomes available to the roots of a plant (Figure 2), e.g. by fertilization with sulfur, sulfate is initially taken up in amounts exceeding the plants requirements for sulfur. As a consequence, excess sulfate is translocated to the leaves with the transpiration stream, thereby leading to an increase in the sulfate content of the leaf tissue (cf. 4) and to an expansion of the glutathione pool(s) of the cells (49). The latter observation indicates that the presence of high concentrations of sulfate in the leaves results in a reduction of sulfate to sulfide that exceeds the plants needs for reduced sulfur. Under these conditions glutathione, the predominant long distance transport form of reduced sulfur (43,44), may be translocated from the leaves to the roots at an extent exceeding the needs of the roots for reduced sulfur in protein synthesis. As a consequence, glutathione may accumulate in the roots. Recent experiments with cultured tobacco cells (Rennenberg, unpublished results) show that the sulfate carrier in chloroplast free cells is inhibited by glutathione at physiological concentrations, whereas the sulfate carrier of green cells is not affected. Glutathione, accumulating in the roots, may therefore eventually block further uptake of sulfate by the roots. However, the involvement of two long-distance transport processes will cause a significant delay between the availability of excess sulfate in the soil and the inhibition of sulfate uptake of root cells by glutathione (Figure 2). In the meantime, transient hydrogen sulfide emission by the leaves may be an important process in plants to prevent the accumulation of toxic amounts of reduced sulfur. Clearly, further experiments are necessary to test this hypothesis.

Sulfur metabolism in plants goes through changes during ontogenesis. These changes also effect the availability of cysteine in a way that emission of hydrogen sulfide may be assumed. When sulfur-poor storage proteins are synthesized from sulfur-rich leaf proteins during seed-ripening of cereals (50), when sulfur-rich storage proteins of oil-seed are mobilized upon germination (51), when glutathione is degraded in spruce needles in spring (52), the generation of excess cysteine and its degradation to ammonium, pyruvate, and hydrogen sulfide is very likely to occur. However, none of these developmental processes has been sufficiently studied for the emission of hydrogen sulfide into the atmosphere.

Hydrogen Sulfide Emission in the Field

Investigations by several authors have shown that the emission of hydrogen sulfide is not a laboratory artifact, but also occurs in nature (3,9,53). Emission of volatile sulfur under field conditions was first observed with crops growing without addition of sulfur containing fertilizers and without significant gas mixing ratios of sulfur compounds in the atmosphere (3). In the light, substantial amounts of volatile sulfur were found to be released by all species tested, whereas only minute amounts were emitted in the dark. Even the amount of volatile sulfur emitted in the light was still too small to be analyzed by the gas chromatographic system available to the authors (3). Thus it is not known whether hydrogen sulfide or/and other volatile sulfur compounds were emitted by the plants in these studies.

The first investigations demonstrating that hydrogen sulfide is emitted by plants under field conditions were reported by Hällgren and Fredriksson (9). In these experiments pine trees were exposed in the field to low concentrations of sulfur dioxide. Under these conditions hydrogen sulfide emission was observed to proceed at a maximum rate of 14-20 nmol per mg chlorophyll and hour and was found to continue for several hours after termination of prolonged exposure to sulfur dioxide. Recently, experiments with ^{35}S-sulfate in the field revealed that volatilization is also a strategy of deciduous trees to deal with excess sulfur in their environment (53). The few field data available have been used for an extrapolation of hydrogen sulfide emission from leaves to all species in all

Figure 2. Fluxes of sulfur in higher plants. Open arrows: fluxes of sulfate; closed arrows: synthesis and translocation of reduced sulfur compound. CYS, cysteine; MET, methionine, GSH, reduced glutathione; X-SH, carrier-bound sulfide.

climates (3). From this calculation it appears that sulfur emission by higher plants may account for 7.4 million metric tons of sulfur per year, or approximately 7% of the total biogenic sulfur emitted into the atmosphere. In other studies it was estimated that the amount of hydrogen sulfide produced from photosynthetic reduction of sulfate on a global scale constitutes 50% of the biogenic sulfur in the atmosphere (24). However, these calculations were based on experiments with soybean plants exposed to physiological concentrations of sulfate in growth cabinets under conditions where the roots of the plants are trimmed. As hydrogen sulfide emission can be stimulated considerably when roots are injured (12), these experiments do not seem to be appropriate to calculate the impact of hydrogen sulfide emission by plants in the global biogeochemical cycle of sulfur. The finding that the rates of hydrogen sulfide emission were 1000-fold higher in the laboratory than the spontaneous emissions in the field suggests, on the other hand, that higher plants are capable of contributing much more sulfur to the atmosphere than indicated by the data so far obtained in the field. Although the data presently available indicate that hydrogen sulfide emission by higher plants is a significant contribution to the biogeochemical cycle of sulfur, further field measurements and experiments are urgently needed for a proper quantification of this process on a global scale.

Literature Cited

1. DeCormis, L. C. R. Acad. Sci. 1968, 266, 683-85.
2. DeCormis, L. Proc. 1st Eur. Congr. Influence Air Pollut. on Plants and Animals; 1969; p 75.
3. Filner, P.; Rennenberg, H.; Sekiya, J.; Bressan, R. A.; Wilson, L. G.; LeCureux, L.; Shimei, T. In Gaseous Air Pollutants and Plant Metabolism; Butterworth: London, 1984; pp 291-312.
4. Rennenberg, H. Ann. Rev. Plant Physiol. 1984, 35, 121-53.
5. Schmidt, A. Z. Pflanzenphysiol. 1982, 107, 301-12.
6. Rennenberg, H.; Arabatzis, N.; Grundel, I. Phytochemistry 1987, 26, 1583-89.
7. Schmidt, A.; Rennenberg, H.; Wilson, L. G.; Filner, P. Phytochemistry 1985, 24, 1181-85.
8. Brown, K. A.; Kluczewski, S. M.; Bell, J. N. B. Environ. Exptl. Bot. 1986, 26, 355-64.
9. Hällgren, J.-E.; Fredriksson, S.-A. Plant Physiol. 1982, 70, 456-59.
10. Sekiya, J.; Wilson, L. G.; Filner, P. Plant Physiol. 1982, 70, 437-41.
11. Taylor, G. E.; Tingey, D. T. Plant Physiol. 1983, 72, 237-44.
12. Wilson, L. G.; Bressan, R. A.; Filner, P. Plant Physiol. 1978, 61, 184-89.
13. Bressan, R. A.; Wilson, L. G.; Filner, P. Plant Physiol. 1978, 61, 761-67.
14. Ziegler, I. Residue Rev. 1975, 56, 79-104.

15. Hällgren, J. E. In Sulfur in the Environment, Nriagu; J. O., Ed.; Wiley: New York, 1978; Part II, pp 163-209.
16. Wellburn, A. R. In Sulfur Dioxide and Vegetation; Winner, W. E.; Mooney, H. A.; Goldstein, R. A., Eds; Stanford University: Stanford, 1985; pp 133-47.
17. Silvius, J. E.; Baer, C. H.; Dodrill, S.; Patrick, H. Plant Physiol. 1976, 57, 799-801.
18. Rennenberg, H. Plant Physiol. 1983, 73, 560-65.
19. Rennenberg, H.; Filner, P. Plant Physiol. 1982, 69, 766-70.
20. Rennenberg, H.; Filner, P. Plant Physiol. 1983, 71, 269-75.
21. Rennenberg, H.; Reski, G.; Polle, A. Z. Pflanzenphysiol. 1983, 111, 189-202.
22. Sekiya, J.; Schmidt, A.; Rennenberg, H.; Wilson, L. G.; Filner, P. Phytochemistry 1982, 21, 2173-78.
23. Spaleny, J. Plant Soil 1977, 48, 537-61.
24. Winner, W. E.; Smith, C. L.; Voch, G. W.; Mooney, H. A.; Bewley, J. D.; Krouse, H. R. Nature 1981, 289, 672-73.
25. Harrington, H. M.; Smith, I. K. Plant Physiol. 1980, 65, 151-55.
26. Rennenberg, H. Phytochemistry 1983, 22, 1557-60.
27. Rennenberg, H.; Grundel, I. J. Plant Physiol. 1985, 120, 47-56.
28. Sekiya, J.; Schmidt, A.; Wilson, L. G.; Filner, P. Plant Physiol. 1982, 70, 430-36.
29. Rennenberg, H.; Sekiya, J.; Wilson, L. G.; Filner, P. Planta 1982, 154, 516-24.
30. Giovanelli, J.; Mudd, S. H.; Datko, A. H. In The Biochemistry of Plants; Miflin, B. J., Ed; Academic: New York, 1980; Vol. 5, pp 454-505.
31. Kredich, N. M.; Keenan, B. S.; Foote, L. J. J. Biol. Chem. 1972, 247, 7157-62.
32. Meister, A. Biochemistry of the Amino Acids; Academic: New York, 1965; Vol. 1, p 411.
33. Bisswanger, H. J. Biol. Chem. 1981, 256, 815-22.
34. Hall, I. D.; Smith, I. K. Plant Physiol. 1983, 72, 654-58.
35. Schmidt, A. Z. Pflanzenphysiol. 1982, 107, 301-12.
36. Hamamoto, A.; Mazelis, M. Plant Physiol. 1986, 80, 702-706.
37. Ogawa, T.; Kawasaki, Y.; Sasaoka, K. Phytochemistry 1978, 17, 1275-76.

38. Kluczewski, S. M.; Bell, J. N. B.; Brown, K. A.; Minski, M. J. In <u>Ecological Aspects of Radionuclide Release</u>; Coughtrey, P. J.; Bell, J. N. B.; Roberts, T. M., Eds.; Blackwell: Oxford, 1983; pp 91-104.

39. Kluczewski, S. M.; Brown, K. A.; Bell, J. N. B. <u>Rad. Prot. Dosimetry</u> 1985, <u>11</u>, 173-77.

40. Kluczewski, S. M.; Brown, K. A.; Bell, J. N. B. <u>Atmos. Environ</u>. 1985, <u>19</u>, 1295-99.

41. Brown, K. A.; Bell, J. N. B. <u>Atmos. Environ</u>. 1986, <u>20</u>, 537-40.

42. Fall, R.; Kuster, W. C.; Fehsenfeld, F. C.; Goldan, P. D. This Volume.

43. Rennenberg, H.; Schmitz, K.; Bergmann, L. <u>Planta</u> 1979, <u>147</u>, 667-72.

44. Bonas, U.; Schmitz, K.; Rennenberg, H.; Bergmann, L. <u>Planta</u> 1982, <u>155</u>, 82-88.

45. Smith, I. K. <u>Plant Physiol</u>. 1975, <u>55</u>, 303-307.

46. Brunhold, C.; Schmidt, A. <u>Plant Physiol</u>. 1978, <u>61</u>, 342-47.

47. Fahey, R. C.; Di Stefano D. L.; Meier, G. P.; Bryan R. N. <u>Plant Physiol</u>. 1980, <u>65</u>, 1062-66.

48. Jocelyn, P. C. <u>The Biochemistry of the SH Group</u>; Academic: London, 1972; pp 247, 280.

49. De Kok, L. J.; De Kan, P. J. L.; Tanczos, O. G.; Kuiper, P. J. C. <u>Physiol. Plant</u>. 1981, <u>53</u>, 435-38.

50. Raybould, C.; Unsworth, M. H.; Gregory, P.-J. <u>Nature</u> 1977, <u>267</u>, 146-47.

51. Youle, R. J.; Huang, A. H. C. <u>Am. J. Bot</u>. 1981, <u>68</u>, 44-48.

52. Esterbauer, H.; Grill, D. <u>Plant Physiol</u>. 1978, <u>61</u>, 119-21.

53. Garten, Jr., C. T. This Volume.

RECEIVED August 11, 1988

Chapter 5

Sulfur Emissions from Roots of the Rain Forest Tree *Stryphnodendron excelsum*

Ecosystem, Community, and Physiological Implications

Bruce Haines[1], Marilyn Black[2], and Charlene Bayer[2]

[1]Botany Department, University of Georgia, Athens, GA 30602
[2]Analytical and Instrumentation Branch, Georgia Tech Research Institute, Georgia Institute of Technology, Atlanta, GA 30332

Roots of *Stryphnodendron excelsum* trees in a lowland rain forest in eastern Costa Rica emit sulfur gases. Extrapolated annual estimates of emissions, based on *S. excelsum* tree density, are on the order of 0.29 kg $S.ha^{-1}.yr^{-1}$. At the ecosystem level, this flux is too small to account for the 11 kg $S.ha^{-1}.yr^{-1}$ SO_4-S input-output discrepancy and acid rain reported earlier. At the physiological level, emission of CS_2 is stimulated by disturbance to the roots of *S. excelsum*. Considering the known toxicity of CS_2 to nematodes, root rot fungi, insects, and nitrifying bacteria we suggest that CS_2 emission may, at the community level, be a defensive mechanism against root predators, and pathogens and a nitrogen conserving mechanism.

Sulfur gas flux from living vascular plants to the atmosphere is a little studied part of the global sulfur cycle. Sulfur fluxes from other parts of the biosphere to the atmosphere are more studied. For example, sulfur gas emissions are known from marine systems ([1-6]), from coastal estuarine and marsh systems ([7-16]), from fresh water floodplain lakes ([17]), from temperate soils ([18-20]), and from tropical forests ([20-22]). Laboratory studies have demonstrated sulfur gas emissions from soils ([23-26]), from intact higher plants ([27-30]), and plant parts (see review by Rennenberg [31]). An inventory of reduced sulfur gas emissions from soil, crops, and trees in the United States has been started ([32,33,34]). General reviews of the sulfur cycle are provided by Smil ([35]) and by Ivanov and Freney ([36]). From these studies, the geographic distribution of biogenic sulfur gas source strengths on a global basis and the phylogenetic distribution of sulfur gas emissions within the plant kingdom are still relatively unknown.

During a survey of sulfur gas emissions from a central American rainforest ([37]), *Stryphnodendron excelsum* Harms (Mimosaceae) was found to be a sufficiently strong sulfur emitter that its location in the forest could be detected by odor. The present study attempted to quantify sulfur emissions from *S. excelsum* to the atmosphere.

0097–6156/89/0393–0058$06.00/0

Study Site and Methods

Sampling was performed at the La Selva Biological Station of the Organization for Tropical Studies, Puerto Viejo de Sarapiqui, Provincia Heredia, Costa Rica, 10°24-26 'N lat, 84° 00-02 'W long. Three *S. excelsum* trees ranging from 0.87 to 0.96 m dia and from 22 to 32 m in height were selected within 100 m of the laboratory to facilitate rapid processing of gas samples.

Preliminary sampling suggested that roots of one of the trees extended more than 16 m from the trunk. Later sampling was performed at 4 m intervals on transects on each of the compass quadrants centered on two trees. For one tree near a forest edge, one transect line was run toward the forest interior for 16 m. At each sampling point on each transect a 25 x 25 cm template was used to cut leaf litter and roots. Litter was removed to a 250 ml Nalgene polypropylene wide mouth centrifuge jar (Nalge Co. Rochester, NY). Soil and associated roots were excavated to a depth of 10 cm and removed to the lab where they were separated by 4 mm sieves without addition of water. Dry sieving avoided the possibility of flooding and anaerobism changing the quality and amount of reduced sulfur gas emission. The total root mass was sorted by odor into sulfur emitting roots and non-emitting roots. Sulfur emitting roots were placed in 250 ml polypropylene centrifuge jars and incubated at ambient rainforest temperature. The time course of sulfur gas accumulation in the centrifuge jars was quantified by use of a Perkin-Elmer Sigma 4B gas chromatograph (Perkin-Elmer, Norwalk, CT) fitted with a sulfur specific flame photometric detector, a 2 mm internal dia. 9 m teflon column packed with 5% polyphenyl ether + 0.5% H_3PO_4 on 40/60 teflon, at 90C, with 99 mls N_2 carrier min^{-1}. Reference hydrogen sulfide, carbon disulfide, dimethyl sulfide, and ethyl mercaptan were supplied from permeation tubes (VICI Metronics, Santa Clara, CA) inserted into a Tracor Model 432 Tri-Perm Permeation Calibration system (Tracor, Inc., Austin, TX). Standards and unknown samples were pulled either from sample incubation bottles or from the permeation system to the gas chromatograph through multiposition zero dead volume sampling valves and a 2 ml teflon sample loop using a 10 ml syringe. Sample loop, valves, and connecting teflon lines were heated to 65°C to minimize surface adsorption of sulfur gases.

Verification of CS_2 as the principal sulfur gas from the roots of each of the three trees sampled in this study was performed with a Finnigan OWA 3B gas chromatograph-mass spectrograph (Finnigan Corp., Sunnyvale, CA).

Three incubation regimes were used for sulfur emitting roots.

1) Field moist roots. All root samples were physically pulled from field collected soil and incubated at existing moisture conditions.

2) Vigorously washed roots. Following incubations of field moist roots, they were removed from the bottles and squeezed repeatedly under running water to dislodge adhering soil particles in preparation for dry weight determinations. When the odor of increased sulfur emissions was noted during washing, one subset of samples was re-incubated and reanalyzed by gas chromatography to determine the magnitude of change in emission rate.

3) Gently rinsed roots. For another set of 12 root samples, following incubation at field moist conditions, the roots were kept in the incubation jars and the gas was displaced by 3 changes of water in immediate succession. This was done to determine if wetting alone or both wetting and squeezing were required to stimulate sulfur gas emissions.

Following the various incubation regimes, all roots were washed free of soil and dried to constant weight at 65°C. Dry roots were sorted into diameter classes of 0-1.9, 2.0-4.9, 5.0-9.9 and greater than 10 mm. Means, standard deviations, and correlations were computed with the Statistical Analysis System (38).

Results

Field Moist Samples. Emissions of the sulfur gases hydrogen sulfide, carbon disulfide, dimethyl sulfide + ethyl mercaptan from forest floor litter, roots, and soil calculated from the initial slopes of time course incubations are given in Table I. Rates are given for distances of 4, 8, 12, and 16 m from trunks of *S. excelsum* trees as g S. $m^{-2}.min^{-1}$. Summing outward from the tree trunk, the annual extrapolated emission rate multiplied by the areas of each successive circle (Figure 1), estimated an overall emission of 100 g S. $tree^{-1}.year^{-1}$ from a 200 m^2 area. The fractional contribution of *S. excelsum* trees to the stem cross sectional area (basal area) of the forest is 0.06 (39). From the extrapolated annual emission rate and the proportional contribution of *S. excelsum* to the cross sectional area, an annual emission of 0.29 kg S. $ha^{-1}.yr^{-1}$ was calculated. This assumes constant emission rates instead of possible diurnal and seasonal cycling of rates.

Vigorously Washed Roots. The time courses of CS_2 at field moist conditions and after two vigorous washings are given in Figure 2a. Results from each of the two washings are replotted in Figures 2b and 2c synchronizing the beginnings of the re-incubations in order to facilitate comparisons of initial slopes. For this particular set of roots, CS_2 was detectable in 5 incubations of field moist roots, but detectable in 7 incubations after the first wash and in 8 after the second wash. Following rapid initial accumulation of CS_2, concentrations either increased more slowly, remained the same or decreased.

Gently Rinsed Roots. The field moist incubations showed rapid CS_2 accumulation followed by declining concentration. CS_2 production was not continuous. The initial slopes for the CS_2 time courses were greater for the field moist incubations (Figure 3a) than for post-rinse incubations (Figure 3b). CS_2 was detectable in 7 of the field moist incubations but continued in only 5 of them following rinsing.

Root Weight and S Emission per Unit Root Weight. Dry weights of non-sulfur emitting and of sulfur emitting roots (Table II) show about the same total amount of roots but decreasing amounts of sulfur emitting roots with increasing distance away from *S. excelsum* trunks. The dry weights of 0-1.9 mm dia. non-sulfur emitting roots were negatively correlated (r -0.3, P<0.049, n=36) with the dry weights of 2-4.9 mm diameter sulfur emitting roots. We do not interpret this as evidence for allelopathy. Emission rates first calculated as $g.m^{-2}$ forest in Table I are recalculated using individual sample root weights to give sulfur emission rates per gram root dry weight in Table III to facilitate comparison of data with data from future studies.

Discussion

Results of this study have implications for ecosystem ecology, community ecology, and for physiological-evolutionary ecology.

Ecosystem Level. This study was designed to estimate the potential proportional contribution of sulfur gas emissions from *S. excelsum* to the 11 $kg.ha^{-1}.yr^{-1}$ SO_4-S input-output discrepancy of the rainforest (40). A gas

Table I. Potential Sulfur Emission, Mean (Standard Deviation) g S x 10^{-9} $m^{-2}.min^{-1}$ from Roots, Litter and Soil at 4 Distances from *S. excelsum* Trees. Sample Size = 9 at Each Distance

Source	Distance, m			
Sulfur Compound	4	8	12	16
Roots				
CS_2-S	2959(2010)	1663(1417)	1049(1242)	281(416)
H_2S-S	41.6(83.2)	3.5(10.7)	0(0)	0(0)
Litter				
$(CH_3)_2S$-S + C_2H_5SH-S	0.006(0.22)	0.23(0.62)	0.18(0.49)	0.17(0.38)
Soil	0(0)	0(0)	0(0)	0(0)
Total	3000(1981)	1666(1420)	1049(1242)	281(410)

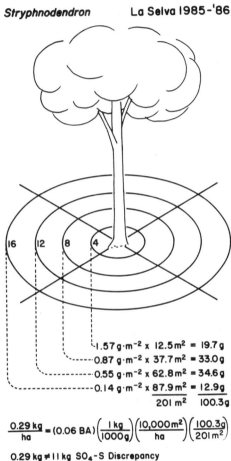

Figure 1. Estimation scheme for sulfur emission from an average *Stryphnodendron* tree where emissions from roots, leaf litter, and soil were sampled at 4 m intervals on transects radiating out from trunk. Emissions at each distance (Table I) were multiplied by the area of circular sampling band between that distance and the previous distance, and the emissions summed through the circular bands for the whole tree.

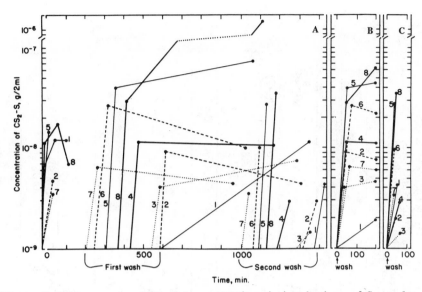

Figure 2. Time course of CS_2-S concentrations in incubations of *S. excelsum* roots. A) time course for individual incubations (identified by numbers) starting as field moist roots and later subjected to two episodes of vigorous washing (W). B) time courses following the first washing replotted synchronizing the starting times of re-incubation. C) replot for the second wash as in B.

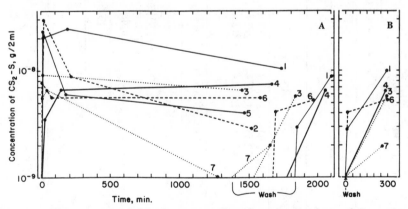

Figure 3. Time course of CS_2-S gas concentrations incubations of *S. excelsum* roots. A) time courses for individual incubations starting as field moist roots then gases displaced by 3 gentle rinsings with water (W). B) time courses after washing re-drawn with starting times synchronized.

Table II. Root Dry Weights, Means (Standard Deviation), in Grams by Diameter Classes, and Total for Non-sulfur Emitting Roots (R) and Sulfur Emitting Roots (RS) from 25 X 25 X 10 cm Deep Soil Volumes Sampled on Transects Radiating out From 3 S. *Excelsum* Trees. Sample Size = 9 at Each Distance

Diameter classes, mm	Distance from tree, m							
	4		8		12		16	
	R	RS	R	RS	R	RS	R	RS
0-1.9	4.2(2.9)	2.3(1.3)	5.2(2.3)	1.2(0.76)	6.7(3.2)	1.0(0.5)	6.4(2.6)	0.39(.49)
2.0-4.9	2.9(2.5)	1.6(2.4)	4.2(3.8)	0.57(0.65)	4.1(2.3)	0.8(1.2)	4.1(3.7)	0.24(0.55)
5.0-9.9	2.0(2.1)	1.1(2.4)	3.3(3.4)	0(0)	1.9(1.3)	0(0)	1.2(2.8)	0(0)
> 10.0	4.2(9.9)	0(0)	4.6(4.9)	0(0)	7.1(9.0)	0(0)	1.2(2.8)	0(0)
Total	13.4(13.4)	5.0(5.5)	17.3(7.8)	1.8(1.1)	19.8(12.5)	1.8(1.3)	14.3(7.4)	0.63(1.0)

Table III. Potential Sulfur Emissions, Mean (Standard Deviation) g S x 10^{-9} . g Root Dry Weight^{-1} min^{-1}. Sample Size = 9 at Each Distance

Compound	Distance from S. *excelsum* tree, m			
	4	8	12	16
CS_2-S	63.2(53.5)	78.7(120)	32.3(51.2)	30.6(40)
H_2S-S	0.67(1.7)	0.49(1.5)	0(0)	0(0)

emission of 11 kg S. ha^{-1}.yr^{-1} to the atmosphere its subsequent oxidation to SO_4^{-2} and dilution in rain, might explain the acid rain (41) reported for this forest (40). A literal interpretation of emissions estimates calculated from roots incubated at field moist conditions and extrapolated to annual estimates, then adjusted for the proportion of *S. excelsum* trunk cross sectional area of the forest indicates annual emission rates of 0.29 kg S.ha^{-1}.yr^{-1}. Emissions from *S. excelsum* may be incorrectly estimated for three reasons.

First, maximum emission rates may have been missed. Emission rates were rapid at first, then declined. For individual samples, between 15 and 25 min were needed to carry the soil from the forest to the laboratory, to extract sulfur emitting roots from the soil, and to place the roots in the incubation bottles. Incubations for determining rates were usually between 5 and 10 min duration.

Second, the decrease in concentration of CS_2 in the headspace of the incubation jars, Figure 2, suggests that two reactions namely CS_2, production and CS_2 consumption were simultaneously in progress. As long as production was greater than consumption, the concentration of CS_2, appeared to increase. When consumption exceeds production, the concentrations appears to decrease. With both reactions in progress, the gross gas production would be the sum of the two rates. Whether CS_2 was simply adsorbed to the surface of the incubation vessel or was oxidized to SO_2 and SO_4^{-2} is unknown, thus the magnitude of the consumption rate is unknown. Potential CS_2 emission rates may thus have been underestimated in these incubations.

Third, the stimulation of gas emission from roots by washing raises the question about the effect of the passage of wetting fronts through the rainforest soil with each of the numerous rain showers.

Future quantification of sulfur gas emissions from *S. excelsum* roots should avoid possible mechanical damage to roots caused by their excavation from soils as in the present study. Instead, plants need to be grown in soil in containers to which simulated rainfall additions are made while sampling for sulfur gases above the soil surface. Once the relations of sulfur gas emission rates to the frequency and amounts of soil wetting are determined, these data can be coupled with a rainfall frequency and quantity model to simulate potential sulfur gas emission for *S. excelsum* on an annual basis.

Perhaps more important than the exact quantities of these emissions with the three sources of uncertainty, are the qualities of the sulfur emissions. Qualitatively, CS_2 emissions have community and physiological implications.

Community Level. Sulfur gas emissions from *S. excelsum* roots may influence the species composition of biological communities through actions as anti-bacterial, anti-fungal, anti-nematode, anti-herbivore, anti-plant (allelopathy) agents. Anti-bacterial properties of CS_2 can inhibit nitrification, the oxidation of ammonia to nitrate (42-44). The anti-fungal properties of CS_2 have been used to control root rot fungi in agriculture and in forestry (45-47). It is a registered fungicide and nematicide (48). Its anti-nematode properties have been used in soil fumigation in agriculture (49). Its anti-herbivore properties have been used to kill insects in stored grains (50). The anti-plant or allelopathic effects have not been investigated as far we know.

Physiological and Evolutionary Level. In field and pot studies *Acacia pulchella* roots suppressed the soil fungus *Phytophthora cinnamomi* and promoted the survival of *Eucalyptus marginata* trees. This fungus is an important pathogen for *Eucalyptus*. Investigating the *Acacia-Phytophthora-Eucalyptus* interaction, Whitfield et al. (51) found CS_2 to be a major constituent of the *Acacia* root volatile compounds.

Carbon disulfide emissions are known from only five plant species: minced leaves of *Brassica oleraceae* (52), intact leaves of *Medicago sativum* L., *Zea mays*, L., *Quercus lobata* Nee (27) and steam distilled roots of *Acacia pulchella* (51). Continuation of surveys of plant species for sulfur gas emission will expand our understanding of the phylogenetic and biogeographic distribution of the sulfur emission phenomenon.

The physiological controls of CS_2 emission from plants apparently are unknown. We offer two interrelated suggestions about controls to CS_2 emission. Removal of CS_2 from around the roots either by extracting roots from soil (Figures 2 & 3) or by vigorous washing (Figure 2) or by gentle rinsing (Figure 3) leads to rapid CS_2 emission followed by relatively steady concentration (Figures 2 & 3). This suggests a feedback mechanism in which lowered CS_2 concentration at the root surface leads to rapid CS_2 release from the root.

A related control of CS_2 release from *S. excelsum* roots is disturbance. In the present study the gentle disturbance of rinsing with water stimulated CS_2 release (Figure 3). More vigorous washing stimulated higher rates of release (Figure 2). In the field, the odor of CS_2 became stronger during the beginning of a rain storm. Given the toxicity of CS_2, to nematodes and insects and the stimulation of CS_2 release by disturbance, we suggest that disturbance to *S. excelsum* roots by root-eating nematodes and insects stimulates CS_2 production as a defense mechanism.

The emission of CS_2 from these legumes may also be a mechanism removing excess SO_4^{-2} from the rhizosphere where SO_4^{-2} might otherwise inhibit the uptake of the Mo required for nitrogen fixation. Rennenberg (31) argues that H_2S emission from plants is a mechanism against excess sulfur accumulation in plants. While Rennenberg's study (31) dealt with H_2S, emission of CS_2 and other sulfur gases could have similar roles. Cole et al (53) showed that SO_4^{-2} can inhibit Mo uptake by algae and bacteria. Sulfur emissions from these trees may have multiple selective advantages.

Acknowledgements

Research supported by U.S. National Science Foundation grants BSR8104700, BSR8012093 and by U.S.D.A. Forest Service, Southeastern Forest Experimental Station, Coweeta Hydrologic Lab., Otto, North Carolina to the University of Georgia. We thank the Organization for Tropical Studies for logistical support. We thank L. Boring, D. Coleman, B. Michel, C. Monk, and M. Watwood for criticism of the manuscript. Literature on the pathological effects of CS_2 was provided by D. Freckman, S. Garabedian and R. Gross to whom we are grateful.

Literature Cited

1. Andreae, M. O.; and Raemdonck, H. Science 1983, 221, 744-7.

2. Kritz, M. A. Journal of Geophysical Research 1982, 87(C11), 8795-803.

3. Barnard, W. R.; Andreae, M. O.; Watkins, W. E.; Bingemer, H.; Georgii, H. W. Journal of Geophysical Research 1982, 87(C11) 8787-93.

4. Andreae, M. O. In Biogeochemistry of Ancient and Modern Environments; Trudinger, P. A.; Walter; M. R; Ralph, B. J., Eds; Australian Academy of Science: Canberra, 1980; pp 253-9.

5. Graedel, T. E. Geophysical Research Letters 1979, 6(4), 329-31.

6. Slatt, B. J.; Natusch, D. F. S.; Prospero, J. M.; Savoie, D. L. Atmospheric Environment 1978, 12, 981-91.

7. Steudler, P. A.; Peterson, B. J. Atmospheric Environment 1985, 19(9), 1411-6.

8. Steudler, P. A.; Peterson, B. J. Nature 1984, 311(5985), 455-7.

9. Jorgensen, B. B.; Okholm-Hanson, B. Atmospheric Environment 1985, 19, 1737-49.

10. Hitchcock, D. R.; Black, M. S. Atmospheric Environment 1984, 18, 1-17.

11. Ingvorsen, K.; Jorgensen, B. B. Atmospheric Environment 1982, 16(4), 855-65.

12. Bandy, A. R.; Maroulis, P. J.; Bonsang, B.; Brown, C. A. In Estuarine Comparisons; Kennedy, V. S., Ed.; Academic Press: New York, 1982; pp 303-12.

13. Aneja, V. P.; Overton, J. H., Jr.; Cupitt, L. T.; Durham, J. L.; Wilson, W. E. Tellus 1979, 31, 174-8.

14. Aneja, V. P.; Overton, J. H., Jr.; Cupitt, L. T.; Durham, J. L.; Wilson, W. E. Nature 1979, 282(5738), 493-6.

15. Zinder, S. H.; Doemel, W. N.; Brock, T. D. Applied and Environmental Microbiology 1977, 34(6), 859-60.

16. Maroulis, P. J.; Bandy A. R. Science 1977, 196, 647-8.

17. Brinkmann, W. L. F.; Santos, U. de M. Tellus 1974, XXVI(1-2), 261-7.

18. Adams, D. F.; Farwell, S. O.; Pack, M. R.; Robinson, R. Journal of the Air Pollution Control Association 1981, 31(10), 1083-9.

19. Adams, D. F.; Farwell, S. O.; Pack, M. R.; Bamesberger, W. L. Journal of the Air Pollution Control Association Journal 1979, 29(4), 380-3.

20. Delmas, R.; Baudet, J.; Servant, J.; Baziard, Y. Journal of Geophysical Research 1980, 85(C8) 4468-74.

21. Delmas, R.; Servant, J. Tellus 1983, 35B, 110-20.

22. Delmas, R.; Baudet, J.; Servant, J. Tellus 1978, 30, 158-68.

23. Farwell, S. O.; Sherrard, A. E.; Pack, M. R.; Adams, D. F. Soil Biology & Biochemistry 1979, 11, 411-5.

24. Banwart, W. L.; Bremner, J. M. Soil Biology & Biochemistry 1976, 8, 19-22.

25. Banwart, W. L.; Bremner, J. M. Soil Biology & Biochemistry 1976, 8, 439-43.

26. Banwart, W. L.; Bremner, J. M. Soil Biology & Biochemistry 1975, 7, 359-64.

27. Westberg, H; Lamb, B. In Environment Impact of Natural Emissions; Aneja, V. P., Ed.; Air Pollution Control Association, Specialty Conference, Transactions, TR-2: Pittsburgh, PA, 1984; pp 41-53.

28. Hällgren, J.; Fredriksson, S. Plant Physiology 1982, 70, 456-9.

29. Winner, W. E.; Smith, C. L.; Koch, G. W.; Mooney, H. A.; Bewley, J. D.; Krouse, H. R. Nature, 1981, 289(5799), 672-3.

30. Spaleny, J. Plant and Soil 1977, 48, 557-63.

31. Rennenberg, H. Annual Review of Plant Physiology 1984, 35, 121-53.

32. Mac Taggart, D.; Adams, D.; Farwell, S. Journal of Atmospheric Chemistry 1987, 5, 417-37.

33. Goldan, P.; Kuster, W.; Albritton, D.; Fehsenfeld, F. Journal of Atmospheric Chemistry 1987, 5, 439-67.

34. Lamb, B.; Westberg, H.; Allwine, G.; Bamesberger, L.; Guenther, A. Journal of Atmospheric Chemistry 1987, 5, 469-91.

35. Smil, V. Carbon-Nitrogen-Sulfur, human interference in grand biospheric cycles; Plenum Press: New York, 1985.

36. Ivanov, M. V.; Freney, J. R The Global Biogeochemical Sulfur Cycle; John Wiley and Sons: New York, 1983.

37. Haines, B.; Black, M.; Fail, J., Jr.; McHargue, L.; Howell, G. In Effects of Atmospheric Pollutants on Forests, Wetlands, and Agricultural Ecosystems; Hutchinson, T. C.; Meema, K. M. Eds.; Springer-Verlag: New York, 1987; pp 599-610.

38. SAS Institute, Inc., Cary, NC, 1985.

39. Hartshorn, G. S. In Costa Rican Natural History; Janzen, D. E., Ed.; The University of Chicago Press: Chicago, IL 1983; pp 118-350.

40. Johnson, D. W.; Cole, D. W.; Gessel, S. P. Biotropica 1979,11, 38-42.

41. Haines, B. L. Oikos 1983, 41, 139-43.

42. Rodgers, G. A.; Ashworth, J.; Walker, N. Zentralblatt für Bakteriologie, Parasitenkunde, Infektionskrankheiten und Hygiene 1980, 135, 477-83.

43. Malhi, S. S.; Nyborg, M. Plant and Soil 1982, 65(2), 203-18.

44. Bremner, J. M.; Bundy, L. G. Soil Biology & Biochemistry 1974, 6, 161-5.

45. Bliss, D. E. Phytopathology 1951, 41, 665-83.

46. Munnecke, D. E.; Kolbezen, M. J.; Wilburn, W. D. Phytopathology 1973, 63, 1352-7.

47. Filip, G. M.; Roth, L. F. Canadian Journal of Forest Research 1977, 7, 226-31.

48. Carter, E. P.; Betz, D. O. Jr.; Mitchell C. T. Annotated index of registered fungicides and nematicides; their uses in the United States; Pesticides Regulation Division, Agricultural Research Service; U. S. Department of Agriculture, 1969.
49. Chandler, W. A. Plant Disease Reporter 1969, 53(1), 49-53.
50. Punj, G. K.; Girish, G. K. Journal of Stored Products Research 1969, 4, 339-42.
51. Whitfield, F. B.; Shea, S. R.; Gillen, K. J.; Shaw, K. J. Australian Journal of Botany 1981, 29, 195-208.
52. Bailey, S. D.; Bazinet, M. L.; Driscoll, J. L.; McCarthy, A. I. Journal of Food Science 1961, 26, 163-70.
53. Cole, J. J.; Howarth, R. W.; Nolan, S. S.; Marino, R. Biogeochemistry 1986, 2, 179-96.

RECEIVED December 27, 1988

TERRESTRIAL AND FRESHWATER SYSTEMS: TRANSFORMATIONS

Chapter 6

Origin of Hydrogen Sulfide in Freshwater Sediments

D. A. Dunnette

Portland State University, Portland, Oregon 97207

Hydrogen sulfide originating from putrefaction of sedimentary protein ranged from 0.02 to 1.51 mg S $L^{-1}d^{-1}$ in 10°C laboratory incubations of sediment samples form two small eutrophic freshwater lakes with an average of approximately 0.4 mg S $L^{-1}d^{-1}$ while sulfate reduction ranged from 0.05 to 3.2 mg S $L^{-1}d^{-1}$ with a mean of approximately 1.5. Although hydrogen sulfide production via putrefaction correlated strongly with numbers of proteolytic bacteria ($r = 0.9$), associations with protein or organic carbon were weak ($r = 0.1$-0.3).

Putrefaction contributions to total hydrogen sulfide production averaged 23% with a range of 5-57% for the two small lakes investigated. It is estimated that over a six month primary production period, putrefaction contributes 5-50% to total hydrogen sulfide production in freshwater systems similar to the ones described in this study.

Hydrogen sulfide is a chemically and biochemically very active component of the sulfur cycle and an important determinant of the Eh and pH limits of the natural environment. Dissimilative sulfate reduction by Desulfovibrio, Desulfobacter, Desulfobulbus and other genera of bacteria has generally been accepted as the only significant source of hydrogen sulfide in the aquatic environment. It is common knowledge, however, that the terminal sulfhydryl amino acid cysteine may undergo microbially mediated anaerobic decomposition or putrefaction to also produce hydrogen sulfide. Putrefaction has largely been ignored because most of the earlier work performed to quantify hydrogen sulfide production was conducted on high mineral content systems in which sulfate reduction would predominate. For example it has been estimated that 97 to 99.5 percent of hydrogen sulfide produced in marine systems originates from sulfate reduction (1,2). Another reason is that there are no generally accepted methods for evaluating the importance of proteinaceous sulfur in hydrogen sulfide production as there are with sulfate reduction where the use of labeled sulfate has been widely used for 30 years using planchet (3) or scintillation (4) techniques.

Investigations suggesting the importance of putrefaction as a route of hydrogen sulfide formation in aquatic systems are sparse but significant and include evidence related to mass balance anomalies (5-7,9,10), bacterial

0097–6156/89/0393–0072$06.00/0

enumeration anomalies (7,8), and ^{35}S studies utilizing labeled cysteine or methionine (11,12). In neither of the latter two cases, however, were actual rates of hydrogen sulfide production via putrefaction determined. This paper describes work undertaken to evaluate the magnitude and relative importance of putrefaction and sulfate reduction in two small monomictic lakes and the factors which may influence these routes of hydrogen sulfide production.

<u>Experimental</u>

Samples were taken from 2 lakes, Third Sister, a moderately eutrophic glacially formed lake of 4 hectares with sediment consisting of approximately 40 percent silt and 45 percent clay and Frains, an 8 hectare lake containing approximately 65 percent silt and 20 percent clay. Frains Lake is a more eutrophic lake with massive algal blooms occurring in early summer. No significant quantities of anthropogenic contaminants are known to enter the lakes.

   The procedures and analytical methods used in this study are described elsewhere (3). Briefly, freshly acquired 50 mL samples of sediment protected from the atmosphere were inoculated with labeled substrate, homogenized, and incubated in the laboratory for 1.5 to 2 hours at 10 °C. [35]S labeled sodium sulfate and cysteine were used to estimate sulfate reduction and putrefaction respectively. One mL aliquots of sample were taken at 30 min intervals and added to measured volumes of cadmium acetate which fixed all hydrogen sulfide produced in the form of CdS. The CdS was then acid distilled under a nitrogen purge, reprecipitated as CdS and deposited on planchets for counting. All operations prior to fixing were conducted anaerobically. Enumerations of proteolytic and sulfate reducing bacteria were conducted using the anaerobic culture method of Hungate (13). Analyses for sulfate and cysteine were determined by chemical reduction of sulfate and column chromatography respectively as described elsewhere (3). Combined reproducibility for incubation and distillation was 11% for homogenized sediment samples (3). If sediment samples are not homogenized, reproducibility is only 80% (4). Reproducibility for cysteine and sulfate determinations was 16% and 9% respectively.

<u>Results</u>

Sediment chemical data including values for organic carbon, protein, free amino acids, cysteine and sulfate are summarized in Table I. The range of variation of principal chemical constituents for hypolimnetic waters of the sampling sites were 2.7-4.7 mg S L^{-1} for sulfate, 3.5-4.6 mg S L^{-1} for protein (as leucine equivalents), 2.5-5.5 mg S L^{-1} for organic carbon and cysteine 0.8-3.6 mg S L^{-1}. Oxidation-reduction potentials during the sampling period varied from -150 to -250 mv vs Ag/AgCl 0.1 KCl. Oxygen concentration in the hypolimnion decreased to 0.0 at both stations by June. Three sets of results are presented to describe the temporal and spatial variation of sulfide production from putrefaction and sulfate reduction: 1) experiments to assess seasonal variation at two different sampling points of varying depth in a single lake, 2) a comparison of sulfide producing characteristics of sediment at different lake depths in each of two lakes and 3) a study in which hydrogen sulfide production rates were measured at distinct depth horizons within the sediment core at a single station.

<u>Seasonal Variation</u>. Twenty two samples were collected from Third Sister Lake over the period April-September at 2 stations of depths 15 m and 10 m. A summary of data obtained from these studies is presented in Tables II, III and

Table I. Sedimentary Chemical Data, Third Sister and Frains Lakes (3)

		Third Sister	Frains
		mgL^{-1}	
Organic carbon	(0.45 μ filtered) mean	6-12	--
Organic carbon	(pH 10 extract) mean	190	--
Protein	(pH 12 extract) mean	120	--
Free ammino acids	(pH 12 extract) mean	45	46
Cysteinne	(pH 12 extract) mean	2.1	1.1
Sulfate	(centrifuged only) mean	2.3	2.2
Sulfate	(pH 12 extract) mean	10.5	--

Table II. Hydrogen Sulfide Production Via Sulfate Reduction
Third Sister and Frains Lakes

	mg S L^{-1} d^{-1} @ 10 °C		
	West(15m)	East(10m)	Frains(10m)
Number of determinations	10	10	3
Maximum	3.2	2.5	--
Minimum	0.7	0.7	--
Mean	1.72	1.72	1.04
Standard deviation	0.84	0.59	0.58

Table III. Hydrogen Sulfide Production Via Putrefaction
Third Sister and Frains Lakes

	mg S L^{-1} d^{-1} @ 10°C		
	West(15m)	East(10m)	Frains(10m)
Number of determinations	10	12	3
Maximum	1.51	0.90	--
Minimum	0.25	0.15	--
Mean	0.65	0.37	0.13
Standard deviation	0.41	0.18	0.03

Table IV. Percent of Hydrogen Sulfide Production Attributable to
Putrefaction, Third Sister and Frains Lakes

		Percent	
	West	East	Frains
Number of determinations	10	10	3
Maximum	49.2	53.0	--
Minimum	7.5	5.1	--
Mean	27.6	18.3	11.1
Standard deviation	13.8	13.1	--

IV. Table II summarizes sulfate reduction rate data at both Third Sister and Frains Lakes and indicates a range of 0.7 to 3.2 mg S L^{-1} d^{-1} and a mean of 1.7 at both stations for 23 determinations. Table III includes data from putrefactive hydrogen sulfide production at the two lakes showing that rates varied from 0.13 at Frains Lake to 1.51 mg S L^{-1} d^{-1} at Third Sister with a mean of approximately 0.4 mg S L^{-1} d^{-1}. Hydrogen sulfide contributions from putrefaction, summarized in Table IV, were found to range from 5.1 to 53.0 percent with means of 27.6, 18.3 and 11.1 percent for Third Sister West, East and Frains sites respectively. Although the sulfate reduction values are similar to those obtained by others (0.01-15 mg S L^{-1} d^{-1}:14), there are no other putrefactive data with which to compare the values of Table III. Microbial enumeration data averaged 5 x 10^2 and 2 x 10^4 cells mL^{-1} for sulfate reducers and proteolytic bacteria respectively. These values are similar to those obtained by others ([14]).

Large seasonal fluctuations were observed for sulfate reduction and putrefaction at both sampling stations of Third Sister Lake. Proteolytic bacteria were strongly correlated with putrefaction rate at both the 15 m and 10 m stations of Third Sister ($r = 0.86$, 0.94 respectively) and inversely associated with sulfate reducing bacteria at both sites ($r = 0.69$, 0.32). No other correlation coefficients (Pearson product moment) associating enumeration results, sulfide production rates or chemical constituents exceeded 0.6. Associations of hydrogen sulfide production via putrefaction and protein or carbon content of hypolimnetic or core waters were not significant. The fact that correlations were not more significant is common in microbial studies which are similar to the study described here ([1,7,12,14]) and is confirmation of the complexity of the environmental factors controlling biogenic sulfur transformations.

Variations Between Lakes. Results of a study to evaluate sulfide production variation with water depth is given in Table V. In this experiment, samples were taken from five different sediment depths over a two-day period at each lake in early October. At both lakes sulfate reduction exceeded putrefaction by a factor of approximately 2 with overall mean rates of 0.55 and 0.29 mg S $L^{-1}d^{-1}$ respectively. Sulfate reduction exceeded cysteine decomposition in all samples except one collected from Third Sister Lake at 17 m. Results of this study show a good correlation at Third Sister Lake between percent hydrogen sulfide production attributable to putrefaction and depth of sampling station ($r = 0.94$) and oxidation-reduction potential ($r = 0.98$). This correlation was not observed at Frains Lake. A possible factor in differences observed may be the physical nature of the sediment at Frains which was less dense and more flocculent than that of Third Sister.

The mean contributions to hydrogen sulfide production due to putrefaction were 30.1% for Frains and 30.6% for Third Sister for the two-day lake evaluation. Total mean hydrogen sulfide production was 1.02 mg S $L^{-1}d^{-1}$ for Third Sister and 0.67 mg S $L^{-1}d^{-1}$ for Frains Lake. The range of hydrogen sulfide production attributable to putrefaction was 5-57%. The greater contribution from putrefaction during this study period (October 7-8) as compared to the seasonal studies of April-September was due to a decrease in sulfate reductive activity while putrefaction was relatively unchanged. The mean numbers of sulfate reducers and proteolytic bacteria at the lakes during this period were 1.8 x 10^2 and 1.6 x 10^4 respectively.

Variation with Sediment Depth. This investigation was undertaken to determine the variation of hydrogen sulfide production and associated microorganisms within the upper 8 cm of sediment from East station. Five fractions were examined representing depths of 0-1, 1-2, 3-4, 5-6 and 8-9 cm. Results are summarized in Table VI. Maximum values for both sulfate

Table V. Data From Third Sister and Frain's Lakes Surveys

Sample Number	Depth (m)	Sulfate Reduct. (mg SL⁻¹d⁻¹)	Cysteine Decomp. (mg SL⁻¹d⁻¹)	Percent Cysteine Decomp.	SO₄ Reducers (mL⁻¹)	Proteolytic (mL⁻¹)	Eh(mv) vsAg-AgCl 0.1Cl	pH
1-3rdS	3.5	0.27	0.02	7.0	7.8×10^2	0.5×10^4	-157	6.7
2-3rdS	7.5	0.59	0.09	13.2	0.3 "	0.7 "	-164	6.7
3-3rdS	12.5	0.94	0.60	36.0	0.9 "	6.0 "	-177	6.7
4-3rdS	13.0	1.07	0.61	39.0	1.6 "	2.6 "	-180	6.7
5-3rdS	17.0	0.40	0.54	57.5	2.0 "	4.8 "	-204	6.7
1-Frai	4.0	0.40	0.16	28.5	1.1 "	2.1 "	-134	6.8
2-Frai	6.5	0.44	0.26	37.2	0.9 "	1.6 "	-197	6.7
3-Frai	7.5	0.44	0.24	35.2	0.4 "	0.6 "	-230	6.8
4-Frai	7.5	0.66	0.09	12.0	1.6 "	1.9 "	-255	6.8
5-Frai	9.5	0.35	0.21	37.4	0.9 "	1.7 "	-235	6.7

Table VI. Summary of Sediment Depth Profile Data
Third Sister East Station (10m) October

Depth cm	Sulfate Reduction mg S L⁻¹d⁻¹	Putrefaction mg S L⁻¹d⁻¹	Sulfate reducers cells mL⁻¹x10⁻²	Proteolytic bacteria cells mL⁻¹x10⁻⁴
0.5	1.2	0.39	1.8	1.1
1.5	1.7	0.75	0.89	1.7
3.5	0.40	0.32	0.31	4.1
5.5	0.26	0.08	0.17	1.6
7.5	0.05	0.03	0.18	0.20

reduction and putrefaction were obtained at a depth of 1-2 cm. Activity dropped off quickly below 2 cm so that at 5-6 cm, rates of hydrogen sulfide production decreased to approximately 10% of the maximum observed for both processes. The average rate of hydrogen sulfide production measured for putrefaction (0.31 mg S L^{-1} d^{-1}) was approximately 40% of that for sulfate reduction (0.75 mg S L^{-1} d^{-1}). Total hydrogen sulfide production averaged 1.1 mg S L^{-1} d^{-1} for the five depths.

Sulfate reducing bacteria decreased exponentially from the sediment-water interface while proteolytic bacteria were most numerous at 3-4 cm with minima at 0-1 cm and 7-8 cm. Proteolytic bacteria ranged from 0.5 to 4.1 x 10^3 cells mL^{-1} while sulfate reducers varied from 0.2 to 1.7 x 10^3 cells mL^{-1}. Results of this study show that more than 80 percent of the hydrogen sulfide produced originates in the vertical horizon between 0 and 4 cm. Maximum numbers of sulfate reducers occurred between 0 and 1 cm and decreased sharply so that 90 percent of the sulfate reducers were found in the upper 2 cm. Proteolytic bacteria occurred in greatest numbers between 3 and 4 cm, dropping off steeply to 1 and 7 cm at 15-25% of their peak values. Maximum sulfate reduction observed was 1.8 mg S L^{-1} d^{-1} while the putrefaction maximum was 0.7 mg S L^{-1} d^{-1}.

The distribution of these data are the result of many chemical, physical and biological factors including pH, Eh, organic carbon, trace substances, possible toxic effects of HS^- and competition for nutrients in the upper 3 cm. Since Eh decreases with decreasing depth, this parameter does not limit distribution of sulfate reducing organisms since they occur in highest numbers near the sediment water interface. The controlling factors may include nutrient, sulfate and carbon availability which would be expected to decrease with increasing depth due to microbial activity and proximity to overlying sources. The factors which affect distribution of putrefactive bacteria appear equally complex possibly including Eh and toxic effects from sulfate reduction near 1.5 cm and competition for nutrients above 3 cm.

<u>Summary/Discussion</u>

Hydrogen sulfide production via putrefaction of endogenous protein at 10 °C ranged from 0.02 to 1.51 mg S L^{-1} d^{-1} in laboratory incubations of sediment from two small eutrophic freshwater lakes with an average of approximately 0.4 mg S L^{-1} d^{-1} whereas sulfate reduction ranged from 0.05 to 3.2 mg S L^{-1} d^{-1} with a mean of approximately 1.5. Sulfate reduction rates are similar to those from other freshwater sites with a range of 0.02 to 15 mg S L^{-1} d^{-1} (<u>2,3,14</u>). No putrefactive data are available with which to compare the results of this investigation. For a larger lake, however, Nriagu (<u>5</u>) estimated the amount of organic sulfur which should be present in sediment based on sulfur content of aquatic macrophytes and microorganisms. Since considerably less was actually found in the sediment, it was suggested that the difference (45%) might represent that fraction of hydrogen sulfide formed from putrefactive processes.

Although hydrogen sulfide production via putrefaction correlated strongly with numbers of proteolytic bacteria (r=0.94, 0.86), none was found for protein (r=0.16) or organic carbon (r=0.34) in Third Sister Lake. The significance of this is not known due to the many associated uncertainties, but it seems reasonable to assume that protein input would result in enhanced hydrogen sulfide production via putrefaction. Results of experiments to test this in laboratory bioreactor studies indicated that protein (egg albumin) added at the rate of approximately 20 ppm d^{-1} increased the population of putrefying bacteria by 90 percent and the rate of sulfide production via putrefaction by approximately 100 percent over a period of 30 days in laboratory incubations at room temperature with an established retention time of 45 days.

A study of the variation of sulfate reduction and putrefaction with sediment depth of 0-8 cm indicated maximum putrefactive and sulfate reducing activity at a depth of 1-2 cm. The data also suggest that oxidation-reduction potential plays an important part in determining the role of putrefaction. However, the significance of this association must be tempered with the understanding that redox equilibrium is never reached in the aquatic environment and that Eh measurements are of value empirically but not thermodynamically.

This study demonstrated that, contrary to popular perceptions, putrefactive contributions to total hydrogen sulfide production can be significant, averaging 23 percent with a range of 5-57% for the two small lakes investigated. Based upon results from these investigations it is estimated that in monomictic freshwater lake systems similar to the ones described in this investigation, putrefaction contributes 5 - 50% to total hydrogen sulfide production in the upper 5 cm of sediment over a six month primary production period, which in this investigation, extended from April through early October. Efforts to assess sulfide production and transformations in such lakes should include a consideration of putrefaction as a sulfide production route.

Literature Cited

1. Jorgensen, B. B. Limnol. Oceanogr. 1977, 22, 814-32.

2. Sokolova, G. A.; Karovoika, G. I. Microbiological Processes in the Formation of Sulfur Deposits; Israeli Program for Scientific Translation, Jerusalem, 1968.

3. Dunnette, D. A., Mancy, K. H.; Chynoweth, D. P. Water Res. 1985, 19, 875-84.

4. Jorgensen, B. B. Geomicrobiol. J. 1978, 1, 11-27.

5. Nriagu, J. O. Limnol. Oceanogr. 1968, 13, 430-9.

6. Koyama, T.; Sugawara, K. J. Earth Sci. Japn. 1953, 1, 24-34.

7. Gunkel, W.; Oppenheimer, C. H. In Marine Microbiology; Oppenheimer, C. H., Ed.; Charles Thomas: 1963.

8. Novozhilova, M. A.; Berezina, F. S. Microbiology 1968, 37, 436-9.

9. Nedwell, D. B.; Floodgate, G. D. Marine Biology 1972, 14, 18-24.

10. Jones, J. G.; Simon, B. M. J. Ecology 1980, 68, 493-512.

11. Zinder, S. H.; Brock, T. D. Appl. Environ. Microbiol. 1978, 35, 344-52.

12. Jones, J. G.; Simon, B. M.; Roscoe, J. V. J. Gen. Microbiol. 1982, 128, 2833-9.

13. Hungate, R. E. In Methods in Microbiology; Norris, J. R.; Ribbons, D.W., Eds.; Academic Press: 1979.

14. The Global Biogeochemical Sulphur Cycle; Ivanov, M. V.; Freney, J. R., Eds.; New York, 1984.

RECEIVED September 23, 1988

Chapter 7

Sulfur Cycling in an Experimentally Acidified Seepage Lake

L. A. Baker, N. R. Urban, P. L. Brezonik, and L. A. Sherman

Department of Civil and Mineral Engineering,
University of Minnesota, Minneapolis, MN 55455

Recent interest in sulfur biogeochemistry of softwater lakes has been generated by the need to understand lake acidification processes. This study examines sulfur biogeochemistry in an experimentally acidified seepage lake in northern Wisconsin. For the pre-acidified lake, direct atmospheric deposition to the lake surface accounted for 93% of total input; seepage inflows and leaf litter accounted for the remaining 7%. Half of the input sulfur was retained by in-lake processes, and seston deposition was the dominant sulfate sink. Even though 70-80% of sediment trap sulfur was recycled, seston-derived sulfur accounts for 70% of sulfur accumulation in cores. Dissimilatory reduction in surficial sediments accounted for 30% of net S retention in the pre-acidified lake and reduction rates appear to be increasing in the acidified basin. Laboratory experiments with 35-S show that sulfur diagenesis is more complex than shown by the conventional model and that organic-inorganic transformations of sulfur in sediments are important diagenetic processes.

Until recently, most investigators studying lake acidification believed that sulfate was conservative in dilute aquatic systems. Attempts to model lake acidification were focused on developing an understanding of terrestrial biogeochemical processes, particularly mineral weathering. Over the past five years, however, there has been a growing awareness that in-lake processes, particularly those involving transformations of sulfate, are important in regulating alkalinity in many softwater lakes.

Hongve (1) first postulated that sediment proceseses may be involved in acid neutralization in softwater lakes. Greater interest in in-lake neutralization processes was generated by the whole-lake acidification experiment of Lake 223. In this experiment, 66-81% of the H_2SO_4 added over a 7-year period was neutralized by in-lake processes and sulfate retention accounted for 85% of the internal alkalinity production. Approximately 30% of the sulfate input was lost by reduction in the anoxic hypolimnion and 70% was lost by reduction within epilimnetic sediments (2). The latter finding was particularly important, because many of the small, oligotrophic lakes that are sensitive to acidification do not have anoxic hypolimnia, and many investigators had discounted the potential role of sulfate reduction in alkalinity regulation of softwater lakes in the belief that reduction would occur only in anoxic hypolimnia.

0097–6156/89/0393–0079$06.50/0
© 1989 American Chemical Society

Several whole-lake ion budgets have shown that internal alkalinity generation (IAG) is important in regulating the alkalinity of groundwater recharge lakes and that sulfate retention processes are the dominant source of IAG (3-5)); and synoptic studies (6-9) have shown that sulfate reduction occurs in sediments from a wide variety of softwater lakes. Baker et al. (10) showed that net sulfate retention in lakes can be modeled as a first-order process with respect to sulfate concentration and several "whole ecosystem" models of lake acidification recently have been modified to include in-lake processes (11).

The dominant role of sulfate reduction in neutralizing acid inputs within lakes has generated considerable interest in developing a better understanding of sulfur biogeochemistry in dilute lakes. Recent studies have been directed toward identifying endproducts of sulfate reduction and determining recycling rates (12-15), determining seasonal patterns of sulfate reduction (15,16), and determining the relative importance of seston deposition versus dissimilatory reduction in removing sulfate from the water column (17-19).

This paper describes the sulfur cycle for Little Rock Lake, Wisconsin, with particular emphasis on processes that remove sulfate from the water column and on subsequent diagenesis and regeneration. Our work and other recent studies in this area show that the classical paradigm of sulfur cycling (20) in lakes is incorrect, or at least incomplete.

Study Site. Little Rock Lake is an 18 hectare seepage lake in northern Wisconsin. The lake is dilute (specific conductance = 11 μS/cm @ 25°C) because 99% of its water input comes from direct precipitation to the lake surface and only 1% comes from groundwater. Water leaves the lake by groundwater recharge and evaporation, and the lake has an outflow-based water residence time of approximately 10 years. Littoral sediments, which comprise 42% of the lake area, are sandy with < 20% organic matter. The central 58% of the basin is gyttja (< 10% dry weight; organic content > 40%) (Figure 1).

Little Rock Lake is currently the subject of a split-basin experimental acidification study. In 1984, the lake was divided into two basins with a reinforced polyvinyl curtain: the north basin is being experimentally acidified and the south basin serves as a control. The pre-acidified lake had a pH of 6.1 and an alkalinity of 25 μeq/L. H_2SO_4 was added to the north basin to lower the pH to 5.5 in 1985 and 1986, and to pH 5.0 in 1987. The north basin will be maintained at pH 5.0 in 1988 and will be acidified to pH 4.5 in 1989 and 1990. Details of experimental design and hypotheses are presented in Brezonik et al. (21).

Sulfate Mass Balance

Inputs and outputs to the lake have been measured to calculate net retention for the pre-acidified lake. Precipitation inputs of sulfate were based on data from wet collectors (1980-1983) compiled by the National Atmospheric Deposition Program (NADP). SO_2 inputs were calculated from regional ambient air concentrations (22) using a deposition velocity of 0.5 cm/sec. Aerosol sulfate was estimated from NADP dry bucket measurements and from dry bucket and snow core measurements made in this study (23). Groundwater inputs occur largely at the southeast corner of the lake and were calculated from modeled inseepage (24) and measured sulfate concentrations in a well located in the major inseepage area. Sulfate output was estimated from mean lakewater sulfate concentration and modeled outflows.

Internal cycling was examined by measuring rates of accumulation in sediment cores, seston deposition rates, and diffusive fluxes to the sediments.

Total sulfur, carbon, and sulfur species were measured in four ^{210}Pb-dated cores. Lakewide sulfur accumulation was calculated from the depositional area (determined from sediment mapping) and area-weighted sulfur accumulation represented by the cores. Cores were taken at depths of 5, 7, and 9 m in the north basin (NB-5, NB-7, and NB-9, respectively), and at a depth of 5 m in the south basin (SB-5) (Figure 1). Three of the cores (NB-5, NB-7, and SB-5) were collected in epilimnetic, depositional areas and one (NB-9) was collected from the hypolimnion (the south basin does not exhibit seasonal stratification). Sedimentation rates were measured by sediment traps, and diffusive fluxes to the sediments were estimated from Fick's Law calculations using 7 pairs of measured porewater gradients. Methodological details are presented elswhere (16,23,25).

For the pre-acidified lake, atmospheric inputs (19.3 mmole/m^2-yr) accounted for 93% of total sulfur input. Wet deposition accounted for 76% of the atmospheric inputs and dry inputs (SO$_2$ + SO$_4^{2-}$ aerosol) accounted for the remaining 24%. Groundwater contributed 1.2 mmole/m^2-yr (5.8% of the total), and leaf inputs (26) provided an additional 0.3 mmole S/m^2-yr to the lake. Fluxes shown in Figure 2 are expressed per unit lake area.

Net sulfur retention was 10.2 mmole/m^2-yr, representing 49% of total input (Figure 2). The high proportion of total sulfur inputs represented by atmospheric inputs and the high net retention of sulfur reflect the preponderance of precipitation in the water budget for Little Rock Lake and its long water residence time. Sulfur accumulation rates in the top 6 cm of sediment were very similar among the three epilimnetic cores (18-19 mmole/m^2-yr) and somewhat higher in the hypolimnetic core (28 mmole/m^2-yr). Our lake-wide estimate of net sediment S accumulation (11.6 mmole/m^2-yr) agrees well with the sink of 10.2 mmole/m^2-yr calculated from inputs minus outputs (Figure 2). The close agreement between sulfur sinks calculated by two independent methods suggests that we have accounted for all major inputs and outputs in our mass balance. In particular, we conclude that there is little emission of volatile sulfur compounds from the lake surface.

Net sulfate retention in Little Rock Lake is comparable with sulfate retention rates observed in other seepage lakes. Net sulfate retention (as a fraction of total input) was 43-46% in acidic McCloud Lake, Florida (4), 75% and 82%, respectively, in Lowery and Magnolia lakes (3), and 40% in nearby Vandercook Lake, Wisconsin (5). These ion budgets, together with ion enrichment calculations for lakes in Florida and the Upper Midwest (27,28) show that sulfate retention occurs in nearly all seepage lakes. By contrast, although sulfate reduction occurs in lakes with short residence times (6), net retention generally is low (< 10% of input; see 10).

Mechanisms of Sulfate Removal from the Water Column

Mass balance calculations clearly show that sulfate is removed from the water column by in-lake processes. Three processes are potentially important: 1) diffusion of sulfate into sediments and subsequent reduction, 2) sedimentation of seston, and 3) dissimilatory sulfate reduction in the hypolimnion.

Dissimilatory Reduction in Surficial Sediments.
Porewater profiles from a number of sites throughout Little Rock Lake show that sulfate is always depleted below the sediment-water interface (Figure 3). Sulfate depletion in porewaters occurs not only in the soft gyttja but also in sandy, littoral sites with organic contents < 10%. The observed depletion of sulfate and the occurrence of H$_2$S indicate that the sediments are anoxic immediately below the sediment-water interface and that sulfate reduction occurs in surficial sediments.

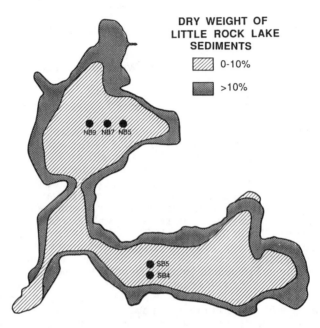

Figure 1. Sediment map of Little Rock Lake. Sediment characteristics were determined from 10-cm cores obtained at 5-m intervals along 28 transects extending from the shoreline towards the center of the lake, plus several samples from deep stations. Individual cores were extruded on site and samples were analyzed for wet weight, dry weight, and loss on ignition. This information was augmented by sonar mapping by Bill Rose (U.S. Geological Survey, Madison, WI).

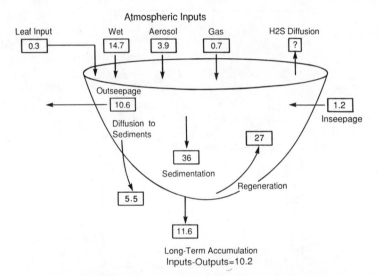

Pre-Acidification Sulfur Budget for Little Rock Lake.

Figure 2. Sulfur balance for Little Rock Lake prior to experimental acidification. All fluxes are in mmole/m²-yr. Details for construction of the sulfur balance are presented elsewhere (References 16,18,25)

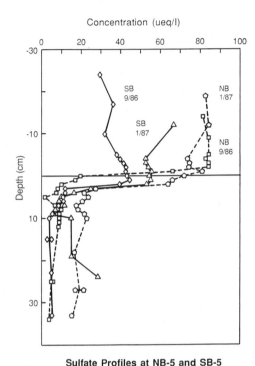

Sulfate Profiles at NB-5 and SB-5

Figure 3. Porewater sulfate profiles at 5-meter sites in the north (acidified) and south (control) basins of Little Rock Lake during September, 1986, and January, 1987. Profiles were determined by porewater equilibrators placed in the lake for three weeks. Both sites are gyttja with > 90% water content.

Numerous laboratory studies involving $^{35}SO_4{}^{2-}$ additions to mixed sediments and intact sediment-water cores from all areas of the lake support the conclusion that sulfate reduction is an important process in Little Rock Lake sediments.

In most of the 30 porewater profiles examined between 1984 and 1987, sulfate concentrations decreased to < 20 μeq/L within 5 cm of the sediment-water interface and remained relatively constant below this depth. These data indicate that sulfate reduction occurs primarily in the upper 5 cm, a contention which is supported by results from laboratory studies in which $^{35}SO_4{}^{2-}$ was added to intact sediment-water cores. We generally observed steeper sulfate gradients in summer than in winter, and hypothesize that winter gradients are not as steep because microbial activity is reduced. So far, however, we have not found a statistically significant relationship between temperature and sulfate flux.

Sediment sulfate reduction rates appear to be higher in the North Basin than in the South Basin. Porewater sulfate profiles from 1986 (north basin acidified to pH 5.5) show that water column sulfate concentrations were higher in the north basin than in the south (control) basin but reach similar values within 5 cm of the interface (Figure 3). A similar result was obtained during experimental acidification of sediment-water mesocosms (29). We anticipate that sulfate fluxes will increase with increasing acidification for two reasons. First, microbial activity of sulfate reducers near the sediment-water interface is likely to increase because lakewater sulfate concentrations are similar to half-saturation values for sulfate reduction. We determined half-saturation values of 26 μM and 70 μM in two separate laboratory studies; these values are similar to values determined by several other investigators (30-32). Second, as sulfate concentrations increase, the rate-limiting step should shift from microbial reduction to diffusion. Several studies show that sulfate fluxes to sediments are first-order with respect to sulfate concentration (8,18,29,33,34). Porewater data for Little Rock Lake support this hypothesis. During the first two years of acidification (to pH 5.5), the mean diffusive flux (calculated from seven porewater profiles in each basin) was higher at the north basin 5-meter site (NB-5)(11.4 ± 5.6 mmole/m^2-yr) than at the comparable south basin 5-meter site (SB-5)(5.5 ± 2.7 mmole/m^2-yr), but a paired difference t-test of fluxes at these sites showed that the differences were not significantly different at the p = 0.05 level. The lack of statistically significant difference probably reflects the fact that sulfate concentrations in the north basin during this phase of acidification were only 20% higher than in the control basin; we expect that divergence in sulfate fluxes between the two basins will increase as the acidification experiment progresses.

Seston Deposition. Sulfur is a minor but essential nutrient for algal production, accounting for 0.15-1.96% of dry weight (35). Sulfur is present as proteins, sulfolipids, ester sulfates, and free sulfate (18,19,36-38) that occur in varying proportions depending upon species and environmental conditions.

Sulfate assimilated by plankton is subsequently removed from the water column by sedimentation. Sediment trap data show that this is an important process in Little Rock Lake (Table I). The mean sulfur content of seston was 0.9-1.0%, comparable with values reported by other investigators (Table I). Annual seston deposition, extrapolated to a whole-lake basis using a depositional area of 58% of total lake area (area of gyttja), was 121 g/m^2-yr (dry weight basis), and sulfur deposition was 36 mmole/m^2-yr. Thus, gross seston deposition (before considering recycling) is a larger flux than diffusion to surficial sediments.

Table I. Sedimentation Rates of Carbon and Sulfur in Little Rock Lake

Site	%S	%C	C/S (wt/wt)	Sedimentation g/m²-yr	S deposition, mmole/m²-yr
Little Rock Lake[1]					
SB-5	0.92	36	37	194	59
NB-9	0.85	46	54	277	74
South Lake[2]	0.86	39	45	71	19
Wintergreen Lake[3]	0.83	24	29	1,030	266
Sudbury Lakes[4]	0.8-1.4	11-26	9-23	----	----

[1]Little Rock Lake sediment trap measurement period was 7/85-12/86. Since two summer periods but only one winter were included, deposition rates are seasonally weighted to represent the best estimate of annual deposition.
[2]Reference (19). Data are for 15.5 m (deepest) trap.
[3]Reference (18).
[4]From Table 7 in (13). Data are for four acid-polluted lakes near Sudbury, Ontario.

Hypolimnetic Sulfate Reduction. Little Rock Lake exhibits dimictic circulation, but the volume of the hypolimnion is only about 3% of the total lake volume. Oxygen depletion occurs in late summer, and a buildup of H_2S and Fe^{2+} is observed following the development of anoxia. Sulfate depletion is observed (Figure 4), but hypolimnetic reduction is a minor sulfate sink for the whole lake because the hypolimnion is small and because much of the hypolimnetic H_2S entrained in the water column during turnover probably is reoxidized. Using an oxidation- diffusion model we estimate that about half of the hypolimnetic H_2S entrained in the water column during turnover is oxidized and the other half is lost by diffusion to the atmosphere; no more than 5% of annual S input leaves by emission of H_2S produced within the hypolimnion.

Recycling of Sulfur to Overlying Water

Seston Sulfur. Much of the sulfur that is immobilized by assimilatory uptake or dissimilatory reduction is oxidized and reenters the water column as sulfate. We have approached the question of seston-S recycling by comparing carbon/sulfur (C/S) ratios in seston and sediment and by following the fate of ^{35}S in labeled algae added to laboratory sediment-water microcosms.

The first approach is based on the hypothesis that C/S ratios in pre-cultural sediments represent thoroughly decomposed seston. If this is true, one can partition sediment sulfur into a "seston" component and an "excess" component representing sulfur incorporated by dissimilatory reduction:

$$XS_i = TS_i - C_i/(C/S)_{pc} \qquad (1)$$

where XS_i = excess S at depth i,
  TS_i = total sulfur at depth i,
  C_i = total carbon at depth i, and
  C/S_{pc} = pre-cultural C/S ratio.

This approach invokes three assumptions. The first assumption is that modern seston has a similar C/S ratio to precultural seston. This assumption cannot be validated directly, but there is no evidence to show that the carbon or sulfur content of algae is any different now than it was in precultural times. The second assumption is that there have been no major changes in the decomposition process that would cause C/S ratios of decomposed seston to change. The absence of abrupt changes in physical characteristics or concentrations of major elements in sediment profiles supports this assumption. Finally, it is assumed that dissimilatory reduction rates in the precultural lake were minor compared to seston deposition rates. This assumption is supported by the following line of reasoning. First, it is likely that the sulfate concentration of precultural precipitation was 10-20 μeq/L (28), compared with a current precipitation sulfate concentration of 34 μeq/L at the Trout Lake NADP station. If one assumes a proportional decrease in lakewater $[SO_4^{2-}]$, precultural concentrations were probably between 15 and 30 μeq/L. Data from a dozen lakes in the Upper Midwest show that the average minimum porewater $[SO_4^{2-}]$ is about 25 μeq/L, implying that sulfate reduction (or at least net sulfate reduction), does not occur when sulfate concentrations are below this level (34). If these conclusions are valid, it is reasonable to assume that there was little or no dissimilatory reduction in sediments of the precultural lake.

Equation 1 can be used to segregate total sediment S into seston-S and excess-S, and recycling can be calculated by subtracting seston-S accumulation rates from seston-S sedimentation rates. In the south basin 5-m core (SB-5), the mean C/S ratio in the precultural lake was 80:1; C/S ratios decrease to 56:1

in the upper 6 cm (Figure 5). Precultural C/S ratios were similar in two other cores from sites with oxic overlying water, but somewhat lower from one site in the anaerobic hypolimnion. Equation 1 was used to apportion total sulfur at SB-5 into seston S (71% of total S, or 13.3 mmole/m^2-yr) and excess S (29% of total S, or 5.5 mmole/m^2-yr). Calculated recycling of seston (sediment trap S deposition minus seston-S accumulation) shows that 80% of seston S is recycled (Figure 6a); similar recycling rates occur at NB-5 and NB-9 (Table II). These recycling rates compare favorably with the 50% recycling rate obtained in experiments in which ^{35}S-labeled algae were added to sediment-water microcosms from Little Rock Lake. Calculated carbon recycling at SB-5 was 61% (Figure 6b), so it appears that recycling of seston sulfur is somewhat more efficient than recycling of seston carbon. Previous investigators have reported lower recycling rates for seston S (Table II). The discrepancy occurs in part because earlier investigators have assumed that all organic S originates from seston. Since H$_2$S produced by dissimilatory reduction also can be sequestered as organic S, earlier investigators may have underestimated seston S recycling.

The recycling rate for ester-S, calculated from ester-S deposition minus ester-S accumulation, is much lower ($< 30\%$ at SB-5) than reported by King and Klug ([18]) or David and Mitchell ([19]), but calculated ester-S recycling rates are predicated on the assumption that ester-S is not formed by diagenetic processes. However, laboratory experiments by Landers and Mitchell ([39]) and Wieder et al. ([40]) indicate that diagenetic formation of ester-S may occcur, so recycling of seston ester-S is probably higher than our calculated value.

<u>Diagenesis of Microbially Reduced Sulfur.</u> Postdepositional transformations play an important role in controlling the extent of recycling of microbially reduced S. Pore water profiles from many freshwater systems clearly show that H$_2$S is a short-lived intermediate in sulfate reduction which does not accumulate in sediments ([14,16,41-43]). However, the conventional paradigm for sulfur diagenesis, in which H$_2$S is initially immobilized by iron monosulfides that later are diagenetically altered to pyrite and elemental S (e.g., [20]), does not apply to all freshwater systems. Instead, organic S and CRS (chromium reducible S, which is believed to represent pyrite + S^o after preliminary acid distillation to remove AVS), are important initial endproducts of dissimilatory reduction.

The importance of iron monosulfides (measured as acid-volatile sulfide, or AVS) varies as a function of the iron content of the sediments. Sediments high in iron (> 20 mg/g) typically have a large fraction of the reduced S bound in iron monosulfides that diagenically alter with time to greigite and/or pyrite ([43-45]). However, even in lakes with significant sedimentary iron, and to a much greater extent in lakes with low sedimentary iron content, organically bound S and CRS are dominant initial endproducts of sulfate reduction. In short-term incubations of sediments from eight softwater lakes, Rudd et al. ([14]) found that 2-90% of the ^{35}SO$_4$$^{2-}$ reduced within the first 24 hours in situ in lake sediments was organically bound, 2-83% was AVS, and 9-46% was CRS.

Little Rock Lake sediments have relatively low iron content (~ 10 mg/g). As a result, even in short (1-3 hr) incubations of pelagic sediments with ^{35}SO$_4$$^{2-}$, 20-30% of the reduced S-35 cannot be recovered by acid distillation. After a 24-hour incubation with north basin gyttja, 32% of reduced S-35 was recovered as AVS, 7% as CRS, and 58% as organic-S. In the same experiment, addition of Fe^{2+} (as FeCl$_2$) or lowering the pH of the sediments (with HCl) resulted in a dramatic increase in the fraction of reduced S bound by Fe (Table III).

The nature of the solid phase S species initially formed by reaction with microbially produced H$_2$S determines subsequent transformations and the extent of recycling, since different solid S species vary in susceptibility to

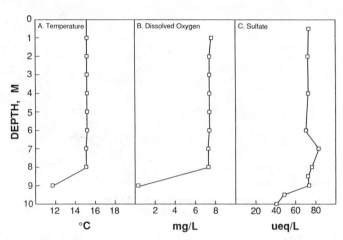

Figure 4. North basin vertical profiles for temperature, D. O., and SO_4^{2-} on September 19, 1986.

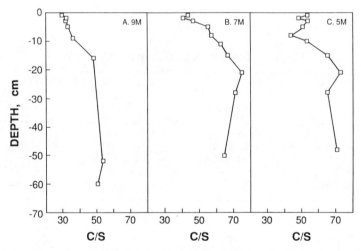

Figure 5. Carbon: sulfur ratios (wt/wt) in three cores from the north basin of Little Rock Lake. Carbon and sulfur concentrations were determined by LECCO analyzers. Cores were collected at 5 m (NB-5), 7 m (NB-7), and 9 m (NB-9). The two shallow cores are from epilimnetic sites at which the overlying water remains oxic throughout the year. The bottom water at NB-9 is anoxic in late summer.

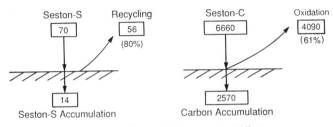

Seston Sulfur and Carbon Recycling
at 5 M Site

Figure 6. Recycling of seston sulfur and carbon at SB-5. Recycling rates were calculated from the difference between measured sediment trap deposition and measured sediment accumulation of seston-S. See text for approach used to calculate seston-S accumulation in sediments.

Table II. Sulfur recycling in Little Rock Lake

| | | % recycled | |
Site	Reference	Total S	Ester S
SB-5	This study	77	27
NB-5	This study	76	--
NB-9	This study	82	--
South Lake	(19)	26	43
Wintergreen Lake	(18)	46	42

Table III. Endproducts of Sulfate Reduction in Little Rock Lake Gyttja in Short-Term Lab Incubations With $^{35}SO_4^{2-}$ Under Various Conditions

Treatment[1]	% of added label			
	SO_4^{2-}	AVS[2]	CRS[3]	Organic S[4]
None	2.9	32.2	6.9	58.0
pH 5.0	2.3	57.5	9.0	31.2
pH 4.5	2.6	57.4	15.8	24.2
2 μM Fe^{2+}	2.7	54.7	44.0	0.0

[1]Sediments were acidified with 1 N HCl, and sediment pH was measured at the end of 24-hour incubations. Reduced iron was added as FeCl$_2$.
[2]AVS determined by one-hour acid distillation with HCL (pH < 1) into 0.17 M NaOH.
[3]CRS determined by chromium reduction method (63).
[4]Organic ^{35}S determined by difference.

oxidation. In in situ chamber incubations of sediments in Lake 302 with $^{35}SO_4^{2-}$, Rudd et al. (14) measured endproducts of sulfate reduction after 19 days and again after 9 months. Organic S was more efficiently retained than inorganic S (57% versus 37% net retention). Greater losses of inorganic S probably occurred because of greater susceptibility of the inorganic S pool to oxidation and because some inorganic S may have been slowly converted to organic S.

Laboratory microcosms with Little Rock Lake sediments inoculated with $^{35}SO_4^{2-}$ also show a gradual increase in organically bound ^{35}S (< 1% to > 30% of reduced S over three months) and ^{35}CRS (20% to > 50%) whether incubated anaerobically or under oxic water columns. Sediments incubated under oxic water columns showed increasing incorporation of ^{35}S into fulvic and humic acids after a one-month delay (up to 30% of reduced S). Whether incorporation into fulvic and humic acids followed partial oxidation to polysulfides or elemental S (cf. 46-49) is not known. However, AVS accounted for < 10% of reduced ^{35}S in our microcosms. Recent marine studies have also shown that H_2S (50-52) can be directly incorporated into acrylate, a breakdown product of β-dimethylsulphoniopropionate (DMSP), but the significance of this reaction in freshwater sediments has not been examined.

Overall recycling rates of sediment sulfur in sediment-water microcosms were comparable when ^{35}S was added as SO_4^{2-} (to promote dissimilatory reduction) or as labeled algae. Of the ^{35}S incorporated by microbial reduction into sediments maintained under oxic water columns nearly 50% was reoxidized within three months.

<u>Summary of Sulfur Diagenesis.</u> The diverse pathways involved in sulfur diagenesis in Little Rock Lake are conceptualized in Figure 7. The emerging picture of sulfur diagenesis resulting from our study and other recent studies is far more complex than previously envisioned. Our awareness of transformations between the organic and inorganic S pools is recent, and we have an incomplete understanding of environmental factors that regulate these transformations and the efficiency by which various species are recycled. Developing a better understanding of sulfur diagenesis in acid-sensitive lakes is essential to fully understand alkalinity regulation in these systems.

<u>Relative Importance of Seston Deposition and Dissimilatory Reduction as In-Lake Sulfate Sinks</u>

In Little Rock Lake, seston deposition appears to be a more important sulfate sink than does dissimilatory reduction. Several previous studies (2,4) have concluded that dissimilatory reduction is the major mechanism for sulfate retention, and Cook et al. (2) concluded that seston deposition was a minor sulfate sink in experimentally acidified Lake 223. The C/S ratio calculations discussed above show that approximately 29% of the total S in recent sediments at SB-5 is excess-S derived from dissimilatory reduction and the remaining 71% originated from seston deposition.

The rate of net sulfate reduction calculated by this approach (5.5 mmole/m²-yr) is in excellent agreement with the average sulfate flux calculated from porewater profiles at this site (5.4 mmole/m²-yr). This suggests that sulfate fluxes calculated from porewater gradients approximate net fluxes, i.e., the integration of gross sulfate reduction and reoxidation of reduced endproducts. We are currently conducting experiments to measure gross annual sulfate reduction with short-term $^{35}SO_4^{2-}$ incubations of intact cores at in situ temperatures throughout the year.

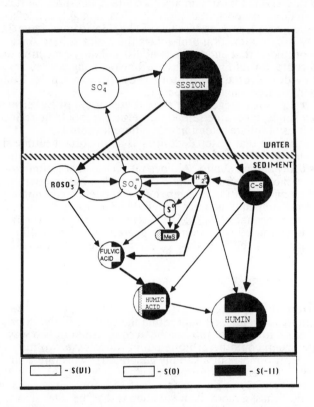

Figure 7. Sulfur transformations of seston-S and microbially reduced S in sediments of Little Rock Lake. This diagram is based on information collected from the lake (cores and sediment traps) and from laboratory sediment-water microcosms to which ^{35}S was added as labeled algae or as $^{35}SO_4^{2-}$. Sizes of circles and arrows are roughly proportional to magnitudes of pools and fluxes, respectively. $ROSO_3^-$ = ester-S; MeS = metal sulfide, C-S = carbon-bonded S.

Seston-S deposition probably is a more important process than dissimilatory reduction in lakes with low $[SO_4^{2-}]$. As lakewater sulfate concentrations increase, seston deposition reaches a plateau limited by the overall primary production rate and the maximum algal S content, but diffusive fluxes continue to increase in direct proportion to $[SO_4^{2-}]$. Thus, in highly acidic lakes (pH ≤ 5; $[SO_4^{2-}] > 100 \mu$eq/L), such as McCloud Lake, Florida and Lake 223, Ontario, dissimilatory sulfate reduction probably is the major sulfate sink. Nriagu and Soon (13) concluded that endproducts of dissimilatory reduction and elevated sediment S content would not be observed below 5 mg/L (240 μeq/L), but we see clear evidence of dissimilatory reduction in Little Rock Lake at concentrations of approximately 50 μeq/L.

These results also suggest that the sulfate model of Baker et al. (10,53), which uses a single, lake-wide retention coefficient should be refined to reflect two mechanisms of sulfate retention with different controls.

The observation that seston deposition is an important sulfate sink in some lakes brings up the subject of how removal of sulfate from the water column by algae alters lakewater alkalinity. Although it is clear that assimilatory reduction consumes two moles H^+/mole sulfate consumed, it has frequently been stated that formation and retention of ester-sulfate (thought to originate primarily in algae) generate only half as much alkalinity as do formation and retention of reduced sulfur compounds (14,39,54). This statement is not strictly correct, and we believe it has caused confusion with regard to in-lake alkalinity regulation. Rudd (14) described the formation of ester sulfate as follows (Equation 4 in Reference 14):

$$ROH + 2H^+ + SO_4^{2-} \text{------>} ROSO_3^- + H^+ + H_2O \tag{2}$$

Any reaction to represent assimilation must be written such that electroneutrality is maintained both within the cell and in the external medium. For ester sulfate formation, this can be accomplished in two ways: First, the ester sulfate group could be protonated, in which case two moles of H^+ are consumed (i.e., two moles of alkalinity generated). This fate is unlikely, since the sulfated polysaccharides formed by algae (e.g., 55,56) as well as smaller sulfate esters likely have pKa values < 2 (57,58) and act as exchange sites in algal cell walls. Thus, a more likely situation is that the ester sulfate groups are dissociated, and charge balance would be achieved by cation exchange. Equations 3 and 4 show the two-reaction sequence using Ca^{2+} as the base cation:

$$| \text{ 2ROH } | + 4H^+ + 2SO_4^{2-} \text{------->} | \text{ 2[ROSO}_3^-\text{]-2H } | + 2H_2O \tag{3}$$

$$| \text{ 2[ROSO}_3\text{]-2H } | + Ca^{2+} \text{-------->} | \text{ 2[ROSO}_3^-\text{]-Ca } | + 2H^+ \tag{4}$$

Net reaction:

$$| \text{ 2ROH } | + 2H^+ + Ca^{2+} + 2SO_4^{2-} \text{------>} | \text{ 2[ROSO}_3^-\text{]-Ca } | + 2H_2O \tag{5}$$

where the boxes represent algal cells.

Although the net reaction (Equation 5) consumes only one mole of alkalinity/mole sulfate utilized, it is important to recognize that two distinct reactions are involved: 1) the formation of the ester sulfate group, which consumes two moles alkalinity/mole sulfate (Equation 3), and 2) the subsequent dissociation/ion exchange reaction (Equation 4), in which two moles of H^+ are generated by the exchange reaction with one mole of Ca^{2+}. Thus, while the net reaction shows one mole of H^+ lost/mole SO_4^{2-} lost, the reaction must be written with a concomittent loss of cations from solution in order to maintain electroneutrality (Equation 5).

<u>Limitations on Sulfate Retention in Softwater Lakes</u>

Sulfate retention in softwater lakes appears to increase in proportion to sulfate loadings (<u>10</u>). This raises the issue of factors that limit the magnitude of sulfate retention. At least three potential conditions might limit sulfate reduction rates: 1) supply of Fe^{2+} to sequester reduced S, 2) supply of carbon to support microbial reduction, and 3) inhibition of sulfate reduction by acidification.

In seepage lakes, external iron supply may not be sufficient to sequester all of the S that accumulates in sediments. For example, at Little Rock Lake, most of the iron probably enters by atmospheric deposition. Based on measurements of ambient air aerosol concentrations and wet deposition of iron for a number of Midwestern sites (<u>59-61</u>), we estimate that iron input to Little Rock Lake is on the order of 2-6 mmole/m²-yr. Our estimate of lake-wide Fe accumulation in sediments (12 mmole/m²-yr) is somewhat higher, and the current Fe:S ratio in sediments is about 1:1 (molar). These data show that the current iron supply is potentially sufficient to sequester all of the S retained in sediments as pyrite or other iron sulfide minerals with higher FeS ratios, but in reality, only an unknown fraction of the total sedimentary iron is available for sequestering reduced S, as illustrated by the effect of soluble Fe^{2+} addition on endproduct formation (above). Factors that determine whether H_2S is sequestered by Fe^{2+} or by organic compounds are not understood. Nevertheless, at much higher sulfate input rates characteristic of the northeast United States, gross iron supply may place an absolute limit on formation of Fe-S endproducts. This conclusion is opposite the conclusion of Rudd et al. (<u>14</u>), who reported that extractable iron concentrations far exceeded total S in cores. The probable reason for this discrepancy is that Rudd et al. were looking at drainage lakes, which presumably receive iron-rich soil material from their watersheds. By contrast, Little Rock Lake and other seepage lakes receive very little soil material (hence little iron input) from the surrounding basin.

A second potential limitation is carbon supply for microbial reduction. Carbon budget calculations for Little Rock Lake show that sulfate reduction uses a small fraction of available carbon. Carbon recycling, estimated as the difference between sediment trap C and sediment C accumulation, is 3310 mmole/m²-yr at SB-5. Our estimate of diffusive flux (i.e., net sulfate flux) at this site (5.4 mmole/m²-yr). If the recycling rate is 80% (similar to the observed recycling of seston-S), lake-wide gross sulfate reduction would be < 30 mmole/m²-yr. Assuming a 1:1 stoichiometry between sulfate reduction and carbon utilization, we conclude that sulfate reduction uses < 1% of total recycled carbon. Thus, on the basis of gross carbon availability, there appears to be plenty of carbon to support sulfate reduction. A more difficult question, and one we have not yet answered, is whether sufficient carbon is supplied in a form available for sulfate reduction. Kelly and Rudd (<u>33</u>) concluded that the occurrence of methane production below the zone of sulfate reduction indicates adequate carbon supply for sulfate reduction. Their conclusion is based on the assumptions that methane production would occur only after all carbon requirements for sulfate reduction are met and that similar substrates are used by methanogens and sulfate reducers.

Finally, there is a potential for inhibition of sulfate reduction by sediment acidification in highly impacted sites. In the first two years of experimental acidification of Little Rock Lake there is no evidence of decreased pH in porewater 1 cm below the interface. It is not clear, however, whether sediment acidification will occur with further increases in acid loadings to the lake. Rudd et al. (6) showed that porewaters from lakes Hovattn and Little Hovattn were acidic at fall turnover and postulated that this may occur by oxidation of reduced sulfur compounds. Although sediments from 223 showed no evidence of acidification after 10 years of experimental lake acidification, the pH of porewater from Lake 114 declined by > 0.5 units after just three years of experimental acidification (6).

In addition to limitations on sulfate reduction, seston deposition of S is limited by algal-S content and primary productivity. The relationship between primary productivity and lake acidification is unclear (62), but limited evidence suggests that primary productivity is not particularly sensitive to moderate lake acidification. Further, there is little evidence to indicate that the S content of seston changes much with acidification (Table III). Hence, within a given lake, the loss of sulfate from the water column from seston deposition probably changes little during the acidification process.

Finally, changes in recycling rates would alter net sulfate retention. Specifically, if acidification resulted in decreased decomposition, one would expect recycling of sulfate to the water column to decrease, and net sulfate retention to increase, during the acidification process. However, the evidence for reduced decomposition is mixed (62), and there are no specific data to evaluate the hypothesis that acidification alters recycling of sulfur.

Conclusions

In-lake processes remove approximately half of the sulfate inputs from the water column of Little Rock Lake. Two processes, seston deposition and dissimilatory reduction, are responsible for sulfate retention. For the pre-acidified lake, seston deposition probably is the dominant sink, accounting for 70% of net retention. Preliminary data and theoretical considerations suggest that the diffusive flux of sulfate to sediments will increase during experimental acidification, and we believe that dissimilatory reduction is the dominant sulfate sink in lakes with elevated sulfate concentrations.

Although both processes result in an equivalent ratio of alkalinity production: net sulfate loss, the factors regulating gross removal from the water column and subsequent diagenesis and recycling are very different for the two processes. Studies in our lab and elsewhere show that sediment diagenetic processes are far more complex than envisioned by earlier investigators; in particular, we now recognize that organic compounds are important in the sequestering of H_2S produced by dissimilatory reduction, but the exact mechanisms remain largely unknown. These mechanisms must be unraveled in order to completely understand factors regulating dissimilatory reduction and recycling of sulfur.

Acknowledgments

This work was supported by two cooperative agreements from the Corvallis and Duluth EPA Laboratories, and we would like to thank the project officers (R. Church and J. Eaton) for their cooperation. The following individuals made significant contributions to the study: Dan Engstrom (dating of cores), T. Perry (porewater equilibrators in 1984 and 1985), C. Conners and S. King (analysis of sulfur species), J. Tacconi (ion budgets), E. Weir, N. Detenbeck, C. Mach, and

J. Wachtler (sampling). We would also like to thank C. Watras and others at the Trout Lake Research Lab of the University of Wisconsin for logistic support during field studies.

Literature Cited

1. Hongve, D. Verh. Internat. Verein. Limnol. 1978, 20, 743-8.

2. Cook, R. B.; Kelly, C. A., Schindler, D. W.; Turner, M. A. Limnol. Oceanogr. 1986, 31, 134-48.

3. Baker, L. A.; Pollman, C. D.; Brezonik, P. L. In Florida Acid Deposition Study, Phase IV; ESE 83-152-0602-0120, Environmental Sciences and Engineering, Inc., Gainesville, FL, 1984, pp. 163-214.

4. Baker, L. A.; Brezonik, P. L.; Edgerton, E. S. Water Resources Res. 1986, 22, 715-22.

5. Lin, J. C.; Schnoor, J. L.; Glass, G. E. In Chemistry of Aquatic Pollutants; Hites, R.; Eisenreich, S. J., Eds.; Advances in Chemistry 216, American Chemical Society: Washington, D. C., 1987; pp 209-28.

6. Rudd, J. W. M.; Kelly, C. A.; St. Louis, V; Hesslein, R. H.; Furutani, A. Limnol. Oceanogr. 1986, 31, 1267-80.

7. Perry, T. E.; Brezonik, P. L.; Pollman, C. D. In Impact of Acid Deposition on Aquatic Biological Systems; Isom, B. G.; Dennis, S. D.; Bates, J. M., Eds.; ASTM STP 928; American Society of Testing and Materials, 1986; pp 67-83.

8. Perry, T. E. M.S. Thesis, University of Minnesota, Minneapolis, 1988.

9. Baker, L. A.; Perry, T. E., Brezonik, P. L. In Lake and Reservoir Management: Practical Applications; Taggart, J.; Moore, L., Eds; North American Lake Management Society: Washington, D. C., 1985; pp 356-60.

10. Baker, L. A.; Brezonik, P. L.; Pollman, C. D. Water, Air, Soil Poll. 1986, 30, 89-94.

11. Gherini, S; Mok, L.; Hudson, R. J. M.; Davis, G. F. Water, Air, Soil Poll. 1985, 26, 425-59.

12. Brezonik, P. L.; Baker, L. A.; Perry, T. E. In Chemistry of Aquatic Pollutants; Hites, R. A.; Eisenreich, S. J., Eds.; Advances in Chemistry Series 216, American Chemical Society: Washington, D. C., pp 229-60.

13. Nriagu, J. O.; Soon, Y. K. Geochem. Cosmichim. Acta 1985, 49, 823-34.

14. Rudd, J. W. M.; Kelly, C. A.; Furutani, A. Limnol. Oceanogr. 1986, 31, 1281-91.

15. Mitchell, M. J., David, M. B.; Uutala, A. J. Hydrobiologia 1985, 21, 121-7.

16. Sherman, L. A. M.S. Thesis, University of Minnesota, Minneapolis, 1988.

17. Baker, L. A., Tacconi, J.; Brezonik, P. L., Verh. Internat. Verein. Limnol. 1988, 23, 346-50

18. King, G. M.; Klug, M. J. Appl. Environ. Microbiol. 1982, 43, 1406-12.

19. David, M. B.; Mitchell, M. J. Limnol. Oceanogr. 1985, 30, 1196-1207.

20. Wetzel, R. G. Limnology; W. B. Saunders and Co.: Philadelphia, 1975.

21. Brezonik, P. L.; Baker, L. A.; Eaton, J.; Frost, T.; Garrison; Kratz, T.; Magnuson, J; Perry, J.; Rose, W.; Shepard, B.; Swenson, W.; Watras, C.; Webster; K. Water, Air, Soil Poll. 1986, 30, 115-22.

22. Atmospheric Precipitation in Ontario Study, Ontario Ministry of Environment, Annual statistics of concentration cumulative ambient air monitoring network, 1985, APIOS-002-87, Ontario Ministry of Environment, 1987.

23. Tacconi, J. M S. Thesis, Univeristy of Minnesota, Minneapolis, 1988.

24. Rose, W., U. S. Geological Survey, Madison, WI, personal communication, 1987.

25. Baker, L. A.; Engstrom, D. E.; Brezonik, P. L., submitted for publication in J. Paleolimnology.

26. Perry, J., Forestry Department, University of Minnesota, personal communication, 1986.

27. Baker, L. A.; Pollman, C. D.; Eilers, J. E., Water Resources Research 1988, 24, 1069-82.

28. Rogalla, J. A.; Brezonik, P. L.; Glass, G. E. Water, Air, Soil Poll. 1986, 31, 95-100.

29. Baker, L. A.; Perry, T. E.; Brezonik, P. L., submitted for publication in Biogeochemistry.

30. Ingvorsen, K.; Jorgensen, B. B. Arch. Microbiol. 1984, 139, 61-6.

31. Nethe-Jaenchem, R.; Thauer, R. K. Arch. Microbiol. 1984, 137, 236-40.

32. Nielsen, P. H. Appl. Environ. Microbiol. 1987, 27-32.

33. Kelly, C. A.; Rudd, J. W. M. Biogeochemistry 1984, 1, 63-77.

34. Grebb, S.; Sherman, L. A.; Perry, T. E.; Baker, L. A., in preparation.

35. Goldman, J. C.; Porcella, D. B.; Middlebrooks, E. J.; Toerin, D. F. Water Res. 1972, 6, 637-79.

36. Schiff, J. A. In Physiology and Biochemistry of Algae; Lewin, R. A., Ed.; Academic Press: New York, 1962; Chapter 14.

37. O'Colla, P. S. In Physiology and Biochemistry of Algae; Lewin, R. A., Ed.; Academic Press: New York, 1962; Chapter 20.

38. O'Kelly, J. C. In Algal Physiology and Biochemistry; Stewart, W. D. P., Ed.; Blackwell Scientific: London, 1974; Chapter 22.
39. Landers, D. H.; Mitchell, M. J. Hydrobiologia, in press.
40. Wieder, R. K.; Lang, G. E.; Granus, V. A. Soil Biol. Biochem. 1987, 19, 101-6.
41. Cook, R. B. Ph.D. Dissertation, Columbia University, 1981.
42. Behr, R. S. M.S. Thesis, University of Manitoba, 1985.
43. Cook, R. B., Can. J. Fish. Aquat. Sci. 1984, 41, 286-93.
44. Davison, W.; Lishman, J. P.; Hikkon, J. Geochem. Cosmichim. Acta 1985, 49, 1615-20.
45. Brown, K. A., Soil Biol. 1986, 18, 131-40.
46. Francois, R. Limnol. Oceanogr. 1987, 32, 964-72.
47. Mango, F. D. Geochim. Cosmichim. Acta, 1983, 47, 1433-41.
48. Casagrande, D. J.; Ng, L. Nature 1979, 282, 598-9.
49. Casagrande, D. J.; Gronli, K.; Sutton, N. Geochim. Cosmichim. Acta 1980, 44, 25-32.
50. Vairavamurthy, A.; Mopper, K. Nature 1987, 329, 623-5.
51. Kiene, R. P.; Taylor, B. F. Nature 1988, 332, 148-50.
52. Vairavamurthy, A.; Mopper, K. Nature, this volume.
53. Baker, L. A.; Brezonik, P. L. Water Resources Research 1988, 24, 65-74.
54. Cook, R. B.; Kelley, C. A., Kingston, J. C.; Kreis, R. G., Jr. Biogeochemistry 1987, 4, 97-117.
55. Mackie, W.; Preston, R. D. In Algal Physiology and Biochemistry; Stewart, W., Ed.; Blackwell Scientific: London, 1974; Chapter 2.
56. Percival, E.; McDowell, R. H. Chemistry and Enzymology of Marine Polysaccharides; Academic Press: New York, 1967.
57. Streitwiese, A.; Heathcock, C. H. Introduction to Organic Chemistry; Macmillan: New York, 1976, pp 498-501.
58. Metzler, D. E. Biochemistry; Academic Press: New York, 1977.
59. Eisenreich, S. J.; Hollad, G. J.; Langevin, S. Precipitation chemistry and atmospheric deposition of trace elements in northeastern Minnesota; prepared for the Minnesota Environmental Quality Council and Minnesota State Planning Agency, 1978.
60. Atmospheric Precipitation in Ontario Study, Annual statistics of concentration and deposition--cumulative precipitation monitoring network, 1985, APIOS-013-87, 1987.
61. Pratt, G., Minnesota Pollution Control Agency, 1987.

62. Dillon, P. J.; Yan, N. D.; Harvey, H. H. In <u>CRC Reviews in Environmental Control, Vol. 13</u>; CRC Press: Boca Raton, FLA, 1984.

63. Zhabina, N. N.; Volkov, I. I. In <u>Environmental Biogeochemistry and Geomicrobiology, Vol. 3, Methods, Metals, and Assessment</u>; Krumbein, W. E., Ed.; Ann Arbor Science: Ann Arbor, MI, pp. 735-45.

RECEIVED August 11, 1988

Chapter 8

Reactivity of Sulfur Nucleophiles Toward Halogenated Organic Compounds in Natural Waters

Jack E. Barbash and Martin Reinhard

Environmental Engineering and Science, Department of Civil Engineering, Stanford University, Stanford, CA 94305

Data from the literature have been assembled to examine the extent to which different sulfur nucleophiles may be responsible for the abiotic dehalogenation of haloaliphatic compounds in natural waters (either pristine or contaminated) or in biodegradation experiments which employ sulfide salts to scavenge molecular oxygen. Information on the abundance of sulfur nucleophiles in hypoxic (i.e., oxygen-deficient) natural waters have been combined with experimental data on both the S_N2 and E2 reactivity of various haloaliphatic substrates toward sulfur nucleophiles and H_2O. From the compiled data, it appears that these reactions are most likely to be significant for the degradation of primary bromoaliphatic compounds, particularly at elevated pH or in the presence of organic cosolvents. Reactions involving chloroaliphatic substrates, though slower, may yield products which are of concern to public health.

The detection of significant levels of halogenated aliphatic contaminants in groundwater resources in the United States (1,2) has spurred a considerable effort to understand the rates and pathways for the various chemical mechanisms by which these compounds may be transformed in natural waters. The dehalogenation mechanisms which are generally believed to be of greatest environmental significance for haloaliphatic compounds in surface and groundwaters include nucleophilic substitution (termed hydrolysis when the nucleophile is H_2O or OH^-), dehydrohalogenation, reductive dehalogenation, and photolysis. Among these, hydrolysis has received the most extensive attention (3). Recent work by a number of investigators, however, has demonstrated that the displacement of halide from bromoaliphatic compounds (4-7), as well as from chloroaliphatic compounds (5) by sulfur nucleophiles may

0097–6156/89/0393–0101$10.50/0
© 1989 American Chemical Society

take place at rates which exceed those of hydrolysis under conditions which are common in reducing environments.

In this discussion, we will examine the potential importance of different sulfur nucleophiles as agents for the dehalogenation of haloaliphatic compounds in natural waters. This assessment will be made by combining experimental data on the reactivity of sulfur nucleophiles toward halogenated organic compounds with information from field studies on the occurrence and abundance of sulfur nucleophiles in natural waters. For the purposes of our discussion, a "sulfur nucleophile" is not only a nucleophile which contains sulfur, but one in which the atom that donates electrons to form a bond with the electrophile is a sulfur atom (8). The natural occurrence or nucleophilic reactivity of sulfate will therefore not be examined. The term "halogenated aliphatic substrate" will be used to denote any organic compound in which a halogen atom is bound to an aliphatic carbon, regardless of whether the molecule also contains an aromatic moiety (e.g., benzyl chloride or DDT). Finally, by "natural waters," we will be referring to all ground-, interstitial, and surface water environments, regardless of whether an anthropogenic component is evident ("contaminated") or absent ("pristine").

OCCURRENCE AND ABUNDANCE OF SULFUR NUCLEOPHILES IN NATURAL WATERS

Environmental Setting

The most common sulfur nucleophiles encountered in natural waters are the following species: H_2S, HS^-, S^{2-}, thiolate anions (RS^-), polysulfides (S_n^{2-}, where $n > 1$) $S_2O_3^{2-}$, $S_4O_6^{2-}$, and SO_3^{2-}. Unlike sulfate, all of these nucleophiles are thermodynamically unstable in the presence of O_2. Hence, they are most commonly found in hypoxic environments (i.e., natural waters containing very low or undetectable levels of dissolved O_2), usually in the presence of high concentrations of reduced organic matter. Environments of this type include marine, estuarine, and marsh porewaters, groundwaters containing high concentrations of leachate derived from domestic refuse (e.g., beneath landfills), and hydrogeologic zones located far down-gradient from their recharge areas.

The speciation of sulfur among its various forms in a particular environment arises from a balance among several simultaneous chemical and biochemical processes, including oxidation and reduction (usually with microbiological mediation), hydrolysis, dissolution, and precipitation. The bacterial reduction of sulfate yields hydrogen sulfide (9), and hence HS^- and S^{2-}, as well (hereafter referred to collectively as $H_2S[aq]$). Species which are partially reduced (i.e., forms in which sulfur exhibits an oxidation state between those of SO_4^{2-} [+ VI] and $H_2S(aq)$ [-II]), such as S_n^{2-}, $S_2O_3^{2-}$, $S_4O_6^{2-}$, and SO_3^{2-} may be produced by the oxidation of $H_2S(aq)$ and sulfide minerals, the reductive dissolution of goethite by HS^-, the hydrolysis of polysulfides, and microbial processes (9-12). Thiols may be produced biologically as a result of sulfate reduction, biodegradation of organic matter or lithotrophic oxidation (i.e., the biological oxidation of inorganic compounds to obtain energy (13)); or abiotically through the reactions of dissolved organic matter with $H_2S(aq)$ or elemental sulfur (11,14).

The concurrent operation of several of these processes in a particular environment may give rise to a complex assemblage of sulfur nucleophiles. For example, Bouleague et al. (11) detected $H_2S(aq)$, S_8, $S_2O_3^{2-}$, SO_3^{2-}, S_n^{2-}, SO_4^{2-}, thiols, and organic polysulfides in the porewaters of a salt marsh along the Delaware estuary. They attributed the observed sulfur speciation to a steady-state interaction between the various reduced sulfur species diffusing upward from the hypoxic sediments (chiefly $H_2S[aq]$, thiols, and organic polysulfides) and O_2 diffusing downward from the sediment-water interface.

Abundance of Sulfur Nucleophiles in Natural Waters

Tables I, II, and III summarize data from several studies on the distribution and abundance of reduced sulfur species in natural waters. The tables give only the maximum concentration (in μM) for each species observed. The data provided in the literature on this subject do not appear to be very extensive (12,15). Table I summarizes data obtained from studies which examined the concentrations of a broad range of chemical parameters, including "total sulfide," which may have included S_n^{2-} as well. For the studies listed in Table II, however, the analytical techniques employed were specifically designed to differentiate among the various sulfur species encountered. Nevertheless, some difficulties still remain with regard to the specific structures of the species whose concentrations are given in Table II.

One such difficulty is that, while it appears that $\Sigma[S_n^{2-}]$ levels may be as high as 400 μM in some hypoxic environments, the precise number (n) of sulfur atoms in these polysulfide species, or even the range of n, is uncertain. Bouleague (16) discusses some of the thermodynamic equilibrium calculations which may be used to estimate the distribution of $\Sigma[S_n^{2-}]$ among the various polysulfide species. Another difficulty with these data is that the structures of the thiols (RSH) detected are not well known. Mopper and Taylor (14) identified 13 different thiols in slurries of intertidal sediments from Biscayne Bay (FL), and found at least 20 more thiols whose structures could not be determined. Their observations (Table III) indicate that thiols encountered in natural waters will probably exhibit a broad variety of structures in any one location, but that a relatively small number of compounds may dominate the mixture.

Table I. Occurrence of $H_2S(aq)$[a] in Groundwaters

Environment	pH	$H_2S(aq)_{max}$[a] (μM)	Reference
Pristine			
groundwater up-gradient from a waste disposal site	5.0	0.2	(53)
limestone aquifer	8.4	6	(53)
sandy aquifer	7.8	2	(53)
sandy aquifer beneath a swamp	7-8	3.4	(54)
Contaminated			
groundwater beneath a municipal landfill	7.8	468	(55)
groundwater beneath a waste disposal site	7.4	593	(53)
groundwater beneath a municipal landfill	6.2	3.6	(54)
groundwater beneath an industrial/municipal landfill	7.7	5.3	(50)
groundwater beneath a chemical plant	8	20	(4)

[a]Sum of $[H_2S]$, $[HS^-]$ and $[S^{2-}]$ concentrations. Tabulated data represent the maximum concentrations observed during each study.

Table II. Occurrence of Reduced Sulfur Species in Groundwaters and
Porewaters. Maximum Observed Concentrations (μM) Given

Environment	pH	$H_2S(aq)$[a]	RSH	$S_2O_3^{2-}$	SO_3^{2-}	$\Sigma[S_n^{2-}]$[b] (n)	Ref.
Groundwaters (discharge areas)							
limestone aquifer	7.4	775		250		355 (3-6)	(16)
black marl	6.9	2021[c]				d	(56)
Porewaters							
salt marsh		2700					(57)
salt marsh	6.27[e]	700[c]			32	d	(12)
(Great Sippewissett, MA)	6.90[e]	3630[f]	<10		624		(15)
salt marsh	6.83	3360[f]	2411		104		(15)
(Great Marsh, DE)	7.17	5060	165[g]	530	177	400[h]	(11)
subtidal porewater (Orleans, MA)	7.76[e]	2500[c]			26	d	(12)
intertidal sediment (slurry[i]; Biscayne Bay, FL)	7.4[j]	4240[k]	100[m]		71		(14)

[a]Sum of $[H_2S]$, $[HS^-]$ and $[S^{2-}]$.
[b]Sum of all polysulfide sulfur species concentrations; n = number of sulfur atoms per polysulfide molecule.
[c]Sum of $[H_2S(aq)]$ and $\Sigma[S_n^{2-}]$.
[d]Included in $[H_2S(aq)]$. See note c.
[e]Sulfur species concentrations obtained at different sites. pH value given was that measured at the location where $[H_2S(aq)]$ value was measured.
[f]$[HS^-]$ in pH 10 buffer.
[g]$\Sigma([RSH] + [RS_mH])$.
[h]Computed from the published data, using the equation:
$$\Sigma[S_n^{2-}] \approx \Sigma[S^{II-}] - [H_2S(aq)].$$
[i]Slurries prepared using two parts sediment to one part deaerated seawater (v/v). Thus, sulfur species concentrations are probably underestimates of aqueous levels.
[j]Measured in porewater directly (?).
[k]$[HS^-]$ in slurry.
[m]See Table III for specific thiols identified in slurries.

Table III. Thiols Identified in Slurries of Marine Sediments (from 14)

Major Thiols (0.5 - 20 μM)[a]
3-mercaptopropionate (dominant)
methanethiol (dominant)
ethanethiol
3-mercapto-1,2-propanediol (monothioglycerol)
2-mercaptoethanol
2-mercaptopyruvate

Minor Thiols (< 0.5 μM)[a]
mercaptoacetate (thioglycollate)
glutathione
2-mercaptoethanesulfonate (Coenzyme M)
l-propanethiol
2-propanethiol
N-acetylcysteine
2-mercaptopropionate

Unknowns
At least 20 unknown thiols, including 4 or 5 which are major.

[a]Slurries prepared using two parts sediment to one part deaerated seawater (v/v). Thus, sulfur species concentrations are probably underestimates of concentrations in original porewaters.

REACTIVITY OF SULFUR NUCLEOPHILES IN MODEL SYSTEMS

Prediction of Nucleophilic Reactivity

Use of Correlation Equations

Limited success has been achieved in predicting absolute rates of reaction between electrophiles and nucleophiles through the correlation of nucleophilic reactivity with the pK_a ([17]) and the electrode potential of the nucleophile ([18,19]), or solvent activity coefficients ([20]). A summary and critical analysis of the correlation equations which have been devised for this purpose has been provided by Duboc ([21]).

Two different correlation equations were used to estimate some of the kinetic data summarized here. One was the two-parameter expression introduced by Swain and Scott ([22]):

$$\log(k_S/k_{H2O})_{SN} = s*n \tag{1}$$

where

k_S $\equiv$ second-order rate constant for the nucleophilic substitution reaction between a particular substrate (RX) and the nucleophile of interest (in the present case, a sulfur nucleophile);

k_{H2O} $\equiv$ second-order rate constant for the nucleophilic substitution reaction between RX and H_2O; and

s, n $\equiv$ empirical constants characteristic of the substrate and the nucleophile, respectively.

The second equation used for the data compiled here was proposed by Edwards ([18]):

$$\log(k_S/k_{H2O})_{SN} = \alpha E_n + \beta H \tag{2}$$

where

E_n $\equiv$ electron donor constant for the nucleophile;

H $\equiv$ basicity of the nucleophile of interest toward a proton (relative to H_2O); and

α, β $\equiv$ substrate constants.

In those instances in which an equation was used, the equation is given in a footnote in the appropriate table.

The Hard and Soft Acids and Bases (HSAB) Model

The various equations proposed for the prediction of nucleophilic reactivity have been found to hold only among similar nucleophiles or similar substrates. Their generality is limited by the fact that even the relative order of reactivity among nucleophiles toward a substrate may be dependent upon substrate structure (23), solvent composition (20), solvent dielectric constant (24), pH, and leaving group (25). Most of these apparent inconsistencies, however, may be explained through an approach which has its origins in the work of Pearson (26), who proposed the "hard and soft acids and bases" (HSAB) model. This model has been criticized for being insufficiently quantitative (27,28). However, Klopman (24) has shown that the predictions made by the HSAB model with respect to relative reactivities among different nucleophiles and different substrates are consistent with the results from frontier orbital energy calculations.

Model Definition. The HSAB model classifies Lewis acids (electrophiles) and bases (nucleophiles) as either "hard" or "soft." Hard acids and bases are relatively small, and exhibit low polarizability and a comparatively low tendency to form covalent bonds. Soft acids and bases have the opposite characteristics (24). Stated simply, the model postulates that hard acids react most readily with hard bases, and soft acids react most readily with soft bases (26).

Using the HSAB terminology, the bimolecular elimination of hydrogen halide from a halogenated alkane with two or more carbon atoms is the result of an attack on a hard electrophilic center (a proton vicinal to the leaving halide). Conversely, using the same substrate, the nucleophilic displacement of halide arises from the attack on a soft electrophilic center (saturated carbon). The HSAB model therefore predicts that in the reaction between a halogenated aliphatic moiety and a hard base, dehydrohalogenation would be the dominant pathway, while for the reaction with a soft base, nucleophilic displacement would dominate, assuming that steric hindrance to attack at carbon is not significant.

Theoretical Basis. Klopman (24) placed Pearson's HSAB model into a more fundamental and quantitative perspective by introducing the concept of "charge- vs. frontier-controlled reactions." Klopman demonstrated that changes in the order of relative reactivities among nucleophiles toward different electrophilic centers (peroxide oxygen, saturated carbon, and carbonyl carbon) could be accurately predicted from differences between the energies of the frontier orbitals of the nucleophile ("donor") and the electrophile ("acceptor"). In the terminology used by Fleming (29), these orbitals correspond to the "highest occupied molecular orbital" (HOMO) of the nucleophile and the "lowest unoccupied molecular orbital" (LUMO) of the electrophile. These energies are, in turn, calculated from fundamental properties of the interacting species, such as ionization potentials (IP), electron affinities (EA), electrostatic charges, desolvation energies, and solvent dielectric constants (24).

Parr and Pearson (30) defined a parameter η, which they called "absolute hardness" ($\eta \equiv \frac{1}{2}$[IP-EA]), and calculated η for a variety of neutral and ionic Lewis acids and bases possessing from one to four atoms. These authors showed that the qualitative predictions of the HSAB model regarding the relative reactivities of these species toward one another may be obtained using the results from simple calculations of stabilization energies using η and electronegativity values.

At present, these theoretical calculations are either too demanding with regard to their data requirements (24) or too simplistic (30) to be used to predict absolute rates of reaction between sulfur nucleophiles and haloaliphatic substrates in natural waters. However, they provide a solid theoretical justification for using Pearson's HSAB model to explain trends in reactivity among different nucleophiles toward haloaliphatic substrates.

Reactivity of Sulfur Nucleophiles

The predictive value of the HSAB model may be demonstrated using the example of the relative reactivities of HS$^-$ and its oxygen analogue (OH$^-$) toward a haloaliphatic compound. Sulfur lies directly below oxygen in the periodic table, indicating that HS$^-$ is more polarizable and has a HOMO which is higher in energy than its oxygen counterpart. HS$^-$ is thus a "softer" nucleophile than OH$^-$, and consequently would be expected to show the greater reactivity with respect to nucleophilic displacement at a saturated carbon. Conversely, OH$^-$ would be expected to show a stronger affinity toward hard electrophiles and hence a higher rate of dehydrohalogenation of haloaliphatic compounds than its sulfur analogue. These predictions are consistent with experimental data on the reactivities of HS$^-$ and OH$^-$ toward 1,2-dichloroethane in aqueous solution (5,31).

Influence of Solution Properties on Reactivity

pH

Among the various sulfur nucleophiles which are encountered in natural waters, most are Bronsted acids or bases, i.e., they can accept or lose a proton. The sulfhydryl functional group is a monoprotic Bronsted acid, while H$_2$S and H$_2$SO$_3$ are diprotic Bronsted acids. The rate of displacement of halide from a given substrate in aqueous solution containing a nucleophilic Bronsted acid depends upon the relative concentrations among the different protonated forms of the acid (of which there are two for a monoprotic acid, three for a diprotic acid), as well as the respective rates of reaction of the different protonated forms with the substrate. Equation 3, derived using an approach identical to that employed to calculate ionization fractions for a diprotic acid (9), depicts the influence of $[H^+]$ on $k_{P1,SN}$, the overall pseudo-first-order rate constant for the nucleophilic attack of a reactive substrate by the three forms of a diprotic acid (NuH_2) in solution. This approach assumes that all activity coefficients are unity.

$$k_{P1,SN} = [\Sigma Nu(aq)] \left[\frac{k_{NuH2}}{1 + K_1/[H^+] + K_1K_2/[H^+]^2} \right.$$

$$\left. + \frac{k_{NuH^-}}{1 + [H^+]/K_1 + K_2/[H^+]} + \frac{k_{Nu2^-}}{1 + [H^+]/K_2 + [H^+]^2/K_1K_2} \right] \qquad (3)$$

where

$[\Sigma Nu(aq)] \equiv [NuH_2] + [NuH^-] + [Nu^{2-}]$;

$K_1, K_2 \equiv$ first and second dissociation constants for NuH_2, respectively,

and

k_{NuH2}, k_{NuH^-} and $k_{Nu2^-} \equiv$ second-order rate constants for reaction with NuH_2, NuH^-, and Nu^{2-}, respectively.

For a monoprotic acid, the third term on the right-hand side is not present, and the third term in the denominator of the first two vanishes, because $K_2 \equiv 0$.

As the tabulated data on the S_N2 reactivity of sulfur species demonstrate (see - Influence of Nucleophile Structure on Reactivity), the second-order rate constant for a nucleophile which is a Bronsted base will always be several orders of magnitude (for these data, from 2.7 to 4.6 log units) larger than that for its conjugate acid. Consequently, as pH increases, a plot of log $(k_{P1,SN}/[\Sigma Nu(aq)])$ versus pH exhibits a smooth, sigmoid step increase for each Bronsted acid-base pair of nucleophiles (two such inflections for a diprotic acid, one for a monoprotic acid). The inflection point for each of these steps corresponds to the pK_a of the conjugate acid of the pair. Although very little data have been published on the pH-dependence of substitution reactions involving sulfur nucleophiles, Haag and Mill (6) demonstrated that the reaction between n-hexyl bromide and $H_2S(aq)$ exhibits this type of pH-dependence.

Concentrations of Organic Co-Solvents

Many bimolecular reactions are accelerated by decreasing the solvent polarity, as well as by changing from a protic to a dipolar aprotic solvent. (According to Parker ([20]), a protic solvent is defined as one in which hydrogen is bonded to oxygen or nitrogen, rather than to carbon. Parker defines a dipolar solvent as one with a dielectric constant greater than 15.) These phenomena arise largely from the influence of the reaction medium on the relative stabilities of ionic reactants, intermediates, and transition states. As the reaction medium becomes more polar, ionic species are stabilized to a greater extent than neutral ones. Similarly, species possessing a highly concentrated charge are more extensively stabilized than those with a more dispersed charge when the solvent polarity is increased. Protic solvents influence the rates of bimolecular reactions primarily (though not exclusively) through their ability to form hydrogen bonds. Dipolar aprotic solvents, which form only very weak hydrogen bonds, are considerably less effective than protic ones at stabilizing ionic species in solution, but are better able to stabilize neutral species ([20,28]).

The reactions of principal interest for the present discussion involve the attack of either an anionic (4) or a neutral species (5) on a neutral substrate.

$$Y^- + RX \longrightarrow [YRX^-]^{\ddagger} \longrightarrow \quad \text{products of substitution} \\ \text{or elimination} \tag{4}$$

$$Y + RX \longrightarrow [YRX]^{\ddagger} \longrightarrow \quad \text{products of substitution} \\ \text{or elimination} \tag{5}$$

Accounting for the solvent effects described above ([20,28]), one would expect that a decrease in solvent polarity or a change from a protic to a dipolar aprotic solvent would accelerate Reactions 4, but might have the opposite effect on Reactions 5.

Data assembled by Parker ([20]) demonstrate these effects for bimolecular reactions involving sulfur nucleophiles and haloaliphatic substrates. As an illustration for the case of Reactions 4, the S_N2 displacement of iodide from CH_3I by SCN^- at 25^oC is accelerated relative to its rate in water by 0.2 log units in methanol, by 1.1 log units in 10% aqueous dimethyl sulfoxide (v/v), and by approximately 2.4 log units in dimethyl formamide (DMF). Furthermore, the rates of bimolecular elimination and substitution of cyclohexyl bromide in the presence of thiophenolate at 25^oC both increase by 2.7 log units when the solvent is changed from ethanol to dimethylformamide ([20]).

With respect to Reactions 5, a change from a protic to a dipolar aprotic solvent appears to influence molecule-molecule reactions to a relatively minor degree: The S_N2 reaction between dimethyl sulfide and CH_3Br at 25^oC is accelerated by 0.5 log units when the reaction medium is changed from methanol to dimethylacetamide ([20]). Additional examples of the effects of solvent change on the reactivity of sulfur nucleophiles are provided by the data assembled in Tables IV through VII, as described below in the section entitled Solvent Composition.

Concentrations of Electrolytes

Nucleophilic Catalysis. The presence of other ions in solution can influence the rates of substitution and elimination reactions in two principal ways. First, various anions may increase the rates of substitution reactions in proportion to their respective nucleophilicities (22,32), an effect referred to here as "nucleophilic catalysis." This phenomenon has been documented in a number of laboratory studies of nucleophilic substitution and dehydrohalogenation reactions which employed a phosphate buffer to maintain constant pH in solution (5,33). Using the hydrolysis of a halo-organic compound by HPO_4^{2-} as an example, this mechanism may be portrayed as follows (3,34):

$$RX + HPO_4^{2-} ---> ROPO_3H^- + X^-$$

$$ROPO_3H^- + H_2O ---> ROH + H_2PO_4^-$$

Although this mechanism appears to be plausible, neither Mabey and Mill (3) nor Jungclaus and Cohen (34) provide any experimental evidence for it. Nevertheless, Perdue and Wolfe (35) have devised a mathematical model for estimating the maximum extent to which several different nucleophilic anions (i.e., those used in common buffer salts) might be expected to enhance the rates of pollutant hydrolysis in aqueous solution.

Salt Effects. Dissolved salts may also affect the rates of nucleophilic substitution and elimination in aqueous solution through their influence on the relative stabilities of the reactants, transition states, and other reactive intermediates. The nonspecific effects of increasing ionic strength are therefore analogous to those arising from increasing solvent polarity (28), and are sometimes referred to as "salt effects."

Influence on Rates of Bimolecular Substitution. Increases in ionic strength have been shown to retard a variety of substitution reactions involving sulfur nucleophiles and haloaliphatic substrates (36-38). In one of these studies, Crowell and Hammett (37) found that the apparent second-order rate constant for the displacement of bromide ion from isopropyl bromide by thiosulfate diminished with increasing ionic strength. This effect was even more pronounced when n-propyl bromide was used as the substrate (37). Parker (20) has noted that a change from a dipolar aprotic solvent to a protic one is likely to have a greater retarding effect on substitution reactions involving primary substrates than on those involving displacement at a secondary carbon. The effect described by Parker arises from the greater stabilization in the transition state of the positive charge at a secondary α-carbon (the carbon atom from which the leaving group departs) than at a primary α-carbon. Rates of substitution at a primary carbon should therefore also be more susceptible to retardation by increases in ionic strength than those at a secondary carbon. The observations by Crowell and Hammett (37) support this conclusion.

Influence on Rates of Bimolecular Elimination. One of the few studies to examine the influence of ionic strength on the rates of bimolecular elimination involving sulfur nucleophiles and haloaliphatic substrates yielded results which are not easily interpreted in terms of the simple stabilization effects discussed above. Bunnett and Eck (39) found that the reaction of 2-chloro-2,3,3-trimethylbutane in methanol, which yielded both a solvolysis and an elimination product (2,3,3-trimethyl-2-butyl methyl ether and 2,3,3-trimethyl-1-butene, respectively), was accelerated by $NaClO_4$ to a greater extent than by $NaSC_2H_5$. Furthermore, the proportion of alkene among the products from the reaction with ethanethiolate was seen to diminish as the concentration of ClO_4^- was raised. According to Bunnett and Eck, these results suggest that the reaction of 2-chloro-2,3,3-trimethylbutane with ethanethiolate in the presence of ClO_4^- in methanol possesses a substantial E2 component, in addition to the E1 and S_N1 pathways (39).

Rates of Reaction Between Sulfur Nucleophiles and Haloaliphatic Substrates

Scope and Limitations of Tabulated Data

Tables IV through IX summarize the data that are currently available on the rates of bimolecular substitution and dehydrohalogenation reactions between sulfur nucleophiles and halogenated aliphatic substrates in aqueous solution (i.e., either measured in water or extrapolated to water from a non-aqueous or partially aqueous solvent). The sulfur nucleophiles considered in these tables are HS^-, S^{2-}, S_4^{2-}, S_5^{2-} (Table IV), $S_2O_3^{2-}$ (Tables V and VIII), SO_3^{2-}, HSO_3^- (Table VI), thiolate anions (Tables VII, VIII, and IX), thiols, thioethers, and thioacids (Table VII). In Tables IV through VII, for each substrate considered, the logarithm of the second-order rate constant for the substitution reaction with the sulfur nucleophile of interest ($\log[k_{S,SN}]$) is tabulated along with the corresponding value for the reaction with H_2O ($\log[k_{H2O,SN}]$) and the logarithm of the ratio between the two rate constants ($\log[k_S/k_{H2O}]_{SN}$). Tables VIII and IX summarize kinetic data for reactions in which both nucleophilic substitution and dehydrohalogenation may occur, and hence include rate constants for both pathways. Also included in Tables IV through IX are some of the experimental conditions under which these rate constants were measured, the solvents

Table IV.　Second-order Rate Constants ($M^{-1}s^{-1}$) for Nucleophilic Substitution of Halogenated Substrates by HS^-, S^{2-}, S_4^{2-}, and S_5^{2-} ($k_{S,SN}$), Compared to $k_{H2O,SN}$. (T=25°C, Unless Otherwise Indicated)

Substrate	Solvent[a] (v/v)	log ($k_{S,SN}$)	log ($k_{H2O,SN}$)	log (k_S/k_{H2O})$_{SN}$	T(°C)	Refs.[b]
Nucleophile: HS^-						
MeI	MeOH	-1.9[c]	-8.88[d]	7[e]		([43,3])
MeBr	f			5.1	0°(h)[f]	([22])
$BrCH_2CH_2Br$	H_2O	-3.5	-10	6.5		([5])
$BrCH_2CH_2Br$	H_2O	-3.4	-10	6.6		([6,59])
$BrCH_2CH_2Br$	H_2O	-3.7[g]	-9.2	5.5		([7,58])
$ClCH_2CH_2Cl$	H_2O	-5.2	-11	5.8		([5])
H_3CCCl_3	H_2O	≤-6[h]	-9.5[i]	≤4		([6,59])
i-PrBr	H_2O	-3.7	-7.2	3.5		([6,59])
$(c-H_2COCH)CH_2Cl$ [j]			-6.0[d]	4.7		([22,3])
n-HexBr	H_2O	-3	-8.5	6		([4])
n-HexBr	H_2O	-2.8	-8.5	5.7		([6,4])
mustard cation	H_2O/EtOH(19:1)			4.8[k]		([42])
$C_6H_5CH_2Cl$	H_2O/diox(39:61)			4.4	50°	([22])
Nucleophile: S^{2-}						
MeBr		-0.3[c]	-8.13[d]	7.8[m]		([18,3])
$(c-H_2COCH)CH_2Cl$ [j]			-6.0[d]	8.1[m]		([18,3])
mustard cation				8.6[m]		([18])
$C_6H_5CH_2Cl$		2.4[c]	-6.64[d]	9.0[m]		([18,3])
Nucleophile: S_4^{2-}						
n-HexBr	H_2O	-0.971	-8.5	7.5		([6,4])
Nucleophile: S_5^{2-}						
n-HexBr	H_2O	-0.955	-8.5	7.5		([6,4])

[a]Solvents shown were those used for the determination of $k_{S,SN}$; $k_{H2O,SN}$ determined in H_2O, unless otherwise specified. Numbers in parentheses are the volumetric proportions for solvent mixtures. Abbreviations are as follows: MeOH = methanol; ace = acetone; diox = dioxane; EtOH = ethanol; DMF = dimethylformamide; DMSO = dimethyl sulfoxide. Solvent not specified when value of log(k_S/k_{H2O})$_{SN}$ was estimated solely from a correlation equation (e.g., those provided in ([18]) and ([22])).

Table IV footnotes—*Continued*

[b]References separated by a comma denote the two references from which the $k_{S,SN}$ (or $\log[k_S/k_{H2O}]_{SN}$) and the $k_{H2O,SN}$ data were obtained, respectively. Both rate constants obtained from the same reference if only one citation given.

[c]Log($k_{S,SN}$) value calculated from the $\log(k_{H2O,SN})$ and $\log(k_S/k_{H2O})_{SN}$ data.

[d]Log($k_{H2O,SN}$) obtained from Mabey and Mill ([3]), and included for purposes of comparison, but not used in calculation of $\log(k_S/k_{H2O})_{SN}$.

[e]Estimated by Pearson et al. ([43]) from data of Swain and Scott ([22]). Adjusted to H_2O as solvent using the following formula ([19]):

$$\log(k_S/k_{H2O})^{H2O} \approx \log(k_S/k_{H2O})^{MeOH} - 1.4$$

[f]Log(k_S/k_{H2O}) calculated by Swain and Scott ([22]), using data for attack of mustard cation by HS⁻ ([42]) in 95% aqueous ethanol at 25°C, in conjunction with their own correlations for reactions involving MeBr. Value of k_{H2O} measured by Swain and Scott at 0°C in 99% aqueous acetone.

[g]Value of $k_{S,SN}$ displayed was that obtained using a sulfide concentration (0.7mM) closest to those used in the other two studies ([5,6]) of the reaction between HS⁻ and 1,2-dibromoethane.

[h]Computed by Haag and Mill ([6]) using data from Klecka and Gonsior ([60]).

[i]Assumes 80% conversion to substitution product (CH_3COOH) during reaction with H_2O ([59]).

[j]1-chloro-2,3-epoxypropane

[k]Value obtained by multiplying the competition factor, F, of Ogston et al. ([42]; $F = k_{S,SN}/k_{H2O,SN}[H_2O]$) by 55.508 M ([$H_2O$(aq)]) and taking the log of the result.

[m]Calculated using the Edwards equation ([18]):

$$\log(k_S/k_{H2O})_{SN} = \alpha E_n + \beta H$$

where E_n = donor constant for nucleophile; H = relative basicity of nucleophile (here, the sulfur nucleophile) toward a proton; and α and β are substrate constants.

Table V. Second-order Rate Constants ($M^{-1}s^{-1}$) for Nucleophilic Substitution of Halogenated Substrates by $S_2O_3^{2-}$ ($k_{S,SN}$), Compared to $k_{H2O,SN}$ (T = 25°C, Unless Otherwise Indicated)

Substrate	Solvent(s)[a] (v/v)	log ($k_{S,SN}$)	log ($k_{H2O,SN}$)	log ($k_s/k_{H2O})_{SN}$	T(°C)[b]	Refs.[c]
MeI	H_2O	-1.5	-8.88	7.4		(41,3)
MeI	H_2O	-2.81	-11.26	8.45[d]	0o	(32)
MeI	MeOH	-0.943		7.6[e]		(43)
MeBr	H_2O	-1.42	-8.13	6.71		(61,3)
MeBr	H_2O	-1.5	-8.13	6.6		(41,3)
MeBr	H_2O/ace (99:1)[f]			6.36	50o;0o	(22)
MeBr	H_2O/EtOH (1:1)	-1.05[g]	-8.13	7.09		(62,3)
MeCl	H_2O	-3.1	-9.36	6.2		(41,3)
$MeClO_4$	H_2O	-0.808[h]	-6.041[h]	5.23[d]	0o	(32)
EtI	H_2O	-2.78	-8.53	5.75		(41,3)
EtBr	H_2O/EtOH (1:1)	-3.02	-8.33	5.31		(37,3)
EtBr	H_2O	-2.99	-8.33	5.34		(41,3)
$BrCH_2CH_2Br$	H_2O	-4.02	-10.0	5.98		(6,59)
$BrCH_2CH_2Br$	H_2O	-3.6[i]	-10.2	6.6		(63)
$ClCH_2CH_2Cl$	H_2O	-5.2[j]	-11.3	6.1		(63)
H_3CCCl_3	H_2O	<-7	-9[k]	< 2		(6,59)
n-PrBr	H_2O/EtOH (1:1)	-3.32	-8.26	4.94		(37,3)
i-PrBr	H_2O	-4.17	-7.16	2.99		(6,59)
i-PrBr	H_2O/EtOH	-4.53	-7.16	2.63		(37,3)
(c-$H_2COCH)CH_2Cl$[m]			-7.8[n]	5.9		(22,3)
i-BuBr	H_2O/EtOH (1:1)	-4.69	-9.05[p]	4.36		(62,3)
n-PēntBr	H_2O/EtOH (1:1)[q]	-3.32[r]	-8.5[s]	5.2		(38,4)
n-PentBr	H_2O/EtOH (39:61)[q]	-3.34[r]	-8.5[s]	5.2		(38,4)
n-HexBr	H_2O	-3.03	-8.5	5.5		(6,4)
mustard cation	H_2O/EtOH (19:1)			6.2[t]		(42)
$C_6H_5CH_2Cl$	H_2O/diox (39:61)			5.5	50o	(22)

[a]See Table IV, note a.
[b]Temperatures separated by a semicolon denote the two temperatures at which the $k_{S,SN}$ and the $k_{H2O,SN}$ data were obtained, respectively. Both rate constants obtained at the same temperature if only one value given.
[c]See Table IV, note b.

Table V footnotes—*Continued*

[d]$Log(k_S/k_{H2O})_{SN}$ calculated from original data at 0°C, rather than from correlation equation given in reference ([32]).

[e]Adjusted to H_2O as solvent using the following formula ([19]):

$$log\ (k_S/k_{H2O})^{H2O} \approx log\ (k_S/_{H2O})^{MeOH} - 1.4$$

[f]99% aqueous acetone used for reactions of MeBr with both $S_2O_3^{2-}$ and H_2O.

[g]$k_{S,SN}$ obtained by extrapolation of data obtained by Dunbar and Hammett ([62]) at 20°C to 25°C, using $E_a = 15.5$ kcal/mole ([62]).

[h]Substitution reactions involving methyl perchlorate in water may have a substantial S_N1 component ([32]).

[i]Rate constant extrapolated to 25°C from 37°C. In the absence of any Arrhenius data for the reaction between $S_2O_3^{2-}$ and $BrCH_2CH_2Br$, an activation energy of 20 kcal/mole, obtained for the reaction between $S_2O_3^{2-}$ and isobutyl bromide over the temperature range of 12.5°-37.5°C ([62]), was used for this calculation.

[j]Rate constant extrapolated to 25°C from 37°C. In the absence of any Arrhenius data for the reaction between $S_2O_3^{2-}$ and $ClCH_2CH_2Cl$, an activation energy of 20.6 kcal/mole, obtained for the reaction between $S_2O_3^{2-}$ and CH_3Cl over the temperature range of 15°-35°C ([41]), was used for this calculation.

[k]Assumes 80° conversion to substitution product (CH_3COOH) during reaction with H_2O ([59]).

[m]1-chloro-2,3-epoxypropane.

[n]$Log(k_{H2O,SN})$ ([3]) included for purposes of comparison only; not included in calculation of $log(k_S/k_{H2O})_{SN}$.

[p]Obtained by extrapolation from 80°C, using Arrhenius data from Mabey and Mill ([3]).

[q]Solvent proportions given on a mass basis (w/w) in original reference ([38]). Converted to volumetric basis (v/v) using 0.789 g/cm^3 for the density of ethanol ([64]).

[r]Rate constants estimated from graphs in ([38]).

[s]Rate constant for n-hexyl bromide ([4]) used as a surrogate. Rationale for this approach comes from the observation that $k_{H2O,SN}$ for MeBr, EtBr, n-PrBr and n-HexBr all agree within a factor of two at 25°C.

[t]See Table IV, note k.

Table VI. Second-order Rate Constants ($M^{-1}s^{-1}$) for Nucleophilic
Substitution of Halogenated Substrates by SO_3^{2-} and HSO_3^- ($k_{S,SN}$),
Compared to $k_{H2O,SN}$ (T = 25°C, Unless Otherwise Indicated)

Substrate	Solvent(s)[a] (v/v)	log ($k_{S,SN}$)	log ($k_{H2O,SN}$)	log ($k_S/$ $k_{H2O})_{SN}$	T(°C)[b]	Refs.[c]
Nucleophile: SO_3^{2-}						
MeI	MeOH	-1.6[d,e]	-8.88[e]	7.1[f]		(43,3)
MeBr		-3.0[g]	-8.13[e]	5.1[h]		(22,3)
MeClO$_4$	H$_2$O	-1.304	-6.041	4.74[i]	0[o]	(32)
BrCH$_2$CH$_2$Br	H$_2$O	-3.6	-10.0	6.4		(6,59)
(c-H$_2$COCH)CH$_2$Cl [j]				4.7		(22)
n-HexBr	H$_2$O	-2.7	-8.5	5.8		(6,4)
mustard cation				4.8		(22)
C$_6$H$_5$CH$_2$Cl				4.4		(22)
Nucleophile: HSO_3^-						
MeClO$_4$	H$_2$O	≤-4.1	-6.041	≤1.9[i]	0[o]	(32)

[a]See Table IV, note **a**.
[b]See Table V, note **b**.
[c]See Table IV, note **b**.
[d]Corrected for transfer from methanol to H$_2$O using
$$\log(k^{H2O}/k^{MeOH})_{SN} = -0.2$$
derived from the reaction between MeI and SCN$^-$ (20).
[e]Not used for the calculation of log ($k_S/k_{H2O})_{SN}$. Included for purposes of
comparison only.
[f]Adjusted to H$_2$O as solvent using the following formula (19):
$$\log(k_S/k_{H2O})^{H2O} \approx \log(k_S/k_{H2O})^{MeOH} - 1.4$$
[g]Calculated from log($k_{H2O,SN}$) and log($k_S/k_{H2O})_{SN}$ data.
[h]Methods used by Swain and Scott (22) to calculate log($k_S/k_{H2O})_{SN}$ for SO_3^{2-}
were identical to those described in Table IV (note **f**) for HS$^-$.
[i]See Table V, notes **d** and **h**.
[j]1-chloro-2,3-epoxypropane

Table VII. Second-order Rate Constants ($M^{-1}s^{-1}$) for Nucleophilic Substitution of Halogenated Substrates by Thiols, Thiolate Anions, Thioacids and Thioethers ($k_{S,SN}$), Compared to $k_{H2O,SN}$ ($T = 25°C$, Unless Otherwise Indicated)

Nucleophile	Substrate	Solvent[a] (v/v)	log ($k_{S,SN}$)	log ($k_{H2O,SN}$)	log (k_S/k_{H2O})$_{SN}$	Refs.[b]
Thiols						
$HSCH_2CH_2SH$	mustard cation	H_2O/EtOH (19:1)			4.7[c]	(42)
$HOCH_2CH$- $(SH)CH_2SH$	mustard cation	H_2O/EtOH (19:1)			4.6[c]	(42)
$HOOCCH$- $(NH_2)CH_2SH$ (cysteine)	mustard cation	H_2O/EtOH (19:1)			4.8[c]	(42)
C_6H_5SH	MeI	MeOH	-4.2	-8.88[d]	4.3[e]	(43,3)
Thiolates						
n-PrS$^-$	MeI	MeOH			8.6[e]	(19)
i-PrS$^-$	DMIH[f]	DMF	-5.9[g,h]	-8.54[i]	2.6	(40,65,66)
$H_3COOCCH_2S^-$	n-BuCl	DMSO	-1.7[j]	-8.4[k]	6.7	(17,65,66)
n-BuS$^-$	n-BuBr	EtOH	-1.7[m]	-8.27[n]	6.6	(36,3)
n-BuS$^-$	n-BuBr	MeOH	-2.6[h,m]	-8.27[n]	5.7	(36,3)
n-BuS$^-$	n-BuCl	DMSO	-0.69[j]	-8.4[k]	7.7	(17,65,66)
t-BuS$^-$	n-BuCl	DMSO	-0.81[j]	-8.4[k]	7.6	(17,65,66)
n-HexS$^-$	n-HexBr	H_2O	-1	-8	7	(4)
$C_6H_5S^-$	MeI	MeOH	0.0294	-8.88[d]	8.5[e]	(43,3)
$C_6H_5S^-$	EtBr	MeOH	-2.4[h,p]	-8.33	5.9	(67,3)
$C_6H_5S^-$	$BrCH_2CH_2Br$	MeOH	-3.2[h,p]	-9.2	6.0	(67,58)
$C_6H_5S^-$	n-PrBr	MeOH	-2.6[h,p]	-8.27	5.7	(67,3)
$C_6H_5S^-$	i-PrBr	MeOH	-4.0[h]	-7.17	3.2	(20,3)
$C_6H_5S^-$	n-BuBr	MeOH	-2.6[h,p]	-8.27[n]	5.7	(67,3)
$C_6H_5S^-$	n-BuBr	MeOH	-2.5[h,m]	-8.27[n]	5.8	(36,3)
$C_6H_5S^-$	n-BuBr	EtOH	-1.9[m]	-8.27[n]	6.4	(36,3)
Thioacids						
H_3CCSS^-	mustard cation	H_2O/EtOH(19:1)			5.5[c]	(42)
$BuCSS^-$	mustard cation	H_2O/EtOH(19:1)			5.4[c]	(42)
Thioethers						
$(CH_3)_2S$	MeI	MeOH	-4.3		4.1[e]	(43)
$(CH_3)_2S$	MeBr	MeOH	-4.8[h]	-8.13	3.3	(20,3)
$(CH_3CH_2)_2S$	MeI	MeOH	-4.5		3.9[e]	(43)
$(C_6H_5CH_2)_2S$	MeI	MeOH	-5		3.4[e]	(43)
c-$(CH_2)_4S$[q]	MeI	MeOH	-4.2		4.3[e]	(43)
c-$(CH_2)_5S$[r]	MeI	MeOH	-4.5		4.0[e]	(43)

Table VII footnotes—*Continued*

[a]See Table IV, note **a**.

[b]Pairs of references denote the two sources from which the $k_{S,SN}$ (or $\log[k_S/k_{H2O}]_{SN}$) and the $k_{H2O,SN}$ data were obtained, respectively. Both rate constants obtained from the same reference if only one citation given. If three references are given, the second and third citations are those from which rate constants and Arrhenius data were obtained, respectively, for the hydrolysis reaction.

[c]See Table IV, note **k**.

[d]Not used for the calculation of $\log(k_S/k_{H2O})_{SN}$. Included for purposes of comparison only.

[e]Adjusted to H_2O as solvent using the following formula ([19]):

$$\log(k_S/k_{H2O})^{H2O} \approx \log(k_S/k_{H2O})^{MeOH} - 1.4$$

[f]2,2-dimethyl-1-iodo-5-hexene

[g]Measurements made at "room temperature." $k_{S,SN}$ for the reaction in DMF estimated from the data given by Ashby et al. ([40]), using the methods described by Moore and Pearson ([68]) for the second-order reaction between equal concentrations of reagents. Data from Parker ([20]) then used to obtain a rough estimate of $\log(k_{S,SN})$ for this reaction in methanol. Parker ([20]) provides values of $\log(k^S/k^M)$, where k^i = S_N2 rate constant for a specific reaction in solvent **i** at 25°C, **S** = solvent of interest and **M** = methanol. For the present purposes, the most appropriate value provided by Parker ([20]) for $\log(k^{DMF}/k^M)$ was the one determined for the reaction between n-BuI and SCN⁻. Using this value yielded: $\log(k_{S,SN}) \approx \log(k_{S,SN}^{DMF}) - \log(k^{DMF}/k^M) = (-3.8) - (1.9) = -5.7$
The result (-5.7) was then further corrected to account for the transfer from methanol to H_2O, according to note **h**.

[h]Corrected for transfer from methanol to H_2O using $\log(k^{H2O}/k^{MeOH}) = -0.2$, derived from the reaction between MeI and SCN⁻ ([20]).

[i]Second-order rate constant for hydrolysis of ethyl iodide by H_2O ([65,66]) used as a surrogate for $k_{H2O,SN}$ for DMIH. Rationale for this approach comes from an analogy with the hydrolysis of n-alkyl bromides, since $k_{H2O,SN}$ for MeBr, EtBr, n-PrBr and n-HexBr all agree within a factor of two at 25°C (see Table V).

[j]Data from Parker ([20]) used to estimate rate constants in H_2O in a manner similar to that described in note **g** above. Datum used was the value of $\log(k^{DMSO/H2O}/k^{H2O})$, where "DMSO/$H_2O$" denotes 10% aqueous DMSO, and the reference reaction was that between MeI and SCN⁻. Thus,
$$\begin{aligned}\log(k_S^{H2O}) &\approx \log(k_S^{DMSO}) - \log(k^{DMSO/H2O}/k^{H2O})\\ &\approx (-0.600) - (1.1) = -1.7 \quad \text{(for S = } H_3COOCCH_2S^-\text{)};\\ &\approx (0.412) - (1.1) = -0.69 \quad \text{(for S = n-BuS}^-\text{); and}\\ &\approx (0.290) - (1.1) = -0.81 \quad \text{(for S = t-BuS}^-\text{).}\end{aligned}$$

[k]Second-order rate constant for hydrolysis of ethyl chloride ([65,66]) used as a surrogate for $k_{H2O,SN}$ for n-butyl chloride, for reasons analogous to those described in note **i**.

[m]Units for rate constants apparently not given by Quayle and Royals ([36]), but assumed to be $M^{-1}s^{-1}$.

[n]Second-order rate constant for hydrolysis of n-propyl bromide used as a surrogate for $k_{H2O,SN}$ for n-BuBr, for reasons described in note **i**.

[p]Interpolated to 25°C using Arrhenius data from Hine and Brader ([67]).

[q]Tetrahydrothiophene.

[r]Tetrahydrothiopyran.

employed and the temperatures at which the reactions were conducted. All rate constants are expressed in units of $M^{-1}s^{-1}$.

Although benzenethiolate has not yet been observed in natural waters, this nucleophile was included in the tables in order to extend the range of substrates examined. This inclusion was justified on the basis of the similarity in reactivity between benzenethiolate and n-butanethiolate in their nucleophilic attack on n-butyl bromide (Table VII), as well as that between benzenethiolate and ethanethiolate in their dehydrohalogenation of DDT at 25°C (Table VIII).

Corrections for Solvent Transfer

Nearly all k_{H2O} values in the tables were measured in aqueous solution in the absence of organic co-solvents. The majority of k_S data, however, were measured in alcohols or mixed-solvent systems. (Non-aqueous solvents are used in these experiments to increase the solubility of the reactants, to increase reaction rates through reduced solvation of ionic reactants, or to influence product distributions). Wherever possible, rate constants measured in organic solvents or water-solvent mixtures were adjusted to estimate their corresponding values in water. Each adjustment was made either on $\log(k_{S,SN})$ if this value was used to calculate $\log(k_S/k_{H2O})_{SN}$, or on $\log(k_S/k_{H2O})_{SN}$ itself, if the latter was taken directly from the original reference or calculated using a correlation equation (e.g., Equations 1 or 2). Because experimental data regarding the effect of solvent changes on the reactivity of sulfur nucleophiles are scarce, not all of the solvent systems encountered could be corrected in this manner.

Parker ([20]) tabulated values of $\log(k^{s1}/k^{s2})$ for a large number of different reactions, where k^{s1} and k^{s2} are the rate constants at 25°C for the reaction of interest in two different solvents (or solvent systems). Three different $\log(k^{s1}/k^{s2})$ values were used in adjusting the $\log(k_{S,SN})$ data ([20]):

$$\log(k^{MeOH}/k^{DMF}) = -1.9$$

$$\log(k^{H2O}/k^{MeOH}) = -0.2$$

$$\log(k^{DMSO/H2O}/k^{H2O}) = +1.1$$

where DMF, MeOH, and DMSO/H_2O denote dimethylformamide, methanol, and 10% aqueous dimethyl sulfoxide, respectively. All three solvent transfer values were derived from the S_N2 reaction between SCN^- and methyl iodide. This was the reaction among those examined by Parker ([20]) which was most closely related to the substitution reactions of interest here.

The following example illustrates the manner in which Parker's ([20]) data were used to correct for solvent transfer. The rate constant for the nucleophilic displacement of iodide from 2,2-dimethyl-l-iodo-5-hexene by 2-propanethiolate

BIOGENIC SULFUR IN THE ENVIRONMENT

Table VIII. Competition Between Dehydrohalogenation and Substitution Pathways for Reactions Between Halogenated Aliphatic Substrates and Sulfur Nucleophiles (Rate Constants Expressed in Units of $M^{-1}s^{-1}$.)

Nucleophile	Substrate	Solvent	$\log(k_E)$	$\log(k_{SN})$	$\dfrac{k_E}{k_E+k_{SN}}$	T(°C)	Refs.[a]
$S_2O_3^{2-}$	$BrCH_2CH_2Br$	H_2O	ND[b]	--[c]	0	21-27	[69]
$S_2O_3^{2-}$	$CH_3CHBrCH_2Br$	H_2O	ND	--	0	21-27	[69]
$S_2O_3^{2-}$	$CH_3CHBrCHBrCH_3$	H_2O	--	--	0.086	100	[69]
$S_2O_3^{2-}$	$(CH_3)_2CBrCH_2Br$	H_2O	--	--	0.192	100	[69]
$S_2O_3^{2-}$	$(CH_3)_2CBrCHBrCH_3$	H_2O	--	--	0.324	100	[69]
EtS⁻	Me_3BuI^d	MeOH	--	ND	1	69.9	[39]
EtS⁻	Me_3BuBr^d	MeOH	--	ND	1	69.9	[39]
EtS⁻	Me_3BuCl^d	MeOH	(-2.9)[e]	ND	1	69.9	[39]
EtS⁻	Me_2BuBr^f	MeOH	-4.54	≤-6.5	≥0.99	69.9	[70]
EtS⁻	Me_2BuBr	MeOH	-4.8[g]	-4.9[g]	0.54	70.0	[71]
EtS⁻	i-PrBr	MeOH	≤-3.9	-2.07	0.015	69.9	[70]
EtS⁻	$C_6H_5C(CH_3)_2Cl$	MeOH	-2.8	ND	1	75.8	[72]
EtS⁻	DDT[h]	MeOH	-4.69	negligible[i]	1	45	[73]
EtS⁻	DDT	EtOH	-4.15	negligible[i]	1	45	[73]
EtS⁻	DDT	EtOH	-5.27	negligible[i]	1	25[j]	[73]
$C_6H_5S^-$	c-$C_6H_{11}Br$	EtOH	-3.50	-3.59	0.550	55	[44]
$C_6H_5S^-$	c-$C_6H_{11}Br$	EtOH	-2.52	-2.70	0.60	75	[74]
$C_6H_5S^-$	c-$C_6H_{11}Br$	DMF (EtOH)	(-4.9)[k]	(-5.1)[m]	(0.60)	25	[74]
$C_6H_5S^-$	c-$C_6H_{11}Cl$	EtOH	-5.65	-5.52	0.427	55	[44]
$C_6H_5S^-$	1,1-c-$C_6H_{10}Br_2$	EtOH	-3.25	ND	1	55	[44]
$C_6H_5S^-$	1,1-c-$C_6H_{10}Cl_2$	EtOH	-4.96	ND	1	55	[44]
$C_6H_5S^-$	DDT	MeOH	-6.0	negligible	1	25[j]	[73]
$C_6H_5S^-$	DDT	EtOH	-5.9	negligible	1	25[j]	[73]

[a]References.

[b]Expected products (alkenes for dehydrohalogenation, thioethers for substitution) not detected.

[c]"--" indicates that a rate constant could not be obtained from the original reference.

[d]2-(I,Br,Cl)-2,3,3-trimethylbutane.

[e]k_E calculated from Bunnett and Eck's data [39], even though strict second-order kinetics not observed. k_E estimated from their data by subtracting the methanolysis rate (4.8 x 10^{-4} s⁻¹) from their observed pseudo-first-order rate constant, dividing the result by 0.368M ([NaSC$_2$H$_5$]) and multiplying by 0.818 to account for the 81.8% yield of the elimination product (2,3,3-trimethylbutene). Only other product observed was the methanolysis product (2=methoxy-2,3,3-trimethylbutane). This estimate of k_E, though not strictly valid, was included only for the purpose of comparison with other values in the table.

[f]2-bromo-3,3-dimethylbutane.

[g]Second-order rate constants estimated from data of Paradisi et al. [71] by dividing pseudo-first-order rate constant by 0.756M ([NaSC$_2$H$_5$]) and multiplying by product yields (0.46 for the substitution product, 0.54 for the sum of all elimination products). Not corrected for methanolysis rate, which appeared to be less than 10% of the overall rate in the presence of EtS⁻.

[h]1,1,1-trichloro-2,2-di-(p-chlorophenyl) ethane.

[i]Product analyses not performed but, by analogy with results from benzenethiolate (which yielded 96% elimination product, DDE [73]), no substitution products expected.

[j]Rate constant extrapolated to 25°C using Arrhenius parameters provided in original reference [73].

[k]Corrected for transfer from DMF to ethanol, using log $(k^{DMF}/k^{EtOH})_E$ = 2.7, derived from the reaction between c-$C_6H_{11}Br$ and $C_6H_5S^-$ [20].

[m]Corrected for transfer from DMF to ethanol, using log$(k^{DMF}/k^{EtOH})_{SN}$ = 2.7, derived from the reaction between c-$C_6H_{11}Br$ and $C_6H_5S^-$ [20].

in DMF (40) was used to estimate the corresponding rate constant in H_2O in the following manner:

$$\log(k_{S,SN}^{H2O}) \approx \log(k_{S,SN}^{DMF}) + \log(k^{MeOH}/k^{DMF}) + \log(k^{H2O}/k^{MeOH})$$

$$= (-3.8) + (-1.9) + (-0.2)$$

$$= -5.9$$

Other instances in which Parker's $\log(k^{s1}/k^{s2})$ data (20) were used for solvent transfer corrections are listed in Tables VII and IX. $\log(k_S/k_{H2O})_{SN}$ values obtained for methanol were corrected for transfer to H_2O using the following relationship, given by Pearson (19):

$$\log(k_S^{H2O}/k_{H2O}^{H2O}) \approx \log(k_S^{MeOH}/k_{H2O}^{MeOH}) - 1.4$$

Data to which this correction was applied have been marked by footnotes in the tables.

Discussion

The principal focus of our examination of the experimental data in Tables IV through IX will be the dependence of the quantity $\log(k_S/k_{H2O})_r$ on the structures of both the sulfur nucleophile and the haloaliphatic substrate. $\log(k_S/k_{H2O})_r$ (referred to here as the "substrate selectivity" for reaction r) is a measure of the selectivity of a particular substrate between each of the sulfur nucleophiles and H_2O with respect to halogen displacement via substitution (if r = S_N) or dehydrohalogenation (if r = E). The use of $\log(k_S/k_{H2O})_r$ for this purpose serves two functions: It "normalizes" sulfur nucleophile reactivity data to a common point of reference, and it gives an indication of the importance of each sulfur nucleophile in reactions with the substrate of interest, relative to the most abundant nucleophile in natural waters, H_2O.

Sources of Uncertainty in Reactivity Data. The experimental conditions employed to obtain the data assembled in Tables IV through IX varied considerably among, and sometimes within the individual studies. Most notable among the parameters whose values showed substantial variability were solvent composition and temperature. Additional uncertainty is introduced when correlation equations are used to estimate $\log(k_S/k_{H2O})$ values for those reactions where actual laboratory data are lacking (18,22).

Solvent Composition. The use of reactivity data obtained from systems containing non-aqueous solvents is probably the major source of uncertainty in individual rate constants, and hence in the tabulated values of $\log(k_S/k_{H2O})$.

Table IX. Second-order Rate Constants ($M^{-1}s^{-1}$) for Dehydrohalogenation and Substitution Reactions of Halogenated Aliphatic Substrates with Sulfur Nucleophiles ($k_{S,E}$ and $k_{S,SN}$, Respectively), Compared to the Corresponding Rate Constants for H_2O ($k_{H2O,E}$ and $k_{H2O,SN}$, Respectively)

Nu[-]	Subst.[a]	Solv.(k_S)[b]	$\log(k_S/k_{H2O})_E$	$\log(k_S/k_{H2O})_{SN}$	$T(^oC)$ (S/H)[d]	Refs.[c] (k_{H2O})
EtS[-]	Me_3BuCl	MeOH	3.1[e]	$(k_S{\approx}0)$[f]	69.9/70	(65,66)
EtS[-]	Me_2BuBr	MeOH	$(k_{H2O}{\approx}0)$[g]	$\leq$-0.3[h,i]	69.9/70	(65,66)[j]
EtS[-]	Me_2BuBr	MeOH	$(k_{H2O}{\approx}0)$	1.3[h,i]	70.0/70	(65,66)[k]
EtS[-]	i-PrBr	MeOH	$(k_{H2O}{\approx}0)$	2.5[i]	69.9/70	(75,66)
EtS[-]	DDT	MeOH	≥ 5[m]	$(k_S{\approx}0)$	45/?[n]	(76)
EtS[-]	DDT	EtOH	≥ 6[m]	$(k_S{\approx}0)$	45/?[n]	(76)
EtS[-]	DDT	EtOH	≥ 5[m]	$(k_S{\approx}0)$	25/?[n]	(76)
C_6H_5S[-]	$c\text{-}C_6H_{11}Br$	EtOH	$(k_{H2O}{\approx}0)$	1.96	55/55	p
C_6H_5S[-]	$c\text{-}C_6H_{11}Br$	EtOH	$(k_{H2O}{\approx}0)$	1.88	75/75	p
C_6H_5S[-]	DDT	MeOH	≥ 4[m]	$(k_S{\approx}0)$	25/?[n]	(76)
C_6H_5S[-]	DDT	EtOH	≥ 4[m]	$(k_S{\approx}0)$	25/?[n]	(76)

[a]Haloaliphatic substrate. Compound abbreviations identical to those used for Table VIII.

[b]Solvents used for determining rate constants for sulfur nucleophile ($k_{S,E}$ and $k_{S,SN}$).

[c]References for data on reactions with H_2O only; references for data on reactions with sulfur nucleophiles are the same as those given in Table VIII. References separated by commas are those from which rate constants and Arrhenius data were obtained, respectively.

[d]Temperature used to measure k_S/temperature used to measure k_{H2O}.

[e]$k_{H2O,SN}$ for hydrolysis of 2-chloropropane used as a surrogate.

[f]Rate of substitution by sulfur nucleophile deemed negligible (see Table VIII).

[g]Rate of dehydrohalogenation by H_2O deemed negligible (78).

[h]$k_{H2O,SN}$ for hydrolysis of 2-bromobutane used as a surrogate.

[i]$k_{S,SN}$ corrected for transfer from methanol to H_2O using $\log(k^{H2O}/k^{MeOH})_{SN}$ = -0.2, derived from the reaction between MeI and SCN[-] (20).

[j]Reference for k_S data: (70).

[k]Reference for k_S data: (71).

[m]Assumes $t_{1/2} \geq 3$ yr for DDT in H_2O ((76), cited in (77)), and that this occurs solely via elimination to DDE. This is probably a substantial underestimate of $t_{1/2}$ for the abiotic dehydrohalogenation of this compound in natural soils.

[n]Temperature unknown. Data obtained in natural soils.

[p]R.E. Robertson, unpublished data, cited in (78).

Due to the limitations on the available data for estimating the effects of solvent transfer (19,20), many of the data listed in Tables IV through IX are reported for reaction in pure ethanol, aqueous ethanol, or 39% aqueous dioxane, and have not been adjusted for transfer to pure water. Some indications of the differences between the rates of reaction in ethanol (or aqueous ethanol), and those in pure water are demonstrated by the following examples (all at 25°C, unless otherwise indicated).

(a) The rate of reaction between $S_2O_3^{2-}$, and methyl iodide at 15°C increases threefold (0.5 log units) when the solvent is changed from pure H_2O to a 2.5:1 (v/v) ethanol/H_2O mixture (41).

(b) When the reaction medium is changed from pure H_2O to 1:1 H_2O/ethanol, $\log(k_{S,SN})$ for the attack by $S_2O_3^{2-}$ increases by 0.5 for methyl bromide, but decreases by 0.03 and 0.36 for ethyl bromide and isopropyl bromide, respectively (Table V).

(c) Changing the reaction medium from methanol to ethanol increases $\log(k_{S,SN})$ for the attack of $C_6H_5S^-$ on n-butyl bromide by +0.6, and increases $\log(k_{S,SN})$ for the attack of n-butanethiolate on the same substrate by +0.9 (Table VII). (For the purpose of relating these data to pure water, it is noted that $\log[k^{H2O}/k^{MeOH}] = -0.2$ for the reaction between SCN⁻ and methyl iodide (20).)

It is therefore expected that the use of $k_{S,SN}$ data from experiments employing pure ethanol or aqueous ethanol as the reaction medium may introduce an error in $\log(k_{S,SN})$ values of one log unit or less, relative to the value anticipated for pure H_2O.

The kinetic data for the reactions of benzyl chloride with HS⁻ and $S_2O_3^{2-}$ were obtained in 39% aqueous dioxane (22). These data were not adjusted for transfer to pure water, due to the apparent absence of any correction parameters for solvent transfer from dioxane (or aqueous dioxane) in the published literature.

Temperature. The majority of the data in Tables IV through IX were obtained at 25°C. For several of the studies, however, the experiments were conducted at other temperatures. In those instances where Arrhenius data were available for the reactions of interest, rate constants (k[T]) obtained at other temperatures were adjusted to 25°C using the Arrhenius equation:

$$\ln(k[T]) = \ln A - (E_a/RT)$$

where A, E_a, and R denote the pre-exponential factor, activation energy, and universal gas constant, respectively, and temperature (T) is expressed in degrees Kelvin. Each of these adjustments has been noted in the table footnotes.

Use of Correlation Equations. Many of the $\log(k_S/k_{H2O})_{SN}$ data listed in the tables were obtained through the use of semi-empirical equations relating

$\log(k_S/_{H2O})_{SN}$ to various physical and empirical constants (e.g., Equations 1 and 2). Although the magnitudes of the deviations between measured and predicted values of $\log(k_S/k_{H2O})_{SN}$ depend to a large extent upon the structures of the nucleophile and substrate, Swain and Scott (22) found that their two-parameter correlation (Equation 1) predicted $\log(k_S/k_{H2O})_{SN}$ to within ±0.6 for most of the 47 substitution reactions which they examined. The four-parameter Edwards equation (Equation 2) was able to predict $\log(k_S/k_{H2O})_{SN}$ to within ±0.3 for the 13 reactions examined (18).

From the above discussion, it appears that the tabulated values of $\log(k_S/k_{H2O})_{SN}$ are reliable to within roughly one log unit, and hence that differences among $\log(k_S/k_{H2O})_{SN}$ values which exceed two log units (i.e., a difference of two orders of magnitude between the actual ratios) probably reflect genuine differences in reactivity.

Influence of Nucleophile Structure on Reactivity. The substrate selectivity of the primary n-alkyl bromides toward the various sulfur nucleophiles listed in Tables IV through VII shows little if any dependence on the length of the alkyl chain (see below). Thus, the data in Tables IV through VII may be used to construct the following approximate order of reactivity of sulfur nucleophiles with respect to the displacement of halide from a primary n-alkyl bromide in H_2O:

$$S_5^{2-} \approx S_4^{2-} \approx \text{n-HexS}^- > \text{n-BuS}^- \approx C_6H_5S^- > HS^-$$

$$\approx SO_3^{2-} \gtrsim S_2O_3^{2-} > RSR' >> H_2O$$

Similar trends have been published elsewhere for the nucleophilic attack at saturated carbon (22,23,32,42,43) and have, in fact, provided most of the data upon which the above trend was based. The main area where these trends disagree is in whether thiosulfate or sulfite is the more reactive nucleophile, probably because the reactivities of these two species are so similar. The remainder of the above trend is consistent among different studies, and thus provides some indication (in conjunction with other data from Tables IV through VII) of the structural features which influence the reactivity of these nucleophiles.

Length of Alkyl Chain. The reactivity of an aliphatic thiolate appears to increase with increasing length of the alkyl chain. In keeping with this trend, the nucleophile with no carbon atoms, HS^-, is the least reactive. At least two factors may be responsible for this trend: The increase in polarizability and the decrease in the degree of solvation with increasing alkyl chain length. Furthermore, from the three most reactive nucleophiles listed in the trend above, it seems to make little difference in reactivity whether the terminal sulfur ($-S^-$) is attached to a chain of sulfur atoms or a chain of methylene groups

($^-CH_2^-$). By contrast with the aliphatic thiolates, the length of the alkyl chain appears to exert little influence on the nucleophilicity of a thioacid. This is indicated by the fact that the substrate selectivities of mustard cation (β-chloroethylethylenesulfonium ion) are virtually identical for the attack by CH_3CSS^- and $C_4H_9CSS^-$ (Table VII).

σ- versus π-Electron Systems. 1-Hexanethiolate is a stronger nucleophile than benzenethiolate. This is what one would expect if there were a higher electron density on the terminal sulfur atom in the former species. A difference of this nature is quite possible for at least two reasons: σ systems are better electron donors than π systems, and a benzene ring is better able to delocalize the negative charge on the sulfur atom, via resonance stabilization, than an n-alkyl group.

Protonation. The protonation of a sulfur nucleophile causes a substantial reduction in its nucleophilicity. This effect is in accordance with the general theory of nucleophilic substitution reactions (28,29). Pairs of nucleophiles (and their associated substrates) which demonstrate the effect of protonation on nucleophilicity are S^{2-}/HS^- (RX = methyl bromide, 1-chloro-2,3-epoxypropane, mustard cation, and benzyl chloride), SO_3^{2-}/HSO_3^- (RX = methyl perchlorate) and $C_6H_5S^-/C_6H_5SH$ (RX = methyl iodide). Among the reactions examined here, the influence of protonation is most pronounced for the reaction between benzyl chloride and either S^{2-} or HS^-, for which the difference between $\log(k_{S(2-)}/k_{H2O})_{SN}$ and $\log(k_{HS^-}/k_{H2O})_{SN}$ is 4.6. By contrast, the influence of protonation is least pronounced for the reactions between methyl bromide and the same pair of nucleophiles, for which the corresponding difference is only 2.7.

Alkylation. The reactivity of an anionic sulfur nucleophile is also greatly diminished when a second alkyl (or arylalkyl) group is attached to the terminal sulfur atom. Although the data in Table VII do not afford comparisons among precisely equivalent nucleophiles, we see that $\log(k_S/k_{H2O})_{SN}$ diminishes by 4.7 for the attack on methyl iodide when the nucleophile is changed from n-propanethiolate to diethyl sulfide. Similarly, $\log(k_S/k_{H2O})_{SN}$ is reduced by 5.1 when the attacking nucleophile is changed from benzenethiolate to dibenzyl sulfide for the same substrate. This effect is probably a result of both the neutralization of the negative charge (thereby reducing the electron density) and an increase in crowding at the sulfur atom of the nucleophile. Some evidence of the latter effect is provided by the relative reactivities of the thioethers toward methyl iodide (43):

$$c\text{-}(CH_2)_4S > (CH_3)_2S > c\text{-}(CH_2)_5S > (CH_3CH_2)_2S > (C_6H_5CH_2)_2S$$

This series suggests that the nucleophilicity of a thioether toward a haloaliphatic substrate increases both as the size of the attached substituents decreases and as the structure becomes more rigid via cyclization. It should be noted, however, that the differences among the $\log(k_S/k_{H2O})_{SN}$ values used to deduce the above trend are quite small. Thus, these structural effects on the nucleophilicity of thioethers, if they exist at all, are quite subtle.

Branching at the α-Carbon. The values of $\log(k_S/k_{H2O})_{SN}$ for the reactions of n-butyl thiolate and t-butyl thiolate with l-chlorobutane are nearly identical (Table VII). This indicates that any reduction in nucleophilic reactivity which might arise from the increased crowding near the sulfur atom in t-butyl thiolate is more than compensated for by the increased electron density on sulfur in this species (due to the presence of three methyl groups, rather than one propyl group, on the α-carbon), relative to its straight-chain analogue. Crowding near the sulfur atom also appears to exert a relatively minor influence on the relative E2 reactivities of benzenethiolate and ethanethiolate toward DDT (Table IX). Despite the presence of the bulky phenyl ring attached to sulfur, benzenethiolate causes dehydrohalogenation of this hindered substrate at only a slightly lower rate than does the comparatively unencumbered ethanethiolate.

Influence of Substrate Structure on Reactivity. *Length of n-Alkyl Chain.* From the tabulated data, it appears that substrate selectivity shows little if any dependence upon the length of the carbon chain in primary, n-alkyl bromides. $\log(k_S/k_{H2O})_{SN}$ values for $S_2O_3^{2-}$ (the sulfur nucleophile for which these data are most extensive) in its reactions with ethyl, n-propyl, n-pentyl, and n-hexyl bromides all lie within the relatively narrow range of 4.74-5.9, with no obvious dependence on chain length. The reaction of $S_2O_3^{2-}$ with methyl bromide yields a $\log(k_S/k_{H2O})_{SN}$ value which lies slightly, but probably not significantly above this range (6.36). Similarly, in the reactions involving HS⁻, $\log(k_S/k_{H2O})_{SN}$ for the reaction with methyl bromide (5.1) is not appreciably different from the two values observed for the reactions with n-hexyl bromide (5.5 and 5.7).

Branching at the Electrophilic Center. The data indicate that branching at the α-carbon on the substrate tends to reduce differences in reactivity between sulfur nucleophiles and H_2O. This conclusion is based upon the following trends among the tabulated data.

(i) Values of $\log(k_S/k_{H2O})_{SN}$ for the reactions of HS⁻, $S_2O_3^{2-}$, and $C_6H_5S^-$ with alkyl bromides all show a marked decrease of two to three log units when the substrate is changed from a primary to a secondary bromide. By comparison with the range of 4.74-5.9 mentioned above for the primary n-alkyl bromides, the range of $\log(k_S/k_{H2O})_{SN}$ for these three nucleophiles reacting with either isopropyl bromide or bromocyclohexane is 1.96-3.4.

(ii) The substrate selectivity drops by 6.0 log units for the displacement of iodide by propanethiolate anions when the substrate is changed from methyl iodide to the secondary substrate 2,2-dimethyl-1-iodo-5-hexene. (This comparison involves two different structural isomers of the nucleophile, but n-propyl thiolate and isopropyl thiolate should exhibit similar reactivity, based on our examination of structural effects on nucleophilicity.)

(iii) The sulfur nucleophiles are essentially unreactive with respect to the displacement of halide from any of the tertiary halides examined (Tables VIII and IX), as would be generally expected for S_N2 reactions.

Leaving Group. The data summarized in Tables IV through VIII suggest that $\log(k_S/k_{H2O})_r$ decreases in the following order for most of the sets of reactions between a given sulfur nucleophile and a series of alkyl halides (RX):

$$RI > RBr > RCl$$

Examples of this behavior include the reactions of HS^- with MeX or XCH_2CH_2X (S_N2), SO_3^{2-} with MeX (S_N2), and $C_6H_5S^-$ with either $c\text{-}C_6H_{11}X$ or $1,1\text{-}c\text{-}C_6H_{10}X_2$ (E2, with EtO^-, rather than H_2O as the competing oxygen nucleophile (44)). Furthermore, in his study of the competition between E2 and S_N2 pathways in the reactions between $C_6H_5S^-$ and halocyclohexanes in ethanol, McLennan (44) found that the proportion of olefin among the reaction products was larger for the brominated substrates than for the corresponding chlorinated ones (Table VIII). The substrate selectivity trend noted above is not supported by all of the tabulated data, however. It appears that many of the difficulties in drawing conclusions regarding the influence of leaving group on $\log(k_S/k_{H2O})_{SN}$ arise from trying to compare results obtained under substantially different experimental conditions.

Selectivity versus Reactivity. The "reactivity-selectivity principle" (RSP) states that "in a set of similar reactions, the less reactive the reagent, the more selective it is in its attack" (45). This principle is widely held, but its generality has recently been questioned (28,46-48). The reactivity data assembled in Tables IV through VIII appear to provide an opportunity to examine the validity of the RSP, using $\log(k_A/k_B)_{SN}$ as a measure of the selectivity of a haloaliphatic compound between nucleophiles A and B.

In general, the various reactivity trends discussed in the previous sections are in direct opposition to the RSP: The data in Tables IV through VIII indicate an <u>increase</u> in selectivity with increasing reactivity. Two examples of these apparent violations of the RSP (restated from earlier sections) are given below.

(i) Benzyl chloride is more reactive than methyl bromide toward H_2O, HS^- or S^{2-} (Table IV), yet $\log(k_{S(2-)}/k_{HS-})_{SN}$ for the former substrate (4.6) is also larger than that for methyl bromide (2.7).

(ii) Among primary haloaliphatics in their reactions with HS^-, SO_3^{2-}, $S_2O_3^{2-}$ and $C_6H_5S^-$, both reactivity and substrate selectivity ($\log[k_S/k_{H2O}]_{SN}$) decrease in the following order:

$$RI > RBr > RCl$$

Koskikallio (49) observed this trend for values of $\log(k_S/k_{H2O})_{SN}$ in the attack of MeI, MeBr and MeCl by a variety of species, including the sulfur nucleophiles $S_2O_3^{2-}$ and SO_3^{2-}. The data assembled here extend this observation to HS^- and $C_6H_5S^-$, as well as to several additional primary haloaliphatic substrates.

IMPLICATIONS FOR NATURAL WATERS

Reaction Products

The products arising from the displacement of halogens from haloaliphatic compounds by sulfur nucleophiles may be quite reactive, giving rise to a complex assemblage of end-products. This has already been observed in the reactions between HS^- and brominated aliphatic compounds (4,7). Among the sulfur nucleophiles considered here, HS^-, S^{2-}, and $S_2O_3^{2-}$ are most likely to exhibit this behavior. Some of the products of these reactions may also be quite toxic. The most prominent examples of toxic substances among the potential products are the β-halogenated diethyl sulfides, a group of compounds which includes mustard gas.

The nucleophilic attack of a haloaliphatic substrate by HS^- or S^{2-} gives rise to a thiol or a thiolate anion, the latter of which is quite a reactive nucleophile (Table VII). The work of Schwarzenbach et al. (4) and Weintraub and Moye (7) has shown that the further reaction of the thiols and/or thiolates generated by the initial attack leads to a mixture of thioethers among the final products. The thioethers are also nucleophilic (Table VII), but they are substantially less reactive than thiols or thiolates. Thus, the thioethers are more likely to dominate the product assemblage than their thiol or thiolate precursors. Scheme I illustrates some of the pathways by which thioethers may be produced by the reaction between $H_2S(aq)$ and 1,2-dibromoethane in aqueous solution, following the initial displacement of one bromide ion by HS^- (5). The displacement of halide from a haloaliphatic compound by thiosulfate produces a Bunte salt, which can react with H_2O to form the corresponding thiol plus HSO_4^- (37). The thiol may then react with additional haloaliphatic substrate to give rise to thioethers, as described above.

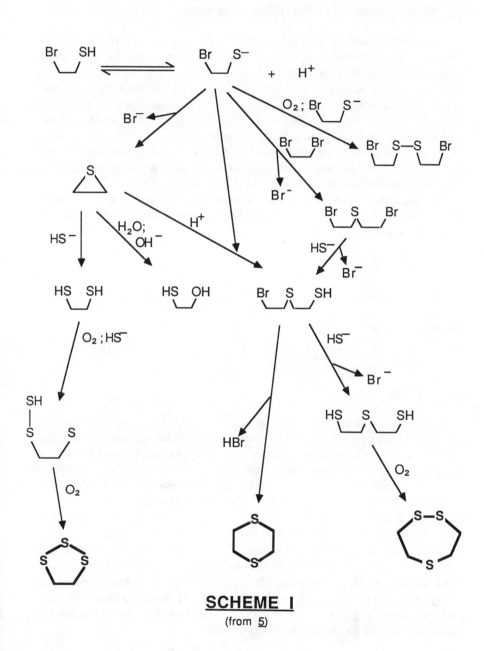

SCHEME I
(from <u>5</u>)

Reaction Rates and the Influence of Environmental Parameters

Sulfur Nucleophile and Haloaliphatic Substrate

The primary objective of this discussion has been to determine whether there are any sulfur nucleophiles in natural waters which are sufficiently reactive and present in high enough concentrations to dehalogenate haloaliphatic compounds *in situ* at rates which are environmentally significant. Data on the abundance of sulfur nucleophiles in natural waters (Tables I through III) may be combined with laboratory-derived estimates of the reactivity of these species toward haloaliphatic compounds (Tables IV through IX) in order to identify any halogenated substrates for which this might be the case. Here, we will examine a few illustrative examples, rather than attempt to provide exhaustive lists of these substrates for each nucleophile.

A haloaliphatic compound whose rate of reaction with a sulfur nucleophile (Nu) exceeds its rate of reaction with H_2O is one for which the following inequality is satisfied:

$$k_{Nu}[Nu] > k_{H2O}[H_2O] \qquad (6)$$

Rearranging Expression 6 and taking the logarithm of both sides, we obtain a form for which the substrate selectivity data (Tables IV through VII and IX) may be used directly:

$$\log \left[\frac{k_{Nu}}{k_{H2O}} \right] > \log \left[\frac{[H_2O]}{[Nu]} \right] \qquad (7)$$

Order-of-magnitude estimates of [Nu] may be obtained from Tables I through III. Substituting an estimate of [Nu] for a given sulfur nucleophile into the right-hand side of Expression 7 then yields an approximate lower bound for $\log(k_{Nu}/k_{H2O})$. The substrate selectivity data may then be scanned to determine those haloaliphatic substrates which may react faster with the sulfur nucleophile in question (at the specified concentration) than with H_2O.

Table X lists a few of the haloaliphatic substrates which were found to satisfy Expression 7 for some of the sulfur nucleophiles encountered (or closely related to those found) in natural waters. The results in the table indicate that among the substrates listed in Tables IV through IX, it is principally the brominated compounds for which the reactions with sulfur nucleophiles are likely to compete effectively with hydrolysis. However, due to the long residence times of groundwaters in hydrogeologic systems, reactions which are not as rapid as hydrolysis may also be significant.

Table X. Results from Sample Calculations used to Identify Haloaliphatic Substrates Which may React Faster with Sulfur Nucleophiles than with H_2O in Natural Waters at 25°C (see Text for Details)

Nucleophile	Anticipated Concentration[a] (μM)	Haloaliphatic Substrate(s) for which log (k_{Nu}/k_{H2O}) > log$([H_2O]/[Nu])$[b]
HS^-	1000	$BrCH_2CH_2Br$, $ClCH_2CH_2Cl$, n-HexBr
S^{2-}	10^{-13}	NF[c]
S_4^{2-},S_5^{2-}	100	n-HexBr
$S_2O_3^{2-}$	100	MeBr, $BrCH_2CH_2Br$, n-PentBr
SO_3^{2-}	100	$BrCH_2CH_2Br$
HSO_3^-	100	NF
EtS^-	10	DDT (elimination)
n-BuS$^-$	10	NF
n-PrS$^-$	0.1	MeI

[a]Order-of-magnitude estimate, obtained from data in Tables I through III (pH = 7).

[b]See Expression 7 in text. Selected examples only, obtained from Tables IV through IX. Not intended to be an exhaustive list. Refers to nucleophilic substitution reactions, unless otherwise specified.

[c]None found in Tables IV through IX.

Other Solution Properties

From the section entitled Influence of Solution Properties on Reactivity, it is apparent that the rate of bimolecular displacement of halide from a haloaliphatic substrate by an anionic sulfur nucleophile will be enhanced in natural waters exhibiting elevated pH, high concentrations of organic solvents, or low ionic strength. Hypoxic aqueous environments, where sulfur nucleophiles are most common, tend to exhibit neutral pH (Tables I and II). Organic solvents are most likely to be encountered in groundwaters contaminated by landfill leachate, leaking chemical storage facilities or chemical spills. Contaminated groundwaters also tend to exhibit high electrolyte concentrations.

Occurrence of Reaction Products in Natural Waters

Recent field evidence indicates that the occurrence of these reaction products may be relatively common in hypoxic groundwaters which have been contaminated by bromoaliphatic compounds. Schwarzenbach et al. (4) reported the presence of a complex mixture of alkyl and chloroalkyl sulfides in a hypoxic groundwater polluted by a variety of bromo- and chloroaliphatic compounds. Ethanethiol (at a concentration of approximately 2 μM) was detected by Jackson et al. (50) in groundwater contaminated by a variety of chlorinated and brominated solvents beneath a municipal/industrial landfill. In addition, Watts and Brown (51, cited in 7) have reported the presence of ethanethiol, diethyl disulfide and triethyl disulfide in Florida groundwaters contaminated by 1,2-dibromoethane.

Final Comments

The data assembled here suggest that the products of dehalogenation reactions involving sulfur nucleophiles are likely to be detected with increasing frequency in hypoxic natural waters contaminated by halogenated aliphatic compounds. Furthermore, these reactions represent abiotic degradation pathways which, if ignored, may lead to inaccurate estimates of the biotransformation of haloaliphatic compounds in laboratory studies which employ Na_2S as part of an anaerobic medium (52). In light of the potential importance of these reactions as mechanisms for the abiotic transformation of haloaliphatic compounds in hypoxic laboratory cultures and natural waters, as well as the possible toxicity of some of the reaction products, it is clear that additional laboratory data on the reactivity of sulfur nucleophiles toward halogenated organic compounds in aqueous solution are needed.

ACKNOWLEDGMENTS

The authors would like to express their appreciation to Werner Haag (SRI, Menlo Park, CA) for calling their attention to several important references used for this paper, as well as to Professor Lisa McElwee-White (Stanford University) for her helpful comments. This work was supported in part by the R.S. Kerr Environmental Research Laboratory of the U.S. Environmental Protection Agency in Ada, Oklahoma, through CR-812462 (Dermont Bouchard, Project Officer); by a grant from the Shell Companies Foundation; and by the Achievement Rewards for College Scientists Foundation, Inc., through a scholarship to J.B.

LITERATURE CITED

1. Westrick, J. J.; Mello, J. W.; Thomas, R. F. J. Am. Water Works Assoc. 1984, 76, 52-9.

2. Barbash, J. E.; Roberts, P. V. J. Water Poll. Cont. Fed. 1986, 58, 343-8.

3. Mabey, W.; Mill, T. J. Phys. Chem. Ref. Data 1978, 7, 383-415.

4. Schwarzenbach, R. P.; Giger, W.; Schaffner, C.; Wanner, O. Env. Sci. Tech. 1985, 19, 322-7.

5. Barbash, J. E.; Reinhard, M. Env. Sci. Tech., in press.

6. Haag, W. R.; Mill, T. Env. Tox. Chem., in press.

7. Weintraub, R. A.; Moye, H. A. Abstr. Div. Env. Chem. Am. Chem. Soc. 1987, 27, 236-40.

8. Streitwieser, A., Jr.; Heathcock, C. H. Introduction to Organic Chemistry, 2nd ed.; Macmillan Publishing Co., Inc.: New York, NY, 1981.

9. Stumm, W.; Morgan, J. J. Aquatic Chemistry, 2nd ed.; Wiley-Interscience: New York, 1981.

10. Pyzik, A. J.; Sommer, S. E. Geochem. Cos. Acta 1981, 45, 687-98.

11. Boulegue, J.; Lord, C. J., III; Church, T. M. Geochim. Cos. Acta 1982, 46, 453-64.

12. Luther, G. W., III; Giblin, A. E.; Varsolona, R. Limnol. Oceanogr. 1985, 30, 727-36.

13. Brock, T. D.; Smith, D. W.; Madigan, M. T. Biology of Microorganisms, 4th ed.; Prentice-Hall: Englewood Cliffs, NJ, 1984.

14. Mopper, K.; Taylor, B. F. In Organic Marine Geochemistry; Sohn, M. L., Ed.; ACS Symposium Series No. 305; American Chemical Society: Washington, DC, 1986; pp 324-39.

15. Luther, G. W., III; Church, T. M.; Howarth, R. W. In Organic Marine Geochemistry; Sohn, M. L., Ed.; ACS Symposium Series No. 305; American Chemical Society: Washington, DC, 1986; pp 340-55.

16. Boulegue, J. Geochim. Cos. Acta 1977, 41, 1751-8.

17. Bordwell, F. G.; Hughes, D. L. J. Org. Chem. 1982, 47, 3224-32.

18. Edwards, J. O. J. Am. Chem. Soc. 1954, 76, 1540-7.

19. Pearson, R. G. J. Org. Chem. 1987, 52, 2131-6.

20. Parker, A. J. Chem. Rev. 1969, 69, 1-32.

21. Duboc, C. In Correlation Analysis in Chemistry: Recent Advances; Chapman, N. B.; Shorter, J., Eds.; Plenum Press: New York, 1978; pp 313-55.

22. Swain, C. G.; Scott, C. B. J. Am. Chem. Soc. 1953, 75, 141-7.

23. Edwards, J. O.; Pearson, R. G. J. Am. Chem. Soc. 1962, 84, 16-24.

24. Klopman, G. J. Am. Chem. Soc. 1968, 90, 223-34.

25. Modena, G.; Paradisi, C.; Scorrano, G. In Organic Sulfur Chemistry: Theoretical and Experimental Advances; Bernardi, F.; Csizmadia, I. G.; Mangini, A., Eds.; Studies in Organic Chemistry, Volume 19; Elsevier: New York, 1985; pp 568-95.

26. Pearson, R. G. J. Am. Chem. Soc. 1963, 85, 3533-9.

27. Ritchie, C. D. J. Am. Chem. Soc. 1983, 105, 7313-8.

28. March, J. Advanced Organic Chemistry. Reactions, Mechanisms, and Structure, 3rd ed.; Wiley: New York, 1985.

29. Fleming, I. Frontier Orbitals and Organic Chemical Reactions; Wiley: New York, 1974.

30. Parr, R. G.; Pearson, R. G. J. Am. Chem. Soc. 1983, 105, 7512-6.

31. Walraevens, R.; Trouillet, P.; Devos, A. Int. J. Chem. Kin. 1974, VI, 777-86.

32. Koskikallio, J. Acta Chem. Scand. 1969, 23, 1477-89.

33. Burlinson, N. E.; Lee, L. A.; Rosenblatt, D. H. Env. Sci. Tech. 1982, 16, 627-32.

34. Jungclaus, G. A.; Cohen, S. Z. Abstr. Div. Env. Chem. Am. Chem. Soc. 1986, 26, 12-16.

35. Perdue, E. M.; Wolfe, N. L. Env. Sci. Tech. 1983, 17, 635-42.

36. Quayle, O. R.; Royals, E. E. J. Am. Chem. Soc. 1942, 64, 226-30.

37. Crowell, T. I.; Hammett, L. P. J. Am. Chem. Soc. 1948, 70, 3444-50.

38. Bunton, C. A.; Robinson, L. J. Am. Chem. Soc. 1968, 90, 5972-79.

39. Bunnett, J. F.; Eck, D. L. J. Org. Chem. 1971, 36, 897-901.

40. Ashby, E. C.; Park, W. S.; Goel, A. B.; Su, W. Y. J. Org. Chem. 1985, 50, 5184-93.

41. Slator, A. J. Chem. Soc. 1904, 35, 1286-1304.

42. Ogston, A. G.; Holiday, E. R.; Philpot, J. St. L.; Stocken, L. A. Trans. Farad. Soc. 1948, 44, 45-52.

43. Pearson, R. G.; Sobel, H.; Songstad, J. J. Am. Chem. Soc. 1968, 90, 319-26.

44. McLennan, D. J. J. Chem. Soc. (B), Phys. Org. 1966, 705-8.

45. Morrison, R. T.; Boyd, N. B. Organic Chemistry, 3rd ed.; Allyn and Bacon: Boston, 1973.

46. Johnson, C. D. Chem. Rev. 1975, 75, 755-64.

47. Pross, A. Adv. Phys. Org. Chem. 1977, 14, 69-132.

48. Johnson, C. D. Tetrahedron 1980, 36, 3461-80.

49. Koskikallio, J. Acta Chem. Scand. 1972, 26, 1201-8.

50. Jackson, R. E.; Patterson, R. J.; Graham, B. W.; Bahr, J.; Belanger, D.; Lockwood, J.; Priddle, M. Contaminant hydrogeology of toxic organic chemicals at a disposal site, Gloucester, Ontario; NHRI Paper No. 23; IWD Scientific Series No. 141; National Hydrology Research Institute: Ottawa, Canada, 1985.

51. Watts, G.; Brown, N. Groundwater Investigation Report No. 85-10; Groundwater section, Fla. Dept. Env. Reg., 1985.

52. Bouwer, E. J.; McCarty, P. L. App. Env. Mic. 1983, 45, 1286-94.

53. Champ, D. R.; Gulens, J.; Jackson, R. E. Can. J. Earth Sci. 1979, 16, 12-23.

54. Nicholson, R. V.; Cherry, J. A.; Reardon, E. J. J. Hyd. 1983, 63, 131-76.

55. Kunkle, G. R.; Shade, J. W. Ground Water 1976, 14, 11-20.

56. Boulegue, J.; Ciabrini, J.; Fouillac, C.; Michard, G.; Ouzounian, G. Chem. Geol. 1979, 25, 19-29

57. King, G. M.; Klug, M. J.; Wiegert, R. G.; Chalmers, A. G. Science 1982, 218, 61-3.

58. Weintraub, R. A.; Jex, G. W.; Moye, H. A. In Evaluation of Pesticides in Ground Water; Garner, W. Y.; Honeycutt, R. C.; Nigg, H. N., Eds.; ACS Symposium Series No. 315; American Chemical Society: Washington, DC, 1986; pp 294-310.

59. Haag, W. R.; Mill, T. Env. Sci. Tech. 1988, 22, 658-63.

60. Klecka, G. M.; Gonsior, S. J. Chemosphere 1984, 13, 391-402.

61. Moelwyn-Hughes, E. A. Trans. Faraday Soc., 1941, 37, 279-81.

62. Dunbar, P. M.; Hammett, L. P. J. Am. Chem. Soc. 1950, 72, 109-12.

63. Ehrenberg, L.; Osterman-Golkar, S.; Singh, D.; Lundqvist, U. Radiat. Bot. 1974, 14, 185-94.

64. The Merck Index, 9th ed.; Windholz, M.; Budavari, S.; Stroumtsos, L. Y.; Fertig, M. N., Eds.; Merck: Rahway, NJ, 1976.

65. Laughton, P. M.; Robertson, R. E. Can. J. Chem. 1959, 37, 1491-7.

66. Robertson, R. E.; Hepolette, R. L.; Scott, J. M. W. Can. J. Chem. 1959, 37, 803-24.

67. Hine, J.; Brader, W. H., Jr. J. Am. Chem. Soc. 1953, 75, 3964-6.

68. Moore, J. W.; Pearson, R. G. Kinetics and Mechanism, 3rd ed.; Wiley: New York, 1981.

69. Gavrilov, B. G.; Tishchenko, V. E. J. Gen. Chem. U.S.S.R. 1948, 18, 1687-91; Chem. Abstr. 1949, 43, 2569h.

70. Bunnett, J. F.; Eck, D. L. J. Am. Chem. Soc. 1973, 95, 1900-4.

71. Paradisi, C.; Bunnett, J. F.; Eck, D. L. J. Org. Chem. 1980, 45, 2506-7.

72. Bunnett, J. F.; Davis, G. T.; Tanida, H. J. Am. Chem. Soc. 1962, 84, 1606-14.

73. England, B. D.; McLennan, D. J. J. Chem. Soc. (B), Phys. Org. 1966, 696-705.

74. Beltrame, P.; Biale, G.; Lloyd, D. J.; Parker, A. J.; Ruane, M.; Winstein, S. J. Am. Chem. Soc. 1972, 94, 2240-55.

75. Laughton, P. M.; Robertson, R. E. Can. J. Chem. 1956, 34, 1714-8.

76. Menzie, C. M. Ann. Rev. Entomol. 1972, 17, 119-222.

77. Khan, S. U. Pesticides in the Soil Environment; Elsevier: New York, 1980.

78. Ellenrieder, W.; Reinhard, M. Chemosphere 1988, 17, 331-44.

RECEIVED June 13, 1988

THE OCEANS: DISTRIBUTIONS

Chapter 9

Distribution of Dimethyl Sulfide in the Oceans

A Review

William J. Cooper[1] and Patricia A. Matrai[2]

[1]Drinking Water Reseach Center, Florida International University,
Miami, FL 33199
[2]Division of Marine and Atmospheric Chemistry, Rosenstiel School
of Marine and Atmospheric Sciences, University of Miami,
Miami, FL 33149-1098

The distribution of dimethyl sulfide (DMS) in surface oceans and the principle source of DMS, dimethylsulfoniopropionate (DMSP) in phytoplankton, are reviewed. The distribution of DMSP in marine phytoplankton is widespread and shows considerable variation in concentration from species to species. No simple correlation between chlorophyll a and DMS concentration exists in oceanic surface waters. However, average surface water DMS concentrations show relatively small variations over large extents of the oceans. The variations that are observed may be attributted to factors such as phytoplankton variability, seasonal changes in water masses, differences in vertical mixing in the water column, and ventilation of DMS to the atmosphere.

The worlds oceans cover approximately 70% of the surface of the earth, and may play a significant role in the transport of volatile sulfur compounds from marine, coastal and estuarine environments to the atmosphere. These compounds include hydrogen sulfide, H_2S, carbonyl sulfide (COS), carbon disulfide, (CS_2), methyl mercaptan (CH_3SH), dimethyl sulfide, (CH_3SCH_3), (DMS), and dimethyl disulfide (CH_3SSCH_3) (1). Although it was initially thought that H_2S was the major compound involved in this transport (2), more recent evidence indicates that DMS accounts for most of the sulfur transferred from marine environments to the atmosphere (3).

Thus, the purpose of this chapter is to:
a. review the known sources of DMS in oceanic environments,
b. present an overview of the spatial and temporal variability of DMS distribution in the worlds oceans,
c. point out where major information gaps exist, and,
d. put into context the following chapters in this section.

0097–6156/89/0393–0140$06.00/0
© 1989 American Chemical Society

Marine Source of DMS

The first report of DMS associated with a marine alga was that of Haas (4).
Since then, it has been well established that the predominant biological source
of DMS in algae is dimethylsulfoniopropionic acid (DMSP), a compound
believed to be involved in biochemical methylation (5-9) and/or
osmoregulation (10-13). The enzymatic cleavage of DMSP results in the
formation of DMS and acrylic acid (14), as shown in equation 1.

$$(CH_3)_2\text{-S-}CH_2\text{-}CH_2\text{-COOH} \quad \text{------>} \quad (CH_3)_2S \; + \; CH_2\text{=}CH\text{-COOH} \qquad [1]$$

DMSP   DMS   Acrylic Acid

DMSP may also undergo base-promoted hydrolysis to give DMS and acrylic
acid. The base-promoted decomposition of DMSP is first order with respect to
DMSP and OH$^-$, and at seawater pH 8.1 and 10°C DMS has a half-life of 8
years (15). This suggests that this process is of minimal importance in
accounting for surface ocean DMS concentrations.
   DMSP was first isolated as the chloride salt from a red marine alga
Polysiphonia fastigiata (= *P. lanosa*) by Challenger and Simpson (5).
Subsequent studies showed that it was also present in the green marine algae
Enteromorpha intestinalis (8,16) and *Ulva lactuca* as well as numerous other
macro algae (8,17,18), but concentrations were not reported.
   The concentration of DMSP has been established in brackish water
phytoplankton (19), where samples collected during a plankton bloom,
consisting mainly of *Oscillatoria agardhii* Gom., contained 62 ug DMSP g^{-1} wet
weight (the salinity was 4-5 o/oo). A second sample consisting mainly of
Aphanizomenon flos-aquae Rolfs and *Nodularia spumigina* Merrt, upon
centrifugation yielded approximately the same concentration of DMSP as the *O.*
agardhii. Two samples obtained at a later date resulted in DMSP
concentrations of 25 and 62 ug g^{-1} wet weight (19).
   In an attempt to resolve taxonomic arguments, Craige and co-workers (20)
reported the presence of DMSP in 12 of 13 unicellular Prasinophyceae they
studied, and only 1 of the 10 unicellular Chlorophyceae examined. This limited
data set implied a relatively widespread occurrence of DMSP in the marine
members of the families. Reed (12) initiated a study to quantitatively
determine the concentration of DMSP in marine algae. The data from that
study are summarized in Table I.
   The concentration of DMS has also been studied in algal cultures.
Lovelock (21) mentioned that "seawater in equilibrium" with the alga
Polysiphonia lanosa (= *P. fastigiata*) contained 10^5 times the concentration of
DMS as found in open ocean samples. Andreae and co-workers (22,23)
determined the concentration of DMS in several axenic algal cultures. They
found that a culture of the coccolithophorid *Cricosphaera carteri* contained the
highest DMS concentration, 1.3 x 10^3 ug DMS L^{-1}. A culture of *Platyomonas* cf.
suecia was shown to have 2.8 ug DMS L^{-1} while the diatoms *Thalassiosira*
fluviatilis and *Skeletonema costatum* had 0.24 and 0.70 ug DMS L^{-1}, respectively.
The sterile medium was shown to have no DMS. The above studies were
conducted to determine the origin of DMS and it is difficult, if not impossible,
to extrapolate these data to oceanic environments. The concentration of DMS
in cultured phytoplankton is treated in detail by Keller and co-workers in this
volume (24), and the reader is referred to that paper for more details.
   An excellent study of DMSP in maintaining osmotic pressure in algal cells is
that of Reed (25). (It should be noted that the term osmoregulation is probably
not correct as used in some of the recent literature concerning DMSP

concentrations in algal cells. A hydrostatic pressure, cell turgor, above a critical value is necessary for walled algal cell growth. This pressure is in part maintained by the presence of DMSP [(25) and references therein].) In the study of Reed (25), specimens of *Polysiphonia lanosa* from both marine and estuarine sites were studied. He showed that DMSP accounted for 10.4 and 6.6% of the total cell osmotic potential in marine and estuary organisms, respectively. The other chemical species which accounted for the majority of the pressure regulation were K^+ and Cl^-, accounting for 49.4 and 52.7% of the total cell osmotic potential, in cultures of marine and estuarine organisms, respectively. The concentration of DMSP was found to be 258 and 126 umol L^{-1} in the marine and estuarine algal cells, respectively.

A more recent study (26-28) has extended the results of Reed (25) to several other algae. Dickson and Kirst (26-28) have demonstrated an increased intracellular concentration of DMSP in several algae, *Platymonas subcordiformis, Phaeodactylum tricornutum, Tetraselmis chui, Prasinocladus, Ruttnera spectabilis,* and *Prymnesium parvum,* with increased concentration of NaCl in the culture media. It appears that the increase in intracellular DMSP, over NaCl concentrations of 150 - 700 mM, is related to osmoregulation in these marine eukaryotic unicelluar algae. This corroborates and extends the results of Reed (25). It is important to point out that not all algal species examined (26-28) were able to synthesize DMSP. However, of those species which did, the intracellular concentration of DMSP varied over an order of magnitude from species to species, 3.2 to ~60 mM, at equal NaCl concentration. Whereas, the intracellular concentrations of DMSP varied nearly two orders of magnitude over the range of salt concentrations examined, 3.2 to ~275 mM, DMSP.

White (29) has shown that DMS is found in numerous marine macroalgae after alkaline treatment (Table II). He also presented data that indicated two pools of DMS precursors in marine algae. He demonstrated that a relatively stable compound released DMS only after alkaline hydrolysis at 100°C for 2 hours, and speculated that this compound may be S-methylmethionine.

In general, the mechanism(s) and reasons for the release of DMS by phytoplankton are unknown. Direct release of DMS is likely, as is release upon cell disruption. It has been demonstrated that zooplankton grazing on phytoplankton results in the release of DMS to the surrounding water (30). The release of DMS increases with increasing time in cultures of phytoplankton to which zooplankton have been added. This increase was shown to be linear over the time of the experiments (24 hours).

It is also possible that bacterial decomposition of sulfur containing organic compounds may account for some DMS in natural waters (31-36). Conversely, bacteria may also utilize DMS and therefore act as a sink (37-39). The contribution of bacterial processes to the DMS/DMSP cycle in open ocean environments has not been addressed and is as yet not understood. However, studies to better understand the biogeochemistry of DMS can not exclude bacterial processes (e.g. 40,41).

In summary it appears that both DMS and its major oceanic precursor, DMSP, are relatively widespread in the marine environment. However, considerable additional work is necessary to extend the present data base relative to the occurrence of DMSP in marine algae, and more importantly to better understand its biochemical role in cellular metabolism and the mechanisms involved in the release of DMS into seawater.

Oceanic DMS Concentrations

Surface Waters. Dimethyl sulfide was first reported in oceanic waters by Lovelock and co-workers (42). Samples were obtained off the coast of England

and on a transect from Montevideo to England. The average of 23 measurements gave a surface water concentration of 0.19 ± 0.44 nmol L^{-1}. Subsequently, numerous studies have reported DMS in surface waters and these results are briefly summarized in Table III.

Several recent comprehensive studies have appeared which discuss in detail surface water DMS concentrations. Therefore it is not necessary to review these, however, a brief overview will be presented. The reader is referred to these papers for a more detailed account of the results.

Table I. Concentration of Dimethylsulfoniopropionate (DMSP) in Marine Phytoplankton ([12])

	*umol g^{-1}
Chlorophyta (18 species)	n.d. - 39.5
87.5 % of the species above trace[1]	
11.1 % at trace levels	
Phaeophyta (26 species)	n.d. - trace
none above trace levels	
23.1 % at trace levels	
Rhodophyta (27 species)	n.d. - 97.2
14.8 % of the species above trace	
11.1 % at trace levels	

* cells washed in fresh water, DMSP determined on wet weight
[1] trace = less than 1 mmol kg^{-1}

Table II. Dimethyl Sulfide Concentration in Marine Organisms ([29])

	umol g^{-1}
Chlorophyta (5 species)	11-108
Phaeophyta (3 species)	0.06-3.0
Rhodophyta (5 species)	0.17-1.9
Blue-green bacteria (1 species)	0.36
Grass (1 species)	12

Table III. Summary of Dimethyl Sulfide Concentrations
Reported in Oceanic and Coastal Waters

Location	nM (DMS) $\overline{X}$	range	Number of Samples	Ref.
Atlantic Ocean				
Atlantic Transect	2.8	0.56 - 23	231	(43)
Atlantic	0.19		23	(42)
Atlantic/Indian	0.30	0.11 - 0.61	12	(44)
Sargasso Sea	2.2	0.94 - 4.1		(48)
U.K. Coastal and Shelf Waters				
Summer	6.9	0.44 - 34	198	(50)
Winter	0.13		185	(50)
Coastal Waters		0.44 - 16*		(51)
Pacific Ocean				
North Pacific			1090	(54)
Summer	2.2			
Winter	1.3			
South Pacific	4.4	2.2 - 6.6	12	(56)
Coastal Waters		0.03 - 3.13		(55)
Peru	6.9	0.94 - 44		(3)
Peru (Winter)	1.8	0.33 - 4.2	19	(59)
Southern Pacific				
Drake Passage	1.9 ± 0.5	0.7 - 3.2	31	(57)
Coastal Anartica	1.7	0.6 - 8.6	128	(57)
Eastern Pacific		0.41 - 12		(58)
Summer	1.9			(58)
Winter	0.6			(58)
Central Equatorial				
Pacific	3.1	1.6 - 6.3		(46)
Bering Sea	4.7	1.0 - 17		(47)
Mediterranean	1.9	0.18 - 9.3	8	(44)
Hiro Bay, Japan	5.5	1.4 - 10	10	(45)
Straits of Florida				
(Florida Keys)	1.4	0.78 - 1.9	4	(27)
Coral Reefs				
(Florida Keys)	2.1	0.88 - 3.1	6	(27)
Seagrass &				
Lagoon Beds	8.1	4.0 - 12	2	(27)
Gulf of Mexico (Summer)	4.6	2.0 - 6.9	12	(59)

One comprehensive study of surface DMS concentrations was that of Andreae and Raemdonck (49). The concentration of DMS was determined in 628 surface seawater samples, collected at 0.05 and 3 m. From this data base they estimated that a global weighted-mean DMS concentration in surface waters is 3.2 nmol L^{-1}, higher than the values obtained in earlier studies (21,44). They noted that the DMS concentrations varied significantly from one location to the next and suggested that this patchiness might be related to primary production and also species variations from locale to locale. The correlation between chlorophyll a and DMS concentration for 225 samples was r = 0.53, α = 0.001.

The most comprehensive data set reported to date is that for the North Pacific (54). A total of 1090 samples were analyzed for DMS over four years. The area-weighted mean DMS concentrations were determined to be 1.3 and 2.2 nmol L^{-1} for the winter and summer, respectively.

A seasonal trend in DMS concentration in surface waters has also been reported in an earlier study (58). The average concentration was 0.5 to 2.8 nmol L^{-1}, in winter and summer months, respectively. It should be noted that substantial scatter exists in the data. This scatter is probably related to natural variability of DMS in surface waters and as more measurements are reported, it is likely that the absolute concentration will continue to show scatter. The authors noted that the equatorial samples showed much less seasonal variability than samples from higher latitudes.

Several studies have reported detailed investigations in the variability of DMS concentration is surface waters around the UK (50-52). As in other studies they found large spatial variability in the concentration of DMS. A seasonal trend was observed, with a peak in DMS concentration coincident with the second annual plankton bloom in the summer. These studies are summarized in more detail in a subsequent chapter (53).

As one part of a recent investigation of the biogeochemistry of an estuary, DMS was included (55). They found that the concentration of DMS increased from 0.3 nM in the river (fresh water) end-member to 2-3 nM nearer the ocean. They suggested that two mechanisms, microbial consumption and volatilization, could be important removal processes.

There are very few measurements of the DMS concentration in the sea surface microlayer. The first report (44) indicated an enrichment of 5 times relative to underlying water samples. Other reports indicate that no enrichment was observed in the microlayer DMS concentration (23,47,51). These differences may be related to the sampling techniques used. It is possible that chemical and biological processes in the sea surface microlayer may affect the transfer of DMS from the bulk ocean to the atmosphere. However, at present, very little is known about the processes affecting the chemistry of DMS in the microlayer.

Studies on the biogeochemistry of DMS have been conducted in a coastal salt pond (36,60). The concentration of DMS in the surface water ranged from 5 - 10 nmol L^{-1}. These concentrations are somewhat higher than those observed in most open ocean environments, and are probably due to higher productivity. Because of the limited surface area of similar environments, these relatively high concentrations do not affect DMS flux estimates on a global scale. The subject of DMS biogeochemistry in a salt pond is treated in more detail in the following Chapter, and will not be discussed further here (40).

From the above studies it is possible to develop several generalizations regarding surface water DMS concentrations. These generalizations are that:

1. DMS is ubiquitous in the worlds oceans,
2. the spatial DMS heterogeneity is presumably due to phytoplankton species heterogeneity and vertical mixing,
3. seasonal distributions are apparent which may be related to productivity and phytoplankton blooms, with winter values generally lower than summer values (which may be in part due to increased water ventilation by higher winds during the winter),
4. DMS concentrations are highly variable in coastal and upwelling areas, when compared to open ocean environments,
5. the studies to date are mostly phenomenological, and provide little insight into the processes that lead to the formation or control of the DMS concentration in oceanic environments.

Water Column DMS Profiles. Nguyen et al. (44) was the first to report DMS water column profiles. One profile had a maximum at 10 m, while the second profile maximum was at 30-32 m. The third profile showed a continual decrease in DMS with depth (0-130 m). They concluded from the data that DMS maximum could move vertically and might account for locally high surface DMS concentrations.

Andreae et al. (23) reported two profiles from the Florida Straits. These profiles were characterized by subsurface maxima (~20-25 m) and a gradual decrease in DMS to less than 0.1 nmol L^{-1} below the euphotic zone. A profile of DMS obtained in the Atlantic (47) appeared not to have a distinct subsurface maximum (no surface value of DMS was reported). The 5 m sample contained 3.1 nmol L^{-1}. The concentration decreased through 200 m, where it remained approximately 0.16 nmol L^{-1}, to 4000 m where it was 0.03 nmol L^{-1}.

Cline and Bates (46) have recently reported DMS profiles in the equatorial Pacific. The surface concentrations varied from 1.6 - 6.5 nmol L^{-1}. The profiles had a subsurface maximum from 20 - 40 m, the predominance of the maximum decreased from east to west (150°W to 175°E) and was nearly absent at 175°E. As others have reported, the DMS concentration dropped to undetectable levels below 150 m. They noted the presence of CS_2, CH_3SH and DMDS in some samples, but individually, at no more than 10% of the DMS concentration.

The biological source of DMS was initially established based on correlations to chlorophyll a and the fact that DMS occurred primarily in the top 100 - 200 meters of the ocean. However, now that the biological source of DMS has been accepted, more detailed analysis has failed to provide any simple correlation of DMS and chlorophyll a in natural environments.

Barnard et al. (47) observed that the overall correlation of DMS and the abundance of $P.$ $poucheti$ was good (r = 0.81, n = 79), but that the vertical distributions were not well correlated in the water column.

Discussion

The major oceanic source of DMS is phytoplankton and the principle chemical source of DMS is intracellular DMSP. It is thought that one role of DMSP is the maintenance of cell turgor (25) or osmoregulation (25-28). Cell turgor is maintained as a part of normal growth of a walled cell and, release of DMS may be a consequence of normal cell growth (e.g. 61). If the release of DMS is related to salinity changes, then it is not surprising that the average concentration of DMS is relatively constant over wide areas of the worlds ocean.

In near shore and estuarine environments, changes in salinity might be a major factor in the variability observed in DMS concentrations in these areas. It is clear that, in those organisms which synthesize DMSP, changes in salinity affect intracellular DMSP concentrations (13,26-28). However, from the experiments conducted by Reed (25), it appears that additional laboratory studies are required to better establish the rate at which these changes occur. Field studies designed to help elucidate the effect of salinity changes on DMS release should focus on areas where decreases in salinity might be associated with heavy rains or in shifting currents from major fresh water sources, e.g. the area around the mouth of the Mississippi, Amazon, or other rivers.

As stated above, the major source of DMS in the oceans appears to be phytoplankton. However, no simple relationship between chlorophyll *a* and DMS concentrations in surface waters exists which applies over wide areas of the oceans. Initial interest in correlating DMS and chlorophyll *a* was to confirm the biological origin of DMS in the oceanic waters. More recently, the interest in this correlation is that surface ocean chlorophyll *a* concentrations can be estimated using satellite remote sensing. Thus, if a good correlation existed between chlorophyll *a* and DMS, then remote sensing techniques could be used to estimate world ocean DMS concentrations. Surface ocean DMS concentrations could then be used in estimating the flux of DMS to the atmosphere, given that DMS physiology is better understood.

It is well accepted that chlorophyll *a* is not an indicator of primary productivity. Therefore, an important question which remains unanswered is the relationship of DMS to primary productivity.

Several approaches may be used to develop better correlations of DMS and chlorophyll *a*. One approach is to extend surveys of laboratory cultured phytoplankton (24) to identify which organisms biosynthesize DMSP. However, until there is a better understanding of the underlying biochemistry of DMSP, these studies may be largely phenomenological. Another possibility is the empirical approach, where large extents of the surface oceans are sampled and DMS and chlorophyll *a* determined simultaneously (53).

Numerous reports have appeared which show that depth profiles of DMS concentration change on time scales of less than 24 hours. These changes are largely unexplained. One possible explanation that might account for the variations in DMS concentration observed over short time scales, i.e. on the order of 4 - 8 hours, is zooplankton grazing. Dacey and Wakeham (30) demonstrated experimentally that zooplankton grazing results in the release of DMS from algal cells containing DMSP. The relative contribution of this process to DMS in oceanic environments is not known although it is estimated to be six times direct release.

Other processes which may explain changes in DMS concentration with depth, on a short time scale, are those related to bacteria. Studies conducted in sediments (31-39,62) have shown that DMS can be consumed by microorganisms. Based on sulfur requirements of microorganisms, the availability of sulfur from sulfate in these environments far exceeds that of DMS. However, as a carbon source, the concentration of DMS is similar to other compounds of low molecular weight and may be cycled by microorganisms to serve as both a carbon and sulfur source. Although the utilization of DMS aerobically has been reported (63,64), the extent to which similar processes exist in oceanic environments is not known. For futher discussions of the microbial processes related to DMS the reader is referred to another Chapter of this book (65).

There is a paucity of data for DMS concentrations in surface waters in the winter months and at high latitudes. Although some data obtained off the coast of Peru was during the winter (59), it was also during the end of the 1982-83 El Nino Southern Oscillation, and therefore may not be representative of DMS concentrations during normal conditions where upwelling is occurring. The data of Turner and Liss (53) and Bates and co-workers (54) are the most extensive to date, and additional measurements in other areas of the worlds oceans during the winter are suggested. Except for the measurements of Berresheim et al. (57), data on DMS concentrations are lacking for surface waters in extreme latitudes. Because of their high biological productivity these regions may be rich sources of DMS, and further data are needed to refine surface water concentrations and flux estimates.

A major need exists to better understand the processes which control the formation and decomposition (or loss) of DMS from oceanic environments.

Literature Cited

1. Andreae, M. In The Role of Air-Sea Exchange in Geochemical Cycling; Baut-Menard, P., Ed.; D. Reidel Pub. Comp.: 1986; pp 331-62.

2. Eriksson, E. J. Geophys. Res. 1963, 68, 4001-8.

3. Andreae, M.O. In The Biogeochemical Cycling of Sulfur and Nitrogen in the Remote Atmosphere; Galloway, J. N.; Charlson, R.J .; Andreae, M. O.; Rodhe, H., Eds.; D. Reidel Pub., Co.: Dordrecht, 1985; pp 5-25.

4. Haas, P. Biochem. J. 1935, 29, 1297-9.

5. Challenger, F.; Simpson, M. I. J. Chem. Soc. 1948, 1591-7.

6. Maw, G. A.; du Vigneaud, V. J. Biol. Chem. 1948, 174, 381- 2.

7. Challenger, F. Adv. Enzymology. 1951, 12, 429-91.

8. Challenger, F.; Bywood, R.; Thomas, P.; Hayward, B. J. Arch. Biochem. Biophys. 1957, 69, 514-23.

9. Challenger, F. Aspects of the Organic Chemistry of Sulfur; Butterworth, London, 1959; p 253.

10. Dickson, D. M.; Wyn Jones, R. G.; Davenport, J. Planta, 1980, 150, 158-65.

11. Dickson, D. M.; Wyn Jones, R. G.; Davenport, J. Planta, 1982, 155, 409-15.

12. Reed, R. H. Mar. Biol. Lett. 1983, 4, 173-81.

13. Vairavamurthy, A.; Andreae, M. O.; Iverson, R. L. Limnol. Oceanogr. 1985, 30, 59-70.

14. Cantoni, G. L.; Anderson, D. G. J. Biol. Chem. 1956, 222, 171-7.

15. Dacey, J. W. H.; Blough, N. V. Geophys. Res. Lett. 1987, 14, 1246-9.

16. Bywood, R.; Challenger, F. Biochem. J. 1953, 53, 26.

17. Ackman, R. G., Tocher, C. S.; McLachlan, J. J. Fish. Res. Bd. Canada 1966, 23, 357-64.

18. Tocher, C. S.; Ackman, R. G.; McLachlan, J. Can J. Biochem. 1966, 44, 519-22.

19. Granroth, B.; Hattula, T. Finn. Chem. Lett. 1976, 6, 148-50.

20. Craige, F. S.; McLachlan, J.; Ackman, R. G.; Tocher, C. S. Can. J. Bot. 1967, 45, 1327-34.

21. Lovelock, J. E. Nature (London), 197⁄, 248, 625-26.

22. Andreae, M. O. In Biogeochemistry of Ancient and Modern Environments; Trudinger, P. A.; Walter, M. R.; Ralph, B. J., Eds.; Springer-Verlag. Berlin, 1980; 253-9.

23. Andreae, M. O.; Barnard, W. R.; Ammons, J. M. Environ. Biogeochem. Ecol. Bull. (Stockholm) 1983, 35, 167-77.

24. Keller, M. D.; Bellows, W. K.; Guillard R. R. L. In Biogenic Sulfur in the Environment; Saltzman, E. S.; Cooper, W. J., Eds.; ACS Symposium Series 393, Chapter 11; American Chemical Society: Washington, DC, 1988.

25. Reed, R. H. J. Exp. Mar. Biol. Ecol. 1983, 68, 169-93.

26. Dickson, D. M. J.; Kirst, G. O. Planta 1986, 167, 536-43.

27. Dickson, D. M. J.; Kirst, G. O. New Phytol. 1987, 106, 645-55.

28. Dickson, D. M. J.; Kirst, G. O. New Phytol. 1987, 106, 657-66.

29. White, R. H. J. Mar. Res 1982, 40, 529-36.

30. Dacey., J. W. H.; Wakeham, S. G. Science 1986, 233, 1314-6.

31. Wagner, C.; Stadtman, E. R. Arch. Biochem. Biophys. 1962, 98, 331-6.

32. Kadota, H.; Ishida, Y. Ann. Rev. Microbiol. 1972, 26, 127-38.

33. Zinder, S. H.; Doemel, W. N.; Brock, T. D. Appl. Environ. Microbiol. 1977, 34, 859-60.

34. Zinder, S. H.; Brock, T. D. J. Gen. Microbiol. 1978, 105, 335-42.

35. Zinder, S. H.; Brock, T. D. Arch. Microbiol. 1978, 116, 35-40.

36. Wakeham, S. G.; Howes, B. L.; Dacey, J. W. H. Nature(London) 1984, 310, 770-2.

37. Zinder, S. H.; Brock, T. D. Nature(London) 1978, 272, 226-7.

38. Kiene, R. P.; Oremland, R. S.; Catena, A.; Miller, L. G.; Capone, D. G. Appl. Environ. Microbiol. 1986, 52, 1037-45.

39. Zeyer, J.; Eicher, P.; Wakeham, S. G.; Schwarzenbach, R. P. Appl. Environ. Microbiol. 1987, 53, 2026-32.

40. Wakeham, S. G.; Dacey, J. W. H. In Biogenic Sulfur in the Environment; Saltzman, E. S.; Cooper, W. J., Eds.; ACS Symposium Series 393, Chapter 10; American Chemical Society: Washington, DC, 1988.

41. Kiene R. P.; Taylor, B. F. In Biogenic Sulfur in the Environment; Saltzman, E. S.; Cooper, W. J., Eds.; ACS Symposium Series 393, Chapter 14; American Chemical Society: Washington, DC, 1988.

42. Lovelock, J. E.; Maggs, R. J.; Rasmussen, R. A. Nature (London) 1972, 237, 452-3.

43. Barnard, W. R.; Andreae, M. O.; Watkins, W. E.; Bingemer, H.; Georgii, H. W. J. Geophys. Res. 1982, 87, 8787-93.

44. Nguyen, B. C.; Gaudry, A.; Bonsang, B.; Lambert, G. Nature (London) 1978, 275, 637-9.

45. Yamaoka, Y.; Tanimoto, T. Nippon neogeikagaku kaishi 1976, 50, 35-6.

46. Cline, J. D.; Bates, T. S. Geophys. Res. Lett. 1983, 10, 949-52.

47. Barnard, W. R.; Andreae, M. O.; Iverson, R. L. Continental Shelf Res. 1984, 3, 103-13.

48. Andreae, M. O.; Barnard, W. R. Mar. Chem. 1984, 14, 267-79.

49. Andreae, M. O.; Raemdonck, H. Science 1983, 221, 744-7.

50. Turner, S. M.; Malin, G.; Liss, P. S.; Harbour, D. S.; Holligan, P. M. Limnol. Oceanogr. 1988, 33, 364-75.

51. Turner, S. M.; Liss, P. S. J. Atmos. Chem. 1985, 2, 223-32.

52. Holligan, P. M.; Turner, S. M.; Liss, P. S. Cont. Shelf Res. 1987, 7, 213-24.

53. Turner, S. M.; Liss, P. S. In Biogenic Sulfur in the Environment; Saltzman, E. S.; Cooper, W. J., Eds.; ACS Symposium Series 393; Chapter 12; American Chemical Society: Washington, DC, 1988;

54. Bates, T. S.; Cline, J. D.; Gammon, R. H.; Kelly-Hansen, S. R. J. Geophys. Res. 1987, 92, 2930-8.

55. Froelich, P. N.; Kaul, L. W.; Byrd, J. T.; Andreae, M. O.; Roe, K. K. Estuarine Coastal Shelf Sci. 1985, 20, 239-64.

56. Nguyen, B. C.; Bergeret, C.; Lambert, G. In Gas Transfer at Water Surfaces; Brutsaert, W.; Jirka, G. H., Eds.; D. Reidel: Hingham, Mass., 1984,; pp 539-45.

57. Berresheim, H.; Andreae, M. O.; Ayers, G. P.; Gillet, R. W. In Biogenic Sulfur in the Environment, Saltzman, E. S.; Cooper, W. J., Eds.; ACS Symposium Series 393; Chapter 21; American Chemical Society: Washington, DC, 1988.

58. Bates, T. S.; Cline, J. D. J. Geophys. Res. 1985, 90, 9168-72.

59. Zika, R. G.; Saltzman, E. S.; Cooper, D. J.; Cooper, W. J., unpublished data.

60. Wakeham, S. G.; Howes, B. L.; Dacey, J. W. H.; Schwarzenbach, R. P.; Zeyer, J. Geochim. Cosmochim. Acta 1987, 51, 1675-84.

61. Bjornsen, P. K. Limnol. Oceanogr. 1988, 33, 151-4.

62. Kiene, R. P.; Visscher, P. T. Appl. Environ. Micro. 1987, 53, 2426-34.

63. Suylen, G. M. H.; Kuenen, J. G. Antonie van Leeuwenhoek J. Microbiol. Serol. 1986, 52, 281-93.

64. Suylen, G. M. H.; Stefess, G. C.; Kuenen, J. G. Arch. Microbiol. 1986, 146, 192-8.
65. Taylor, B. F.; Kiene, R. P. In Biogenic Sulfur in the Environment; Saltzman, E. S.; Cooper, W. J., Eds.; ACS Symposium Series 393, Chapter 13; American Chemical Society: Washington, DC, 1988.

RECEIVED October 5, 1988

Chapter 10

Biogeochemical Cycling of Dimethyl Sulfide in Marine Environments

Stuart G. Wakeham[1] and John W. H. Dacey[2]

[1]Skidaway Institute of Oceanography, P.O. Box 13687, Savannah, GA 31416
[2]Department of Biology, Woods Hole Oceanographic Institution, Woods Hole, MA 02543

Dimethylsulfide (DMS) plays an important role in the global atmospheric sulfur cycle. This single compound contributes a major portion of the reduced biogenic sulfur transferred from the ocean to the atmosphere. For this reason, there is considerable interest in characterizing the biogeochemical processes by which DMS is produced and consumed. Numerous research projects are currently addressing aspects of DMS cycling in oceanic, coastal, and intertidal environments. This paper provides an overview of the biogeochemistry of DMS in marine systems and synthesizes the current state of knowledge in this area of research.

The production of volatile reduced sulfur compounds in marine ecosystems and the subsequent efflux of these compounds to the marine atmospheric boundary layer is an important source of sulfur to the global atmosphere (1). Independent of its role in the atmospheric sulfur budget, Charlson et al. (2) have suggested that dimethylsulfide (DMS) also plays a major role in cloud formation over oceans. Oxidation products of DMS appear to serve as sites for cloud nucleation.

In the ocean DMS is the predominant volatile sulfur compound (3-6). The range of DMS emission rates of 0.7-13 μmol S/m^2/yr from surface seawater (4,7-9) yields an estimated global flux of 1.1 ± 0.5 Tmol S/yr (Andreae, 1985; Tmol = 32x10^{12} g). This flux is a major component of the 1.1-1.6 Tmol S/yr of biogenic sulfur transferred to the atmosphere from the world ocean and of the 1.9-3.6 Tmol S/yr total global (marine + terrestrial) biogenic sulfur flux. In comparison, coastal wetlands and marshes emit sulfur at significantly higher rates on an aerial basis than does the ocean, e.g. for DMS, 1-246 μmol/m^2/d (10-12). However wetlands occupy a relatively small area (4 x 10^{11} km^2 vs 3 x 10^{14} km^2 of ocean) and their role in the global sulfur cycle is minor, with a contribution of some 2% of the total gaseous flux. Approximately half the sulfur flux is from marshes is DMS and half may be H$_2$S (10).

Interest in biogeochemical processes controlling the emission of DMS from the ocean has led to increased efforts to determine the sources and sinks of DMS. The precursor of DMS in the marine environment is dimethylsulfonio-propionate (DMSP), also known as dimethylpropiothetin (DMPT). DMSP occurs in many species of marine phytoplankton (13-16) and higher plants

0097–6156/89/0393–0152$06.00/0

(17-19). DMSP is involved in the cycling of methionine (20). In saline environments, DMSP also appears to be important in regulating cellular osmotic pressure (21), and it has been suggested that sulfur may be stored in that form (22). Yet another role for DMSP may be as a precursor of acrylic acid, which along with DMS is formed by the enzymatic cleavage of DMSP (23). Acrylic acid is a broad-spectrum bacteriocide (24) which, when excreted by marine phytoplankton, may inhibit colonization of healthy algal cells by bacteria (25).

Our investigations of the biogeochemistry of DMS have focussed on a variety of coastal marine environments. In this paper we present an overview of our research on the processes leading to DMS production and consumption. We discuss our results in the context of obtaining a broader understanding of DMS cycling in the marine environment.

Distribution and Production of DMS in the Ocean

Interest in evaluating the sea-air flux of DMS has resulted in a very large data set (some 3000 measurements) on DMS concentrations in surface seawater in various oceanic biogeographical zones (3,4,6,7,9). Concentrations in surface waters range from about 1 nmol/L in oligotrophic areas, 1-2 nmol/L in coastal areas, to 2-5 nmol/L in upwelling regions. Surface concentrations vary both seasonally and with latitude. The depth distribution of DMS in the water column is characterized by a surface or near-surface concentration maximum of a few nmol/L, sharply decreasing at the base of the euphotic zone to low levels of less than 0.1 nmol/L in the deep ocean. The depth profile in oligotrophic waters of the Cariaco Trench off Venezuela (Figure 1) is typical of the open ocean (see Wakeham et al. (5) for analytical details). We also determined the depth profile for DMSP, the algal precursor to DMS, in the Cariaco Trench. DMSP associated with particulate matter, most likely algal cells, showed a subsurface concentration maximum, while free DMSP concentrations were highest at the sea surface. Total DMSP concentrations were about 2-4x greater than DMS.

Several mechanisms may be involved in the production of DMS in seawater. Phytoplankton may produce DMS as a normal metabolic product (26,27). Neither the biochemical function of DMS production nor the rate at which DMS is released in the ocean are understood. Furthermore there is extensive evidence that distributions of DMS in the ocean are poorly correlated with phytoplankton production or biomass (3,26). This has led to the view that DMSP synthesis and the subsequent release of DMS are highly species-specific and need not correlate with the abundance of phytoplankton. In fact, algal DMSP content and DMS production by various species of algae in culture varies over several orders of magnitude (15,26-28). Early research into DMS production by marine algae led to the discovery that *Polysiphonia fastigiata* produces DMS by enzymatic cleavage of DMSP (23). Several coccolithophorids and dinoflagellates release DMS at particularily high rates. Thus, Barnard et al. (16) suggested that high concentrations of DMS in the Bering Sea result primarily from release of DMS by *Pheaocystis pouchetti*. Similarily, Turner et al. (6) found a strong correlation between DMS in the English Channel and abundances of the dinoflagellate *Gyrodinium aureolum*. Actual release rates in the ocean are unknown, but could be expected to be of the order of 10^{-15} mol/cell/day based on laboratory data (27,28). For the two algal species for which there are published data (*Hymenomonas carterae* (27) and *Gymnodinium nelsonii* (28)), turnover of intercellular DMSP (releasing DMS into seawater) is 1.4% and 0.3% per day respectively. Subsequent work (Figure 2) suggests that

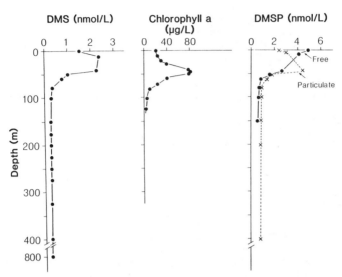

Figure 1. Vertical concentration profiles for DMS, chlorophyll a and DMSP in the western basin of the Cariaco Trench off Venezuela (10° 39'N, 65°30'W). DMS was determined by sparging and gas chromatography with a flame photometric detector. Particulate DMSP was determined by base treatment of material collected on 0.22 μm filters and analysis of the DMS released; free DMSP was determined as DMS released upon base treatment of sparged water samples obtained after initial DMS analysis. Chlorophyll *a* data from W. Cooper and R. Zika (personal communication).

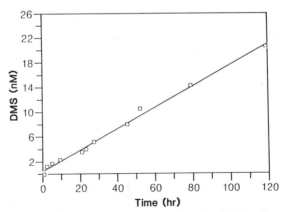

Figure 2. Release of DMS by *Prorocentrum micans* in culture. Experiments were conducted in 1-liter glass bottles with silicone rubber stoppers and with a phytoplankton cell density of 500 cells/mL. The bottles were placed on a rotator (2 rev/min) in low light (2-30 μeinstein/m² sec). DMS increase in the headspace was measured by gas chromatography/flame photometric detector. A linear regression of the data yields a DMS production rate of 2.1 x 10^-10 μmol/cell day, corresponding to a DMSP turnover of 0.26%/day.

DMS production by another dinoflagellate, *Prorocentrum micans*, also occurs at a rate of 0.3% of DMSP per day.

In recent surveys of DMS distributions in seawater, there has been concern artifacts might result from the disruption of algal cells during analysis of DMS (i.e. during filtration or sparging of unfiltered samples), presumably as a consequence of cell lysis and subsequent enzymatic decomposition of DMSP to DMS. These concerns led us to ask the question if natural processes in the ocean lead to DMS release into seawater as a result of loss of integrity of algal cells? We investigated one such process, grazing of marine phytoplankton by zooplankton (28). In a series of experiments (Figure 3), there was a marked enhancement of DMS release when phytoplankton were ingested by zooplankton. In fact, DMS production rates in treatments containing zooplankton grazing on phytoplankton were an order of magnitude higher than rates observed in treatments containing phytoplankton alone.

We have developed a simple model to evaluate the importance of zooplankton grazing in the marine DMS cycle (Figure 4). In our grazing experiments, approximately one-third of the algal DMSP ingested by zooplankton was recovered as free DMS in seawater. Thus, DMS produced by zooplankton grazing on DMSP-containing phytoplankton exceeds the direct release of DMS by phytoplankton if more than 3% of the DMSP-bearing phytoplankton in a parcel of water are ingested per day. In the ocean, however, zooplankton almost surely graze more than 3% of the standing stock of phytoplankton per day. Assuming that phytoplankton biomass is approximately balanced (where zooplankton grazing is equivalent to phytoplankton growth), a phytoplankton production rate of 0.2 per day (e.g., North Pacific Gyre, (29)) would require that about 20% of phytoplankton cells be ingested each day. Under such conditions, the rate of release of DMS during grazing could be six times greater than that released by the phytoplankton alone.

Although our experiments demonstrated that grazing increases DMS production, the mechanism by which DMS is released is unclear. It may be that capture and handling of algal cells leads to cell disruption and to enzymatic decomposition of cellular DMSP. Or DMSP decomposition may occur during digestion in the intestinal tract of the zooplankton or by microbial activity in fecal material. As noted above, about one-third of the DMSP in ingested algal cells is converted to DMS. At present, we are unable to account for the remaining two-thirds. Some may have been oxidixed to dimethylsulfoxide (DMSO) or degraded further. The disappearance of DMS during the zooplankton grazing experiment in Figure 3 appears to have been due to oxidation of DMS by bacteria. On the other hand, some cellular DMSP may have been released unaltered into seawater. Free (dissolved) DMSP has been measured in several studies (5,6; Cariaco Trench reported here, Figure 1). Subsequent decomposition of free DMSP could lead to further production of DMS. The reaction of DMSP with OH⁻ yields DMS and acrylic acid. Abiotic decomposition of DMSP with OH⁻ at the pH of seawater is unlikely to be important; kinetic experiments by Dacey and Blough (30) indicate that DMSP has a 8-year half-life when reacting with OH⁻ in seawater at 10°C. Biotic decomposition appears more likely. For example, Wakeham et al. (5) supplemented unfiltered seawater from a coastal marine pond with DMSP and found complete conversion to DMS in several days. Turner et al. (6) reported similar biotic decomposition of DMSP to DMS in 0.2um-filtered seawater. While both experiments indicate that a biological mechanism is required to decompose DMSP to DMS, with a biological half-life of several days for DMSP in seawater, it is unknown whether the reaction is catalysed by enzymes also released from disrupted algal cells or whether bacteria are responsible. Bacterial fermentation of DMSP has been reported by Wagner and Stadtman

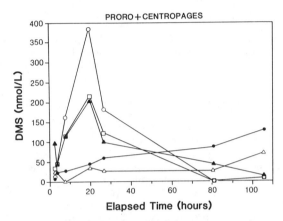

Figure 3. DMS production during zooplankton grazing on phytoplankton: (○ , ■ , ▲ ,) *Centrophages hamatus* grazing on *Prorocentrum micans*; (● and △) *P. micans* alone. Experimental conditions as for Figure 2 (phytoplankon 500 cells/mL) except zooplankton (40/L) added. DMS production is accelerated when the phytoplankton are ingested by zooplankton. However, when experiments are run for extended time intervals, DMS disappears from solution, presumably by microbial oxidation. At the end of this experiment, zooplankton had grazed the phytoplankton to very low levels, and oxidation of DMS that had been created in the process has resulted in a net accumulation of less DMS than in the phytoplankton-alone treatments.

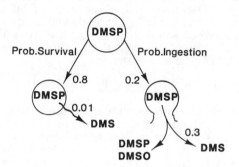

intact,undisturbed phytoplankton:
Prob.Survival x DMSP Turnover = 0.8 x 0.01 = 0.008

ingested phytoplankton:
Prob.Ingestion x DMSP Conversion = 0.2 x 0.3 = 0.06

Figure 4. Conceptual model illustrating the importance of zooplankton grazing on phytoplankton to DMS production in the ocean. For undisturbed phytoplankton, the probability of releasing DMS from intercellular DMSP is estimated as the product of survival against herbivory (0.8 assuming a algal turnover of 0.2/day) times the turnover of intercellular DMSP (<1%/day). For zooplankton grazing, the probability of DMS production is grazing rate (0.2 per day) times the DMSP-to-DMS conversion efficiency (0.3). Thus the production of DMS via herbivory and ingestion of DMSP containing algal cells is approximately 6x the release rate by undisturbed algal cells.

(31). Dacey and Blough (unpub. results) have isolated a bacterium which grows on DMSP as its sole source of carbon, growing on acrylate while quantitatively liberating DMS.
If DMS concentrations at the surface of the ocean are presumed to be at steady state, production must balance loss. The fate of DMS is thought to be evasion across the sea surface into the marine atmospheric boundary layer. However, since rates of DMS production are unknown, it is impossible to compare production with flux to the atmosphere, which is relatively well constrained. An alternative sink for DMS in seawater is microbial consumption. The ability of bacteria to metabolize DMS in anaerobic environments is well documented (32-34). Data for aerobic metabolism of DMS are fewer (there are at present none for marine bacteria), but Sivela and Sundman (35) and de Bont et al. (36) have described non-marine aerobic bacteria which utilize DMS as their sole source of carbon. It is likely that bacterial turnover of DMS plays a major role in the DMS cycle in seawater.

DMS Cycling in a Stratified Coastal Salt Pond

To better characterize processes involved in the production and consumption of DMS in marine environments, we are conducting a study of DMS biogeochemistry in a seasonally stratified coastal marine basin, Salt Pond, Cape Cod, MA (Wakeham et al., 5,37). The rationale for studying a coastal pond is that as it becomes thermally stratified in summer, the water column is effectively partitioned into three biogeochemical zones (Figure 5): an oxygenated epilimnion (0-2 m water depth), an oxygen-deficient metalimnion (2-3.5 m) and a strongly anoxic hypolimnion (3.5-5 m). Thus a variety of both aerobic and anaerobic processes occur simultaneously, but at different depths in the same water column.
The epilimnion of Salt Pond is probably typical of moderately productive and oxic coastal seawater in general. Summer DMS concentrations in the epilimnion are in the range of 5-10 nmol/L. DMSP concentrations were higher than DMS concentrations, and free DMSP was more abundant than particulate DMSP. The mechanisms of DMS production in the epilimnion are also probably the same as those occurring in most oceanic regimes, that is some combination of direct release by phytoplankton and formation of DMS as algal cells are ingested by herbivores.
It is in the oxygen-deficient metalimnion, however, that the highest DMS concentrations were found (Figure 5). Particulate DMSP also showed a strong concentration peak in the metalimnion, but in contrast to the overlying epilimnion, free DMSP was significantly less abundant. It is possible that the particulate DMSP peak in the metalimnion arises from algal cells which settle out of the epilimnion and are concentrated near the metalimnion-hypolimnion boundary, as indicated by the strong peak in chlorophyll a at this depth (Figure 5). Production of DMS in this zone could then result from physiological stress on the algal cells under conditions of reduced oxygen tensions or to bacterial decomposition of algal material. In either case the peak in DMS concentration tracks the seasonal vertical migrations of the oxic-anoxic boundary as the pond mixes down in winter and then restratifies in summer (37).
A marked decrease in the concentration of DMS was always observed at the top of the hypolimnion (Figure 5) where H_2S concentrations begin to increase as a result of sulfate reduction in sediments and bottom waters of the pond. A concentration profile of this type strongly suggests a significant removal process at the top of the hypolimnion. Given the H_2S and DMS profiles, it was suspected that the anaerobic phototrophic bacteria, which are abundant in the hypolimnion (e.g. bacteriochlorophyll d in Figure 5), might

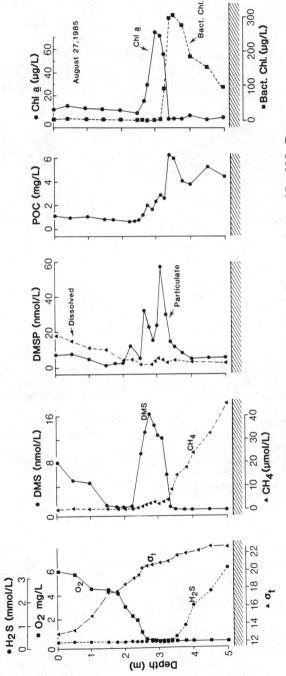

Figure 5. Water column profiles in coastal Salt Pond on August 27, 1985. Bact. chl is primarily bacteriochlorophyll *d*.

consume DMS as well as H_2S. Initial experiments confirmed that DMS is degraded by inocula from the hypolimnion of the pond (5).

In a subsequent study (34), enrichment cultures of phototrophic purple bacteria from Salt Pond oxidized DMS to DMSO, with DMS serving as an electron acceptor for photosynthesis and the oxidation of DMS to DMSO being used to reduce CO_2 to biomass. Furthermore, a pure strain of bacteria has been isolated which is capable of both phototrophic oxidation of DMS to DMSO and chemotrophic reduction of DMSO back to DMS in the dark (J. Zeyer, pers. commun.).

A comparison of estimated rates of DMS production and consumption in Salt Pond allows us to calculate a first-order budget for DMS (Figure 6). We assume that production of DMS is restricted to the epilimnion and metalimnion, and that the major removal processes are tidal exchange with adjacent Vineyard Sound, gas exchange to the atmosphere, and microbial consumption in the hypolimnion. In the case of DMS production, it is impossible to derive realistic estimates of release by phytoplankton because the species composition of the pond is virtually uncharacterized and there are too few data on DMS release rates by algae, certainly none in nature. On the other hand, we do know the inventory of DMSP and the turnover rate of particulate organic carbon (POC) in the pond. If particulate DMSP in the epilimnion and metalimnion (100 $\mu mol/m^2$ in August, 1985) decomposes to produce DMS at a rate similar to that of POC turnover (0.2/day; 38), then a DMS production of 20 $\mu mol/m^2/day$ would result. This production rate includes contributions from both microbial decomposition of algal cells and grazing; direct exudation by phytoplankton would produce some unknown amount of additional DMS.

If the DMS inventory in Salt Pond is at steady state in summer (5), production should approximately balance removal. Tidal removal of DMS to Vineyard Sound is minimal. Outflow from Salt Pond is thought to be primarily surface water, and using a maximum tidal range of 0-0.2 m/d and a mean surface water concentration of 10 nmol/L, we calculate an export rate of less than 2 $\mu mol/m^2/d$. The water-air flux of DMS may be calculated using the two film model of Liss and Slater (39; flux = $-k_1 C_w$). With the same surface water DMS concentration (C_w) and an estimated mass transfer coefficient (k_1) for DMS of 1.5 cm/h, the projected flux of DMS from the pond into the atmosphere would be 4 $\mu mol/m^2/d$. This compares with the range of estimated emissions from the ocean of 5-12 $\mu mol/m^2/d$ (1).

Microbial consumption of DMS in the hypolimnion is more difficult to estimate since our laboratory experiments showing degradation have not been rigorous enough to yield degradation rates applicable to natural conditions. However, if we assume that DMS is transported into the hypolimnion by eddy diffusion, with an arbitrary eddy diffusion coefficient of 1 m^2/d, the observed concentration gradient across the oxic/anoxic boundary would support a sink for 30 $\mu mol/m^2/d$ DMS produced in the hypolimnion. We therefore hypothesize that anaerobic microbial consumption in the hypolimnion of Salt Pond may be the major sink for DMS produced in the metalimnion. At this point we cannot estimate the potential for DMS oxidation in the epilimnion.

DMS Emission from Coastal Salt Marshes

Coastal wetlands have long been noted for their relatively high emission of volatile sulfur gases to the atmosphere; indeed the typical odor of marshes often is due largely to DMS. Several studies have reported emissions of DMS, H_2S, and other sulfur compounds, dimethyldisulfide, carbonyl sulfide, and carbon disulfide (10-12,40-42). DMS and H_2S constitute the bulk of the flux, with DMS predominating in vegetated areas and H_2S in mud flats. Fluxes of DMS and H_2S from coastal wetlands are typically 1-2 orders of magnitude higher than from oceanic regions (e.g. 2-10 μmol/m^2/d for the ocean vs 100-200 μmol/m^2/d for marshes), but on a global scale wetlands contribute only a small fraction (2%) of the total biogenic sulfur emission. Diurnal, tidal, and seasonal variations in DMS flux have been reported (12,40-42).

Since some DMS is produced during the decomposition of sulfur-containing organic matter, primarily amino acids, in waterlogged soils (43), it had been generally assumed that, like H_2S, DMS in marshes was derived from anaerobic decomposition. However, whereas a turnover of approximately 0.1% per day of the H_2S typically present in marsh pore waters (500 μmol/L at 0-2 cm) can support the observed flux (150 μmol/m^2/d), a turnover of 100-30,000% per day of pore water DMS (50 nmol/L) would be needed to give the observed DMS flux (1-246 μmol/m^2/d (19)). This discrepency suggests that the factors regulating emission of H_2S and DMS from saltmarshes must be very different. This observation and results from other investigators (42) suggest that emission of DMS may be controlled by the physiology of marsh grasses.

DMSP is present in a number of marsh plants, but only in *Spartina alterniflora* (19) and *S. anglica* (22) is it particularly abundant. *Spartina alterniflora* is the dominant species in temperate marshes of North America but not in marshes or wetlands at lower latitudes. To date, most DMS emission measurements have been made in marshes dominated by *S. alterniflora*. Emission rates in areas having other marsh grasses, with lower DMSP content, are likely to be considerably lower. Estimates of DMS emission from saltmarshes in general which are based on fluxes from *S. alterniflora* without considering the species of grass are likely to be considerably overestimated.

Salt Pond DMS "Budget"

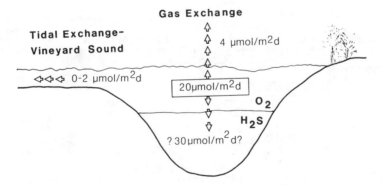

Figure 6. Schematic of the cycle of DMS in Salt Pond. See text for discussion.

Concentrations of DMSP in *S. alterniflora* are highest in leaves and lowest in roots and rhizomes (19, Figure 7). DMSP content of leaves also generally increases with increasing salinity (19,44) and also appears to depend on the nutritional status of the grass (19). These several factors are consistent with a presumed osmoregulatory function for DMSP. An osmoregulatory role for DMSP has been postulated by analogy with the quaternary ammonium analogs of DMSP, most notably glycine betaine, a well-documented osmoticum (21). Dickson et al. (45) reported shifts in the DMSP content of the marine macroalga *Ulva lactuca* in response to changes in salinity. Certainly the concentrations of several 10's of mmol/L in water contained in plant tissue mean that DMSP contributes to the osmotic pressure of the plant. It is yet unknown how dynamic this solute pool is, and recent evidence with a green macroalga *Enteromorpha intestinalis* (46) and the European marsh grass *Spartina anglica* (22) have cast doubt on the capability of plants to adjust their DMSP content in response to shifts in salinity.

Regardless of the physiological role of DMSP in these plants, virtually nothing is known about the factors controlling the *in vivo* decomposition of DMSP which results in the liberation of DMS. Presumably enzymatic decomposition in leaves is the major mechanism resulting in DMS emission to the atmosphere from areas of *S. alterniflora* (Figure 8). If DMSP decomposition and subsequent release of DMS by leaves of marsh grasses are influenced by plant physiology, then understanding the mechanism of this conversion has important implications with respect to design of experiments for measuring DMS emission in the field. It is likely that the production and release of leaf DMSP are influenced by short-term changes in heat, incident solar radiation, and water balance. Clearly, chamber experiments must be designed to minimize unnatural perturbations of the environment around the plant. There is also evidence that soils are a sink for DMS (19, Figure 8). Kiene and Visscher (47) found consumption of DMS in anaerobic saltmarsh sediments. Uptake and consumption of DMS by soil microorganisms will affect the net emission rates and must be considered when extrapolating DMS flux rates to different geographical regions.

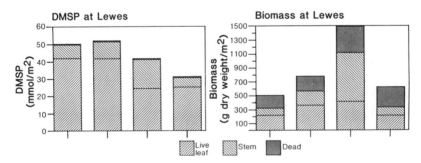

Figure 7. DMSP inventory in *S. alterniflora* and above-ground biomass at four *S. alterniflora* sites in Great Marsh, Lewes, Delaware, in June, 1986. Most above ground DMSP is concentrated in live leaves although stems and dead leaves represent the majority of above-ground biomass at this time. Roots have less DMSP per unit biomass but may represent a larger pool on an aerial basis. Below-ground DMSP is less likely, however, to play a role in DMS emission as discussed by Dacey et al. (19).

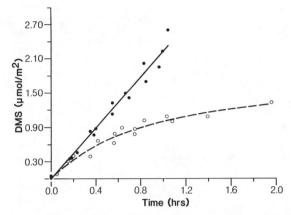

Figure 8. DMS accumulation in headspace above enclosed *S. alterniflora* cores. The straight line (●) is a linear fit of data from three runs with sediment flooded; the curved line (○) is a hyperbolic fit with sediment exposed. These data suggest that once ambient concentrations of DMS in air reach about 4 pmol/mL, the sediment surface has the capacity to consume 1.7 μmol DMS/m^2/h.

By analogy to zooplankton grazing in the ocean, herbivory may also be involved in DMS release in salt marshes. Freshly-deposited Canada goose feces yielded DMS within 48 hours in amounts equivalent to 80-190 μmol/g dry weight. Leaves of *S. alterniflora* were a principal component of material in the feces, and the quantity of DMS released from feces suggests that a major portion of the DMSP contained in the leaves (80-300 μmol/g dry weight; [19]) is decomposed to DMS. However, since the role of herbivory in the turnover of organic matter is less in marshes than in the ocean, it is unlikely that herbivory plays a significant role in DMS flux in these sites. In salt marsh grass, the liberation of DMS is probably controlled primarily by physiological and decompositional processes.

Summary

We propose a model (Figure 9) to illustrate key factors controlling the formation and removal of DMS in marine systems. DMS in the marine environment appears to arise primarily from DMSP, although other sulfonium compounds have been identified. The distribution of DMSP among phytoplankton species and its physiological role need further investigation. We propose that under most circumstances DMS released by phytoplankton represents a small turnover of intercellular DMSP. In situations where cells are stressed, as in the hypoxic zone of Salt Pond, DMS release may be accelerated. Our work with zooplankton grazing and the recent observations of significant soluble DMSP pools suggests that factors other than algal physiology are important in the formation of DMS. The high concentrations of free DMSP are almost certainly the consequence of the physical disruption of cells, most likely during grazing. The apparent chemical stability of DMSP in seawater suggests that the removal of DMSP is controlled by bacteria.

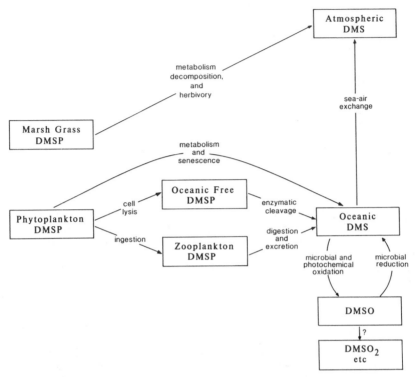

Figure 9. Schematic of biogeochemical processes producing and consuming DMS in marine environments. DMSP in certain phytoplankton and marsh grasses may be slowly metabolized and released directly. Disruption of DMSP-containing cells, either by herbivory or by microbial decomposition, results in DMSP release into solution and enhanced production of DMS. DMS may be oxidized microbially to DMSO, and perhaps DMSO$_2$; photochemical decomposition of DMS in surface waters may also occur. Residual DMS may escape into the atmosphere, where it undergoes further photochemical degradation.

Very little is known about the removal of DMS from seawater. Bacterial and photochemical oxidation undoubtedly result in DMSO and further-oxidized products. It is not clear what percentage of DMS formed in seawater actually reaches the atmosphere. Its flux to the atmosphere certainly plays a major role in the chemistry of the atmosphere. A thorough understanding of the global distribution and dynamics of that flux will depend on obtaining an increased understanding of the processes that control the dynamics of DMS in marine ecosystems.

Acknowledgments

We thank W. Cooper and R. Zika for chlorophyll *a* data from the Cariaco Trench, and M. Hardisky for biomass data from Great Marsh, Lewes, DE. E. Canuel and L. Hare provided laboratory assistance. This research was supported by National Science Foundation grants OCE-84-16203, OCE-87-14170 and OCE-87-19000, and National Aeronautics and Space Administration grant NAGW-606.

Literature Cited

1. Andreae, M. O. In The Biogeochemical Cycling of Sulfur and Nitrogen in the Remote Atmosphere; Galloway, J. N.; Charlson, R. J.; Andreae, M. O.; Rodhe, H., Eds.; Reidel: Dordrecht, 1985; pp 5-25.
2. Charlson, R. J.; Lovelock, J. E.; Andreae, M. O.; Warren, S. G. Nature 1987, 326, 655-61.
3. Andreae, M. O.; Barnard, W. R. Mar. Chem. 1984, 14, 267-79.
4. Cline, J. D.; Bates, T. S. Geophys. Res. Lett. 1983, 10, 949-52.
5. Wakeham, S. G.; Howes, B. L.; Dacey, J. W. H.; Schwarzenbach, R. P.; Zeyer. J. Geochim. Cosmochim. Acta 1987, 51, 1675-84.
6. Turner, S. M.; Malin, G.; Liss, P. S.; Harbour, D. S.; Holligan, P. M. Limnol. Oceanogr. 1988, 33, 364-75.
7. Andreae, M. O.; Raemdonck, H. J. Geophys. Res. 1983, 90, 12,891-900.
8. Nguyen, B. C.; Bonsang, B.; Gaudry, A. J. Geophys. Res. 1983, 88, 10,903-14.
9. Bates, T. S.; Cline, J. D.; Gammon, R. H.; Kelly-Hansen, S. R. J. Geophys. Res. 1987, 92, 2930-8.
10. Steudler, P. A.; Peterson, B. J. Nature 1984, 311, 45-57.
11. Jorgensen, B. B.; Okholm-Hansen, B. Atmos. Environ. 1985, 19, 1737-49.
12. Cooper, W. J.; Cooper, D. J.; Saltzman, E. S.; De Mello, W. Z.; Savoie, D. L.; Zika, R. G.; Prospero, J. M. Atmos. Environ. 1987, 21, 1491-5.
13. Ackman, R. G.; Tocher, C. S.; McLachlan, J. J. Fish. Res. Bd. Canada 1966, 23, 357-64.
14. Andreae, M. O. In Biogeochemistry of Ancient and Modern Environments; Trudinger, P. A.; Walter, M. R., Eds.; Springer Verlag: New York, 1980; pp 235-59.
15. White, R. H. J. Mar. Res. 1982, 40, 529-36.
16. Barnard, W. R.; Andreae, M. O.; Iverson, R. L. Cont. Shelf Res. 1984, 3, 103-13.
17. Reed, R. H. Mar. Biol. Lett. 1973, 4, 173-81.
18. Larher, F.; Hamlin, J.; Stewart, G. R. Phytochem. 1977, 16, 2019-20.
19. Dacey, J. W. H.; King, G. M.; Wakeham, S. G. Nature 1987, 330, 643-5.
20. Maw, G. A. In The Chemistry of the Sulfonium Group; Sterling, C. J. M.; Patai, S., Eds.; J. Wiley: New York, 1981; pp 703-70.
21. Wyn Jones, R. G.; Storey, R. In The Physiology and Biochemistry of Drought Resistant Plants; Paleg, L. G.; Aspinall, D., Eds.; Academic Press: Sydney, 1982; pp 171-204.
22. van Diggelen, J.; Rozema, J.; Dickson, D. M. J.; Broekman, R. New Phytol. 1986, 103, 573-86.
23. Cantoni, G. L.; Anderson, D. G. J. Biol. Chem. 1956, 222, 171-7.
24. Sieburth, J. McN. J. Bacteriol. 1961, 82, 72-9.
25. Davidson, A. T.; Marchant, H. J. Mar. Biol. 1987, 95, 481-7.
26. Andreae, M. O. In The Role of Air-Sea Exchange in Geochemical Cycling; Baut-Menard, P.; Liss, P. S.; Merlivat, L., Eds.; Reidel: Dordrecht, 1986; pp 331-62.
27. Vairavamurthy, A.; Andreae, M. O.; Iverson, R. L. Limnol. Oceanogr. 1985, 30, 59-70.
28. Dacey, J. W. H.; Wakeham, S. G. Science, 1986, 233, 1314-6.
29. Welschmeyer, N. A.; Lorenzen, C. J. Limnol. Oceanogr. 1985, 30, 1-21.

30. Dacey J. W. H.; Blough, N. V. Geophys. Res. Lett. 1987, 14, 1246-9.
31. Wagner, C; Stadtman, E. R. Arch. Biochem. Biophys. 1962, 98, 331-6.
32. Zinder, S. H.; Brock, T. D. Appl. Environ. Microbiol. 1978, 35, 344-52.
33. Kiene, R. P.; Oremland, R. S.; Catena, A.; Miller, L. G.; Capone, D. G. Appl. Environ. Microbiol. 1986, 52, 1037-45.
34. Zeyer, J.; Eicher, P.; Wakeham, S. G.; Schwarzenbach, R. P. Appl. Environ. Microbiol. 1987, 53, 2026-32.
35. Sivela, S.; Sundman, V. Arch. Microbiol. 1975, 103, 303-4.
36. de Bont, J. A. M.; van Dijken, J. P.; Harder, W. J. Gen. Microbiol. 1981, 127, 315-23.
37. Wakeham, S. G.; Howes, B. L.; Dacey, J. W. Nature 1984, 310, 770-2.
38. Lohrenz, S. E.; Taylor, C. D.; Howes, B. L. Mar. Ecol. Prog. Ser., in press.
39. Liss P. S.; Slater, P. G. Nature 1974, 247, 181-4.
40. Adams, D. F.; Farwell, S. O.; Robinson, E.; Pack, M. R.; Bamesberger, W. L. Environ. Sci. Technol. 1981, 15, 1493-8.
41. Aneja, V. P.; Aneja, A. P.; Adams, D. F. J. Air Pollut. Control Assoc. 1982, 32, 803-7.
42. De Mello, W. Z.; Cooper, D. J.; Cooper, W. J.; Saltzman, E. S.; Zika, R. G.; Savoie, D. L.; Prospero, J. M. Atmos. Environ. 1987, 21, 987-90.
43. Bremner, J. M.; Steele, C. G. Adv. Microb. Ecol. 1978, 2, 155-201.
44. Reed, R. H. Mar. Biol. Lett. 1983, 4, 173-81.
45. Dickson, D. M. J.; Wyn Jones, R. G.; Davenport, J. Planta 1980, 150, 158-65.
46. Edwards, D. M.; Reed, R. H.; Chudek, J. A.; Foster, R.; Stewart, D. P. Mar. Biol. 1987, 95, 583-92.
47. Kiene, R. P.; Visscher, P. T. Appl. Environ. Microbiol. 1987, 53, 2426-34.

RECEIVED September 15, 1988

Chapter 11

Dimethyl Sulfide Production in Marine Phytoplankton

Maureen D. Keller, Wendy K. Bellows, and Robert R. L. Guillard

Bigelow Laboratory for Ocean Sciences, McKown Point,
West Boothbay Harbor, ME 04575

Significant dimethyl sulfide (DMS) production is confined to a few classes of marine phytoplankton, mainly the Dinophyceae (dinoflagellates) and the Prymnesiophyceae (which includes the coccolithophores). One hundred and twenty-three individual clones of phytoplankton representing twelve algal classes were examined in exponential growth for intra- and extracellular DMS (and its precursor DMSP). There is a strong correlation between the taxonomic position of the phytoplankton and the production of DMS. Although the Dinophyceae and Prymnesiophyceae predominate, other chromophyte algae (those possessing chlorophylls *a* and *c*) also contain and release significant amounts of DMS, including some members of the Chrysophyceae and the Bacillariophyceae (the diatoms). The chlorophytes (those algae possessing chlorophylls *a* and *b*) are much less significant producers of DMS with the exception of a few very small species. Other classes, including the cryptomonads and the cyanobacteria, are minor producers.

The oceans are a significant source of organic sulfur compounds that are implicated in acid precipitation and the production of atmospheric aerosols which affect global climate. These compounds are largely biogenic in origin, the most important being dimethyl sulfide (DMS) produced by marine phytoplankton (1-4). Although the distribution of DMS is broadly similar to that of primary productivity (5,6), attempts to directly correlate DMS production to primary production have been only moderately successful (e.g., 7). This is mainly because only certain groups of algae are known to produce significant amounts of DMS. Thus, correlations of DMS with chlorophyll *a* measurements are often poor and need to be supplemented with information on species composition. Field observations have implicated the colonial prymnesiophyte, *Phaeocystis* sp., with high levels of DMS (4,8,9). Coccolithophores (members of the Prymnesiophyceae) and some dinoflagellates (Dinophyceae) have also been suspect (4,9).

Until recently, no comprehensive survey of DMS production by phytoplankton has been made. Single clones of various marine species have been examined and of these, the coccolithophore, *Hymenomonas* (ex. *Cricosphaera*, *Syracosphaera*) *carterae* had the highest DMS levels (10,11).

0097–6156/89/0393–0167$06.00/0
© 1989 American Chemical Society

Representatives of the classes Bacillariophyceae (diatoms), Chlorophyceae, Chrysophyceae, Cryptophyceae and the Dinophyceae (dinoflagellates) had much lower concentrations. Freshwater representatives of these classes also showed little DMS production, although significant levels were found in freshwater cyanophytes (12). Marine macroalgae also release substantial quantities of DMS. The Chlorophyceae, especially *Ulva*, *Enteromorpha*, and *Codium*, and the red alga, *Polysiphonia*, are capable of producing large amounts of DMS in contrast to the Phaeophyceae (brown algae) which produce little (13,14).

In algae, DMS is produced by the cleavage of dimethylsulfoniopropionate (DMSP). The by-products of this enzymatic cleavage are DMS and acrylic acid (1:1). In macroalgae, there is a clear relationship between DMSP and osmotic adaptation (14). The intracellular concentrations of DMSP increase with increasing external salt concentrations. A similar response has been observed in a few species of phytoplankton, including the heterotrophic dinoflagellate, *Crypthecodinium* (ex. *Gyrodinium*) *cohnii* (15) and the coccolithophore, *Hymenomonas carterae* (16), but only at extreme salinities or osmotic pressures (using sucrose). The role of DMSP and other osmotica, glycine betaine and homarine, have also been examined for a halotolerant strain of *Platymonas* (17). *Phaeocystis*, the prymnesiophyte most commonly associated with high DMS concentrations *in situ*, has not been examined for this property but it has been speculated that acrylic acid, the other product of DMSP decomposition, plays an important role in this alga's ecology. The acrylic acid produced by *Phaeocystis* may have antibiotic properties (18) and may inhibit bacterial growth or feeding by zooplankton on the colonies of *Phaeocystis*. Ageing cultures of *Phaeocystis* have excretion rates of acrylic acid up to $7 \mu g/1$ (19). In this situation, DMS would be the incidental byproduct of acrylic acid production.

While it is generally accepted that phytoplankton, and most likely only certain groups of phytoplankton, are the most significant source of DMS in the oceans, there is little understanding of the physiology of its production or release. It is unclear whether algae actively excrete DMS or DMSP. Recent studies suggest that a considerable amount of DMSP is present in the water column (4,20), probably as a result of cell disruption. It is unlikely that DMSP passes through intact cell membranes, both because of the size and polarity of the molecule. It has been convincingly demonstrated that zooplankton stimulate the release of DMS (21). Thus, as algae age or are eaten, DMS is released, as well as DMSP. Once in the water column, there is little agreement as to its fate. DMS is soluble in seawater (22) and shows a steep gradient in concentration across the air-sea interface. An unknown proportion is oxidized to DMSO, both by bacteria (23) and photochemically (24). The fate of free DMSP is also unknown, although it is likely that bacteria mediate its breakdown to DMS at unknown rates (25). Processes occurring above the sea surface are also poorly understood. Most of the DMS is thought to be oxidized to SO_2 and subsequently to H_2SO_4, resulting in acid precipitation events, or becomes sulfate aerosols which serve as cloud condensation nuclei (CCN), critical to cloud formation and thus climate (2-4). Essential to a determination of the role of DMS in the global sulfur cycle is an understanding of the phytoplankton, determining which phytoplankters are capable of significant DMS production and the environmental variables affecting their physiology.

Methods

All the phytoplankton clones are maintained at and available from the Provasoli-Guillard Center for Culture of Marine Phytoplankton (Bigelow Laboratory for Ocean Sciences, McKown Point, West Boothbay Harbor, Maine

04575, U.S.A.). Algal cultures were grown in 100 ml batch cultures in appropriate media and, when possible, under identical light conditions (10^{16} quanta·cm^2·sec^{-1}; 14:10 light:dark cycle) and at 20°C. Media selection and growth temperature were dependent upon the environment from which the phytoplankter was isolated. Cell count samples were preserved with Lugol's fixative and counted on a 0.1 mm hemocytometer. Cell volumes were measured with live samples. Because we were interested in the total potential production of DMS by each clone, we chose to measure both the intracellular DMS (in the form of DMSP) and the extracellular DMS and DMSP. To do this, we took advantage of the specific reaction which cleaves DMSP 1:1 into DMS and acrylic acid with the addition of a strong base (10,13). This method is simple and specific, and also eliminates some of the problems of sample preparation. Sampling, filtration and bottle effects undoubtably influence DMS release from DMSP. By measuring all the DMS (and DMSP) in a culture, we obtain levels of total DMS, but are unable to discriminate between free DMS and that produced by the cleavage of DMSP. Thus, data are presented as DMSP-equivalents (Figure 1). While still in mid-exponential growth (as measured by fluorescence and cell counts), each phytoplankton culture was divided into two 50 ml portions. One was placed in a silanized 100 ml serum bottle and sealed. The base (2 mls) was added, the sample incubated and headspace gas analysis was performed 24 h later. All samples were incubated for 24 h at room temperature in the dark, then incubated for 30 min at 40°C immediately before sampling to maximize release to the headspace. Constant headspace:sample ratios were maintained at all times. The number from this sample represents the total DMS produced by the culture without discriminating between intra- and extracellular pools. The other 50 ml portion of the culture was gently (preferably by gravity) filtered through a GF/F filter (Whatman, nominal pore size 0.7 μm). The filter was placed in a 10 ml silanized vial, sealed and base (1 ml) added. After incubation, headspace gas analysis was performed. The filtrate was collected, transferred to a silanized 100 ml serum bottle and treated in the same manner as the whole culture. DMSP standards were prepared by dissolving 1 mg DMSP (Research Plus, Bayonne, N.J., U.S.A.) in 100 mls of 0.2 μm filtered seawater. The primary stock is volatile and it was remade weekly. Appropriate dilutions were made into 50 ml portions for liquid standards and for filter standards, small amounts of the primary stock were applied directly and absorbed by GF/F filters. One ml of the base was added to filter samples, 2 mls to liquid samples. The level of detection is 50 ng DMS for filter standards and 300 ng DMS for liquid standards. Appropriate amounts of headspace gas were removed and injected directly onto a Chromosil 330 Teflon column (6 ft.; 1/8" O.D.); injector 70°; column 70°; He carrier gas 30 ml/min and detected with a FPD (detector temperature 150°). A Varian 3300 gas chromatograph was used. All measurements were performed in triplicate. Standards were run before and after and randomly during sample analysis.

Results and Discussion

One hundred and twenty-three individual phytoplankton clones representing twelve algal classes were examined for total DMS(DMSP) production. DMS was measured in whole cultures, cell and filtrate fractions (as outlined in Figure 1). DMS measurements for whole cultures (cells and filtrate) are given in Table I. Values for DMS levels in all fractions, whole culture, filtered cells and filtrate are not reported here but comparisons of the whole culture sample with the filtered cells and accompanying filtrate revealed that, for the most part, amounts of DMS in the two fractions were equivalent to the amount in the whole culture. Differences can largely be attributed to filtering problems. Even

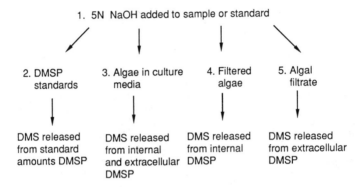

Figure 1. Sampling protocol for DMS measurements.

Table I. Phytoplankton Clones Examined for DMS Release. ng DMSP Corresponds to DMSP - Equivalents Calculated from Standard Curves as Explained in Text. N.D. = Not Detectable. See Text for Limits of Detection.

Class	Genus/Species	Clone	pg DMSP/cell	μm DMSP/cm^3 of cell volume
Bacillariophyceae Centrales (CENTRIC DIATOMS)	Biddulphia sp.	L1474	1.0	0.69
	Chaetoceros affinis	CCUR*	N.D.	
	Chaetoceros decipiens	WTCD	N.D.	
	Chaetoceros didymum	L162*	0.04	18.21
	Chaetoceros simplex	BBSM	N.D.	
	Coscinodiscus sp.	COSC1*	10.4	0.17
	Ditylum brightwellii	L154*	9.81	4.61
	Melosira nummuloides	MEL3	34.5	264.18
	Minidiscus trioculatus	GMe41*	0.10	32.91
	Porosira glacialis	18*	9.37	2.09
	Rhizosolenia setigera	RHIZ0	15.1	0.46
	Skeletonema costatum	SKEL	0.50	21.87
	Skeletonema menzellii	MEN5	0.14	30.30
	Thalassiosira guillardii	7-15*	N.D.	
	Thalassiosira pseudonana	3H	0.08	16.64
	Thalassiosira rotula	MB411	0.25	1.05
	Thalassiosira sp.	PP86A*	5.48	1.82
Bacillariophyceae Pennales (PENNATE DIATOMS)	Amphora coffaeformis	47M	N.D.	
	Asterionella glacialis	A6	N.D.	
	Cylindrotheca closterium	NCLOST	1.49	41.42
	Navicula pelliculosa	O4	N.D.	
	Nitzchia laevis	O7	0.52	7.34
	Phaeodactylum tricornutum	PHAEO	N.D.	

Continued on next page

Table I continued

Class	Genus/Species	Clone	pg DMSP/cell	μm DMSP/cm^3 of cell volume
Chloromonadophyceae (CHLOROMONADS)	*Chattonella harima*	WTOMO	N.D.	
	Chattonella luteus	OLISTH	N.D.	
Chlorophyceae (CHLOROPHYTES)	*Chlamydomonas sp.*	EPT3	N.D.	
	Chlorella capsulata	OPT10	0.61	25.22
	Chlorella sp.	O17	N.D.	
	Chlorococcum sp.	Chloro-1	0.08	0.14
	Dunaliella tertiolecta	DUN	N.D.	
	Nannochloris atomus	GSBNanno	N.D.	
	Stichococcus sp.	NB3-18	N.D.	
	unidentified coccoid	IVP16AX	N.D.	
Chrysophyceae (CHRYSOPHYTES)	*Chrysamoeba sp.*	IG5	2.68	596.27
	Chrysosphaera sp.	UW397	N.D.	
	Ochromonas sp.	IC_1	0.38	200.75
	Ochromonas sp.	VT_1	0.58	529.10
	unidentified coccoid	IVR5ax	0.24	422.39
	unidentified flagellate	MC1	N.D.	
Cryptophyceae (CRYPTOMONADS)	*Chroomonas salina*	3C	N.D.	
	Cryptomonas sp.	ID_2	2.86	345.52
	Cryptomonas sp.	PHI	N.D.	
	Cyanophora paradoxa	CY2	N.D.	
	Rhodomonas lens	RLENS	N.D.	
	unidentified cryptomonad	IVF_3	N.D.	
	unidentified cryptomonad	IVP_8	N.D.	
	unidentified cryptomonad	10C*	N.D.	
Cyanophyceae (BLUE-GREEN ALGAE)	*Synechococcus bacillaris*	SYN	N.D.	
	Synechococcus sp.	DC_2	N.D.	
	Synechococcus sp.	L1602	N.D.	
	Synechococcus sp.	L1604	N.D.	
	Synechocystis sp.	CN0117	N.D.	
	Trichodesmium sp.	MACC0993	N.D.	

Table I continued

Class	Genus/Species	Clone	pg DMSP/cell	μm DMSP/cm^3 of cell volume
Dinophyceae (DINOFLAGELLATES)	Amphidinium carterae	AMPHI	19.3	2201.50
	Cachonina niei	CACH	43.0	192.54
	Ceratium longipes	090201*	2.0	0.23
	Crypthecodinium cohnii	CCOHNII	45.7	377.61
	Dissodinium lunula	L823	116.0	1.94
	Gambierdiscus toxicus	GT200A	160.0	10.08
	Gonyaulax polyedra	GP60e	18.0	4.01
	Gonyaulax spinifera	W1*	145.0	16.49
	Gymnodinium nelsoni	GSBL	244.0	29.55
	Gymnodinium simplex	WT8	32.0	45.75
	Gymnodinium sp.	94GYR*	24.0	124.63
	Gyrodinium aureolum	KT3	0.72	0.65
	Gyrodinium aureolum	PLY497A	0.68	0.36
	Heterocapsa pygmaea	GYMNO	19.5	451.49
	Heterocapsa sp.	GT23*	78.1	190.30
	Oxyrrhis marina	NEPCC534 (w/DUN)	N.D.	
	Prorocentrum minimum	EXUV	21.4	888.06
	Prorocentrum sp.	IIB$_2$b$_1$	16.4	1082.10
	Prorocentrum sp.	86183*	N.D.	
	Prorocentrum sp.	M12-11*	593.0	190.30
	Protogonyaulax tamarensis	GT429*	265.0	139.55
	Pyrocystis noctiluca	CCMP4	6.06	0.01
	Scrippsiella trochoidea	PERI	384.0	350.00
	Symbiodinium microadriaticum	HIPP	24.2	344.78
	Thoracosphaera heimii	L603	26.6	194.03
	unidentified dinoflagellate	DDT*	91.6	83.58
Euglenophyceae (EUGLENOPHYTES)	Eutreptia sp.	EEUI	N.D.	

Continued on next page

Table I continued

Class	Genus/Species	Clone	pg DMSP/cell	μm DMSP/cm^3 of cell volume
Eustigmatophyceae (EUSTIGMATOPHYTES)	*Nannochloropsis*	GSB Sticho	0.02	22.76
Prasinophyceae (PRASINOPHYTES)	*Mantoniella squamata*	PLY189	0.13	29.18
	Micromonas pusilla	DW8	0.03	161.94
	Micromonas pusilla	IB$_4$	0.02	287.31
	Nephroselmis pyriformis	PLY58	1.08	36.57
	Pedinomonas minutissima	VA3	N.D.	
	Pseudoscourfielda marina	IVP$_{11}$	0.02	5.65
	Pyraminonas sp.	13-10PYR	0.02	0.53
	Tetraselmis levis	PLATY 1	1.62	32.76
	Tetraselmis sp.	OPT4	1.09	45.45
	unidentified coccoid	Ω48-23	0.09	47.69
	unidentified coccoid	VH$_2$ax	1.92	126.87
	unidentified coccoid	VB$_1$	1.38	156.72
	unidentified flagellate	BE92*	0.03	484.33
Prymnesiophyceae (PRYMNESIOPHYTES including COCCOLITHOPHORES)	*Chrysochromulina ericina*	NEPCC 109A	3.81	251.49
	Chrysochromulina herdlansis	NEPCC 186	3.62	412.69
	Coccolithus neohelis	CONE	2.53	85.08
	Emiliania huxleyi	BT6	0.75	166.42
	Imantonia rotunda	IIE$_6$	0.18	159.70
	Imantonia rotunda	1197NTA	0.26	87.31
	Isochrysis galbana	ISO	0.50	56.87
	Pavlova lutheri	MONO	0.05	3.28
	Pavlova pingus	IG$_7$	0.71	46.87
	Pavlova sp.	IIB$_3$$^+$	0.22	53.36
	Pavlova sp.	IIB$_3$ax	0.65	156.72
	Pavlova sp.	IIG$_3$$^+$	0.45	51.19

Table I continued

Class	Genus/Species	Clone	pg DMSP/cell	μm DMSP/cm^3 of cell volume
	Pavlova sp.	IIG$_3$ax	0.52	59.18
	Pavlova sp.	IIG$_6$+	0.65	73.66
	Pavlova sp.	IIG$_6$ax	0.73	83.58
	Phaeocystis sp.	677-3	2.29	260.45
	Phaeocystis sp.	1209	1.0	113.43
	Pleurochrysis carterae	COCCOII	12.0	170.15
	Prymnesium parvum	PRYM	1.7	111.94
	Syracosphaera elongata	SE62	19.8	35.30
	Umbilicosphaera sibogae	L1178	13.8	195.52
	unidentified coccolithophore	8613COCCO*	1.1	125.37
	unidentified flagellate	3D*	1.57	179.10
	unidentified flagellate	8610G6*	3.37	139.55
	unidentified flagellate	8610C3*	3.15	358.21
	unidentified flagellate	326-1*	5.55	365.67
Rhodophyceae	*Porphyridium cruentum*	PORPH	N.D.	
	Rhodosorus marinus	RHODO	0.31	35.67
unidentified	*unidentified green flagellate*	LOBD*	2.81	94.78

*Gulf of Maine isolates
+Bacteria isolated from these clones showed no detectable DMS release

distribution and extraction of phytoplankters that grow in clumps or attached to the culture vessel walls is very difficult.

Using the DMS values (calculated from the external DMS standards) and cell count data, we normalized the measurements to pg DMSP-equivalents/cell. Also, because there is a great diversity of size and form among the phytoplankton clones, ranging in the largest dimension from less than 1 μm to several hundred μm's, we calculated DMS levels on a per cell volume basis ($v = 0.5236(d)^3$). We used total cell volumes for these values, not attempting to estimate osmotic volume or to discriminate between vacuolate and non-vacuolate species. Thus, amounts of DMSP-equivalents (μM) per unit cell volume (cm^3) may be underestimated by approximately a factor of 2 (assuming an osmotic volume of 50% of total volume) for many species. Since the range of values is over five orders of magnitude, this discrepancy is probably not significant. When compared with published values of DMSP/cell volume ([4,16,17,21]) the values are comparable though somewhat lower as expected. For example, a *Platymonas* (= *Tetraselmis*) had levels ranging from 50-250 μM/cm^3 ([17]) and a clone of *Gymnodinium nelsoni* had an estimated level of 280 ([21]). Our values for Tetraselmis clones were approximately 50 μM/cm^3; for *Gymnodinium nelsoni*, 30 μM/cm^3. Other Gymnodinium's were considerably higher (up to 150 μM/cm^3).

The DMS (in the form of DMSP) is largely intracellular for the phytoplankton, which were undergoing exponential growth. DMS does not appear to be released into the medium in any quantity and it is likely that the measured DMS is released, presumably in large part as DMSP, only upon cell death and lysis or with some mechanical disruption, such as grazing ([21]). Some members of the Prymnesiophyceae had substantial quantities of DMS (DMSP) in the medium. We believe that this is a real phenomenon and not just a filtering artifact. The significance of the extracellular DMS (DMSP) among the Prymnesiophytes is unclear. Certain members of this class, notably *Phaeocystis*, produce substantial quantities of extracellular mucilage and reports of sulfur odors during *Chrysochromulina* blooms suggest that this phenomenon occurs *in situ* as well ([9]).

Most of the phytoplankters examined were xenic (i.e. not bacteria-free). Because of concern that the presence of bacteria might affect DMS concentrations, we examined for DMS production several clones of algae with and without their accompanying natural bacteria. In the presence of bacteria, DMS concentrations were not significantly less (Table I). We also examined bacteria isolated from the algal cultures separately, both for DMS production and utilization of DMSP as a growth substrate, with negative results in both cases.

An analysis of DMS production by each phytoplankton class follows:

Bacillariophyceae (centric and pennate diatoms). For the most part, the diatoms are not significant producers of DMS. An exception is the estuarine species, *Melosira nummuloides*. In sufficient numbers (i.e. bloom situations) certain other species could be important. Therefore diatoms cannot be summarily dismissed as sources of DMS; some consideration of species composition must be included.

Chloromonadophyceae (Chloromonads). This class of phytoplankton has few known marine representatives but these species can form extensive blooms. They are not however a significant source of DMS.

Chlorophyceae (Chlorophytes-true green algae). These algae do not produce DMS in any quantity. One minor exception is the *Chlorella* isolated from a small salt pond.

Chrysophyceae (Chrysophytes). There is a considerable range of DMS levels within this class. DMS was undetectable in several clones but was very high, based on cell volume, in others.

Cryptophyceae (Cryptomonads). The cryptomonads are insignificant producers of DMS. The small *Cryptomonas* clone, ID2, isolated from an oceanic area, gave the only detectable DMS concentration in this group. It produces a substantial amount.

Cyanophyceae (blue-green algae). The important prokaryote, *Synechococcus*, produces negligible amounts of DMS, as do other marine representatives of this class. By contrast, freshwater representatives apparently do produce DMS (12).

Dinophyceae (Dinoflagellates). The dinoflagellates, along with the prymnesiophytes, are the major producers of DMS, although there is considerable variation when compared by cell volume. The dinoflagellates are a diverse group with a large size range and with auto- and heterotrophic representatives. Some of the very large oceanic dinoflagellates, such as *Pyrocystis* or *Ceratium*, are minor DMS producers. Others, such as the common coastal bloom-forming species, *Amphidinium* or *Prorocentrum*, are major potential sources. The heterotrophic dinoflagellate, *Crypthecodinium cohnii*, as reported (15), produces a substantial quantity of DMS, but there are many other dinoflagellates which produce equivalent or greater amounts. The symbiotic dinoflagellate, *Symbiodinium*, associated with corals, contains a large amount of DMS. Reports of major DMS release from coral reefs have been reported (11) and may be related to this phytoplankter.
Attempts to correlate DMS production with Orders within the Dinophyceae are only moderately conclusive. Certainly the Order Prorocentrales, which includes only the *Prorocentrum*, is significant. The Orders Gymnodiniales (which includes *Amphidinium* and *Gymnodinium* among others) and Peridiniales (which includes *Heterocapsa* and *Protogonyaulax*) are important but also include members which are insignificant DMS producers; *Gyrodinium* and *Oxyrrhis* and *Ceratium*, respectively.

Euglenophyceae; Eustigmatophyceae; Rhodophyceae. These classes have few marine representatives and are insignificant producers of DMS.

Prasinophyceae (Prasinophytes-green algae). For the most part, these algae do not produce much DMS but because of the very small size of representative clones, the intracellular quantities are quite large. This aspect will be discussed in further detail.

Prymnesiophyceae (Prymnesiophytes-includes the coccolithophores). The prymnesiophytes are the second important class of phytoplankton in terms of DMS production. Although much less substantial per cell than the dinoflagellates, they do produce equivalent amounts when compared on a per cell volume basis. The coccolithophores are all significant producers but other bloom-forming prymnesiophytes, such as *Chrysochromulina*, are also important. *Phaeocystis*, the phytoplankter most commonly linked to DMS emissions *in situ*, does not stand out among the prymnesiophytes in terms of DMS production. It

is exceptional however in the magnitude and longevity of the blooms it produces (4,9,19).
It is difficult to extrapolate the values obtained for DMS from laboratory cultures to field situations but the clear taxonomic pattern of DMS production has important ecological implications. Based on this taxonomic relationship, it is evident that DMS will vary temporarily and spatially, dependent on the species composition of the flora. Although areas of high primary productivity may have significant DMS emissions, it is equally possible that they will not or the emissions may vary seasonally. A general understanding of phytoplankton distribution and succession coupled with this taxonomic survey of the DMS production allows for speculation on DMS patterns in the oceans.

Although the diatoms are not major DMS producers, their numerical abundance and diversity precludes their elimination from this discussion of those phytoplankters central to DMS production. Diatoms seasonally dominate the flora in most areas of the oceans, but are especially abundant in temperate and polar regions. Phytoplankton blooms are controlled in large part by the physical state of the environment, which affects light, temperature and nutrients. Diatoms dominate in turbulent waters, including coastal and upwelling areas, and during or immediately after mixing events, such as the spring and fall water column turnover. Dinoflagellates, in contrast, predominate in stable environments, normally stratified, warm water communities (26). The seasonal cycle of phytoplankton is a function of latitude and the timing and duration of the biological seasons are critical to the development of the phytoplankton community. Thus with increasing latitude, the peak of phytoplankton abundance will occur correspondingly later. If the oceans are divided into three macrozones, boreal (subarctic and subantarctic), temperate and tropical, certain generalities can be made. The boreal regions and temperate regions are not dissimilar except for the timing of the blooms. Both are essentially diatom-dominated communities with sporadic and often very large blooms of dinoflagellates and coccolithophores in the summer. Nanoplankton often dominate in the transitional early summer months.

The succession of phytoplankton within the Gulf of Maine serves as an example (see Table I for Gulf of Maine isolates). The spring diatom bloom, consisting of small, rapidly-growing diatoms, develops along the southwest coast of the Gulf and moves eastwards across the shelf. The diatoms are dominated by small centrics, mostly *Thalassiosira*, *Chaetoceros* and *Skeletonema costatum*. This community typically reestablishes itself in a second, smaller fall bloom. As nutrients are depleted and stratification occurs, the spring diatom community diminishes and is replaced by a mixed population of flagellates and dinoflagellates. Its composition is not as predictable. The dominant dinoflagellate off-shore is *Ceratium longipes*; inshore populations consist of *Protogonyaulax tamarensis*, the common "red-tide" organism of temperate waters, *Prorocentrum* spp., *Peridinium* spp. and *Dinophysis* spp. (27-31). Other dinoflagellates, such as *Gyrodinium* sp., occasionally form massive localized patches (29). *Phaeocystis pouchetii* is always a component of the spring flora and occasionally, but not predictably, will disrupt the typical succession and completely dominate surface waters for prolonged periods. There are numerous records of *Phaeocystis* blooms throughout the Gulf with concentrations exceeding 10^6/l (19,27,30). There are also reports of blooms of the coccolithophore, *Emiliania huxleyi* offshore (38,31). We have observed major, extensive coccolithophore (presumably *E. huxleyi*) blooms in mid-summer since 1983.

Thus, for the Gulf of Maine, we would expect to see a strong seasonality of DMS emissions. The winter and early spring months would be periods of negligible release; mid-spring would be less predictable depending on the occurrence of *Phaeocystis*. In years of a *Phaeocystis* bloom, this period might see a major DMS emission. In early summer, near-shore waters would account for a substantial portion of the annual DMS release, with blooms of the common dinoflagellates, *Prorocentrum* and *Protogonyaulax*. Offshore, the picture would be less clear. *Ceratium longipes* is a minor DMS producer but the poorly-characterized nanoplankton population may represent a significant source. In later summer, a major *E. huxleyi* bloom would be responsible for high DMS concentrations. The fall months would show a sharp decline in DMS levels as the diatoms become dominant again. *Rhizosolenia*, common in late summer, is known to form macroscopic mats in many environments (32,33), and may provide an exception. In such cases, cell densities reach 10^3/ml and could represent a significant, if patchy, source of DMS. Also a few coastal forms such as *Melosira* and *Cylindrotheca* that occur in abundance on tidal flats and at the thermocline of stratified coastal water columns may be important.

This scenario is somewhat typical of boreal and temperate regions in all the oceans and applies also to neretic areas in tropical seas. The estuarine embayments in the tropics are often seasonally dominated by *Prorocentrum*, *Scrippsiella* and *Gymnodinium*. In oligotrophic areas in the tropics, dinoflagellates are often abundant and congregate at the pycnocline, forming a deep chlorophyll maximum at approximately the 1% light level (34). This "shade flora" is usually dominated by *Ceratium* or *Pyrocystis*, neither a major DMS producer. However, since the maximum number of dinoflagellates is usually subsurface (at the 10-1% light levels) in all environments, an interesting question is raised. Maximum DMS concentrations have been observed just above the deep chlorophyll maximum (7), presumably released by those phytoplankters comprising the maximum. Does this DMS contribute to the amount of DMS released to the atmosphere at the air-sea interface? Since many dinoflagellates are known to make diel vertical migrations, it is very possible in shallow water columns. For non-migrating species or deep euphotic zones, it is less clear.

Coccolithophores are important in oligotrophic waters. The majority of coccolithophore species are tropical but the few boreal species, notably *E. huxleyi*, can form blooms, with cell numbers of up to 3.5 x 10^6/l recorded (35). *E. huxleyi* is also the species with the greatest biogeographical range and except in tropical waters is the dominant coccolithophore. High numbers of coccolithophores are also seen in equatorial waters and in marginal seas in the tropics, but are dominated by *Gephyrocapsa* (36). *Phaeocystis*, another prymnesiophyte, forms extensive blooms in late spring in the Arctic and Antarctic.

Recent advances in satellite technology make coccolithophore blooms visible from space (37) and there is speculation that coccolithophore blooms are becoming more persistent and regular in certain areas (26). It is also possible that other phytoplankters which are major producers of DMS are becoming more abundant. *Phaeocystis* blooms in particular are becoming more intense and frequent in the North Sea. It is suggested that these phenomena are linked to increasing eutrophication (26,38).

The distribution of *Phaeocystis* is well-known because of its colonial form; likewise *E. huxleyi* is well-characterized because of its easily recognized coccoliths. Both of these species are members of the nanoplankton (phytoplankton less than 20 μm), which in general are poorly described. Many of the nanoplankton are flagellates or coccoid cells which belong to a smaller size fraction, the ultraplankton (those cells passing through a 3 μm filter (39)) and are members of several different algal classes, including the Chrysophyceae, Prasinophyceae and Prymnesiophyceae. They are mentioned here because many produce substantial amounts of DMS. These small algae are present in all environments at densities of 10^6 - $10^7/l$ and often dominate deep chlorophyll maxima in oceanic areas (39-42). Thus, representatives of the Chrysophyceae, Prasinophyceae and Prymnesiophyceae may be important sources of DMS in areas they dominate. Comparisons of production per unit cell volume reveal that DMS levels are roughly equivalent to those of the dinoflagellates and coccolithophores. Since the abundance of these small phytoplankton is usually at least two orders of magnitude higher than their larger counterparts, size differences (which are also approximately two orders of mangitude) are overcome. DMS emissions from oligotrophic regions may be mainly from this size class of phytoplankters. In coastal waters, phytoplankton of ultraplanktonic size are often major bloom-formers; e.g. the prymnesiophyte, *Chrysochromulina* (9), the prasinophyte, *Micromonas pusilla* (39) and the newly-described chrysophyte, *Aureococcus anophagefferens*, responsible for the "brown tide" in Long Island Sound, U.S.A. waters, reaching densities of $10^9/l$ (43). All of these species and other similar forms may be responsible for occasional major DMS emissions. It is also noted that the ultraplankton are quite plastic and easily pass through filter pores much smaller than their smallest dimension (39). Therefore, it is recommended that filters used to either remove particulates from water samples before DMS measurement or collect cells for chlorophyll measurements be no larger than GF/F (nominal pore size 0.7 μm). Correlations of DMS concentrations with chlorophyll *a* may improve with this method.

DMS levels in marine phytoplankton are highly variable, ranging from non-detectable up to 600 pg/cell or intracellular concentrations as high as 2000 μm DMSP (equivalents)/cm^3 cell volume. There is a clear taxonomic pattern, with major production confined to the Dinophyceae and the Prymnesiophyceae. However, representatives of other classes, especially other chromophyte (containing chlorophylls *a* and *c*) plankters, may be as significant. The eukaryotic ultraplankton may be especially important. Based on this taxonomic survey of DMS levels within phytoplankton and known distributions of individual phytoplankters *in situ*, it can be expected that DMS production will be seasonal and regional in nature and highly dependent on the numerical abundance of key species. A consideration of phytoplankton species composition and succession must be included in any area under study for DMS emissions from the oceans.

Acknowledgments

Although the research described in this article has been funded wholly or in part by the United States Environmental Protection Agency through R812662 to R.R.L. Guillard, it has not been subjected to Agency review and therefore does not necessarily reflect the views of the Agency and no official endorsement should be inferred. We thank Patrick M. Holligan and Rhonda C. Selvin for their comments and suggestions. This manuscript is Bigelow contribution no. 88002.

Literature Cited

1. Andreae, M. O.; Raemdonck, H. Science 1983, 221, 744-7.

2. Bates, T. S.; Charlson, R. J.; Gammon, R. H. Nature 1987, 329, 319-21.

3. Charlson, R. J.; Lovelock, J. E.; Andreae, M. O.; Warren, S. G. Nature 1987, 326, 655-61.

4. Turner, S. M.; Malin, G.; Liss, P. S.; Harbour, D. S.; Holligan, P. M. Limnol. Oceanogr. 1988, 33, 364-75
5. Barnard, W. R.; Andreae, M. O.; Watkins, W. E.; Bingmer, H.; Georgii, H. W. J. Geophys. Res. 1982, 87, 8787-93.
6. Bates, T. S.; Cline, J. D. J. Geophys. Res. 1985, 90, 9168-72.
7. Andreae, M. O.; Barnard, W. R. Mar. Chem. 1984, 14, 167-79.
8. Barnard, W. R.; Andreae, M. O.; Iverson, R. L. Cont. Shelf. Res. 1984, 3, 103-13.
9. Holligan, P. M.; Turner, S. M.; Liss, P. S. Cont. Shelf. Res. 1987, 7, 213-24.
10. Ackman, R. G.; Tocher, C. S.; McLachlan, J. J. Fish. Res. Bd. Can. 1966, 23, 357-64.
11. Andreae, M. O.; Barnard, W. R.; Ammons, J. M. Ecol. Bull. 1983, 35, 167-77.
12. Bechard, M. J.; Rayburn, W. R. J. Phycol. 1979, 15, 389-93.
13. White, R. H. J. Mar. Res. 1982, 40, 529-36.
14. Reed, R. H. Mar. Biol. Letts. 1983, 4, 173-81.
15. Kadota, H.; Ishida, Y. Bull. Jap. Soc. Scient. Fish. 1968, 34, 512-8.
16. Vaivaramurthy, A.; Andreae, M. O.; Iverson, R. L. Limnol. Oceanogr. 1875, 30, 59-70.
17. Dickson, D. M.; Kirst, G. O. Planta. 1986, 167, 536-43.
18. Sieburth, J. McN. J. Bact. 1961, 82, 72-9.
19. Guillard, R. R. L.; Hellebust, J. A. J. Phycol. 1971, 7, 330-8.
20. Wakeham, S. G.; Howes, B. L.; Dacey, J. W. H.; Schwarzenbach, R. P.; Zeyer, J. Geochim. Cosmochim. Acta 1987, 51, 1675-84.
21. Dacey, J. W. H.; Wakeham, S. G. Science 1986, 233, 1314-6.
22. Dacey, J. W. H.; Wakeham, S. G.; Howe, B. L. Geophys. Res. Letts 1984, 11, 991-4.
23. Zeyer, J.; Eicher, P.; Wakeman, S. G.; Schwarzenbach, R. P. Appl. Environm. Microbiol. 1987, 53, 2026-32.
24. Brimblecombe, P.; Shooter, D. Mar. Chem. 1986, 19, 343-53.
25. Dacey, J. W. H.; Blough, N. V. Geophys. Res. Letts. 1987, 14, 1246-9.
26. Holligan, P. M. Rapp. P.v. Reun. Cons. Int. Explor. Mer 1987, 187, 9-18.
27. Bigelow, H. B. Bull. Bureau Fish. XL(II) 1924.
28. Gran, H. H.; Braarud, T. J. Biol. Bd. Can. 1935, 1, 279-467.
29. Yentsch, C. S.; Glover, H. E.; Apollonio, S.; Yentsch, C. M. Bigelow Technical Report #78029 1978.

30. Flagg, C. N.; Magnell, B. A.; Frye, D.; Cura, J. J.; McDowell, S. E.; Scarlet, R. I. Final Report, v.1. U.S. Dept. Interior No. AA851-CT1-39 1982.

31. Marshall, H. G. J. Plank. Res. 1984, 6, 169-93.

32. Caron, D. A.; Davis, P. G.; Madin, L. P.; Sieburth, J. McN. Science 1982, 218, 795-7.

33. Martinez, L.; Silver, M. W.; King, J. M.; Alldredge. Science 1983, 221, 152-4.

34. Taylor, F. J. R. In The Biology of Dinoflagellates; Taylor, F. D. R., Ed.; Blackwell Scientific, 1987; pp. 399-502.

35. McIntyre, A.; Be, A. W. H. Deep Sea Res. 1967, 14, 561-97.

36. Okada, H.; Honjo, S. Deep Sea Res. 1967, 20, 355-74.

37. Holligan, P. M.; Viollier, M.; Harbour, D. S.; Camus, P.; Champagne-Philippe. Nature 1983, 304, 339-42.

38. Lancelot, C.; Billen, G.; Sournia, A.; Weisse, T.; Colijn, F.; Veldhuis, M. J. W.; Davies, A.; Wassman, P. Ambio. 1987, 16, 38-46.

39. Murphy, L. S.; Haugen, E. M. Limnol. Oceanogr. 1985, 30, 47-58.

40. Estep, K. W.; Davis, P. G.; Hargraves, P. E.; Sieburth, J. McN. Protistologica 1984, 20, 613-34.

41. Hooks, C. E.; Bidigare, R. R.; Keller, M. D.; Guillard, R. R. L. J. Phycol. 1988, in press.

42. Glover, H. E.; Keller, M. D.; Guillard, R. R. L. Nature 1986, 319, 142-3.

43. Smayda, T. In Proceedings of the Emergency Conference on "Brown Tide" and other unusual algal blooms, Oct. 23-4, New York, 1986.

RECEIVED September 6, 1988

Chapter 12

Dimethyl Sulfide
and (Dimethylsulfonio)propionate
in European Coastal and Shelf Waters

S. M. Turner, G. Malin, and P. S. Liss

School of Environmental Sciences, University of East Anglia,
Norwich NR4 7TJ, United Kingdom

The seasonal and spatial variations in surface water dimethyl
sulphide (DMS) concentrations have been determined, enabling
an estimation of the flux of marine biogenic sulphur to the
atmosphere for the European Shelf system. This natural emission
may account for a significant amount of the excess atmospheric
sulphate over Scandinavia in Spring and Summer. Several
ecologically important, bloom-forming species of phytoplankton
have been identified as DMS producers, and where the algal
population is nearly monospecific, there are good correlations
between DMS and biomass. Concentrations of dimethylsulphoniopropionate (DMSP), the precursor of DMS, are on
average 14 times higher than concentrations of DMS and there is
good correlation between the two compounds. Intracellular levels
of DMSP vary according to algal species.

Over the past few years we have been studying the waters around the United
Kingdom, including the North Sea, Irish Sea and N.E. Atlantic, in order to
characterise dimethyl sulphide (DMS) emissions and assess the significance of
this natural contribution to acidity of rainfall and the sulphur cycle. Biogenic
DMS concentrations in seawater vary considerably both temporally and
spatially and coastal and shelf water systems often contain higher
concentrations of volatile sulphur than the open oceans (1,2).
   In order to establish the sources of DMS we have made measurements of
dimethylsulphoniopropionate (DMSP; the precursor of DMS) and attempted to
relate the two compounds to phytoplankton species and abundance. We have
also investigated variations in DMS and DMSP with depth through the water
column, with respect to diel cycling and monitored the effect of high and low
nitrate concentrations on DMSP levels in laboratory cultures and in the N. Sea.

Experimental

Surface seawater samples were taken using ships' continuous non-toxic pump
supplies, the average depth of intake being 3m. A 500ml bottle was carefully
filled with the fast-flowing, bubble-free water and allowed to overflow for about
a minute and then stoppered (ground glass). Hence, exposure to air and

0097–6156/89/0393–0183$06.00/0
© 1989 American Chemical Society

potential problems of degassing or contamination were minimised. Aliquots for analysis were immediately sub-sampled from the bottom of the bottle using glass syringes, to avoid possible storage effects and degassing by contact with air. One aliquot was purged with scrubbed N_2 for 20 minutes, to extract the dissolved gases which were cryo-focussed at -150°C. The sample was thawed and injected via a six-port valve into a gas chromatograph fitted with two Chromosil 330 columns (SUPELCO) and two flame photometric detectors (Varian 3700 with Aerograph dual flame detector, 365nm optical filter). Two further aliquots were prepared for DMSP analysis; one unfiltered and one filtered using Whatman GF/C filters, with a nominal, initial retention size of 1.2 μm. The samples were reacted with 10M sodium hydroxide (final pH = 12) in ground glass stoppered bottles at room temperature, for at least 6 hours. The time for complete alkaline conversion of DMSP to DMS was determined using natural samples. The DMSP concentration was then determined as DMS by the analysis described above. Dissolved and particulate DMSP were calculated by subtraction :

$$DMSP_d = DMSP_{filtered} - DMS$$
and
$$DMSP_p = DMSP_{total} - (DMS + DMSP_d).$$

(The "dissolved" fraction consists of DMSP that passes through the filter, and the "particulate" fraction is that retained. Hence, the particulate DMSP is associated with algal cells, gut contents of zooplankton and detrital material.)

Standard hydrographic data (temperature, salinity, fluorescence and inorganic nutrients) were generally recorded continuously. Chlorophyll a concentrations of discrete samples were determined fluorometrically. Water samples were preserved with Lugol's iodine solution and 0.5% neutralised formaldehyde for onshore identification and enumeration of the phytoplankton using an inverted microscope.

Area of Study

Figure 1 is a map of the study area and shows the regions covered by the cruises and the position of a sampling station just off the coast of East Anglia. We have tried to maximise geographical and seasonal coverage, but as most of the work has been done on ships of "opportunity", certain areas, such as the Wadden Sea, German Bight and eastern North Sea, have not yet been properly sampled.

Seasonal Variation in Surface Water Concentrations

DMS in seawater is produced by marine phytoplankton (3,4) and in the latitudes delineated by the European Shelf system would be expected to show some seasonality. Figure 2a shows the variation of DMS concentration (log scale) at the fixed station just off the coast of East Anglia. For the two years of monitoring, there was a marked increase in DMS concentration in spring, reaching a maximum in summer and then decreasing. The maxima coincided with blooms of the alga, *Phaeocystis pouchetii*, which occur annually in this area, but vary in timing from year to year.

Figure 2b is a compilation of all the data obtained on cruises showing mean values and ranges of DMS concentration plotted on a logarithmic scale. The mean values for winter and summer are 0.1 and 9.4 nmol DMS (S) l^{-1} respectively, a seasonal difference of two orders of magnitude. As yet, the data set for DMSP is not as comprehensive as for DMS, but as the former is the precursor of the latter, a similar seasonal pattern might be expected. For

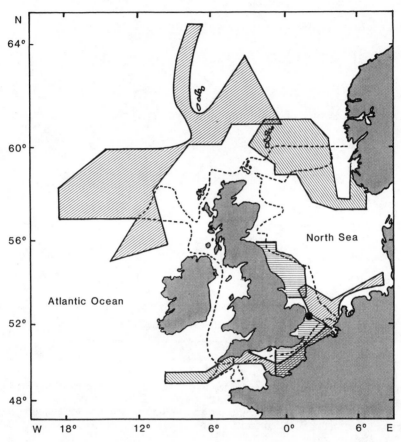

Figure 1. Map to show the areas covered by cruises, 1984 to 1987, and the position of a sampling station, ●, off the East Arglian coast.

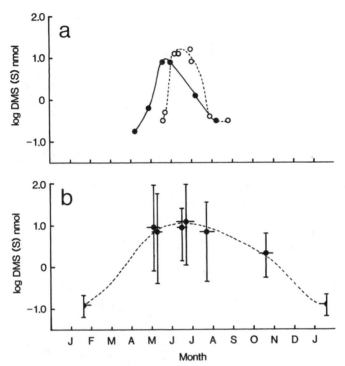

Figure 2. Seasonal variation of DMS in surface water (approx. 3m depth); (a) log DMS concentration at the East Anglian sampling station for 1983 ---O--- and ——●—— 1984. (b) compilation of data from all cruises showing the log values for mean DMS concentrations, ranges of DMS concentration and duration of each cruise.

example, for the 3 cruises falling in the period 22nd April to 3rd July, the average concentration of total DMSP was 134.4 nmol $DMSP_t$ (S) l^{-1}, with a mean value of 30.0 nmol $DMSP_t$ (S) l^{-1} for an October cruise. The mean DMS concentrations for the two periods were about 9.4 and 2.2 nmol DMS (S) l^{-1}. Thus, from Spring to Autumn, $DMSP_t$ and DMS concentrations decreased proportionally (by a factor of 4.3 to 4.5) and appeared to maintain a ratio of about 14:1.

Spatial Variation

In addition to temporal variation in DMS concentration Figure 2b also illustrates the large spatial variations that can be encountered in relatively small sampling areas. As a case study, Figure 3a shows the location of a cruise in April/May 1987. Figure 3b describes the distribution of DMS in surface water on this cruise, when the mean concentration was 9.2 nmol DMS (S) l^{-1}, (n = 162). There were large transverse concentration gradients ranging from 0.8 to nearly 90 nmol DMS (S) l^{-1}. In an area between the Shetland Islands and the Norwegian coast DMS concentrations ranged from 15.6 to 89.8 nmol DMS (S) l^{-1}, with two discrete areas having concentrations greater than 32.8 nmol DMS (S) l^{-1}. The highest DMS levels were of a similar magnitude to those reported for the approaches to the Rio de la Plata estuary (5) and in a coastal salt pond (6). The distribution of DMSP in European Shelf surface waters is similar to that of DMS, in that large ranges of concentrations are found over relatively small distances. Total DMSP concentrations are on average 15.6 times higher than those of DMS (means from 5 cruises, SD = 3.3), but the ratio is very variable for individual samples. The ratio of $DMSP_p$ to $DMSP_d$ is also very variable, but mean values for the different cruises fall in the range of 1 to 2.6.

Figures 3c and d show the distributions of total DMSP and chlorophyll *a* for the April/May cruise. Comparing b, c and d in Figure 3, it can be seen that high DMS was not always associated with high DMSP values and chlorophyll appears to be better correlated with the $DMSP_t$ than the DMS.

Phytoplankton Species and Chlorophyll *a*

Although DMS is ubiquitous in surface seawater and is associated with marine phytoplankton, field studies in coastal and shelf areas have generally shown that DMS is poorly correlated with chlorophyll (2,5,7,8). This is due to the fact that DMS production appears to vary with algal species, some groups being more prolific producers than others (2, and Keller, M. D., Bigelow Laboratory for Ocean Sciences, personal communication). In laboratory cultures, DMS production also varies with the stage of algal growth (2) and this may also be a factor in natural populations.

Figure 4a shows the relationship between DMS and chlorophyll *a* for a suite of samples taken on a circumnavigation of Britain in 1985. The samples were taken from areas with very different hydrographic characteristics. Again, there appears to be little correlation between DMS and chlorophyll. However, if the phytoplankton are identified and enumerated, samples containing a dominant species or group can be abstracted from the total data set. When DMS is plotted against chlorophyll for these sub-sets, some improved correlations appear. For coccolithophores, the dinoflagellate, *Gyrodinium aureolum* and dinoflagellates excluding *G. aureolum* (Figure 4; c, e and f) there is reasonable correlation and the difference in slopes indicates how DMS production per unit chlorophyll varies for different algal species. The plots for flagellates and diatoms (b and f), however, do not show clear correlations.

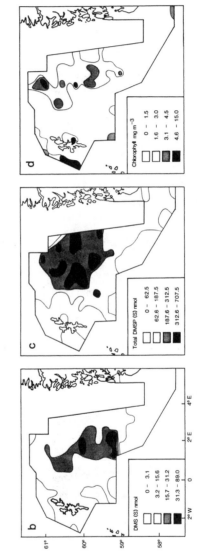

Figure 3. Surface water concentrations of DMS, total DMSP and Chlorophyll *a* in April/May 1987 (n = 162). (a) map to show the area of the North Sea studied. (b) distribution of DMS, (c) distribution of DMSP and (d) distribution of Chlorophyll *a*.

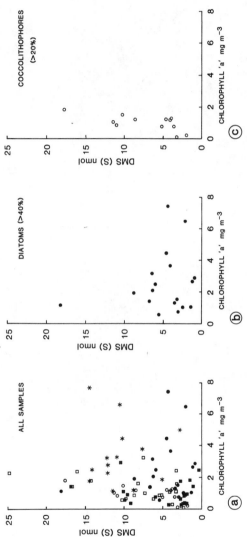

Figure 4. Relationship between DMS and chlorophyll *a*, July/August 1985. After identification and enumeration of phytoplankton, cell numbers were converted to carbon and each particular group or species was expressed as a percentage of total phytoplankton biomass. DMS and chlorophyll data values for samples containing an identifiable dominant group or species were plotted. Where groups were represented by a small number of samples, data sets were expanded by including samples which contained <50% of the total biomass. Thus the figures given in parentheses are the minimum % biomass represented by each group. a) all samples, b) Diatoms; DMS = -2.06 Chl. + 10.31 (r = 0.19, n = 18), c) Coccolithophores; DMS = 9.50 Chl. + 2.68 (r = 0.67, n = 12).

continued on next page.

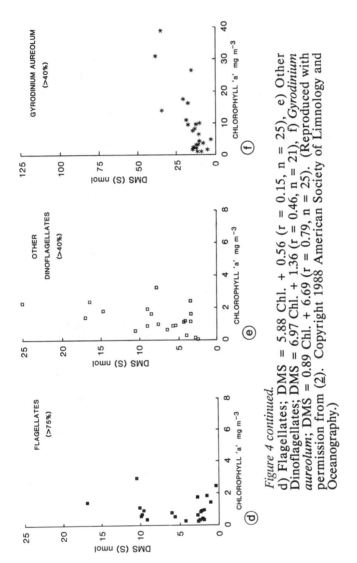

Figure 4 continued.
d) Flagellates; DMS = 5.88 Chl. + 0.56 (r = 0.15, n = 25), e) Other Dinoflagellates; DMS = 6.97 Chl. + 1.36 (r = 0.46, n = 21), f) Gyrodinium aureolum; DMS = 0.89 Chl. + 6.69 (r = 0.79, n = 25). (Reproduced with permission from (2). Copyright 1988 American Society of Limnology and Oceanography.)

Several factors will tend to complicate this type of analysis of field data; phytoplankton other than the dominant named group or species may make a significant contribution to the DMS and chlorophyll concentrations. Also, within a group, individual species may have widely differing abilities to produce DMS (e.g. within the group 'other dinoflagellates', 3 samples stand apart from the rest and these were identified as containing significant numbers of *Scrippsiella trochoidea*). Also, as the mechanisms for DMS production are not yet well understood, external factors (salinity, temperature, nutrients etc.) may influence the amount of DMS produced by phytoplankton, even at the species level (see later discussion on DMS and DMSP relationships). This factor is probably not very significant for *G. aureolum* (Figure 4e) as the samples were restricted to a 'bloom' area where external factors will vary less than where data were accumulated from a number of very different areas e.g. the diatoms (Figure 4f).

As DMS is ultimately derived from intracellular DMSP, it is necessary to quantify this source strength to characterise the potential for DMS production, which may be modified by alternative sinks for DMSP. Figure 5 shows the relationship of DMSP to cell carbon (mmol DMSP (S) /mmol C, l^{-1} seawater) for blooms of two common algal species, *P. pouchetii* and *G. aureoleum*. From this data it appears that *P. pouchetii* produces about seven times more DMSP than *G. aureolum*. This might be taken to suggest that *P. pouchetii* is seven times more prolific a DMS-producer than *G. aureolum*. However, the ratio of DMS concentration (per unit carbon) for the two species is only three, which implies that there must be quite different rates of DMS production and/or DMSP comsumption for the two cases.

Relationship of DMS and DMSP

Laboratory culture experiments have shown that during the exponential growth phase of algal development, the rates of DMS and $DMSP_d$ production are relatively low and fairly constant (2). It is only at some point in the stationary phase (when the number of cells destroyed/dying is equal to the number of cells produced) that concentrations of dissolved sulphur species start increasing rapidly (2). In the case of laboratory cultures, therefore, it could be postulated that during the stationary phase, it is the leakiness of the external membranes of aging or dying cells that is responsible for the surge in DMS and $DMSP_d$ in the medium. As DMS can be produced from DMSP by enzymic action, a proportion of the DMS may be released directly by the cells and some from the $DMSP_d$ in the medium. In natural populations in the sea, however, grazing is the main cause of phytoplankton mortality. Dacey and Wakeham (9) have shown that algal cultures produced more DMS when zooplankton were present. Also, when the zooplankton were isolated from the culture and introduced into water without algae, the DMS concentration of the water increased with time, suggesting DMS production from the guts and faeces of zooplankton. These observations indicate that the majority of DMS in seawater may not be a product of cell metabolism, but is produced by disruption of cells. If this is so then there should be very little correlation between DMS and $DMSP_p$; Figure 6 shows field data which tend to corroborate this. On the whole, DMS is slightly better correlated with total DMSP rather than with its dissolved or particulate fractions. The variation in DMS: DMSP ratios at the individual sample level may not only be attributable to different mechanisms and rates of DMSP decomposition, but also, to loss of DMS through bacterial utilisation, photochemical reactions and variations in the air-sea transfer rate with different

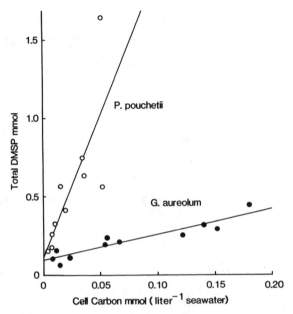

Figure 5. Relationship between total DMSP and cell carbon for *Phaeocystis pouchetii* and *Gyrodinium aureolum*, where the two species constituted >70% and >80% by carbon of the total phytoplankton biomass respectively.

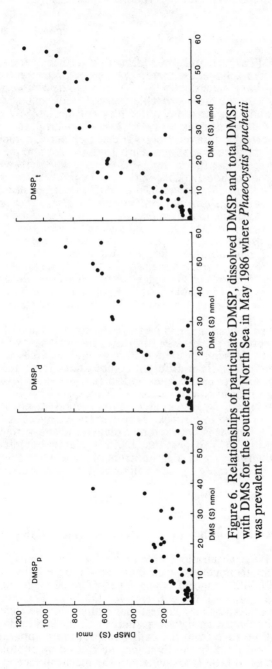

Figure 6. Relationships of particulate DMSP, dissolved DMSP and total DMSP with DMS for the southern North Sea in May 1986 where *Phaeocystis pouchetii* was prevalent.

wind regimes.   Ratios will also vary if the pool of DMSP has already been depleted and the rate of DMS production is more rapid than the rate of DMS loss.

Depth Profiles

Depth profiles for DMS are generally similar, with a surface or near-surface maximum in the first 50m, followed by a steep gradient to low concentrations of the order of 0.1 nmol l^{-1} at greater depths (5).   DMS is still detectable at depths of 1000's of meters, where the most probable source is sedimenting particles, such as faecal pellets and marine snow, which may contain DMSP. Figure 7 gives two examples of depth profiles for DMS, DMSP, chlorophyll a and cell numbers of the numerically dominant algal species. Figure 7a shows near surface maxima for all the parameters, with sharp gradients down to the base of the thermocline. In this case, the numbers of $G.$ $aureolum$ cells were well correlated with chlorophyll and DMSP$_p$, and were in good agreement with DMS and DMSP$_d$. In Figure 7b, however, the DMS subsurface maximum at 10m was situated above the chlorophyll maximum at 22m. The dominant species was, the dinoflagellate, $Ceratium$ $lineatum$, which was associated with significant concentrations of DMS in surface water samples at the same location. This profile suggests one of two things; that a numerically minor species overlying the $C.$ $lineatum$ maximum, was responsible for the production of DMSP$_d$ and DMS, or that zooplankton grazing and bacterial processes were more active in the top 15 metres of the water column. This case indicates that the relationships between DMS, DMSP and phytoplankton are unclear and emphasises the need for studies of processes in laboratory cultures to assist in the interpretation of field observations.

Various attempts have been made to measure the concentration of DMS in the microlayer, the water at the air/sea interface, the thickness of which is defined operationally by the sampler used as 10 to several hundreds of microns (10). Microlayer samples can have different compositions from under-lying water, at least with respect to particulates and surface-active materials (10). Sampling the microlayer for dissolved gases is difficult especially if the air and water are out of equilibrium. Nguyen et al. (11) have reported enrichments of DMS in the microlayer of up to 5 times. However, most investigators have found no significant differences between microlayer and sub-surface water (12,13). Turner and Liss (13) suggest that DMS is slightly depleted at the interface, except in the presence of high numbers of algae when sampling or storage artifacts may be the cause of the small enrichments measured in these cases.

Diel Cycling

We have adopted two approaches to investigate whether DMS production varies on a daily cycle:
    1) DMS and DMSP concentrations were determined for a 'parcel' of water in a bloom of Phaeocystis pouchetii, tracked using a drogue buoy. Figure 8 shows that DMS and DMSP$_d$ concentrations appeared to be higher during the dark than during the light, with the converse for DMSP$_p$. The decrease in internal DMSP during darkness may suggest that synthesis of the compound is suspended or minimised when no light is available and the loss through DMS production or DMSP leakage from the cells becomes more apparent. The decrease in DMSP$_d$ and DMS in the light may be caused by photochemical processes. However, as so little is known about the mechanisms and rates of loss and production of these compounds, any conclusions are highly speculative.

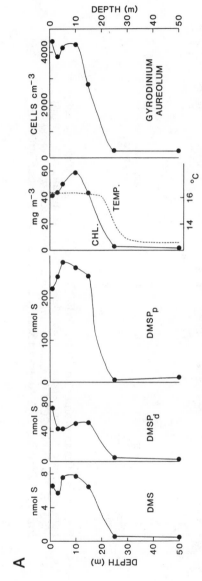

Figure 7A. Depth profiles for DMS, DMSP, chlorophyll, temperature and cell number for two stations where different algal species predominate, July/August 1985; A) 49C28'N, 03C55'W; *Gyrodinium aureolum*. (Reproduced with permission from (2). Copyright 1988 American Society of Limnology and Oceanography.)

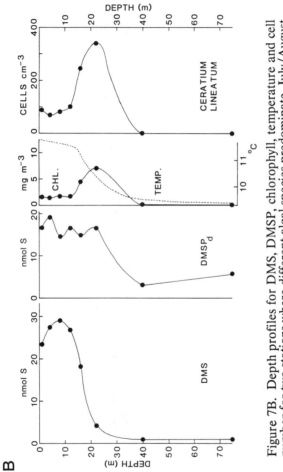

Figure 7B. Depth profiles for DMS, DMSP, chlorophyll, temperature and cell number for two stations where different algal species predominate, July/August 1985; B) 58C58'N, 02C00'W; *Ceratium lineatum*. (Reproduced with permission from (2). Copyright 1988 American Society of Limnology and Oceanography.)

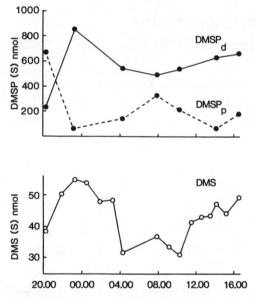

Figure 8. Diel variations of dissolved DMSP, particulate DMSP and DMS in a 'parcel' of water (3m depth) tracked using a drogue buoy off the north of Holland in May 1986 where *Phaeocystis pouchetii* was dominant (average chlorophyll *a* concentration was 30mg m^{-3}.

2) Any attempt to compare results for samples taken during light and dark on our cruises is thwarted by the wide ranges of concentrations we have encountered in surface waters. Therefore we have employed the tentative relationships between DMS and DMSP fractions, described above, to normalise and possibly enhance differences in the light and dark. ie. the ratio DMS + $DMSP_d$ to $DMSP_p$ should be greater in the dark. Mean values for this ratio were calculated for samples taken over periods of about four hours around mid-day and midnight respectively, for 5 cruises. The mean value for dark was 0.68 (n = 104, SD = 0.44) compared with the light value of 0.94 (n = 111, SD = 0.60). Therefore, if there is any diel cycling of these compounds, the variations in concentration must be quite small.

Flux of Sulphur (DMS) to the Atmosphere

In order to assess the atmospheric significance of the biogenic emission of sulphur, we have chosen to relate the natural flux to the anthropogenic emissions of SO_2 over Europe. During winter months the production of DMS in seawater is low and European SO_2 emissions are maximal, therefore, the flux of DMS will be of little significance (2). We will, therefore, describe and contrast emission rates for spring and summer.

The flux of DMS into the air is calculated using the following equation;

$$F = \Delta C \times k$$

where, ΔC = the concentration difference between air and water that drives the flux ($\Delta C = C_a H^{-1} - C_w$, where C_a and C_w are gas concentrations in air and water respectively and H is the Henry's Law constant, the ratio of $C_a:C_w$ at equilibrium),

and   k   = the transfer velocity which quantifies the rate of exchange.

The concentration term for DMS (S) is calculated using the mean of the average DMS concentrations found on cruises between April to September and is equal to 8.8 nmol DMS (S) l^{-1}. Air concentrations are of the order of 3 nmol DMS (S) m^{-3} (1) and are assumed to be negligible, hence the concentration difference is equal to the mean DMS concentration in water. The transfer velocity, k, was calculated from the equations of Liss and Merlivat (14) using the mean wind speeds measured on board ship, which procedure yields a value of 15.4 cm h^{-1}. Therefore DMS emission is equal to;

32 μmoles S m^{-2} d^{-1}

The emission of SO_2 from Europe in 1983 was about 638 Gmoles S yr^{-1} (15). The mean seasonal factor for spring and summer is 0.67 and the area of Europe taken as 9.1 x 10^{12}m^2, so the man-made emission rate is;

128 μmoles S m^{-2} d^{-1}

Thus, the biogenic emission is equivalent to 25% of the anthropogenic emission, expressed on a unit area basis. Owing to the decrease in SO_2 emissions in Europe over the last few years, the relative significance of marine DMS has probably increased.

Conclusions and Discussion

Our field studies in European shelf waters have shown that DMS production is seasonally and spatially very variable. During spring and summer the mean DMS concentration is about 3.5 times higher than the world ocean average. On a global scale, however, the area considered is small and will not significantly influence the calculation of total marine DMS fluxes. However, comparison of the natural and anthropogenic emissions for Europe shows that the marine source is highly significant on a local scale. The natural marine and anthropogenic land emissions calculated here are averages but, as DMS production occurs episodically in hotspots and certain areas of Europe produce little anthropogenic SO_2, biogenic fluxes may be of even greater significance in some areas. As yet there have been no studies that could positively identify acid precipitation events with episodes of high DMS production but modelling exercises have suggested that DMS may be responsible for 40% of the SO_4^{2-} deposited over Scandinavia during summer ([16]).

Several bloom-forming species of phytoplankton have now been identified as DMS producers and their varying abilities to produce DMS have been preliminarily characterised. However, there is still very limited understanding of the processes and factors affecting DMS production and considerably more field observations coupled with laboratory studies are required. There is evidence to show that, over the last two decades, there have been changes in the phytoplankton populations of some areas of the North Sea ([17]). These may be due to climatic changes, eutrophication, land engineering projects or a combination of all or some of these factors. Nevertheless, in some areas, there have been increases in algal biomass and duration of blooms and also a shift in favour of flagellates, rather than diatoms ([18]). This, together with the findings of our DMS study might suggest that there may have been an increase in the emissions of biogenic sulphur from the North Sea.

It has been suggested that DMS is a precursor of SO_4^{2-} condensation nuclei which has implications for climate control via a cloud albedo feedback mechanism ([19]). An increase in cloud condensation nuclei (CCN) would lead to increased cloud cover and cooling, thus counteracting greenhouse warming effects caused by increased CO_2 concentrations. Although the role of DMS in cloud seeding may be of significance over the open oceans, for most of the European shelf system, terrestrial and anthropogenic sources of CCN will probably be dominant. This may, however, not be so for the North East Atlantic.

Since Lovelock et al.'s original discovery of the ubiquity of DMS in the seawater ([20]), interest in this biogenic gas has greatly increased. It is now purported to play important roles in the global cycling of sulphur, the transfer of this element from the sea to the land and in controlling the acidity of rain and marine aerosols, as well as being involved in the climate regulation mechanism through cloud albedo, mentioned above. Future studies will have to consolidate our understanding of these various important roles of DMS, and also address whether analogous gases with halogen and nitrogen as the central atom may play equally significant roles in environmental chemistry.

Acknowledgments

This research was supported by grants from the U.K. Natural Government Research Council, the U.K. Department of the Government and the U.S. National Science Foundation (as part of the SEAREX Program).

Literature Cited

1. Andreae, M. O. In The Role of Air-Sea Exchange in Geochemica Cycling; Buat-Menard, P., Ed.; Reidel: Dordrecht, 1986; pp 331-62.
2. Turner, S. M.; Malin, G.; Liss, P. S.; Harbour, D. S.; Holligan, P. M. Limnol. Oceanogr. 1988, 33(3), 364-75.
3. Ishida, Y. Mem. Coll. Agric. Kyoto 1968, 94, 47-82.
4. Vairavamurthy, A.; Andreae, M. O.; Iverson, R. L. Limnol. Oceanogr. 1985, 31(1), 59-70.
5. Andreae, M. O.; Barnard, W. R. Mar. Chem. 1984, 14, 267-79.
6. Wakeham, S. G.; Howes, B. L.; Dacey, J. W. H. Geochim. Cosmochim. Acta 1987, 51, 1675-84.
7. Holligan, P. M.; Turner, S. M.; Liss, P. S. Continental Shelf Res. 1987, 8, 213-24.
8. Barnard. W. R.; Andreae, M. O.; Watkins, W. E.; Bingemer, H.; Georgii, H.-W. J. Geophys. Res. 1982, 87, 8787-93.
9. Dacey, J. W. H.; Wakeham, S. G. Science 1986, 233, 1314-6.
10. Liss, P. S. In Dynamic Processes in the Chemistry of the Upper Ocean; Burton, J. D.; Brewer, P. G.; Chesselet, R., Eds.; Plenum Press: New York and London, 1986; pp 41-51.
11. Nguyen, B. C.; Gaudry, A.; Bonsang, B.; Lambert, G. Nature 1978, 275, 637-9.
12. Andreae, M. O.; Barnard, W. R.; Ammons, J. M. In Environmental Biogeochemistry; Hallberg, R., Ed.; Ecol. Bull.: Stockholm, 1983; Vol. 35, pp 167-77.
13. Turner, S. M.; Liss, P. S. J. Atmos. Chem. 1985, 2, 223-32.
14. Liss, P. S.; Merlivat, L. In The Role of Sea-Air Exchange in Geochemical Cycling; Buat-Menard, P., Ed.; Reidel: Dordrecht, 1986; pp 113-27.
15. Saltbones, J.; Dovland, H. EMEP/CCC Report; N.I.L.U., 1987; 1986.
16. Fletcher, I. S. In Biogenic Sulfur in the Environment; Saltzman, E. S.; Cooper, W. J., Eds.; American Chemical Society: 1988, this volume.
17. Cadee, G. C. Netherlands J. Sea. Res. 1986, 20(2/3), 285-90.
18. Berg, J.; Radach, G. ICES, Sess.R, Biol. Oceanogr. Committee L:2,C.M. 1985.
19. Charlson, R. J.; Lovelock, J. E.; Andreae, M. O.; Warren, S. G. Nature 1987, 326, 655-61.
20. Lovelock, J. E.; Maggs, R. J.; Rasmussen, R. A. Nature 1972, 237, 452-3.

RECEIVED September 6, 1988

THE OCEANS:
BIOLOGICAL TRANSFORMATIONS

Chapter 13

Microbial Metabolism of Dimethyl Sulfide

Barrie F. Taylor[1] and Ronald P. Kiene[2]

[1]Division of Marine and Atmospheric Chemistry,
Rosenstiel School of Marine and Atmospheric Science,
University of Miami, Miami, FL 33149–1098
Marine Institute, University of Georgia, Sapelo Island, GA 31327

The biogenesis of dimethyl sulfide (DMS) provides a principal input of volatile sulfur to the atmosphere. This contribution has significant effects on the sulfur cycle and on global geochemistry. Natural emissions of biogenic sulfur are roughly equivalent to anthropogenic sources of gaseous sulfur oxides. With a biotic component of such magnitude, the processes involved in the biological turnover of DMS need to be well understood. In saline environments the biodegradation of dimethylsulfoniopropionate (DMSP) can generate DMS, whereas in terrestrial regions methionine and S-methylmethionine are probably the main precursors of DMS. Microorganisms produce DMS from DMSP and methionine under aerobic and anaerobic conditions. DMS is biochemically oxidized to dimethyl sulfoxide (DMSO) which may, in turn, be biologically reduced back to DMS. These interconversions occur in both oxic and anoxic habitats. Under anaerobic conditions where light is present, DMS may be oxidized to DMSO by phototrophic bacteria. DMSO is an electron acceptor which supports the anaerobic metabolism or growth of a variety of microorganisms, with DMS as the reduced endproduct. DMSO is also aerobically used as a carbon and energy source by some species of *Hyphomicrobium*. The biodegradation of DMS proceeds in oxic and anoxic environments. Aerobic metabolism supports the growth of obligate methylotrophs (*Hyphomicrobium* sp.) and obligate Calvin cycle autotrophs (*Thiobacillus* sp.). In anoxic sediments sulfate-reducers and methanogens consume DMS. At low levels of DMS sulfate-reducers outcompete methanogens for DMS, thus CO_2 and H_2S are the endproducts of oxidation in marine sediments. At higher substrate concentrations, or in freshwater sediments (sulfate absent), DMS supports methanogenesis.

The biotransformations of sulfur compounds by microorganisms can have large-scale impacts on global chemistry. As an example, sulfate-reducing bacteria have, throughout history, formed major deposits of elemental sulfur and iron sulfides on Earth, and these processes are continuing today (1). Contemporary sulfate-reduction coupled with the oxidation of reduced inorganic sulfur

0097–6156/89/0393–0202$06.00/0
© 1989 American Chemical Society

compounds by lithoautotrophic bacteria, has important consequences for the mineralization of organic matter in coastal regions and the development of food chains (2-4). Similarly, the microbial generation of volatile organo-sulfur compounds can have a great impact on the sulfur cycle and global geochemistry by transferring sulfur from the biosphere to the atmosphere (5). Volatile sulfur compounds generated by microbial processes are H_2S, alkylated sulfides and disulfides, COS, CS_2 and SO_2. It has been calculated that the global fluxes of biogenic sulfur and anthropogenic SO_2 into the atmosphere are about equal (6). Interest focusses on dimethylsulfide (DMS) because of its quantitative importance as a natural source of sulfur to the atmosphere. DMS is estimated to form approximately half of the biogenic input of volatile sulfur to the atmosphere, with about 75% of the DMS being generated in marine regions (6).

DMS is photochemically oxidized in the atmosphere to methanesulfonic and sulfuric acids. These strong acids contribute, along with nitric and organic acids, to the natural acidity of precipitation. Recent problems with acid rain have aroused interest in the anthropogenic and natural sources of volatile sulfur compounds (7).

In addition to affecting the pH of precipitation, the emission of DMS has been linked with the regulation of global climate. Recent studies (8,9) have indicated that the number of oceanic sulfate aerosols, which form cloud condensation nuclei, may be directly affected by DMS inputs to the atmosphere. Thus the rates of DMS release may influence cloud formation over the ocean, which in turn impacts on the global heat balance, thereby giving the biota a modicum of "control" over the climate (9). This intriguing hypothesis, together with concerns over acid rain implicitly point out the need for a greater understanding of mechanisms for DMS production and consumption, and how environmental factors affect these processes. This review concentrates on the microbiological processes involved in the biogenesis of DMS and its subsequent fate through biotransformation and catabolism.

Biogenesis of Dimethylsulfide

DMS is formed during the biodegradation of organic sulfur compounds and by the biological methylation of sulfide and methanethiol. Precursors of DMS are methionine, S-methylmethionine, dimethylsulfoniopropionate (DMSP), dimethylsulfonioacetate (DMSA), dimethyl sulfoxide (DMSO), methionine sulfoxide and sulfone, trimethylsulfonium salts, S-methylcysteine, homocysteine, dimethyl disulfide (DMDS), 2-keto-4-methiolbutyrate, and 2-mercaptoacetate (thioglycollate).

Methionine and DMSP have been investigated most frequently as progenitors of volatile sulfur compounds in a variety of environments. Methionine is a ubiquitous component of proteins and DMSP is present in many algae and some plants which grow in saline environments (10,11). DMSP appears to serve as an osmolyte in many organisms (12), although for some marine algae and plants, DMSP is less important than other solutes as a short-term osmotic effector (13,14). DMSP probably has other metabolic functions, for example as a methylating agent (15,16) or as a sulfur storage molecule (14).

Methionine as a Precursor of DMS. Methionine was recognized as a precursor of methylated sulfur compounds about 50 years ago (see 17 for references). Subsequent studies have shown the production of volatile methylated sulfides from methionine in many environments. The emphasis has been on anaerobic transformation in studies with natural samples or microcosms. Francis et al. (18) observed the production of methanethiol, DMS and DMDS from methionine in soil incubations. The production of the volatile sulfur compounds required amendment with glucose. In work with rumen fluid, Salsbury and Merricks (19) detected methanethiol and DMS when methionine was added. Carbon tetrachloride, an inhibitor of C-1 metabolism mediated by B_{12} derivatives (20,21), blocked methanogenesis and also DMS production from methionine but had no effect on methanethiol formation. In decaying cyanobacterial mats in a hot spring the principal volatile sulfur product during organic matter degradation was DMDS, presumably resulting from methionine (22). Studies with lake sediments identified methanethiol and DMDS as the major products of methionine biodegradation (23). More recent investigations of anaerobic transformations with salt marsh sediments have confirmed the conversion of methionine into volatile sulfur compounds with a preponderance of methanethiol production over that of DMS (24,25). The unifying conclusion from these investigations of anaerobic methionine catabolism, based on work with radiolabelled substrates and inhibitors, is that demethiolation yields methanethiol which may then be converted to DMS and/or DMDS. Sulfate-reducers (but not methanogens) contribute, along with other bacteria, to the anaerobic demethiolation of methionine or its breakdown product 2-keto-4-methiolbutyrate (25). DMS probably arises via methylation of methanethiol by a thiol S-methyltransferase (see below). Alternative fates for methanethiol involve metabolism by sulfate-reducing and/or methanogenic bacteria to yield CO_2 and CH_4. In freshwater sediments CH_4 rather than CO_2 is produced from methanethiol, even at low concentrations of methanethiol, whereas in marine environments the situation may be reversed (24,26). This observation is in accord with the dominance of sulfate-reduction in anaerobic degradation in coastal sediments (2,4). Methanethiol is a substrate which behaves similar to methanol in marine sediments being monopolized by sulfate-reducers at low concentrations but by methanogens at higher levels (27,28). Kiene and Capone (24) suggested that methanethiol is not directly attacked by methanogens but rather that it must first be converted to DMS via methylation.

With the exception of demethiolation, little is known of the enzymatic mechanisms for methionine degradation, particularly with respect to DMS production. A great variety of microorganisms produce methanethiol from methionine under either oxic or anoxic conditions. Bacteria capable of the transformation include species of *Pseudomonas* and *Clostridium*, strict aerobes and anaerobes respectively, with *Proteus vulgaris*, *Enterobacter aerogenes* and *Escherichia coli* as examples of facultative organisms; fungi are represented by species of *Penicillium*, *Candida*, *Aspergillus* and *Scopulariopsis*. Early data on the topic was summarized by Kadota and Ishida (29) and Figure 1 outlines some of the microbial transformations of organosulfur compounds. Several investigators using cell-free systems showed that methionine was converted to 2-ketobutyrate, methanethiol and ammonia (29), and there is general agreement on these as end-products of methionine catabolism, together with DMDS, by intact cells of many microorganisms. DMDS arises from methanethiol by an oxidation that probably occurs both chemically and biochemically. Segal and Starkey (17) observed variation in the relative rates of methanethiol and DMDS production during the growth of different bacteria. Presumably some species enzymatically catalyzed the conversion of methanethiol to DMDS. There are, also, some reports of DMS production from methionine, for example by the bread mold *Scopulariopsis brevicaulis* (30); a possible explanation is the methylation of methanethiol catalyzed by a thiol S-methyltransferase (31,32). Kiene and Capone (24) more recently observed the anoxic conversion of methanethiol to DMS in slurries of salt marsh sediments and concluded, from inhibitor effects, that sulfate-reducers, methanogens and other anaerobes all participated in the methylation process. Starkey and his colleagues provided good evidence for most of the steps shown in Figure 1, in both bacteria and

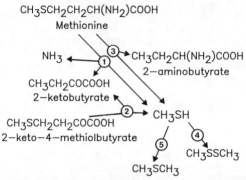

Figure 1. Microbial transformations of methionine which generate methylated sulfides. 1) methionine γ-lyase; 2) and 3) possible demethiolations; 4) chemical and probably biochemical oxidation; 5) thiol S-methyltransferase.

fungi (17,33). Only a few aerobic bacteria isolated from soil used methionine as a sole source of carbon, nitrogen and sulfur. None of the fungi grew on methionine and this was attributed to the inability of the isolates to grow on 2-ketobutyrate (33). Deamination preceded demethiolation and in the fungi the second step required energy and oxygen. Aeration also promoted demethiolation by a bacterial isolate. However, anaerobic demethiolation by cell-free extracts of anaerobes and facultative anaerobes was reported by other workers (29). Except for methionine γ-lyase (see below) which catalyzes a simultaneous deamination and demethiolation of methionine, few of the enzymatic conversions shown in Figure 1 have been characterized in microorganisms. Deamination of methionine involves either amino acid oxidases or aminotransferases, but the enzymatic demethiolation of 2-keto-4-methiolbutyrate has not been studied (34). Challenger and Liu (35) detected methanethiol and DMS production from 2-keto-4-methiolbutyrate by S. brevicaulis; direct demethiolation of 2-keto-4-methiolbutyrate may have occurred, with subsequent methylation of the methanethiol, but amination to methionine may have preceded the demethiolation step.

Methionine γ-lyase, in contrast, has been purified from Aeromonas sp., Pseudomonas putida and Clostridium sporogenes (36). This enzyme effects a α,γ-elimination of methanethiol and ammonia from L-methionine with the direct formation of 2-ketobutyrate:

$$CH_3SCH_2CH_2CH(NH_2)COOH + H_2O = \\ CH_3SH + CH_3CH_2COCOOH + NH_3$$

The Aeromonas and Pseudomonas enzymes have been studied in detail. Both enzymes are very sensitive to sulfhydryl and carbonyl inhibitors and possess 4 moles of pyridoxal 5'-phosphate bound in aldimine linkages to ξ-amino groups of lysine residues. They have similar substrate specificities and, in particular, show about two-fold higher rates with homocysteine as compared to L-methionine. Also attacked, but at slower rates, were ethionine (55% of methionine rate), S-methylcysteine (9%), S-methylmethionine (4%) and the sulfoxide (21%) and sulfone (55%) of methionine. The substrate specificity of methionine γ-lyase presents possibilities for the the environmental generation of DMS and ethyl sulfides from plant residues containing S-methylmethionine or S-methylcysteine, and their ethyl analogs. Methionine sulfone is even a good substrate for methionine γ-lyase. Enzymes similar to the S-alkylcysteinease from P. cruciviae (see below) probably account for the environmental occurrence of a variety of alkylated sulfides and perhaps the formation of methylated sulfides from methionine sulfoxide (5).

Production of DMS from S-Methylmethionine. The methylated derivative of methionine, S-methylmethionine is a sulfonium compound which is present in a variety of terrestrial plants, and is a precursor of DMS (37-39). White (40) also suggested that S-methylmethionine might occur in some marine algae. S-methylmethionine is synthesized in the Jack bean via methyl transfer from S-adenosylmethionine to methionine (41). Mazelis et al., (42) isolated a Corynebacterium sp. from soil that enzymatically degraded S-methylmethionine to DMS and homoserine. The enzyme responsible for the transformation was partially purified and was specific for the L-form of S-methylmethionine. DMSP, DMSA and methionine sulfoxide were not substrates for the sulfonium lyase but S-adenosylmethionine was converted to methylthioadenosine and homoserine. The enzyme had a pH optimum of 7.8, did not require pyridoxal 5'-phosphate, and was not inhibited by sulfhydryl-blocking agents. It was also very thermostable, retaining 40% activity even after 5 minutes boiling. The environmental distributions of S-methylmethionine and the sulfonium lyase are unknown but S-methylmethionine may be a significant precursor of DMS in terrestrial regions. For example, the high DMS emissions associated with corn plants (43) is probably due the the enzymatic breakdown of S-methylmethionine (37).

Methionine and its S-methyl derivative, S-methylmethionine, are probably important precursors of DMS in terrestrial regions. However, plants also produce a variety of nonprotein sulfur amino acids such as S-methylcysteine, its γ-glutamyl and sulfoxide derivatives, and djenkolic acid (44). The biosyntheses and functions of these compounds are poorly understood, and mechanisms for their biodegradation may be of more than academic interest, particularly with respect to the generation of volatile sulfur compounds.

DMSP as a Precursor of DMS. DMSP is a significant precursor of DMS in marine environments with the DMSP being released from phytoplankton by a variety of mechanisms (45,46). An elimination reaction yields DMS and acrylate from DMSP:

$$(CH_3)_2S^+CH_2CH_2COO^- = (CH_3)_2S + CH_2=CHCOO^- + H^+$$

The reaction proceeds abiotically at pH 13 or higher but at common environmental or physiological pH values the chemical conversion is extremely slow. For example, in seawater at pH 8.2 and 10°C the half-life for DMSP is about 8 years (47). Enzymes which promote the transformation occur in aerobic and anaerobic bacteria (47,48), in the red macroalga *Polysiphonia lanosa* (15), and in the marine planktonic dinoflagellate *Gyrodinium cohnii* (29). Wagner and Stadtman (48) isolated a strictly anaerobic spore-forming bacterium, a *Clostridium* species, from river mud using DMSP as the main source of carbon. DMSP was converted to DMS, propionate and acetate:

$$3(CH_3)_2S^+CH_2CH_2COO^- + 2H_2O =$$

$$2CH_3CH_2COO^- + CH_3COO^- + 3(CH_3)_2S + CO_2 + 3H^+$$

The organism also fermented acrylate, as do other clostridia, to a 2:1 molar mixture of propionate and acetate. Bacteria that ferment acrylate, derived from DMSP, probably abound in coastal marine sediments (Kiene and Taylor, this volume). Kiene and Taylor (this volume) observed that acrylate was immediately metabolized in anoxic slurries of coastal marine sediments with the appearance of a 1:2 molar mixture of acetate and propionate. The acetate and propionate were consumed by sulfate-reducing bacteria. Acetate is a quantitatively important substrate in anoxic marine sediments (49) and DMSP may be a significant precursor of acetate in such environments.

The data of Wagner and Stadtman (48) indicated an initial enzymatic cleavage of DMSP into DMS and acrylate, even though the transformation was not directly demonstrated. Likewise, Dacey and Blough (47) did not confirm the enzymatic conversion but showed growth on DMSP with the production of DMS and acrylate. However, the enzymatic cleavage of DMSP was studied with a partially purified preparation from *Polysiphonia lanosa* (15) and with cell-free extracts of *Gyrodinium cohnii* (29). Both enzymes required sulfhydryl groups for activity; the *P. lanosa* enzyme was unstable without reduced glutathione or cysteine and the activity in *G. cohnii* was inhibited by p-chloromercuribenzoate or iodoacetamide. pH optima were on the acidic side; pH 5.1 for *P. lanosa* and 6.0-6.5 for *G. cohnii*. The enzymes may occur extracytoplasmically; the *P. lanosa* enzyme was membrane bound and that from

G. cohnii was optimally active in 0.4M NaCl, an environment similar to seawater. An extracytoplasmic location of the enzyme would allow conservation of energy as a proton gradient during the cleavage of DMSP. With the enzyme from *P. lanosa*, Cantoni and Anderson (15) also reported the slow cleavage of dimethylacetothetin (dimethylsulfonioacetate or DMSA) but no activity with S-methylmethionine, S-adenosylmethionine, betaine (trimethylglycine) or choline. This indicates that DMSP lyase is distinct from the sulfonium lyase for S-methylmethionine described by Mazelis et al. (42). Interestingly, cell-free extracts of G. *cohnii* catalyzed the methylation of homocysteine with DMSP to produce methionine and 3-methiolpropionate, indicating a metabolic as well as osmotic function for DMSP. Recent work by Dacey, Wakeham and King (50) suggests a DMSP cleavage enzyme (lyase) might occur in higher plants. The marsh grass *Spartina alterniflora* had high levels of DMSP and the leaves were a source of DMS whereas the sediments were a sink. Dead and moribund leaves contained acrylate, presumably left over from the enzymatic attack of DMSP. However, DMSP degradation by phyllosphere microbes was not excluded and must be considered a distinct possibility.

Franzman et al., (51) and Gibson et al. (52) recently detected high levels of DMS in the hypersaline Organic Lake located in the Vestfold Hills of Antarctica. The highest concentrations of DMS, about 1.6 μM, occurred in the anoxic bottom waters with a salinity of approximately 174 o/oo. Gibson et al. (52) speculated that DMS catabolism was inhibited at high salinity whereas its biogenesis proceeded. Sulfate reducing and photosynthetic bacteria, potential consumers of DMS, were absent from Organic Lake. High levels of DMS in the anoxic monimolimnion were not correlated with algal biomass but were correlated with bacterial numbers. Gibson et al. (52) suggested that DMS was generated by bacteria. Interestingly, DMSP concentrations were high in the anoxic waters, perhaps indicating its formation by bacteria. Bacteria have not been shown to contain DMSP although *E. coli* will accumulate exogenous DMSP or DMSA when subjected to osmotic stress (53). The presence and persistence of DMSP at depth in sediments from unvegetated areas of coastal marshes (54) suggests that organisms other than algae may naturally synthesize this compound.

Phosphatidylsulfocholine the sulfonium analog of phosphatidylcholine is present in some marine microalgae (55-57) and sulfocholine, $(CH_3)_2S^+CH_2CH_2OH$, liberated from the phospholipids may be a precursor of DMS. Sulfocholine may be oxidized to DMSA which is a substrate for the DMSP cleavage enzyme from *P. lanosa*. Salsbury and Merricks (19) showed DMS formation from DMSA in rumen fluid. Enzymes more specific for DMSA cleavage may occur, especially in bacteria associated with algae or plants, or in the rumen. The relationship between the recently discovered demethylation pathway, which yields 3-mercaptopropionate (58,59), and the cleavage pathway for DMSP breakdown needs investigation in order to understand the factors affecting or controlling the relative contributions of the different routes. A similar demethylation pathway may exist for sulfocholine and/or DMSA which leads to the formation of 2-mercaptoethanol and/or 2-mercaptoacetate both of which have been detected in the porewaters of anoxic coastal sediments (60). DMSP and DMSA may even be a source of methyl groups for methanethiol and DMS production via thiol S-methyltransferases.

<u>Methylation of Sulfide and Methanethiol</u>. Another pathway for DMS biogenesis involves its formation from inorganic sulfide via methylation reactions. Interest in the enzymatic methylation of sulfide and organic sulfides was initially concerned with the detoxification of thiols in animals (61). The

enzyme thiol S-methyltransferase was first purified from rat liver and it catalyzed the following reactions:

$$H_2S + S\text{-adenosylmethionine} = CH_3SH + S\text{-adenosylhomocysteine}$$

$$CH_3SH + S\text{-adenosylmethionine} = (CH_3)_2S + S\text{-adenosylhomocysteine}$$

Also active as substrates for S-methylation were 2-mercaptoacetate (thioglycollate), 2-mercaptopropionate (2-MPA), 2-mercaptoethanol, and the methyl ester of 3-mercaptopropionate (3-MPA). The Vmax values were similar for all the substrates but Km values varied from 64 μM and 240 μM for H_2S and CH_3SH, respectively, to 8 mM for 2-mercaptoethanol. The enzymatic activity was generally detected in animal tissues that are exposed to xenobiotics (gut, liver, kidney, lungs) but was present at low levels in heart and skeletal muscle. Thiol S-methyltransferase activities have recently been detected in aerobic and facultatively aerobic bacteria, plants, a yeast (*Candida lipolytica*), an alga (*Euglena gracilis*) and the ciliate protozoan Tetrahymena thermophila (31,32,62,63). Cell-free extracts of *Tetrahymena thermophila* showed similar rates of methylation with sulfide and CH_3SH; also the methylation of CH_3SH was demonstrated with intact cells (32). In all the studies, animal and microbial, a single enzyme in each organism appeared to be responsible for the methylation of the different substrates, i.e. sulfide, methanethiol and organic xenobiotic thiols. The activity was also present in bacteria grown in the absence of sulfide, and some isolates even generated sulfide from sulfate for the methylation reaction (31). With bacteria the methylation of CH_3SH by intact cells was not demonstrated and there have been questions about thiol methylating activity in strict anaerobes (31,64). However, strictly anaerobic bacteria undoubtedly contribute to the methylation of sulfide and organic thiols, such as the methylthio-derivatives of polychlorinated biphenyls that have been detected in sediments (65). Also, as noted above, Kiene and Visscher (25) and Kiene and Capone (24), observed the formation of DMS from methanethiol in slurries of salt marsh sediments. Methylation of inorganic sulfide as source of methanethiol and DMS needs investigation in various environments, as does the ability of strict anaerobes to effect these methylations.

Reduction of DMSO. DMSO is a waste product from paper mills (66) and is used in medicine and pharmacology as a vehicle to promote the passage of drugs across cell membranes (67). There is therefore, an anthropogenic component of DMSO occurrence in the environment, but probably not of global significance. DMSO occurs in natural aquatic environments especially surface seawater. It is generated by the oxidation of DMS (see below) derived from the enzymatic cleavage of DMSP. Levels of DMSO were reported to be equal to or greater than those of DMS in oceanic waters (68) but this conclusion needs re-examination since the analytical method did not distinguish between DMSO and DMSP. DMSO is reduced chemically to DMS by sulfide (69) and this could occur in natural environments, especially marine sediments. More likely, however, DMSO will be biologically reduced. Many microorganisms, both bacteria and yeasts, reduce DMSO to DMS under anaerobic conditions (69), and in some organisms, but not all, it allows anaerobic growth (70). The enzymes responsible for DMSO reduction have been studied from fungi and bacteria. Normally DMSO reductases are only synthesized or expressed in the absence of oxygen. An exception to this generalization concerns *Hyphomicrobium* EG and *Hyphomicrobium* S which grew aerobically on DMSO and carried out an aerobic reduction of DMSO to DMS catalysed by a DMSO reductase. This enzyme has not been studied in detail but DMSO reductases which function in anaerobes have received considerable attention. Interest in DMSO and trimethylamine N-oxide is keen because the products of their reduction, DMS and trimethylamine, are of concern to the food industry. DMS at the correct concentrations imparts flavor to cheeses and beers (71) and trimethylamine is generated during the bacterial spoilage of fish. Trimethylamine N-oxide (TMAO) is present in the tissues and fluids of some marine fish where it fulfills osmotic functions (72). DMSO is reduced by facultative and strict anaerobes, and also photosynthetic bacteria but not by sulfate-reducing bacteria (24,69,70,73,74). Kiene and Capone (24) observed that molybdate (an inhibitor of sulfate-reduction) slightly stimulated DMS production from DMSO, by slurries of salt marsh sediments, probably because use of DMS by sulfate-reducers was inhibited, whereas DMS production from DMSO was unaffected.

The situation with respect to the different enzymes that reduce organic N- and S-oxides is complex. In general, the enzymes have broad substrate specificities and those which reduce TMAO also reduce DMSO, and *vice versa*; both types usually reduce methionine sulfoxide. For example, a DMSO reductase from *E. coli* reduced DMSO at about 30% of the rate of TMAO; methionine and tetrahydrothiophene sulfoxide were reduced at rates similar to DMSO, and the enzyme catalyzed the reduction of chlorate and a various organic N-oxides at higher rates than DMSO (75). Single enzymes that reduce both DMSO and TMAO have been confirmed in *Proteus vulgaris* and *Rhodobacter capsulatus* (76,77). One exception, is the marine bacterium *Shewanella* sp. NCMB 400, with an inducible TMAO reductase that does not reduce DMSO, chlorate or pyridine N-oxide (78). Since TMAO and DMSO probably occur together in many marine environments an interesting recent technical development involves NMR to follow the simultaneous use of TMAO and DMSO by *Rhodobacter capsulatus* (79). DMSO and TMAO reductases are usually periplasmic and contain molybdenum (75,78,80). Molybdopterin the common molybdenum cofacter for all molybdoenzymes with the exception of nitrogenase (81), was recently identified in the DMSO reductase of *E. coli* (75).

Other Precursors of DMS. Several other organosulfur compounds, in addition to methionine, S-methylmethionine, DMSP and DMSO, have been reported to be involved in the biogenesis of methylated sulfides. Early reports of these investigations were summarized by Kadota and Ishida (29). Challenger's group reported DMS production by fungi from S-methylcysteine (30,35). Bremner and Steele (5) observed the stimulation of methanethiol, DMS and DMDS production in soil, with either aerobic or anaerobic (waterlogged) incubation, by S-methylcysteine but not by S-methylmethionine. DMS was also formed from homocysteine, probably after methylation to methionine but possibly via sulfide formation with subsequent thiol S-methyltransferase activity. More recently, Kiene and Capone (24) noted equivalent stimulatory effects on the anaerobic production of volatile organosulfur compounds by methionine and S-methylcysteine in slurries of salt marsh sediment; methanethiol and DMS were detected with the earlier and greater appearance of methanethiol. Biochemical studies of S-methylcysteine catabolism were carried out by Nomura et al. (82). *Pseudomonas cruciviae* was isolated from soil using S-methylcysteine as the sole source of carbon and energy. An enzyme was purified 30-fold from cells grown on S-methylcysteine which liberated methanethiol from S-methylcysteine; it also attacked S-methylcysteinesulfoxide with the formation of methyl methanethiosulfinate:

$$CH_3SCH_2CH(NH_2)COOH + H_2O = CH_3SH + CH_3COCOOH + NH_3$$

$$2\,CH_3S(O)CH_2CH(NH_2)COOH + H_2O = CH_3S(O)SCH_3 +$$

$$2\,CH_3COCOOH + 2\,NH_3$$

The enzyme resembled methionine γ-lyase in requiring pyridoxal phosphate for activity but it did not attack methionine. The enzyme was described as an S-alkylcysteinase because it used a range of S-alkylcysteines and S-alkylcysteinesulfoxides as substrates. Similar rates of attack were obtained with methylated and ethylated substrates whereas with propyl, butyl, isobutyl, allyl, amyl and isoamyl derivatives of cysteine the rates were between 40% and 75% lower.

Wagner et al. (83) isolated *Pseudomonas* MS which grew on trimethylsulphonium salts as a sole source of carbon and energy, with the evolution of DMS. A partially purified enzyme preparation catalyzed the transfer of a methyl group from trimethylsulfonium chloride to tetrahydrofolate. Neither S-adenosylmethionine nor DMSP functioned in the methyl transfer reaction. The methyltransferase was devoid of vitamin B_{12}, had a molecular weight of about 100,000 with a pH optimum of 7.8.

Challenger and Liu (35) found that 3-methiolpropionate or 3-mercaptopyruvate caused DMS formation by *Scopulariopsis brevicaulis*. Labarre and Bory (84) observed the conversion of 2-mercaptoacetate to a mixture of H_2S and methanethiol by *Clostridium oedematicus* but mechanisms were not investigated; probably a thiol S-methyltransferase catalyzes methanethiol synthesis from sulfide or via an S-methyl derivative of 2-mercaptoacetate.

Catabolism of Dimethylsulfide

DMS is degraded aerobically and anaerobically by microorganisms and its catabolism by pure and mixed cultures, or consortia, has recently been studied.

Anaerobic Catabolism.   Anaerobic DMS degradation has been examined with samples from freshwater, estuarine, alkaline/hypersaline and salt marsh sediments (25-27,54). Zinder and Brock (26) using sediments from Lake Mendota, observed the predominant conversion of DMS to CH_4 with little formation of CO_2. They were, however, unable to secure methanogenic enrichments using DMS and three pure cultures of methanogens (*Methanosarcina barkeri, Methanobacterium ruminantium, M. thermoautotrophicum*) did not metabolize either DMS or methanethiol. In salt marsh sediments, and a variety of other sediments, DMS at low concentrations was converted primarily to CO_2 by sulfate-reducing bacteria (25,27,54). Only at high substrate levels, similar to the situation with methionine (methanethiol) metabolism in salt marsh sediments, was methanogenesis stimulated by DMS. It should be noted that sulfate reducers which use DMS have yet to be isolated in pure culture.

DMS catabolism by pure cultures of microorganisms has been predominantly examined with aerobes; there are two reports of growth on DMS by anaerobes. One is the methanogenic isolate of Kiene et al. (27) and the other is the anoxygenic phototrophic bacterium of Zeyer et al. (85) (see below). Kiene et al. (27) isolated a pure culture of a methanogen, from an estuarine sediment, that grew on DMS. The organism was obligately methylotrophic and grew on DMS, methanol or trimethylamine but not on H_2 plus CO_2, formate, acetate, methionine or DMDS. No growth occurred on methanethiol, even though it was postulated as an intermediate in DMS metabolism, possibly because of toxicity problems. The biochemistry of DMS conversion to methane still needs investigation as it appears to be different than that of methanol or trimethylamine (86).

Aerobic Catabolism. Isolates which grow aerobically on DMS have proved to be either strictly methylotrophic species of *Hyphomicrobium* or strictly autotrophic species of *Thiobacillus* (87). There was a report of methanethiol oxidation catalyzed by methane monooxygenase (88) but the possible metabolism of methylated sulfides by aerobic methanotrophs has not been further investigated. *Hyphomicrobium* S was obtained from a batch enrichment with DMSO as sole carbon and energy source, and the only substrates it grew on were DMSO and DMS (89). *Hyphomicrobium* EG was later isolated from a chemostat enrichment on DMSO, and it also grew on DMS, methylamine, dimethylamine, and trimethylamine N-oxide (TMAO) and weakly on formate or trimethylamine (90). Unlike many hyphomicrobia neither organism grew anaerobically with nitrate, nitrite or fumarate as an electron acceptor (90). The catabolic pathway for DMSO and DMS metabolism is shown in Figure 2. A DMSO reductase was formed under aerobic conditions and functioned better with NADH than NADPH in *Hyphomicrobium* S whereas in *Hyphomicrobium* EG the preferred coenzyme was NADPH (89,90). The subsequent pathway was the same in both organisms with oxidation of DMS by a monooxygenase to give methanethiol and formaldehyde. A methanethiol oxidase generated sulfide and formaldehyde, with the sulfide being further oxidized to sulfate. NAD-dependent formaldehyde and formate dehydrogenases catalyzed the oxidation of formaldehyde to CO_2. Carbon assimilation was probably at the level of formaldehyde via the serine pathway (91) as indicated by the presence of high levels of hydroxypyruvate reductase and low levels of ribulose bisphosphate carboxylase (89,92). The methanethiol oxidase of *Hyphomicrobium* EG was purified and studied in detail (93). The enzyme catalyzed the following reaction:

$$CH_3SH + O_2 + H_2O = HCHO + H_2S + H_2O_2$$

with the direct involvement of molecular oxygen. The molecular weight of the enzyme was 40-50,000; there was no obvious coenzyme as judged by spectral characteristics, and no metal requirement. The enzyme is of particular interest because methanethiol and all the products (sulfide, HCHO, peroxide) are toxic molecules. The importance of the enzyme in the physiology of the organism is apparent from the fact that it constituted 8-10% of the total cellular protein. In addition to methanethiol the enzyme also catalyzed the oxidation of ethanethiol (Km 18 μM) and sulfide (Km 60 μM) but not methanol or methylamine.

Experiments with *Hyphomicrobium* EG growing in a methylamine-limited chemostat proved that the organism gained energy from the oxidation of sulfide or thiosulfate. However, both *Hyphomicrobium* S and *Hyphomicrobium* EG depended on a methylated substrate for cellular carbon, i.e. they are strict methylotrophs. In contrast to this situation the bacteria isolated aerobically on DMS, and DMDS, by Kelly and his colleagues have proved to be strict autotrophs, fixing CO_2 via the Calvin cycle and possessing high levels of ribulose bisphosphate carboxylase (87,94,95). *Thiobacillus thioparus* TK-m was derived from a strain of Thiobacillus thioparus that was isolated from activated sludge (94). The organism only obtained energy from the oxidation of the sulfur moiety of DMS which is probably equivalent to sulfide (Figure 2). The pathway for DMS catabolism resembles that employed by the hyphomicrobial isolates (87; Figure 2).

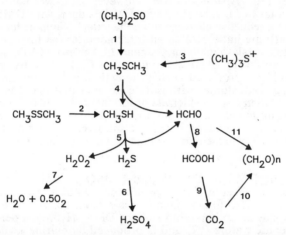

Figure 2. Aerobic catabolism of methylated sulfides (adapted from Kelly, 1988). 1) DMSO reductase (*Hyphomicrobium* sp.); 2) DMDS reductase (*Thiobacillus* sp.); 3) trimethylsulfonium-tetrahydrofolate methyltransferase (*Pseudomonas sp.*); 4) DMS monooxygenase; 5) methanethiol oxidase; 6) sulfide oxidizing enzymes; 7) catalase; 8) formaldehyde dehydrogenase; 9) formate dehydrogenase; 10) Calvin cycle for CO_2 assimilation (*Thiobacillus* sp.); 11) serine pathway for carbon assimilation (*Hyphomicrobium* sp.).

Smith and Kelly (95) isolated bacteria from a variety of habitats (soil, peat, freshwater and marine sediments) which grew on DMDS as a sole source of energy and carbon. All of the isolates were thiobacilli that grew on DMDS, DMS, and inorganic sulfur compounds, and were strictly autotrophic, showing no heterotrophic or methylotrophic growth. There was variation in the inorganic sulfur compounds oxidized; all used thiosulfate, tetrathionate and sulfide but not all grew on elemental sulfur or thiocyanate. One organism grew on thiourea but none used DMSO. Smith and Kelly (95) examined one isolate (E6) in detail. Cells grown on DMDS oxidized DMDS, sulfide, thiosulfate, formaldehyde and formate in accord with the operation of a similar pathway to that proposed for *Hyphomicrobium*. However, DMS was not oxidized thereby indicating a lack of involvement of a DMS oxidase, with DMDS metabolism probably via a reductive cleavage to methanethiol. Support for catabolism by a pathway with methanethiol as an intermediate was further provided by a sensitivity to a catalase inhibitor, 3-amino-1,2,4-triazole (AT), during growth on DMDS, as noted for *Hyphomicrobium* EG (92). Sensitivity to AT can be ascribed to H_2O_2 produced during the action of methanethiol oxidase (92). Chemostat growth studies with isolate E6 indicated that energy was derived from the oxidation of formaldehyde or formate but, unlike *Thiobacillus versutus* (96), the organism did not grow autotrophically on these organic compounds. A requirement for a reduced inorganic sulfur compound by the isolate may indicate an inability to assimilate sulfate as noted for some photosynthetic sulfur bacteria.

Oxidation of DMS to DMSO and DMSO₂. DMS is chemically and biochemically oxidized to dimethylsulfoxide (DMSO). Mechanisms for the *in situ* oxidation of DMS to DMSO in seawater have received little attention, even though this may be an important sink for DMS. Hydrogen peroxide occurs in surface oceanic waters (97) and is produced by marine algae (98). It may participate in a chemical oxidation of DMS, since peroxide oxidizes sulfides to sulfoxides (99). Photochemical oxidation of DMS to DMSO occurs in the atmosphere and DMSO is found in rain from marine regions (68). DMS is also photo-oxygenated in aqueous solution to DMSO if a photosensitizer is present; natural compounds in coastal seawater catalyzed photo-oxidation at rates which may be similar to those at which DMS escapes from seawater into the atmosphere (100).

In addition to chemical and photochemical processes, the formation of DMSO from DMS may be enzymatically catalyzed in aqueous environments. The anaerobic oxidation of DMS to DMSO by a photosynthetic purple sulfur bacterium, probably a species of *Thiocystis*, was recently reported (85). Nothing is known about the mechanism of the oxidation although it cannot involve a molecular oxygen. Biochemical mechanisms requiring molecular oxygen can generate sulfoxides and sulfones from organic sulfides (101). The transformations are performed by monooxygenases that catalyze the following general reaction:

$$RSR + NADPH + H^+ + O_2 = RS(O)R + NADP^+ + H_2O$$

Two different enzyme systems have been described, one uses cytochrome P-450 to activate oxygen whereas the other employs flavin adenine dinucleotide (FAD). Both the cytochrome P-450 and the flavin monooxygenase systems have broad substrate specificities and oxidize and oxygenate a variety of organic nitrogen or sulfur compounds. The enzymes have a widespread distribution and have been detected in animals, plants, fungi and bacteria. Their function appears to be primarily one of detoxification of xenobiotics. Microbial enzymes have received little attention with respect to their sulfoxidizing activities or potential. DMS is a substrate for the purified flavin monooxygenase from hog liver with a Km value of 8 μM (102). The enzyme also converts DMSO to dimethylsulfone but there is doubt about the role of the enzyme in sulfone formation *in vivo*, because of the high concentration of DMSO required (102). Sulfones are, however, produced during normal metabolism and, for example, urinary excretion in man amounts to 4-11 mg per day of dimethylsulfone (103). Worth noting with respect to dimethylsulfone production in natural environments is the oxidation of DMSO to dimethylsulfone by a chloroperoxidase enzyme from the fungus *Caldariomyces fumago* (104). This observation may have particular relevance to the oceans because of the common presence of haloperoxidases in marine microorganisms (105). The active species in the oxidation of organic nitrogen and sulfur compounds by flavin monooxygenase is probably a peroxyflavin, as is the case in the luciferin-luciferase system of luminescent marine bacteria (106). Luminescent bacteria abound in seawater 107) and they might conceivably contribute to a biochemical oxidation of DMS to DMSO.

Summary and Conclusions

Microorganisms can both generate and consume DMS. Enzymes that catalyze the cleavage of methylated sulfides from organic sulfur compounds are compiled in Table 1. Figure 3 summarizes current knowledge about the catabolism and biotransformations of DMS.

The production of DMS in the biosphere is balanced by losses to chemical and biological oxidation as well as by emissions to the atmosphere. The relative amount of DMS that escapes to the atmosphere, compared to that formed biologically, is unknown and will probably be the subject of considerable research in the years to come.

In order to assess the roles of microbes as global consumers of DMS more knowledge is needed about the biochemical pathways of catabolism under aerobic and anaerobic conditions. Also, the ecological factors which influence the distribution of DMS degraders and their activities. Parallel studies on DMS biogenesis are also needed. The comparative importance of various precursors must be unequivocally established in different environments (i.e. soils, forests, sediments, lakes, oceans). Since the bulk of DMS arises from plant-derived molecules, some interesting questions remain concerning why some species produce significant quantities of DMS precursors (e.g. DMSP, S-methylmethionine), whilst others of similar taxonomic background or ecological niche do not.

Implicit to interest in the transformations of sulfur compounds in Nature is the question of human activities promoting the production of sulfur species (or their precursors) which have undesirable environmental impacts. A well-established example, albeit on a relatively small scale, is the leaching of mine dumps causing local problems with acidity and metal toxicity. On larger scale, the burning of sulfur-containing fossil fuels, has affected the global sulfur cycle by injecting huge amounts of previously sequestered sulfur into the atmosphere. The effects of the combustion era are now being seen in the form of acid rain.

The global nitrogen cycle has been altered by the use of fertilizers and cultivation of nitrogen-fixing legumes. This has contributed to problems of eutrophication and groundwater contamination with nitrate. Increased nitrogen inputs with higher algal and plant productivity may also affect the sulfur cycle. Elevated inputs of organic matter into salt marshes, estuaries and nearshore regions will promote anoxia and stimulate sulfate reduction and the sulfur cycle in general; probably resulting in greater releases of volatile sulfur compounds.

Table 1. Enzymes that Generate Methylated Sulfides
from Organosulfur Compounds

Enzyme	Substrate	Products
DMSP lyase	DMSP	DMS
	DMSA	DMS
Methionine γ -lyase	Methionine	Methanethiol
	Homocysteine	H$_2$S
	S-methylcysteine	Methanethiol
	Methionine sulfoxide	
	Methionine sulfone	
	S-methylmethionine (weak)	DMS
S-alkylcysteinase	S-methylcysteine	Methanethiol
	S-methylcysteine sulfoxide	Methyl methane-thiosulfinate
Thiol S-methyltransferase	Sulfide	Methanethiol
	Methanethiol	DMS
	Thioglycollate	
Sulfonium lyase	S-methylmethionione	DMS
	S-adenosylmethionine	Methylthioadenosine
Trimethylsulfonium-tetrahydrofolate methyl-transferase	Trimethylsulfonium salts	DMS

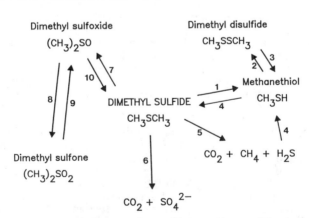

Figure 3. Summary of DMS transformations. 1) aerobic and anaerobic
bacteria; 2) chemical and probably biochemical oxidation; 3) aerobic and
anaerobic bacteria; 4) thiol S-methyltransferase; 5) sulfate reducers and
methanogens; 6) aerobic bacteria (hyphomicrobia and thiobacilli); 7) chemical
and biochemical (aerobic and anaerobic); 8) chloroperoxidase; 9) mechanism
unknown; 10) aerobic and anaerobic bacteria.

The microbial biogenesis and catabolism of DMS, as well as other biogenic sulfur compounds, and the environmental factors affecting these processes deserve attention in order to understand the relationships between acid rain, climate and the global cycle of sulfur.

Acknowledgments

Financial support from the National Science Foundation (Grant OCE-8516020) is gratefully acknowledged. Contribution No. 623 of the University of Georgia Marine Institute.

Literature Cited

1. Trudinger, P. A. Phil. Trans. R. Soc. Ser. 1982, B 298, 563-81.

2. Jorgensen, B. B. Nature 1982, 296, 643-5.

3. Jorgensen, B. B. Symp. Soc. Gen. Microbiol. 1988, 42, 31-63.

4. Howarth, R. W. Biogeochemistry 1984, 1, 5-27.

5. Bremner, J. M.; Steele, C. G. Adv. Microbial Ecol. 1978, 2, 155-201.

6. Andreae, M. O.; Raemdonck, H. Science 1983, 221, 744-7.

7. Nriagu, J. O.; Holdway, D.A.; Coker, R.D. Nature 1987, 237, 1189-92.

8. Bates, T. S.; Charlson, R. J.; Gammon, R. H. Nature 1987, 329, 319-21.

9. Charlson, R. J.; Lovelock, J. E.; Andreae, M. O.; Warren, S. G. Nature 1987, 326, 655-61.

10. Blunden, G.; Gordon, S. M. In Progress in Phycological Research; Round, F. E.; Chapman, D. J., Eds.; Biopress Ltd.: Bristol, England, 1986; Vol. 4, pp 39-80.

11. Keller, M. D.; Bellows, W. K.; Guillard, R. F. In Biogenic Sulfur in the Environment; Saltzman, E. S.; Cooper, W. J., Eds.; ACS Symposium Series, American Chemical Society, Washington DC, 1988.

12. Vairavamurthy, A.; Andreae, M. O.; Iverson, R. L. Limnol. Oceanogr. 1985, 30, 59-70.

13. Edwards, D. M.; Reed, R. H.; Stewart, W. D. P. Mar. Biol. 1988, 98, 467-76.

14. Van Diggelen, J.; Rozema, J.; Dickson, D. M. J.; Broekman, R. New Phytol. 1986, 103, 573-86.

15. Cantoni, G. L.; Anderson, D. G. J. Biol. Chem. 1956, 222, 171-7.

16. Durell, J.; Cantoni, D. G.; Anderson, G. L. Biochim. Biophys. Acta. 1957 26, 270-82.

17. Segal, W.; Starkey, R. L. J. Bacteriol. 1969, 98, 908-13.

18. Francis, A. J.; Duxbury, J. M.; Alexander, M. Soil Biol. Biochem. 1975, 7, 51-6.

19. Salsbury, R. L.; Merricks, D. L. Plant Soil 1975, 43, 191-209.
20. Bauchop, T. J. Bacteriol. 1967, 94, 171-5.
21. Wood, J. M.; Moura, I; Moura, J. G.; Santos, M. H.; Xavier, A. V.; LeGall, J.; Scandellari, M. Science 1982, 216, 303-5.
22. Zinder, S.H.; Doemel, W. N.; Brock, T. D. Appl. Environ. Microbiol. 1977, 34, 859-60.
23. Zinder, S. H.; Brock, T. D. Appl. Environ. Microbiol., 1978, 35, 344-52.
24. Kiene, R. P.; Capone, D. G. Microbial Ecol. 1988, 15, 275-91.
25. Kiene, R. P.; Visscher, P. T. Appl. Environ. Microbiol. 1987, 53, 2426-34.
26. Zinder, S. H.; Brock, T. D. Nature 1978, 273, 226-8.
27. Kiene, R. P.; Oremland,R. S.; Catena, A.; Miller, L. G.; Capone, D. G. Appl. Environ. Microbiol. 1986, 52, 1037-45.
28. King, G. M. Geomicrobiol. J. 1984, 3, 275-306.
29. Kadota, H.; Ishida, Y. Ann. Rev. Microbiol. 1972, 26, 127-38.
30. Challenger, F.; Charlton, P. T. J. Chem. Soc. 1947, 1591-7.
31. Drotar, A. M.; Burton, G. A., Jr.; Tavernier, J. E.; Fall, R. Appl. Environ. Microbiol. 1987, 53, 1626-31.
32. Drotar, A. M.; Fall, L. R.; Mishalanie, E. A.; Tavernier, J. E., Jr.; Fall, R. Appl. Environ. Microbiol. 1987, 53, 2111-8.
33. Ruiz-Herrera, J.; Starkey, R. L. 1969. J. Bacteriol. 1969, 99, 544-51.
34. Soda, K. In Sulfur and Sulfur Amino Acids, Methods in Enzymology; Jacoby, W.B.; Griffith, O. W., Eds.; vol 143. Academic Press: New York, 1987; Vol. 143, pp. 453-9.
35. Challenger, F.; Liu, Y. C. Rec. Trav. Chim. Pays Bas 1950, 69, 334-42.
36. Esaki, N.; Soda, K. In Methods in Enzymology; Jakoby, W. B.; Griffith, O. W., Eds.; Academic Press: New York, 1987; Vol. 143, pp. 459-65.
37. Bills, D. D.; Keenan, T. W. J. Agric. Food Chem. 1968, 16, 643-5.
38. Challenger, F. Aspects of the organic chemistry of sulfur; Butterworth, London, 1959.
39. Hattula, T.; Granroth, B. J. Sci. Food Agric. 1974, 25, 1517-21.
40. White, R. H. J. Mar. Res. 1982, 40, 529-36.
41. Greene, R. C.; Davis, N. B. Biochim. Biophys. Acta 1960, 43, 360-2.
42. Mazelis, M.; Levin, B.; Mallinson, N. Biochim. Biophys. Acta 1965, 105, 106-14.
43. Goldan, P. D.; Kuster, W. C.; Albritton, D. L.; Fehsenfeld, F. C. J. Atmos. Chem. 1987, 5, 439-67.

44. Giovanelli, J. In Methods in Enzymology; Jakoby, W. B.; Griffith, O. W., Eds.; Academic Press: New York, 1987; Vol. 143, pp. 419-26.

45. Dacey, J. W. H.; Wakeham, S. G. Science 1986, 233, 1314-6.

46. Nguyen, B. C.; Belviso, S.; Mihalopoulos, N.; Gostan, J.; Nival, P. Mar. Chem. 1988, 24, 133-41.

47. Dacey, J. W. H.; Blough, N. V. Geophys. Res. Lett. 1987, 14, 1246-9.

48. Wagner, C.; Stadtman, E. R. Arch. Biochem. Biophys. 1962, 98, 331-6.

49. Sorensen, J.; Christensen, D.; Jorgensen, B. B. Appl. Environ. Microbiol. 1981, 42, 5-11.

50. Dacey, J. W. H.; King, G. M.; Wakeham, S. G. Nature 1987, 330, 643-5

51. Franzmann, P. D.; Deprez, P. P.; Burton, H. R.; van den Hoff, J. Aus. J. Mar. Freshw. Res. 1987, 38, 409-17.

52. Gibson, J. A. E.; Garrick, R. C.; Franzmann, P. D.; Deprez, P. P.; Burton, H. R. The production of reduced sulfur gases in saline lakes of the Vestfold Hills; Submitted for publication.

53. Chambers, S. T.; Kunin, C. M.; Miller, D.; Hamada, A. Appl. Environ. Microbiol. 1987, 169, 4845-7.

54. Kiene, R. P. FEMS Microbiol. Ecol. 1988, 53, 71-8.

55. Anderson, R.; Kates, M.; Volcani, B. E. Nature 1976, 263, 51-3.

56. Bisseret, P.; Ito, S.; Tremblay, P-A.; Volcani, B. E.; Dessort, D.; Kates, M. Biochim. Biophys. Acta 1984, 796, 320-7.

57. Bisseret, P.; Dessort, D.; Nakatani, Y.; Kates, M. Chem. Phys. Lipids 1985, 36, 309-18

58. Kiene, R. P.; Taylor, B. F. Nature 1988, 332, 148-50.

59. Kiene, R. P.; Taylor, B. F. Appl. Environ. Microbiol. 1988, 54, 2208-12.

60. Mopper, K.; Taylor, B. F. In Organic Marine Geochemistry; Sohn, M. Ed.; ACS Symposium Series. No. 305.; American Chemical Society: Washington, DC, 1986; pp. 324-339.

61. Weisinger, R. A.; Jakoby, W. B. In Enzymatic Basis of Detoxification; Jakoby, W. B., Ed.; vol. 1. Academic Press: New York, 1980; Vol. 1 pp. 131-40.

62. Drotar, A. M.; Fall, R. Plant Cell Physiol. 1985, 26, 847-54.

63. Lamoureux, G. L.; Rusness, D. G. Pest. Biochem. Physiol. 1980, 14, 50-61.

64. Larsen, G. L. Xenobiotica 1985, 15, 199-209.

65. Buser, H-R.; Muller, M. D. Environ. Sci. Technol. 1986, 20, 730-5.

66. Sivela, S.; Sundman, V. Arch. Microbiol. 1975, 103, 303-4.

67. Jacob, S. W.; Herschler, R. Ann. New York Acad. Sci. 1975, 243, 1-480.

68. Andreae, M. O. Limnol. Oceanogr. 1980, 25, 1054-63.
69. Zinder, S. H.; Brock, T. D. J. Gen. Microbiol. 1978, 105, 335-42.
70. Zinder, S. H.; Brock, T. D. Arch. Microbiol. 1978, 116, 35-40.
71. Anness, B. J.; Bamforth, C. W. J. Inst. Brew. 1982, 88, 244-52
72. Yancey, P. H.; Clark, M. E.; Hand, S. C.; Bowlus, R. D.; Somero, G. N. Science 1982, 217, 1214-22.
73. Yen, H. C.; Marrs, B. Arch. Biochem. Biophys. 1977, 181, 411-8.
74. Richardson, D. J.; King, G. F.; Kelly, D. J.; McEwan, A. G.; Ferguson, S. J.; Jackson, J. B. Arch. Microbiol. 1988, 150, 131-7.
75. Weiner, J. H.; MacIsaac, D. P.; Bishop, R. E.; Bilhous, P. T. J. Bacteriol. 1988, 170, 1505-10.
76. McEwan, A. G.; Wetzstein, H. G.; Meyer, O.; Jackson, J. B.; Ferguson, S. J. Arch. Microbiol. 1987, 147, 340-5.
77. Styrvold, O. B.; Strom, A. R. Arch. Microbiol. 1984, 140, 74-8.
78. Clarke, G. J.; Ward, F. B. J. Gen. Microbiol. 1988, 133, 379-86.
79. King, G. F.; Richardson, D. J.; Jackson, J. B.; Ferguson, S. J. Arch. Microbiol. 1987, 149, 47-51.
80. McEwan, A. G.; Wetzstein, H. G.; Ferguson, S. J.; Jackson, J. B. Biochim. Biophys. Acta 1985, 806, 410-7.
81. Nichol, C. A.; Smith, G. K.; Durch, D. S. Ann. Rev. Biochem. 1985, 54, 729-64.
82. Nomura, J.;Nishizuka, Y.; Hayaishi, O. J. Biol. Chem. 1963, 238, 1441-6.
83. Wagner, C.; Lusty, S. M., Jr.; Kung, H-F.; Rogers, N. L. J. Biol. Chem. 1967, 242, 1287-93.
84. Labarre, C.; Dory, J. 1969. Ann. Inst. Pasteur 1969, 117, 222-5.
85. Zeyer, J.; Eicher, P.; Wakeham, S. G.; Schwarzenbach, R. P. Appl. Environ. Microbiol. 1987, 53, 2026-32.
86. Oremland, R. S.; Kiene, R. P.; Mathrani, I.; Whiticar, M. J.; Boone, D. R. Appl. Environ. Microbiol, submitted.
87. Kelly, D. P. Symp. Soc. Gen. Microbiol. 1988, 42, 65-98.
88. Colby, J.; Stirling, D. I.; Dalton, H. Biochem. J. 1977, 165, 395-402.
89. De Bont, J. A. M.; van Dijken, J. P.; Harder, W. J. Gen. Microbiol. 1981, 127, 315-23.
90. Suylen, G. M. H.; Kuenen, J. G. Antonie van Leeuwenhoek 1986, 52, 281-93.
91. Anthony, C. The biochemistry of methylotrophs; Academic Press: New York, 1982;

92. Suylen, G. M. H.; Stefess, G. C.; Kuenen, J. G. Arch. Microbiol. 1986, 146, 192-8.

93. Suylen, G. M. H.; Large, P. J.; van Dijken, J. P.; Kuenen, J. G. J. Gen. Microbiol. 1987, 133, 2989-97.

94. Kanagawa, T.; Kelly, D. P. FEMS Microbiol. Lett. 1986, 34, 13-19.

95. Smith, N. A.; Kelly, D. P. J. Gen. Microbiol. 1988, 134, 1407-17.

96. Kelly, D. P.; Wood, A. P. In Microbial Growth on C1 Compounds; Crawford, R. L.; Hanson, R. S., Eds.; American Society for Microbiology: Washington DC, 1984; pp 324-9.

97. Petasne, R. G.; Zika, R. G. Nature 1987, 325, 516-8.

98. Palenik, B.; Zafiriou, O. C.; Morel, F. M. M. Limnol. Oceanogr. 1987, 32, 1365-9.

99. Snow, J. T.; Finlay, J. W.; Kohler, G. O. Ann. New York Acad. Sci. 1976, 243, 228-36.

100. Brimblecombe, P.; Shooter, D. Mar. Chem. 1986, 19, 343-53.

101. Holland, H. L. Chem. Rev. 1988, 88, 473-85.

102. Ziegler, D. M. In Enzymatic Basis of Detoxification; Jakoby, W. B., Ed.; Academic Press: New York, 1980; Vol. 1, pp. 201-27.

103. Williams, K. I. H.; Burstein, S. H.; Layne, D. S. Arch. Biochem. Biophys. 1966, 113, 251-2.

104. Geigert, J.; De Witt, S. K.; Neidleman, S. L.; Lee, G.; Dalientos, D. J.; Moreland, M. Biochem. Biophys. Res. Commun. 1983, 116, 82-5.

105. Neidleman, S. L.; Geigert, J. Biodehalogenation: Principles, Basic Roles and Applications; Wiley, New York., 1986.

106. Hastings, J. W.; Nealson, K. H. Ann. Rev. Microbiol. 1977, 31, 549-95.

107. Hastings, J. W.; Nealson, K. H. In The symbiotic luminous bacteria; Starr, M. P.; Stolp, H.; Truper, H. G.; Balows, A.; Schlegel, H. G., Eds.; The Prokaryotes: Springer-Verlag, New York, 1981; Vol. 2, pp. 1332-1345.

RECEIVED October 13, 1988

Chapter 14

Metabolism of Acrylate and 3-Mercaptopropionate

Decomposition Products of (Dimethylsulfonio)propionate in Anoxic Marine Sediments

Ronald P. Kiene[1] and Barrie F. Taylor[2]

[1]Marine Institute, University of Georgia, Sapelo Island, GA 31327
[2]Division of Marine and Atmospheric Chemistry,
Rosenstiel School of Marine and Atmospheric Science,
University of Miami, Miami, FL 33149–1098

The anaerobic metabolism of acrylate and 3-mercaptopropionate (3-MPA) was studied in slurries of coastal marine sediments. The fate of these compounds is important because they are derived from the algal osmolyte dimethylsulfoniopropionate (DMSP), which is a major organic sulfur compound in marine environments. Micromolar levels of acrylate were fermented rapidly in the slurries to a mixture of acetate and propionate (1:2 molar ratio). Sulfate-reducing bacteria subsequently removed the acetate and propionate. 3-MPA has only recently been detected in natural environments. In our experiments 3-MPA was formed by chemical addition of sulfide to acrylate and was then consumed by biological processes. 3-MPA is a known inhibitor of fatty acid oxidation in mammalian systems. In accord with this fact, high concentrations of 3-MPA caused acetate to accumulate in sediment slurries. At lower concentrations, however, 3-MPA was metabolized by anaerobic bacteria. We conclude that the degradation of DMSP may ultimately lead to the production of substrates which are readily metabolized by microbes in the sediments.

Dimethylsulfoniopropionate (DMSP) is an organic sulfur compound that occurs at high concentrations in many marine algae and plants, where it fulfills an osmotic function (1,2). DMSP is a principal form of organic sulfur in productive marine environments and its microbial catabolism entails either enzymatic cleavage to dimethyl sulfide (DMS) and acrylate (3,4):

$$(CH_3)_2S^+CH_2CH_2COO^- = (CH_3)_2S + CH_2{=}CHCOO^- + H^+$$

or conversion to 3-mercaptopropionate (3-MPA) (5,6). 3-MPA is a novel organosulfur compound, only recently detected in natural environments, which

0097–6156/89/0393–0222$06.00/0

is formed by a chemical addition of sulfide to acrylate (5,7) and by biochemical demethylation of DMSP (5,6) (Figure 1). Both acrylate and 3-MPA have been reported to inhibit the β-oxidation of fatty acids (8,9). The mechanism of action of 3-MPA resides in its ability to form coenzyme A derivatives (Figure 2) which inhibit enzymes involved in fatty acid oxidation (10). However, in spite of these reported effects, we observed the metabolism of both acrylate and 3-MPA in anoxic slurries of coastal marine sediments.

Experimental

Sediments were collected in polybutyrate tubes from nearshore regions in Biscayne Bay, Florida. Sediment slurries were prepared by mixing the upper 15 cm of the cores with equal volumes of deoxygenated seawater under a stream of N_2. The slurries were sieved through 1.5 mm screens to remove large particles and then 50 ml portions were dispensed into serum bottles (60 ml volume). The bottles were sealed with butyl rubber stoppers, held in place by aluminum crimps, and incubated under N_2 in the dark at 23-26°C with rotary shaking. The preparation of the slurries promoted the release of 3-MPA into the seawater and 24-48 hours preincubation was usually necessary to allow the endogenous levels to decrease. Samples for HPLC analysis were removed with disposable plastic syringes (18 gauge needles) and centrifuged at 13,000 x g for about 1 min to remove sediment.

3-MPA was determined as its isoindole derivative using aminoethanol and o-phthalaldehyde (11). 0.5 ml samples were derivatised, immediately after centrifugation of the slurries, with 10 μl of o-phthalaldehyde (20 mg/ml in methanol) and 10 μl of aminoethanol (20 μl/ml of sodium borate buffer, pH 9.2). After allowing 1 minute for reaction 20 μl of the sample was introduced into the HPLC via a Rheodyne 7125 valve (Rheodyne, Cotati, CA) with an injection loop. A C18 reverse-phase column (Waters Radial-Pak, 5μm particles, 10 cm by 1 cm i.d.) was used to separate isoindoles using a binary gradient of 0.05M sodium acetate buffer (pH 5.7) (solvent A) and methanol (solvent B). The gradient protocol was: isocratic at 10% B for 1 min; 10% to 25% B in 1 min; isocratic at 25% B for 4 min; 25% to 50% B in 7 min; isocratic at 50% B for 6 min; 50% to 60% B in 2 min; 60% to 67.5% B in 4 min; 67.5% to 100% B in 1 min; isocratic at 100% for 3 min; 100% to 10% B in 2 min; isocratic at 10% B for 3 min. Solvent was pumped at 1 ml/min with an Autochrom M500 pump. The isoindoles were detected with a Hitachi Model F1000 fluorimeter using excitation and detection wavelengths of 340 and 450 nm respectively. Peak areas were integrated with a Chromatopac C-R3A Data Processor (Shimadzu, Kyoto, Japan). 3-MPA eluted after about 15 min and its detection limit was about 1.0 nM.

Acrylate, propionate and acetate were separated by HPLC on a Benson carbohydrate column (30 cm by 0.6 cm i.d.) (Chromtec, Fort Worth, FL). The HPLC system used a Waters UK6 injector and a Waters Model 6000A pump (Waters Associates, Milford, MA) with a Conductomonitor III Detector (Laboratory Data Control, Riviera Beach, FL) and a Shimadzu Data Processor. The solvent was 0.15 mM H_2SO_4 at a flow rate of 0.5 ml/min. Typical retention times (min) were: acrylate, 13.3; acetate, 12.3; propionate, 14.5. The detection limits for acrylate, acetate and propionate were 10, 30 and 40 μM respectively.

20 mM molybdate was used to inhibit sulfate-reduction (12,13). Chloramphenicol (125 μg/ml) and tetracycline (50 μg/ml) were used as general inhibitors of microbial activity and slurries were pre-incubated for 2-24 hr with these antibiotics before beginning an experiment.

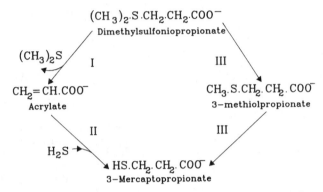

Figure 1. Routes for the production of acrylate and 3-MPA from DMSP. I = enzymatic cleavage of DMSP; II = Michael addition; III = biochemical demethylations.

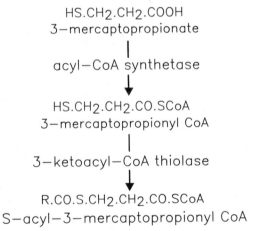

Figure 2. Metabolism of 3-MPA in mitochondria. Abstracted from Cuebas et al. (10).

Organic solvents were purchased from Burdick and Jackson (Muskogee, WI). Acrylic acid was bought from Fluka Chemicals (Happaugue, NY); other chemicals and biochemicals were obtained from either the Aldrich Chemical Co. (Milwaukee, WI) or the Sigma Chemical Co. (St. Louis, MO).

Results

Acrylate was readily consumed and without a lag when it was added at about 700 μM (Figure 3). This activity was blocked by autoclaving and strongly inhibited by antibiotics. Acetate and propionate accumulated transiently in samples which received acrylate (Figure 4A & B). Molybdate had no effect on the rate of consumption of acrylate (Figure 3) but it promoted the accumulation of acetate and propionate (Figure 4). Sediments treated with molybdate alone showed a substantial, linear accumulation of acetate. When this endogenous rate of acetate formation (molybdate present) was subtracted from that observed with acrylate plus molybdate, the molar ratio of propionate to acetate formed from acrylate was about 2 to 1.

In contrast to micromolar levels, millimolar concentrations of acrylate were only utilized after a lag of about 3 days (Figure 5A). Again autoclaving but not molybdate inhibited acrylate consumption. Acetate and propionate were produced from acrylate, and acetate concentrations continued to increase after propionate levels began to decrease (Figure 5B & 5C). Molybdate prevented the disappearance of both acetate and propionate.

3-MPA was formed from 12 mM acrylate at similar rates in either autoclaved or uninhibited sediments (Figure 6). However, 3-MPA persisted in autoclaved systems but not in the untreated slurries.

When 3-MPA was added at millimolar levels we observed the accumulation of acetate (Figure 7). 1 mM 3-MPA caused a small, transient accumulation of acetate whilst 10 mM 3-MPA caused acetate to accumulate for a period of 5 days. The initial rate of acetate accumulation was faster with a higher level (20 mM) of 3-MPA but its total production after 5 days was only about half that seen with 10 mM 3-MPA. Acetate concentrations did not change significantly upon further incubation (data not shown).

Discussion

Acrylate and 3-MPA were readily metabolised by sediment microbes when they were present at micromolar concentrations; but were not so easily degraded at millimolar levels. Acrylate is metabolised by a variety of bacteria. *Escherichia coli* converts acrylyl-CoA to pyruvate via lactyl-CoA, and some clostridia ferment acrylate, via CoA intermediates, to a mixture of acetate and propionate. Of particular relevance to the fate of DMSP was the isolation by Wagner and Stadtman (14) of a species of *Clostridium* (probably *C. propionicum*) from river mud that fermented DMSP as follows:

$$3(CH_3)_2S^+CH_2CH_2COO^- + 2H_2O = 3\,(CH_3)_2S + 2CH_3CH_2COO^-$$
$$+ \; CH_3COO^- + CO_2 + 3H^+$$

The organism also fermented acrylate to acetate and propionate:

$$3CH_2=CHCOO^- + 2H_2O = 2CH_3CH_2COO^- + CH_3COO^- + CO_2$$

Presumably an enzyme similar to that described for marine algae converted DMSP to DMS and acrylate with subsequent fermentation of the latter compound (3,4). Clearly, similar fermentative bacteria may be common in

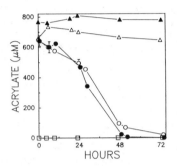

Figure 3. Metabolism of 700 μM acrylate in sediment slurries. Treatments: acrylate alone, ◯ ; acrylate plus autoclaved slurry, ▲ ; acrylate plus 20 mM molybdate, ● ; acrylate plus antibiotics, △ ; endogenous (no additions),☐ .

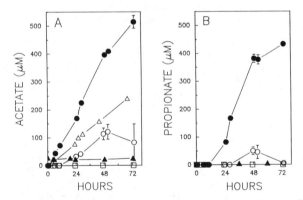

Figure 4. Accumulation of acetate (panel A) and propionate (panel B) in sediment slurries. Treatments: 700 μM acrylate, ◯ ; 700 μM acrylate plus 20 mM molybdate, ● ; 700 μM acrylate and autoclaved slurry, ▲ ; 20 mM molybdate alone, △ ; endogenous (no additions), ☐ .

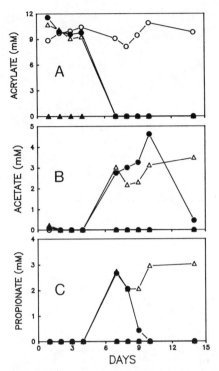

Figure 5. Metabolism of 10 mM acrylate in sediment slurries. Panel A shows acrylate concentrations; panel B shows acetate concentrations; panel C shows propionate concentrations. Treatments: uninhibited slurries, ● ; autoclaved slurries, ○ ; 20 mM molybdate added, △ ; endogenous (no acrylate added), ▲ .

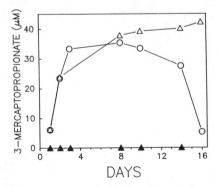

Figure 6. 3-Mercaptopropionate production and metabolism in sediment slurries. Treatments: 10 mM acrylate added, ○ ; 10 mM acrylate with autoclaved slurries, △ ; endogenous (no acrylate added), ▲ .

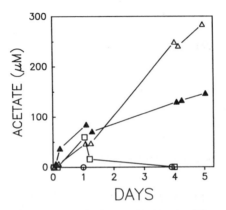

Figure 7. Effects of millimolar concentrations of 3-mercaptopropionate on acetate accumulation in sediment slurries. Symbols: no 3-MPA, ◯ ; 1 mM 3-MPA, ● ; 10 mM 3-MPA, △ ; 20 mM 3-MPA, ▲ .

coastal marine sediments since micromolar levels of acrylate was rapidly metabolized in our experiments to acetate and propionate. Fermentation rates for acrylate exceeded the oxidation rates for acetate and propionate, thereby allowing these products to accumulate transiently (Figure 4 & 5). Millimolar concentrations of acrylate generated about equimolar amounts of acetate and propionate (Figure 3). At the lower substrate level, and with molybdate present to block further metabolism, the molar ratio of acetate to propionate was about 1:2 after correcting for acetate production from endogenous substrates (Figure 4A). The recovery of fermentation products was about 108% after three days. These results are in agreement with the stoichiometry reported by Wagner and Stadtman (14) for acrylate fermentation.

The inhibition by molybdate of propionate and acetate utilization implicates sulfate-reducing bacteria in their catabolism. Acetate is an important substrate for sulfate-reducing bacteria in marine sediments (15-17) but acrylate has not previously been recognized as a precursor of acetate in marine sediments. The propionate formed from acrylate may also be a precursor of acetate because it is converted to acetate by Desulfobulbus species, which have been identified as the dominant propionate-oxidising sulfate-reducers in marine and estuarine sediments (18,19). This is suggested in our experiments by the continued production of acetate after propionate began to decrease (Figure 5), and by the blockage with molybdate of propionate consumption. Thus DMSP may be a precursor for several volatile fatty acids in marine sediments. The importance of DMSP as a source of metabolisable substrates in natural environments is not known.

The chemical addition of sulfide to acrylate to yield 3-MPA occurred in the slurries. However, consumption of acrylate via the addition reaction was probably secondary to its fermentation to acetate and propionate. Kiene and Taylor (6) showed that 10 μM acrylate gave very little accumulation of 3-MPA in sediments, probably because both acrylate and 3-MPA were rapidly metabolised.

Millimolar concentrations of 3-MPA caused acetate accumulation in the sediment slurries (Figure 7). This might be expected because 3-MPA inhibits fatty acid oxidation in rat heart mitochondria (9). In mitochondria, 3-MPA is converted into 3-mercaptopropionyl-CoA and S-acyl-3-mercaptopropionyl-CoA (Figure 2; 10) which respectively inhibit the medium and long-chain acyl-CoA dehydrogenases of β-oxidation (9). Clearly different enzymes will be affected in the inhibition of acetate oxidation but 3-MPA may prove useful as a selective inhibitor of fatty acid catabolism in environmental studies. Thiols such as 3-MPA bind significantly to sediment particles. In sterile slurries about half of the added 3-MPA leaves solution within 24 hours (6). The transient accumulation of acetate with 1 mM 3-MPA may be due to eventual lowering of the dissolved concentration to non-inhibitory levels during the incubation. At lower concentrations 3-MPA was consumed biologically (Figure 4; 6). Its lack of use in autoclaved slurries suggests biotransformation and this raises biochemical questions about its metabolism. We are currently studying the metabolism of 3-MPA by bacterial cultures and in sediments.

Acknowledgments

We thank Ken Mopper and A. Vairavamurthy for helpful discussions, Marcia House for help with organic acid analysis, and the National Science Foundation for financial support (Grant No. OCE-8516020). Contribution No. 622 from the University of Georgia Marine Institute.

Literature Cited

1. Vairavamurthy, A.; Andreae, M.O.; Iverson, R.L. Limnol. Oceanogr. 1985,
    30, 59-70.
2. White, R.H. J. Mar. Res. 1982, 40, 529-36.
3. Cantoni, G.L.; Anderson, D.G. J. Biol. Chem. 1956, 222, 171-7.
4. Kadota, H.; Ishida, Y. Ann. Rev. Microbiol. 1972, 26, 127-38.
5. Mopper, K.; Taylor, B. F. In Organic Marine Chemistry; Sohn, M., Ed.;
    ACS Symposium Series No. 305; American Chemical Society: Washington,
    D.C., 1986.
6. Kiene, R. P.; Taylor, B.F. Nature 1988, 332, 148-50.
7. Vairavamurthy, A.; Mopper, K. Nature 1987, 329, 623-5.
8. Thijsse, G. J. E. Biochim. Biophys. Acta 1964, 84, 195-7.
9. Sabbagh, E.; Cuebas, D.; Schulz, H. J. Biol. Chem. 1985, 260, 7337-42
10. Cuebas, D.; Beckmann, J. D.; Frerman, F. E.; Schulz, H. J. Biol. Chem.
    260, 7330-6.
11. Mopper, K.; Delmas, D. Analyt. Chem. 1984, 56, 2557-60.
12. Oremland, R. S.; Taylor, B. F. Geochim. Cosmochem. Acta 1978, 42,
    209-14.
13. Taylor, B. F.; Oremland, R. S. Curr. Microbiol. 1979, 3, 101-3.
14. Wagner, C.; Stadtman, E. R. Arch. Biochem. Biophys. 1962, 98, 331-6.
15. Banat, I. M.; Lindstrom, E. B.; Nedwell, D. B.; Balba, M. T. Appl. Environ.
    Microbiol. 1981, 42, 985-92.
16. Sorensen, J.; Christensen, D.; Jorgensen, B. B. Appl. Environ. Microbiol.
    1981, 42, 5-11.
17. Winfrey, M. R.; Ward, D. M. Appl. Environ. Microbiol. 1983, 45, 193-9.
18. Laanbroek, H. J.; Pfennig, N. Arch. Microbiol. 1981, 128, 330-5
19. Widdel, F.; Pfennig, N. Arch. Microbiol. 1982, 131, 360-5.

RECEIVED August 22, 1988

Chapter 15

Mechanistic Studies of Organosulfur (Thiol) Formation in Coastal Marine Sediments

Appathurai Vairavamurthy and Kenneth Mopper

**Division of Marine and Atmospheric Chemistry,
Rosenstiel School of Marine and Atmospheric Science,
University of Miami, Miami, FL 33149-1098**

Studies were carried out to understand the influence of environmental variations, in particular pH and salinity, on organosulfur (thiol) formation in marine sediments via the Michael addition mechanism. Reactions of HS⁻ and a polysulfide species, S_4^{2-}, with simple organic molecules containing activated, unsaturated bonds (acrylic acid and acrylonitrile) revealed that polysulfide ions were more reactive than bisulfide ion under conditions typical of marine sediments. Furthermore, polysulfide ion reactions were especially more important for organic Michael acceptors having a terminal carboxyl group (e.g. acrylic acid) than for neutral molecules. Variation of pH influenced reaction rates probably by its effect on the speciation of the reactants. The reaction of HS⁻ with acrylonitrile (a neutral organic molecule) showed maximum reaction rates around neutral pH. However, with acrylic acid the pH maximum was shifted to a lower pH value, close to its pK_a, because of the higher reactivity of the undissociated acid. High ionic strength influenced the rate of the addition reaction positively, depending on the polarizability of the organic molecule. This effect suggests that hypersaline palaeoenvironmental conditions would have favored organosulfur formation by the Michael addition mechanism.

Sulfur transformations in marine sediments constitute a significant part of the global sulfur cycle. Most past studies have emphasized inorganic aspects such as pyrite formation (1-6). Recently, several geochemical studies have focused on organosulfur compounds because of their importance in the origin of sulfur in petroleum as well as to better understand organic matter diagenesis and microbial transformations (7-14). The major classes of organosulfur compounds in marine sediments and in petroleum include, thiols, organic sulfides, disulfides, and thiophene derivatives (15). The presence of high concentrations of organically-bound sulfur in some sediments and petroleums (up to 10% in some high sulfur crude oils) contrasts with the relatively low organic sulfur content in biomass (16-19). Several studies have conclusively shown that during early diagenesis H_2S formed from bacterial sulfate reduction becomes incorporated into organic matter and forms the major fraction of organically-bound sulfur in sediments (17-22). However, until recently, the exact chemical

0097–6156/89/0393–0231$06.00/0

mechanisms by which H_2S reacts with sedimentary organic matter have not been well understood.

Recently, 3-mercaptopropionic acid (3-MPA) was identified as a major thiol in anoxic intertidal Biscayne Bay (Florida) sediments ([23,24]). We showed the formation of this thiol by the reaction of HS⁻ with acrylic acid at ambient temperatures in seawater medium and suggested this reaction for the origin of 3-MPA in sediments ([24]). A likely source of acrylic acid in marine sediments is the common osmolyte β-dimethylsulfoniopropionate from marine algae and plants, which upon enzymatic cleavage yields acrylic acid and dimethylsulfide ([25-27]). Formation of 3-MPA was hypothesized to occur by the Michael addition mechanism, whereby the nucleophile HS⁻ adds to the activated double bond in the α,β-unsaturated carbonyl system of acrylic acid ([24]). Recently, a biological route also has been shown for 3-MPA formation from DMSP ([28]). Based on the abiotic mechanism for 3-MPA formation, we suggested that Michael addition of HS⁻ to activated unsaturated bonds in organic molecules (Michael acceptors) could be a major pathway for the incorporation of sulfur into sedimentary organic matter during early diagenesis. Recently LaLonde et al. ([29]) have shown using model systems that reactions of polysulfide ions with conjugated ene carbonyls form thiophenes at ambient temperatures, substantiating further the importance of the Michael addition mechanism for the geochemical origin of organosulfur compounds.

Studies by Francois ([19,30]) indicated that polysulfides (S_n^{2-}) react more rapidly than bisulfide in the incorporation of sulfur into humic substances. Molecular orbital theory considerations, as pointed out by LaLonde et al ([29]), also suggest higher reactivity of polysulfides with organic Michael acceptors, than bisulfide. These polysulfide species, in fact should constitute a significant fraction of the total inorganic reactive sulfur in reducing sediments, particularly when conditions exist for incomplete oxidation of sulfide (e.g. salt marsh and coastal marine sediments) ([8,21,31,32]). In the present study we have compared the reaction kinetics of polysulfide addition to acrylic acid with that of bisulfide addition. These reactions form 3-polysulfidopropionic acid and 3-mercaptopropionic acid respectively. Reaction of sulfur nucleophiles with acrylic acid provides a simple and easily studied system, where the products (thiols) are monitored by HPLC after precolumn fluorometric labelling ([33]).

We also report here the effects of two other important environmental variables, pH and salinity, on the reaction of bisulfide with acrylic acid. Since acrylic acid posesses an ionizable functional group (-COOH), we have also carried out similar studies using acrylonitrile, a neutral analogue, to compare and contrast the results. Knowledge of these variables is important for understanding diagenesis of recently deposited organic matter in reducing sediments as well as for gaining insights into paleoenvironmental conditions that favored incorporation of sulfur into organic matter.

Methods

Kinetic studies were carried out in amber glass bottles using de-aerated, seawater or the appropriate buffered solution at $40 \pm 0.1°C$. The reactions were studied under pseudo first-order conditions using 5 mM NaHS or Na_2S_4 (Morton Thiokol, Danvers, MA) and 100 or 200 μM of the unsaturated compound (acrylic acid or acrylonitrile) (Aldrich Chemical, Milwaukee, WI). All the chemicals were of reagent grade. Appropriate amounts of the reactants were added to reaction bottles from concentrated stock solutions that were made fresh in filtered de-aerated seawater or Milli-Q water. Samples were removed at different time intervals and analyzed for the thiols, 3-mercaptopropionic acid and 3-mercaptopropane nitrile. Thiols were determined by HPLC after precolumn fluorescence labelling using o-phthalaldehyde (33).

Since the reaction products from the addition of S_4^{2-} to acrylic acid (Equation 3) or acrylonitrile cannot be derivatized directly using the o-phthalaldehyde reagent, they were cleaved with tributyl phosphine (TBP) to the corresponding thiols (Equation 4) prior to HPLC analysis. TBP is a specific reagent for the cleavage of -S-S- linkage (34). When used with aqueous samples, it should be added dissolved in an organic phase (e.g. methanol or 1-propanol) (20-50%, v/v) to enhance miscibility (34,35). Furthermore, the molar concentration of TBP should be at least 2-5 times that of the disulfide present in the sample. Under these conditions, reaction times of 20-30 min at room temperature or 5-10 min at 40°C are adequate for quantitative cleavage.

Effect of pH on the addition reactions was studied from pH 4 to 10. For pH values below 7, the reaction was buffered with a 25 mM sodium acetate solution whereas for pH values above 7, a 25 mM disodium tetraborate solution was used; the pH was adjusted by adding either HCl or NaOH. The combination electrode used for the determination of experimental pH, was calibrated with National Bureau of Standards (NBS) buffers for Milli-Q water (36) and tris(hydroxymethyl)aminomethane (TRIS) buffers for seawater (37) and NaCl solutions (38) on the pH_F (free proton) scale. Since the addition of reactants caused a small pH change in the buffered medium, the experimental pH values shown in the results were measured after the reactants were added to reaction bottle. Samples from low pH reaction series were adjusted to pH 9 by addition of a strong borate buffer just prior to HPLC analysis. This was necessary because thiol analysis using o-phthalaldehyde requires this pH for optimum derivatization.

Results and Discussion

Bisulfide and Polysulfide as Nucleophiles. Hydrogen sulfide (H_2S) and bisulfide ion (HS^-) are probably the primary sulfur nucleophiles (i.e. species with a lone pair of electrons on sulfur) in reducing sediments. However, several environmental factors such as diffusion of oxygen, the presence of Fe(III) may cause incomplete oxidation of sulfide to form polysulfide ions (S_n^{2-}, where $n > 1$) (31).

$$2\,HS^- + O_2 ----> (1/4)\,S_8 + 2\,OH^- \tag{1}$$

$$HS^- + (n-1)/8\,S_8 <----> S_n^{2-} + H^+ \tag{2}$$

Bacterially produced elemental sulfur can also react with hydrogen sulfide to form polysulfide ions. Thus, polysulfide ions should constitute a significant fraction of sulfur nucleophiles in reducing sediments especially where sulfide oxidation is incomplete, such as in intertidal and salt marsh sediments (31,32). The polysulfide ions should also be important at redox boundaries (anoxic/suboxic) in the water column of marine anoxic basins, such as the Black Sea.

We carried out the reaction of polysulfide with acrylic acid as a pseudo first-order reaction using excess sodium tetrasulfide at 40°C in seawater medium. The addition product, presumably 3-tetrasulfidopropionic acid (Equation 3) was then cleaved with tributyl phosphine to 3-mercaptopropionic acid just prior to HPLC analysis (Equation 4).

$$^-SSSS^- + CH_2{=}CHCOO^- + H_2O ---> {}^-SSSS\text{-}CH_2CH_2COO^- + OH^- \tag{3}$$

$$^-SSSS-CH_2CH_2COO^- + 3 (C_4H_9)_3P + 3 H_2O \longrightarrow$$

$$HSCH_2CH_2COO^- + 3 (C_4H_9)_3 P=O + 3 SH^- + 2 H^+ \qquad (4)$$

The results obtained for reaction of acrylic acid (200 μM) with tetrasulfide (1.25 mm & 5 mM) as well as bisulfide (5 mM) are shown in Figure 1. In the reaction with HS$^-$, the concentration of 3-MPA did not noticeably increase after treatment with TBP. If polysulfide ion was the actual reactive species in this reaction, then TBP treatment should have caused a significant increase of 3-MPA, in comparison to that without TBP. Therefore, these results indicate that HS$^-$ was the reactive species in the sulfide medium. However, in the reaction of acrylate with the polysulfide ion, S_4^{2-}, determination of 3-MPA concentration after TBP treatment revealed a large increase, indicating that S_4^{2-} reacted with acrylate to produce 3-tetrasulfidopropionic acid. The reactivity of tetrasulfide with acrylate was much higher than that of bisulfide. Similar results were also obtained for reaction of acrylonitrile with HS$^-$ and S_4^{2-} (Figure 2). The kinetic data are summarized in Table I. In the polysulfide series low concentrations of 3-MPA were also observed without TBP treatment. It is possible that this 3-MPA was formed from uncatalyzed dissociation of the polysulfide addition product.

The much higher reactivity of polysulfide ions, compared with HS$^-$, towards organic Michael acceptors suggests that, in coastal marine sediments and salt marshes, polysulfide ions could be more important than HS$^-$ for organosulfur formation. In fact, preliminary studies indicate that TBP treatment of anoxic sediment slurries or extracted pore water from Biscayne Bay (Florida) releases substantial amounts of 3-MPA (23), suggesting the presence of polysufido-propionic acids or the disulfide of 3-MPA. This aspect is currently under study.

Recently LaLonde et al (29) pointed out that the reactivity of sulfur nucleophiles with organic electrophiles in the Michael addition mechanism can be considered from the perspective of frontier molecular orbital theory (FMO) (39). Since the addition reaction results in the formation of a sulfur-carbon covalent bond, interactions of molecular orbitals (MOs) will be more important than Coulombic interactions (40). The significant interactions are those between the highest occupied molecular orbital (HOMO) of one molecule (the nucleophile) and the lowest unoccupied molecular orbital (LUMO) of the other (the electrophile). These orbitals are referred to as the frontier orbitals (39,41). The interaction is stronger when the involved orbitals are closer in energy (41). Several studies have shown that for sulfur nucleophiles the energy of the HOMO increases with the number of sulfur atoms in various catenated forms such as the polysulfides (S_n^{2-}) (29). Therefore, the energy separation between HOMO of the nucleophile and LUMO of the organic electrophile is smaller for the combination of a polysulfide ion and an electrophile than for that of bisulfide and the same electrophile (Figure 3). This probably explains the faster reactions of tetrasulfide with acrylate and acrylonitrile compared to those of bisulfide with the same electrophiles.

It is important to note that for organic Michael acceptors, the energy of the LUMO decreases progressively with increase in the number of conjugated carbon-carbon unsaturated bonds in the molecule. As shown in Figure 3, if conjugation to a C=C bond is provided by an unsaturated bond, which is also electron withdrawing (e.g. -CHO, -CN), it causes the energy of the LUMO to decrease much further than that caused by an another C=C bond (41). This enhanced LUMO- lowering effect probably explains the high reactivity of molecules containing α,β-unsaturated carbonyl system (e.g. acrylic acid) with sulfur nucleophiles.

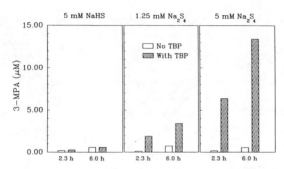

Figure 1. Comparison of reactivities of HS⁻ and S_4^{2-} with acrylic acid in seawater medium (salinity 35 and reaction pH 8.3 ± 0.2) at 40°C. Determinations of 3-MPA were done after 2.3 h and 6.0 h reactions.

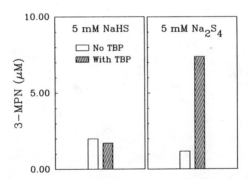

Figure 2. Comparison of reactivities of HS⁻ and S_4^{2-} with acrylonitrile in seawater medium (salinity 35 and reaction pH 8.3 ± 0.2) at 40°C after 6.0 h reaction.

Table I. Pseudo first-order rate constants[*] for Michael addition of sulfur
Nucleophiles (5 mM) to acrylic acid and acrylonitrile in seawater medium
(salinity 35 and reaction pH 8.3 ± 0.2)

Sulfur Nucleophile	Michael Acceptor	k_1 (hr^{-1}) at 40°C
HS$^-$	CH$_2$=CHCOOH	(4.7 ± 0.3) 10^{-4}
S$_4^{2-}$	CH$_2$=CHCOOH	(1.3 ± 0.2) 10^{-2}
HS$^-$	CH$_2$=CHCN	(3.1 ± 0.3) 10^{-3}
S$_4^{2-}$	CH$_2$=CHCN	(1.5 ± 0.3) 10^{-2}

[*]Derived from a minimum of 2 determinations.

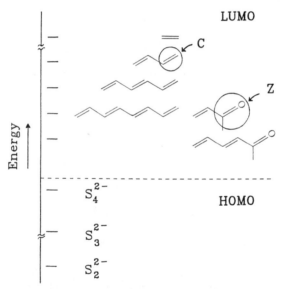

Figure 3. Comparison of the LUMO energy levels for a series of conjugated
unsaturated organic molecules and HOMO energy levels for S_n^{2-}. C = vinyl or
phenyl; Z = electron withdrawing group such as -CHO, -CN etc. (adapted from
LaLonde et al. 1987) (29).

Influences of pH and Salinity Variations. Pseudo first-order rate constants (k_1) for 3-MPA formation from the reaction of HS⁻ with acrylic acid in water as well as in seawater, as a function of pH, are shown in Figure 4. The reaction rate varies with pH and displays a maximum around pH 5. Several studies involving reactions of hydrogen sulfide observed progressively increasing rates from acidic to neutral pH with maxima around neutral pH, indicating that reactivity is related to the dissociation of H_2S to form HS⁻ (42,43). However, in the case of the sulfide-acrylic acid reaction, pH can affect speciation of both reactants. Unionized acrylic acid is expected to be a stronger electrophile and more reactive than acrylate ion. Since acrylic acid has a pKa of about 4.3 (44), the reaction rate of acrylic acid with a hypothetical unionizable nucleophile is expected to be at maximum below this pKa value (pH < 4.3). Therefore, it appears that the maximum rate for the formation of 3-MPA at pH around 5 is a function of speciation of both acrylic acid and hydrogen sulfide.

At basic pH values the rate of 3-MPA formation is reduced, but continues at measurable rates even at a pH value as high as 10. These results indicate that acrylate ion possesses significant reactivity, although the undissociated form is much more reactive. In the acidic pH range, the rate of 3-MPA formation in seawater is similar to that in Milli-Q water, but at basic pH values, the rates in seawater are higher than those in Milli-Q water (Figure 4). In an ionic medium such as seawater, for reactions involving ions, the Bronsted-Bjerrum equation predicts that ionic interactions cause deviations from ideal-solution behaviour (Equation 5) (45).

$$\log(k/k_0) = 2AZ_AZ_B\sqrt{I} \qquad (5)$$

(where k = rate constant; k_0 = ideal solution rate constant; A = constant; Z_A = charge on species A; Z_B = charge on species B; I = ionic strength). This equation shows that ionic strength of the medium positively influences the rate of reaction when both reactants carry the same charge. The higher rates for the reaction in seawater can be explained by this effect because both reactants (HS⁻ and $CH_2=CHCOO^-$) are negatively charged at basic pH values, although other factors, such as ion-pairing also may be involved.

The pseudo first-order rate constant results for the formation of 3-mercaptopropane nitrile (3-MPN) from the reaction of acrylonitrile (a neutral analogue of acrylic acid) with hydrogen sulfide are shown in Figure 5. Here, reaction rates reach maxima around neutral pH. This behaviour, which contrasts with that of 3-MPA formation, probably reflects HS⁻ formation from H_2S dissociation. Unlike acrylic acid which changes speciation with pH, acrylonitrile being a neutral molecule, is probably not strongly affected by pH. At basic pH values reaction rates declined, a trend that was also observed by other workers for reactions involving hydrogen sulfide (42). Formation of S^{2-} cannot account for this behaviour because equilibrium constants for the dissociation of H_2S and HS⁻ (pKa values are about 7.0 and 14 respectively) (46) indicate that HS⁻ is essentially the only species at the basic pH values used in the experiment; thus the cause for this behaviour remains unclear. The pH variations of rates for 3-MPN formation in water and seawater follow each other closely at both acidic and basic pH values, which is in agreement with that expected theoretically.

We also determined the effect of ionic strength on the formation of 3-MPA and 3-MPN in NaCl solutions at pH 8.0 and 40°C. Figure 6 shows plots of log k vs. $I^{1/2}$, where k is the overall rate constant (units: $M^{-1}day^{-1}$), calculated as $k_1/[H_2S]$. The rate of 3-MPA formation shows definite increase (slope from regression is 0.22) with ionic strength, which is in agreement with the Bronsted-Bjerrum equation. The formation of 3-MPN also shows an increase with ionic

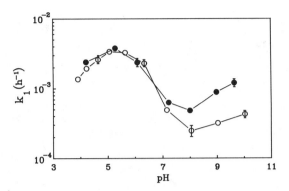

Figure 4. Variation of k_1 with pH for the formation of 3-mercaptopropionic acid at 40°C. (●) seawater medium (I = 0.7); (○) Milli-Q water medium.

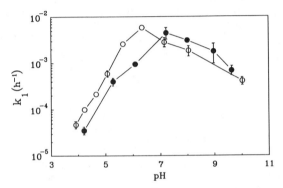

Figure 5. Variation of k_1 with pH for the formation of 3-mercaptopropane nitrile at 40°C. (●) seawater medium (I = 0.7); (○) Milli-Q water medium.

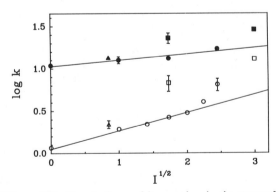

Figure 6. Effect of NaCl solutions of increasing ionic strength (I) on log k for the formation of 3-mercaptopropane nitrile (●), and 3-mercaptopropionic acid (○) at pH 8.1 ± 0.1 and 40°C.   ■ and   □   are values with MgCl2 solutions;   ▲   and   △   indicate rates in seawater.

strength, but the effect is much smaller (slope is 0.07). Since acrylonitrile is not ionized, the Bronsted-Bjerrum equation predicts that this reaction will not be influenced by ionic strength. However, it is possible that the -CN group, because it is highly polarized, may have contributed to the increase with ionic strength. Figure 6 also indicates rates of formation for both 3-MPA and 3-MPN in 1 molal and 3 molal $MgCl_2$ solutions (ionic strength is 3 and 9 respectively) as well as in seawater. A significant increase in rates is observed with magnesium, probably because of its higher positive charge density. It is possible that cations such as magnesium form ion-pairs, in particular with acrylate, which may enhance the electrophilicity of the unsaturated system by partial neutralization of the negatively charged carboxyl group. The presence of polyvalent cations such as magnesium in seawater may explain the somewhat higher value for 3-MPA formation than that predicted for a NaCl solution of corresponding ionic strength.

Summary and Conclusions

The Michael addition mechanism, whereby sulfur nucleophiles react with organic molecules containing activated unsaturated bonds, is probably a major pathway for organosulfur formation in marine sediments. In reducing sediments, where environmental factors can result in incomplete oxidation of sulfide (e.g. intertidal sediments), bisulfide (HS^-) as well as polysulfide ions (S_n^{2-}) are probably the major sulfur nucleophiles. Kinetic studies of reactions of these nucleophiles with simple molecules containing activated unsaturated bonds (acrylic acid, acrylonitrile) indicate that polysulfide ions are more reactive than bisulfide. These results are in agreement with some previous studies (30) as well as frontier molecular orbital considerations. Studies on pH variation indicate that the speciation of reactants influences reaction rates. In seawater medium, which resembles pore water constitution, acrylic acid reacts with HS^- at a lower rate relative to acrylonitrile because of the reduced electrophilicity of the acrylate ion at seawater pH.

Kinetic data show that in seawater medium S_4^{2-} reacts about 20-30 times faster than HS^- with acrylic acid, whereas the reaction of S_4^{2-} with acrylonitrile is only about 4-6 times higher than with HS^-. However, in any environment, the importance of polysulfide versus bisulfide reactions is also dependent on their relative concentrations. In a situation where polysulfide and bisulfide ions are present in similar concentrations, our results imply that polysulfide ions, rather than bisulfide, are the important sulfur nucleophiles for reactions with activated unsaturated molecules having a terminal carboxyl group (e.g. acrylic acid, cinnamic acid). However, for neutral molecules such as fucoxanthin, in addition to polysulfide ions, reactions with bisulfide ions will also be of importance.

Salinity exerts a positive influence on the rate of the addition reaction depending on the polarizability of the organic molecule. This effect is pronounced for unsaturated molecules containing a terminal carboxyl group. These results suggest that hypersaline palaeoenvironmental conditions would have favored organosulfur formation by the Michael addition mechanism.

Acknowledgments

We thank Eric S. Saltzman for his encouragement and B.F. Taylor and F.J. Millero for valuable discussions on this manuscript. Virender K. Sharma is acknowledged with thanks for assistance and helpful suggestions during the study. This work was supported by National Science Foundation grant OCE-8516020.

Literature Cited

1. Goldhaber, M. B.; Kaplan, I. R. In The Sea; Goldberg, E. D., Ed.; Wiley: New York, 1974; Vol. 5, pp 569-655.
2. Goldhaber, M. B.; Kaplan, I. R. Mar. Chem., 1980, 9, 95-143.
3. Sweeny, R. E.; Kaplan, I. R. Econ. Geol., 1973, 68, 618-34.
4. Howarth, R. W.; Jorgensen, B. B. Geochim. Cosmochim. Acta, 1984, 48, 1807-18.
5. Jorgensen, B. B. Limnol. Oceanogr., 1977, 22, 814-32.
6. Berner, R. A. Geochim. Cosmochim. Acta, 1984, 48, 605-15.
7. Bates, T. S.; Carpenter, R. Geochim. Cosmochim. Acta, 1979, 43, 1209-21.
8. Luther III, G. W.; Church, T. M.; Scudlark, J. R.; Cosman, M. Science, 1986, 232, 746-9.
9. Brassel, S. C.; Lewis, C. A.; de Leeuw J. W.; de Lange, F.; Sinninghe Damste, J. S. Nature, 1986, 320, 160-2.
10. Sinninghe Damste, J. S.; ten Haven, H.L.; de Leeuw, J. W.; Schenck, P. A. Org. Geochem., 1986, 10, 791-805.
11. Sinninghe Damste, J. S.; de Leeuw, J. W. Intern. J. Environ. Anal. Chem. 1987, 28, 1-19
12. Sinninghe Damste, J. S.; de Leeuw, J. W.; Kock-van Dalen, A. C.; de Zeeuw, M. A.; de Lange, F; Rijpstra, I. C.; Schenck, P. A. Geochim. Cosmochim. Acta, 1987, 51, 2369-91.
13. Valisolalao, J.; Perakis, N.; Chappe, B.; Albrecht, P. Tetrahedran Lett., 1984, 25, 1183-6.
14. Hussler, G.; Albrecht, P.; Ourisson, G. Tetrahedran Lett., 1984, 25(11), 1179-82.
15. Tissot, B. P.; Welte, D.H. Petroleum Formation and Occurrence, Springer: Berlin, 1984.
16. Thomson, C. J. In Organic Sulfur Chemistry; Freidlina, R. Kh; Skorova, A. E., Eds.; Pergamon Press, 1981, 189-208.
17. Gransch, J.A.; Posthuma, J. In Advances in Organic Geochemistry 1973; Tissot B.; Bienner F.; Eds.; Editions Technip: Paris,1974, 727-39.
18. Nissenbaum, A.; Kaplan, I.R. Limnol. Oceanogr., 1972, 17, 570-82.
19. Francois, R. Geochim. Cosmochim. Acta, 1987, 51, 17-27.
20. Casagrande, D.J.; Idowu, G.; Friedman, A.; Rickert, P.; Siefert, K,; Schlenz, D. Nature, 1979, 282, 599-600.

21. Aizenshtat, Z.; Stoler, A.; Cohen, Y.; Nielsen, H. In Advances in Organic Geochemistry 1981; Bjoroy, M. et al.Eds.; Wiley: Chichester, 1983; 279-88.

22. Dinur, D.; Spiro, B.; Aizenshtat, Z. Chem. Geol., 1980, 31, 37-51

23. Mopper, K.; Taylor, B. F. In Organic Marine Chemistry; Sohn, M. L., Ed.; ACS Symposium Series No. 305; American Chemical Society: Washington DC, 1986; pp 324-39.

24. Vairavamurthy, A.; Mopper, K. Nature, 1987, 329, 623-5.

25. Vairavamurthy, A.; Andreae, M. O.; Iverson, R. L. Limnol. Oceanogr., 1985, 30, 59-70.

26. Charlson, R. J.; Lovelock, J. E.; Andreae, M. O.; Warren, S. G. Nature, 1987, 326, 655-61.

27. Dacey, J. W. H.; Blough, N. V. Geophys. Res. Lett., 1987, 14, 1246-9.

28. Kiene, R. P.; Taylor, B. F. Nature, 1988, 332, 148-50.

29. LaLonde, R. T.; Ferrara, L. M.; Hayes, M. P. Org. Geochem., 1987, 11(6), 563-71.

30. Francois, R. Limnol. Oceanogr., 1987, 32(4), 964-72.

31. Boulegue, J.; Michard, G. In Chemical Modeling in Aqueous Systems; Jenne, E. A., Ed.; ACS Symposium Series No. 93, American Chemical Society: Washington DC, 1979, 25-50.

32. Boulegue, J.; Lord III, C. J.; Church, T.M. Geochim. Cosmochim. Acta, 1982, 46, 453-64.

33. Mopper, K.; Delmas, D. Anal. Chem., 1984, 56, 2558-60.

34. Ruegg, Urs Th.; Rudinger, Josef. In Methods of Enzymology XLVII; Hirs, C. H. W.; Timasheff, S. N. Eds.; Academic Press: New York,1977; 111-16.

35. Humphrey, R. E.; Potter, J. L.; Anal. Chem., 1965, 37, 164.

36. Covington, A. K.; Bates, R. G.; Durst, R. A. Pure Appl. Chem., 1985, 57, 531-42.

37. Millero, F. J. Limnol. Oceanogr., 1986, 31(4), 839-47.

38. Millero, F. J.; Hershey, J. P.; Fernandez, M. Geochim. Cosmochim. Acta, 1987, 51, 707-11.

39. Fukui, K. Accts. Chem. Res., 1971, 4, 57.

40. Klopman, G. J. Am. Chem. Soc., 1968, 90, 223-34.

41. Fleming, I. Frontier Orbitals and Organic Chemical Reactions; Wiley: Chichester, 1976.

42. Chen, K. Y.; Morris, J. C. Environ. Sci. Technol., 1972, 6, 529-37.
43. Millero, F. J.; Hubinger, S.; Fernandez, M.; Garnett, S. Environ. Sci. Technol., 1987, 21, 439-42.
44. Windholz, M. The Merck Index; Merck & Co., Rahway, NJ, 1983; p 19.
45. Stumm, W.; Morgan, J. J. Aquatic Chemistry; Wiley & Sons: New York, 1981, pp 102-104.
46. Morse, J. W.; Millero, F. J.; Cornwell, J. C.; Rickard, D. Earth Sci. Reviews, 1987, 24, 1-42.

RECEIVED July 20, 1988

Chapter 16

Reduced Sulfur Compounds in the Marine Environment

Analysis by High-Performance Liquid Chromatography

Russell D. Vetter[1], Patricia A. Matrai[2,3], Barbara Javor[1], and John O'Brien[1]

[1]Marine Biology Research Division, A−002, Scripps Institution of Oceanography, La Jolla, CA 92093
[2]Institute of Marine Resources, A−018, Scripps Institution of Oceanography, La Jolla, CA 92093
[3]Rosenstiel School of Marine and Atmospheric Science, University of Miami, Miami, FL 33149−1098

The number of oxidation states that sulfur can assume and its ability to act as energy substrate, electron acceptor and structural component make the understanding of the various roles this element plays in the functioning of the marine ecosystem of great importance. Since sulfur exists in a wide variety of highly reactive species, the extraction and measurement of complex mixtures of reduced sulfur compounds from biological samples such as mixed communities, sediments and tissues is extremely difficult and the chance of introducing artifacts is large. To effectively measure such mixtures a protocol should accomplish two things: allow for the fixation of a variety of reduced sulfur compounds to prevent their reaction during extraction and sample preparation; and separate and quantify as many sulfur compounds as possible with a single assay. We have applied the monobromobimane-HPLC method of thiol derivitization to the measurement of reduced sulfur compounds in phytoplankton, sediments and invertebrates. Monobromobimane reacts with the thiol group producing a covalently bonded fluorescent adduct. This process prevents further reaction of the sulfur group and allows the different sulfur compounds to be separated by reverse-phase liquid chromatography at a later time. With the addition of a flow-through scintillation counter it is possible to trace the metabolism of radiolabeled sulfur compounds in highly complex systems. We present the methodology we have developed to measure reduced sulfur compounds in plankton samples, sediment cores and marine invertebrates. The results presented illustrate the variety of ways reduced sulfur compounds participate in the marine ecosystem.

0097−6156/89/0393−0243$06.00/0
© 1989 American Chemical Society

The study of the biogeochemistry of sulfur in the marine environment is currently experiencing rapid growth due to several fundamental discoveries. One is the realization that reduced sulfur compounds produced by marine phytoplankton are a major source of atmospheric sulfate which contributes to acid rain (1, and references therein) and perhaps the cloud cover of the world's ocean (2). The latter function may have a self-regulatory effect on temperature and primary productivity of the earth (3). A second discovery has been that there are tremendous inputs of geothermally derived hydrogen sulfide occurring along ridge-crest spreading centers and subduction zones thoughout the deep ocean (4,5). This hydrogen sulfide is utilized by luxuriant biological communities fueled entirely by the energy contained in this poisonous gas (6-8, and references therein). Most of the organisms in these ecosystems rely upon a symbiotic mode of nutrition whereby sulfide oxidizing chemolithotrophic bacteria housed internally harvest the energy in hydrogen sulfide and use it for the fixation of CO_2 into products available to the animal host as a source of nutrition. These are the only ecosystems not dependent on light as a source of primary productivity that have been found on this planet. Related to this discovery is the recent proposal of Lake (9), that the earliest form of life was probably a thermophilic, acidophilic sulfur oxidizing bacterium. While this hypothesis is still quite controversial it will undoubtedly stimulate further research in the metabolism of sulfur. When considering the metabolism of sulfur in the marine environment it is important to keep in mind the very different uses that organisms have found for this element as inorganic energy source, electron acceptor and structural component.

Covalently-bonded sulfur is a required structural element for all organisms. It is incorporated most commonly as the amino acids cysteine and methionine in proteins and as sulfolipids in membranes. The properties of the thiol group to form disulfide bonds, undergo thiol-disulfide exchange reactions and have a pK_1 of relevance to internal cellular pH makes it useful as a structural element in the disulfide bridges of proteins, as an antioxidant in the glutathione system, and perhaps as a regulatory messenger via addition of small molecular weight thiols to proteins (10).

The structural role of sulfur in marine organisms goes beyond this basic requirement. For example, marine phytoplankton can produce up to millimolar intracellular concentrations of dimethylsulfonioproprionate (11-15), although its physiological significance remains unclear. It has been suggested that this compound serves as a compatible osmolyte since its concentration increases with salinity in a coastal marine coccolithophorid (11). It is tempting to speculate that marine phytoplankton have substituted a sulfur based osmolyte over the more common nitrogen based osmolytes such as trimethylamine-n-oxide or glycine betaine because of the abundance of sulfur over nitrogen in the marine environment, and their ability to reduce sulfate (16). The substitution of sulfur for nitrogen in other cell constituents of marine phytoplankton has not been studied but should be a fruitful area of investigation.

Sulfate reducing bacteria are able to grow heterotrophically in the absence of oxygen by using sulfate as a terminal electron acceptor and releasing hydrogen sulfide. This process is especially important in marine systems because the abundance of dissolved sulfate in seawater allows sulfate reduction to outcompete methanogenesis in the anaerobic remineralization of marine primary production. Howarth and Teal (17) have shown that perhaps twelve times more organic carbon is degraded anaerobically (primarily by sulfate reducing bacteria) than aerobically in a typical New England salt marsh. The resulting sulfide contains large amounts of the energy originally found in the organic carbon. This hydrogen sulfide can interact with organic matter in the sediments to produce organic thiol compounds such as 3-mercaptoproprionate

(18). Natural breakdown of organic matter may also produce organic thiol compounds. Luther et al. (19) demonstrated that there is seasonal variability in the organic thiols of the Delaware marsh. These results have sparked interest in the role of organic thiols in metal and energy cycling in sediments.

The capacity of chemoautotrophic bacteria to oxidize hydrogen sulfide as a source of energy for CO_2 fixation is well known (20). However, the realization that entire bacterial and invertebrate communities can be fueled by geothermally derived sulfide at hydrothermal vents is a comparatively new discovery. The study of how these bacteria and invertebrates act in consort to exploit the energy in hydrogen sulfide while ameliorating its toxic effects on aerobic respiration has resulted in a whole new area of research in the metabolism and detoxification of sulfide (21-24). These results show that invertebrates as well as chemoautotrophic bacteria have specific biochemical mechanisms for the metabolism of hydrogen sulfide. In some cases, the animal may oxidize sulfide to an intermediate form such as thiosulfate and transport this less toxic form to endosymbiotic bacteria; these microorganisms can further oxidize the sulfur to gain energy and fix CO_2. As a result, the blood of these organisms often contains complex mixtures of organic and inorganic reduced sulfur compounds which are difficult to analyze. Furthermore, the recent observation that mitochondria of *Solemya reidi* can produce ATP directly from sulfide (25) raises the extremely important possibility that invertebrates may be able to derive a portion of their energy requirements from the oxidation of inorganic sulfur substrates. Prior to this discovery, such oxidation was thought to be done exclusively by procaryotes.

As our understanding of the way reduced sulfur compounds function in the marine ecosystem has grown so has the need for methods which allow for the sampling and preservation of complex mixtures of organic and inorganic sulfur compounds in the field and their subsequent separation and analysis at a later time.

The Measurement of Complex Mixtures of Reduced Sulfur Compounds

Early studies of reduced sulfur in the environment centered on the role of sulfate reducing bacteria in sediments and the measurement of hydrogen sulfide and other simple inorganic sulfur compounds in pore waters. Simple colorimetric assays are probably still the best for such studies. Typical methods include the methylene blue determination of sulfide (26,27) and the dithio-nitrobenzoic acid (DTNB) method for total thiols (28,29).

The realizations that organic thiols could i) be important in sediment chemistry, ii) contribute to the atmospheric sulfur budget , and iii) fuel entire ecosystems at hydrothermal vents have led to the development of a number of techniques designed to separate and measure complex mixtures of organic and inorganic reduced sulfur compounds. In order to study complex mixtures of reduced sulfur compounds in biological materials and environmental samples, the samples must be protected from further oxidation during extraction and sample preparation. Ideally, all sulfur compounds of interest could be measured by the same automated method. Most of the procedures to be presented here have in common the reaction of the sulfur compound with a hydrophobic reporter group, followed by separation via reversed phase high performance liquid chromatography (HPLC) and detection by either fluorescence or absorbance characteristics. We will briefly discuss the pros's and con's of several methods used by environmental chemists. We emphasize that, at this time, there does not appear to be one ideal method and the choice will depend on the specific needs of the investigator. Such techniques include the polarographic method (19,30), the potentiometric method (31,32), the *o*-phthalaldehyde (OPA-HPLC) method (33), the 2,2'-dithiobis(5-nitropyridine) (DTNP-HPLC) method (34) and the monobromobimane (bimane-HPLC) method (24,35).

The polarographic and potentiometric methods are not HPLC-run. The polarographic method relies upon the measurement of half-wave potentials of various sulfur compounds reacting with a mercury electrode. It is sensitive to submicromolar concentrations (Luther, pers. com.). While sulfide, thiosulfate, polysulfide and polythionates can be measured, the initial sample must be subdivided and pretreated in different ways. The disadvantages are that sample preparation and analysis are time-consuming and there is no way to preserve samples for later analysis nor to study organic thiols with precision.

The potentiometric measurement of hydrogen sulfide *via* a Ag/Ag_2 electrode is well known and such electrodes are commercially available. Boulégue (32) has used the electrode to measure changes in potential during a titration with $HgCl_2$. This method can measure sulfide, polysulfide, thiols, sulfite and thiosulfate. Weaknesses of the method are the following: there is no method of sample preservation; the pH must be adjusted to pH 13 for the measurement of sulfide, thiol and polysulfide then adjusted to pH 7 for the measurement of thiosulfate and sulfite; polysulfide is determined by difference after attack on the polysulfides with added sulfite (if elemental sulfur is present in the sample it will also participate in this reaction); the identity of the organic thiols is unknown; and sulfite can not be measured in seawater samples because of competition with halides (31).

The OPA method is based on the formation of a highly fluorescent isoindole derivative by reaction with *o*-phthalaldehyde and 2-aminoethanol in mildly basic aqueous solution (33). The advantage of this technique is that a large variety of thiols and other reduced sulfur compounds can be detected at subnanomolar to nanomolar concentrations. Its disadvantage is that the fluorescent derivative, which preserves the thiol in its reduced stage, is unstable and must be formed just prior to injection. These characteristics preclude the delayed analysis (e.g., in the laboratory) of large numbers of samples collected in the field.

A DTNP method for HPLC has recently been reported (34). The sulfur compounds are stable after derivatization and the detection limits are quite low. The major disadvantages are that the derivatized thiols are detected by absorbance, and sediment porewaters and biological samples often contain large numbers of absorbing compounds which interfere with authentic thiols. In addition, sulfide is not detected.

The study of sulfide metabolism at hydrothermal vents dictated the development of methods that could process hundreds of samples which contain complex mixtures of sulfur compounds in a variety of blood, seawater and tissues samples. In addition, we needed the capability of using ^{35}S-radiolabeled compounds for the tracing of complex sulfur metabolic pathways in bacteria and animal compartments of the different hydrothermal vent symbioses. In some instances, *in situ* sampling by submersibles at depths of 2500 meters with associated recovery times of two hours necessitated the remote derivatization of samples at depth prior to recovery. None of the above methods completely met our needs. We have adapted the bimane-HPLC method (24,35) for shipboard use and have found it a particularly robust method for studying a number of questions concerning the role of reduced sulfur compounds in the marine environment.

The bimane-HPLC method is not necessarily the most sensitive one but it is highly reproduceable for reduced sulfur compounds ranging in concentration from high nanomolar to tens of millimolar (24,35,36). The technique is potentially more sensitive based on the intrinsic fluorescence of the adducts formed. However, the half time of the bimane reaction is sufficiently slow that concentrations of bimane below 1mM do not result in rapid derivatization. As a result, peaks resulting from thiol concentrations below 500 pM are obscured by baseline noise. Unreacted bimane can be removed (36) but we have not done so routinely. One advantage of the bimane method is that the monobromo form of bimane readily crosses cell membranes so that potentially reactive sulfur compounds can immediately be stabilized during extraction and sample processing. Tissue samples can be stored for over one year without loss (35). The stability of the sulfur-bimane adducts means that the method lends itself to automated sample injection. An extensive list of the organic and inorganic sulfur compounds which form bimane derivatives has been published by Fahey and Newton (35). A further advantage is that the stability and hydrophobicity of the bimane-sulfur derivatives allows for the relatively easy separation and detection of ^{35}S-radiolabeled compounds (24). This will be discussed later in greater detail.

In this paper we review three types of field studies where the method is used to determine very different aspects of sulfur chemistry in the marine environment. These studies include i) the measurement of the intracellular thiol composition of marine phytoplankton in response to light; ii) the reduced sulfur composition of anaerobic sediments; and iii) the metabolism of potentially toxic hydrogen sulfide by sediment dwelling bivalve molluscs housing endosymbiotic sulfur oxidizing bacteria.

Methods

We will present the basic requirements for preparing a bimane derivatized sample for HPLC analysis and the preparation of appropriate blanks. We will then discuss the specific methods used for the three studies presented in the results section: i) phytoplankton studies; ii) sediment core studies; iii) sulfide metabolism studies in invertebrate-bacteria symbiosis.

A good general review of the chemistry of bimane-thiol reactions (37) and HPLC separation of organic bimane-thiol derivatives (35) can be found elsewhere. We prepare a 100 mM bimane stock in acetonitrile which can be diluted as needed with acetonitrile. This stock solution is kept frozen and used over the course of one month. Samples containing 0-500 μM total reduced sulfur are derivatized with 1 mM bimane (final sample concentration) and similarly, samples containing up to 5 mM total reduced sulfur are derivatized with 10 mM bimane. These two bimane concentrations allow us to cover most of the environmental range of interest. As mentioned above, a two-fold excess of bimane to total thiol is the minimum required for rapid sample derivatization but the bimane concentration should not be reduced below 1mM because of an unacceptable increase in the half-time of the reaction. It is the HS$^-$ form of sulfide and thiols which carries out the nucleophilic attack on the bimane, hence pH's of around 8 are optimal for derivatization. This is the approximate pH of seawater and most bloods. Since these two fluids are also well buffered we are able to derivatize these samples without further pH adjustment or buffering. If pH adjustment and buffering is required we use 200 mM HEPES buffer with 5 mM EDTA. Other buffers and metal chelating agents can be used but it is important to run a reagent blank for each new protocol. A sample of water or blood (100 μl) is placed in a 1.5 ml disposable microfuge tube and 10 μl of the appropriate bimane stock concentration are added and allowed to react at pH 8 adjusting with HEPES buffer, if necessary. The samples are allowed to react

10-15 min at room temperature shielded from bright light. Once the bimane reaction is complete, 100 μl of acetonitrile are added and the sample is heated to 60°C for 10 minutes to precipitate protein. The acetonitrile denaturation will precipitate most but not all of the protein. In general, this step is enough to prevent clogging of the guard column during sample analysis. Finally, the added acetonitrile is diluted out so that the injected amount of organic solvent does not affect the chromatography of the initial, primarily aqueous, portion of the HPLC gradient. The 200 μl of sample and acetonitrile are diluted with 300 μl of dilute (25 mM) methane sulfonic acid. The samples are spun to compact the precipitated protein and are stored frozen until analysis. Fahey and Newton have investigated the effects of sample storage conditions (35). They found no significant sample loss in samples stored at 4, -20 and -70°C for periods up to 20 months. Samples repeatedly frozen and thawed did experience losses. Separations are carried out with an increasing hydrophobic gradient of methanol and water using a Gilson dual-pump HPLC apparatus equipped with an Altex C-18 reversed-phase column. Fluorescence detection is done using a broad excitation filter of 305-395 nm and a narrow band emission filter centered at 480 nm. Each analysis protocol should include a reagent blank treated identically to the unknown sample but with the cells or tissue omitted; this sample can be used to identify peaks arising from the reagents. A second blank, the unknown control, is prepared by reacting the thiols present in the cell extract with 2-pyridyl disulfide (PDS) or DTNB prior to derivatization with bimane. These reagents form stable derivatives with thiols which then no longer react with bimane. This blank indicates any intrinsically fluorescent nonthiol components such as plant pigments which may give a chromatographic peak unrelated to the bimane derivatization (Figure 1), and a possible reaction of non-thiol compounds with bimane.

To determine the thiol concentrations in marine phytoplankton, samples are collected on pre-baked glass fiber filters (Whatman GF/F) by filtering two-to-four liters of seawater. Bimane is added and samples are allowed to react as previously described. These particulate samples can have very low thiol concentrations; to decrease any background fluorescence of unreacted bimane, after reaction, the excess bimane is eliminated by gentle filtration and rinsing with filtered seawater. Care is taken to minimize vacuum and exposure to air before adding the bimane. The filters are then homogenized in acetonitrile, placed in disposable 5 ml centrifuge tubes, heated at 60°C for ten minutes, centrifuged and stored frozen until analysis. Prior to analysis, samples are concentrated by evaporation and 25 mM methane sulfonic acid is added to a final volume of 2 ml. All samples are re-centrifuged. Sample volumes of 20-100 μl are injected with a Waters WISP automatic injector. This method is described in more detail in Matrai and Vetter (36).

To determine the concentrations of reduced sulfur species in pore waters, the recovered core is placed in a nitrogen atmosphere in a glove bag and approximately 1 g of the sediment of interest is packed into a 1.5 ml disposable microcentrifuge tube capped with nitrogen. In the case of sediments shown in Table I, the sediment was taken at increments throughout the depth of the core. In the case of the comparison of the sulfur composition of external porewater with the blood composition of sediment dwelling bivalves (Figure 2), the cores were dissected until an animal was found in an undisturbed position and the mud immediately surrounding the animal was sampled. The tubes are spun at 12,000 x g for 2 minutes. After spinning, the supernatant pore water is immediately derivatized as described above. It should be mentioned that the 100 μl sample volume that we prefer is not a minimum sample volume. Sample volumes of as little as 20 μl provide ample material for replicate analyses.

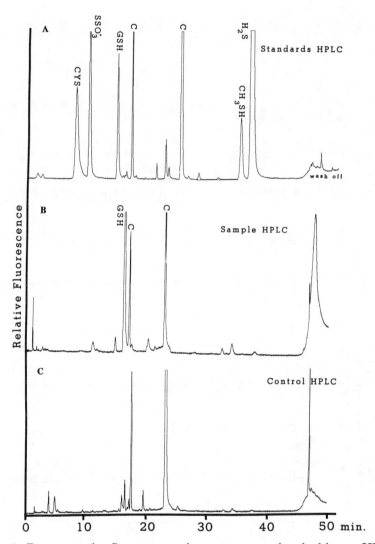

Figure 1. Representative fluorescence chromatograms using the bimane-HPLC method of analyzing sulfur compounds. (A) Standard containing cysteine (cys), thiosulfate (SSO_3^{2-}), methanethiol (CH_3SH), and sulfide (H_2S). (B) Field sample showing glutathione (GSH) and non-thiol (C) peaks present in the control sample. (C) Control sample where authentic thiols in sample were prereacted with PDS rendering them unreactive to bimane. The remaining peaks are fluorescent contaminants in the sample or the reagents. (Reproduced with permission from Reference 36. Copyright 1988 American Society of Limnology and Oceanography, Inc.)

Table I. Comparison Between the Methylene Blue (26), the DTNB (28) and the Bimane-HPLC (24) Procedures. Sediment Core Samples were Obtained from the Mission Bay Saltmarsh and the San Diego Bay Solar Salt Pond, Both Sites are Near San Diego, CA. Depth is Indicated in cm, Salinity in o/oo, Sulfate in mM, and all Thiols in μM

site	depth	salinity	pH	sulfate	methylene blue	DTNB	bimane sulfide	bimane thiosulfate	bimane sulfite	bimane total thiol
salt marsh no plants	0-1	49	7.1	36	1650	1710	1710	45	16	1770
	1-2	51	7.2	37	3530	3250	2880	71	42	2990
	2-3	49	7.3	32	3310	3250	2900	55	56	3010
	3-4	48	7.4	31	3110	2850	2390	2	33	2430
	6-7	43	7.1	29	1420	1480	1200	4	39	1240
	10-11	40	7.1	31	615	1150	642	84	15	740
	14-15	40	7.1	30	262	248	204	78	16	298
	19-20	38	7.3	29	27	20	8	40	3	51
salt marsh Spartina zone	0-2	39	6.4	32	6	0	0	0	0	0
	2-3	39	6.7	32	48	0	0	3	1	4
	3-4	38	6.8	32	247	nd	9	34	0	42
	6-7	37	6.7	30	279	nd	258	41	6	306
	10-11	36	6.6	30	322	250	252	45	4	302
	14-15	35	6.8	30	213	258	207	16	5	228
	19-20	35	6.9	30		nd	131	20	6	156
solar salt ponds core #1	0-1	96		79	70	77	0	48	0	48
	1-2	93		74	169	164	83	51	5	139
	2-3	89		62	198	181	153	79	8	240
	3-4	85		57	328	302	116	152	26	293
	6-7	82		50	1720	1430	1360	270	37	1660
	10-11	81		45	2620	2820	1340	443	64	1850
	14-15	80		53	2340	2340	1980	227	36	2250
	19-20	83		56	2660	2750	2160	228	42	2430
solar salt ponds core #2	0-1	128		121	459	445	289	57	7	350
	1-2	126		130	930	968	764	57	13	843
	2-3	122		109	1160	1210	681	80	11	771
	3-4	117		100	713	668	390	130	8	528
	6-7	110		86	1890	1770	678	156	16	850
	10-11	108		74	702	695	155	279	14	448
	14-15	106		88	1740	2020	706	236	14	955
	19-20	103		81	1100	1120	319	237	16	571

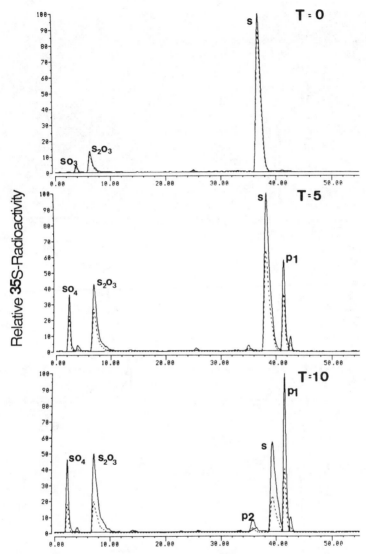

Figure 2. Radioactivity chromatogram of sulfur compounds derivatized with monobromobimane. The reversed-phase HPLC separation is based on the hydrophobic properties of the bimane-sulfur adducts but peak area is based on ^{35}S-radioactivity of the compounds. At time 0 sulfite and thiosulfate impurities are present before addition of the hepatopancrease tissue homogenate. This was a 60 min experiment to determine the sulfide detoxifying functions of the hepatopancrease of the hydrothermal vent crab *Bythograea thermydron*. During this time the proportion of radioactivity in sulfide rapidly decreases and thiosulfate and sulfate accumulate as end products. Two intermediates, p1 and p2 accumulate then decrease during the experiment. The two intermediates are believed to be polysulfides based on similar elution times of polysulfide standards. (Figure is the unpublished chromatograms from the data in Vetter et al. (24).) *continued on next page.*

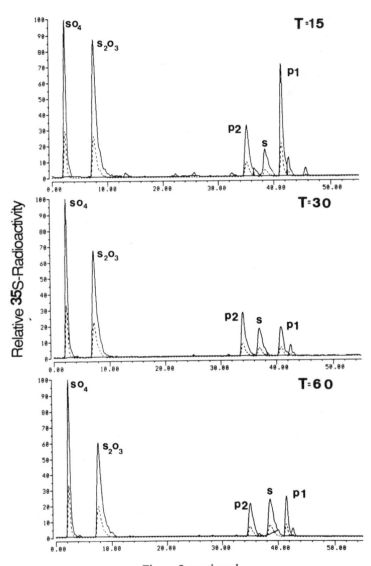

Figure 2 continued.

While pipet error increases at these volumes, it is sometimes necessary when obtaining blood samples from small organisms.

The study of pathways of sulfide metabolism in bacterial-invertebrate symbioses and in benthic invertebrates exposed to potentially lethal concentrations of hydrogen sulfide has benefited greatly from the ability to use ^{35}S-radiolabeled sulfide and thiosulfate in experiments. We have extended the original bimane method to include the use of these compounds by coupling a flow-through scintillation counter in-line with the fluorescence detector (24). After passing through the fluorescence detector, the eluted peak mixes with an equal volume of scintillation cocktail and passes in front of the photomultipliers. Two simultaneous chromatograms are produced: one showing all of the fluorescent thiol peaks, and another showing the radioactivity present in the different peaks.

The combination of radiolabeled sulfide and the bimane-HPLC method is particularly powerful because one of the main obstacles to the use of labeled sulfide is, that aside from radioactive decay, the compound is subject to rapid oxidation in the presence of air. The breakdown products of chemical sulfide oxidation are the same as those of biological oxidation. Previously it has been impossible to check routinely the purity of the purchased isotope and its subsequent purity during a series of experiments. It is our experience that newly purchased sodium sulfide sometimes contains up to 10% thiosulfate as well as traces of sulfite and sulfate (Figure 2), and that the sulfide once hydrated readily oxidizes if stored in a normal refrigerator.

As opposed to normal scintillation counting, the bimane-HPLC scintillation counting method will tell precisely the amount of ^{35}S-sulfide relative to contaminating oxidation products of the newly purchased isotope and at the start of each experiment. By dissolving a shipment of ^{35}S-sulfide in degassed distilled water and immediatly aliquoting the hydrated solution into small volume (100 μl) cryovials stored in liquid nitrogen, we are now able to economically use ^{35}S-sulfide for metabolic experiments while assuring that each day experiments will be done with pure unoxidized isotope (Figure 2). The unused portion of the thawed isotope is discarded. The liquid nitrogen stored sulfide remains unoxidized for over six months. Since the bimane-HPLC method gives a fluorescent measure as well as a radioactive measurement the stoichiometry of reactions in which the product contains more than one sulfur atom can be determined. For example, if radiolabeled sulfide is oxidized to thiosulfate, the thiosulfate could contain one radiolabeled sulfur if the other sulfur comes from sulfite in the medium or two if both sulfurs come from the labeled sulfide. Since the fluorescence of thiosulfate is constant on a per mole basis while the radioactivity signal will double if two radiolabeled sulfurs are present, it is possible to determine the nature of the reaction.

An unexpected advantage of the use of ^{35}S-sulfide is that one of the end products of radiolabelled sulfide metabolism, sulfate, which does not react with bimane, can be detected as a radioactive peak occurring at the beginning of the chromatogram (Figure 2). This peak can be identified as sulfate and not simply unreacted thiols by adding barium chloride to the sample. Barium chloride will precipitate out sulfate as the barium salt and the peak will disappear from the chromatogram. This allows, for the first time, the measurement of the complete metabolism of sulfide to sulfate with a single method (24).

An important consideration in the use of flow-through scintillation counting is that the sample resides in the counting chamber for a short length of time as the peaks move through. This means that, unlike conventional liquid scintillation counting, the counting time can not be varied appreciably. Consequently, the specific activity of the isotope being used must be quite high for good resolution of activity in peaks. The use of 35-S has been most useful in small volume experiments such as the mitochondrial sulfide oxidation experiments presented here.

Results and Discussion

Reduced Sulfur Compounds in Marine Phytoplankton. Marine phytoplankton are the principal producers of sulfur amino acids in the oceans (38) and a primary source of volatile organic sulfur compounds (16). The production of organic sulfur and its concentration in surface ocean waters can depend on biological processes such as the abundance, species composition, and physiological state of phytoplankton (39-41) as well as recycling and lateral transport (42). In addition, zooplankton grazing and bacterioplankton activity may also be important in the processes leading to the production or release of organic sulfur (12,43).

Reduced sulfur compounds have been found in concentrations up to 600 pM of glutathione in the Southern California Bight and in Saanich Inlet (36). Several reduced sulfur compounds were detected of which cysteine, thiosulfate, glutathione, methanethiol and sulfide were identified. Of these thiols, glutathione was, as expected, the most common and abundant thiol in particulate matter, the others having been seen sporadically and in much lower concentrations. Particulate glutathione (PGSH) vertical distributions showed similar trends to those of chlorophyll a (Figure 3) in both areas, sometimes showing a surface or subsurface maximum, and decreasing with depth.

In coastal waters, the vertical distribution of thiols could be due to physical factors, such as light and nutrients, that influence sulfur incorporation, mixing, or the depth-variable taxonomic composition of the algal population. Preliminary results reported by Matrai and Vetter (36) on effects of light and nutrients on the production of particulate glutathione by phytoplanktonic populations in the southern California Bight suggest that light can be an important factor. Low light levels, or exposure to low light intensities, clearly decreased the concentrations of PGSH present in the samples. Phytoplankton show light-dependent sulfur uptake (44). Furthermore, Anderson (45) has indicated that the availability of glutathione as a carrier in sulfate reduction may regulate sulfate uptake and assimilation. However, high light did not necessarily enhance the production of thiols. In the case of nutrients, greater levels of nitrate failed, in general, to elicit a response in thiol production. One possible explanation of these results is that phytoplankton cells were not nitrate-limited, as might be expected in the Bight (46). Fahey et al. (47) also noted higher GSH concentrations for light grown phytoplankton cultures than in darker grown ones.

Reduced Sulfur Compounds in Marine Sediments. To determine the applicability of the bimane-HPLC technique to measure reduced sulfur compounds in sediment porewater samples, we compared the results of the methylene blue method of Cline (26), the DTNB procedure of Ellman (28) and the bimane-HPLC procedure outlined above. Cores included came from a Spartina foliosa marsh in Mission Bay (near San Diego, California), and an evaporation pond for the production of salt in south San Diego Bay (Table I).

The methylene blue technique is a measure of hydrogen sulfide whereas DTNB should measure sulfide and any other organic or inorganic thiol which is present in porewater. Ideally, the bimane sulfide peak should correspond to the methylene blue concentration and the bimane total thiol concentration should correspond to the DTNB estimate. In saltmarsh cores, there was good general

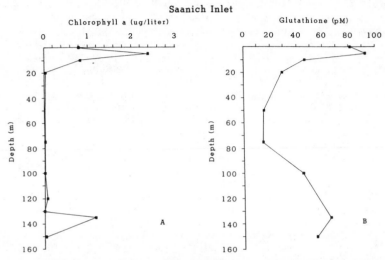

Figure 3. An example of vertical distributions of chlorophyll *a* (A) and particulate glutathione (B) in Saanich Inlet, B.C. (Reproduced with permission from Reference 36. Copyright 1988 American Society of Limnology and Oceanography, Inc.)

agreement in the sediment profiles of the different methods both in the *Spartina* zone and in the unvegetated mud flat. The sulfide concentration as determined by the bimane method was lower than that determined by methylene blue. The total thiol as determined by bimane was very close to the value determined by DTNB but was still slightly lower. Since both methods were standardized against the same sulfide solution there must be differences in the porewater which are reflected in the experimental results. No standard additions were performed. One possible explanation is that the methylene blue method and the DTNB method react with particle and protein bound thiol groups and produce color. No polysulfides were observed. With bimane the fluorescent reporter group remains bound to the sulfur containing compound and precipitates prior to running on the HPLC. If particulate and protein thiols are a large component, it would explain differences in the two methods.

In the solar evaporation ponds, salinities in the cores reached almost four times oceanic values. In these cores the concentration profile of bimane sulfide with depth also tracked that of methylene blue sulfide and bimane total reduced sulfur tracked DTNB. However, the difference between the bimane method and the other two methods is unacceptably large and suggests that there was some inhibition of the bimane reaction. Pore water samples which were diluted to normal seawater salinity with 200 mM HEPES buffer pH 8 were not inhibited. Dilution will of course lead to a loss of sensitivity for trace thiols. Another factor which can effect the yield of the bimane reaction is the unusual pH's that are often encountered in sediment pore waters. Cores from Mono Lake (not shown) had extremely alkaline pH's of 9.8 to 10.1. This high a pH definitely has an effect on the bimane reaction. Pore water samples adjusted to pH 8 gave much higher results which were similar to those obtained from the methylene blue and DTNB method (data not shown). Thiosulfate and sulfite were present in micromolar quantities in cores from all habitats. Thiosulfate was highest in the salt pond cores where it occasionally was more abundant than sulfide. While methane thiol, glutathione, and other organic thiols can be detected by the bimane method, they were not abundant ($< 10 \mu M$) in the core samples we chose to analyze.

We have demonstrated that the bimane-HPLC method gives good results with some sediments. However, there are instances in which the method does not give results equivalent to established methods. At this time we have not undertaken extensive experiments to determine the inhibitory factors, other than pH, which can affect the bimane reaction in cores from extremely alkaline or high salt environments. The chemistry of these types of cores is extremely complex and it is up to investigators working with sediments to determine the applicability of the bimane-HPLC technique to their specific situation. As a first approach, dilution in a buffered pH 8 solution with metal chelating agents is useful although there is a loss of sensitivity due to dilution.

Metabolism of Pore Water Hydrogen Sulfide By the Mitochondria of a Sediment Dwelling Bivalve Mollusc Which Contains Endosymbiotic Sulfur Bacteria. The study of hydrothermal vent animals has stimulated research into how animals in high sulfide habitats resist the toxic effects of hydrogen sulfide (22-24) and potentially harvest the energy released during the oxidation of sulfide (48). As our studies have progressed, it is clear that animals which do not contain symbiotic bacteria, and the animal partners in sulfur symbioses, carry on an active, biochemically mediated, sulfide oxidation pathway (49).

The protobranch mollusc, *Solemya reidi*, is a shallow water bivalve which lives in very high sulfide environments and maintains an autotrophic mode of nutrition *via* symbiotic sulfur oxidizing bacteria located in the gills (50). It can be collected locally and maintained in seawater aquaria in the laboratory. It is our experimental model for sulfur-based bacteria-invertebrate symbiosis. The bimane-HPLC method has been invaluable for the study of sulfur metabolism in this symbiosis. *Solemya* are exposed to high concentrations of sulfide in the sewage sludge that they inhabit. Since the mantle tissue is thin and covers a wide surface area on the inside of the shell it was expected that the blood of the animal might contain equivalent amounts of sulfide. We compared the sulfur constituents in sediment pore water surrounding the organisms with the blood's sulfur composition. While the sediment porewaters did contain high amounts of hydrogen sulfide, the blood of the six different clams contained almost no sulfide (Figure 4). Possible explanations for this observation are that hydrogen sulfide is excluded from the animal or that it enters and is rapidly metabolized.

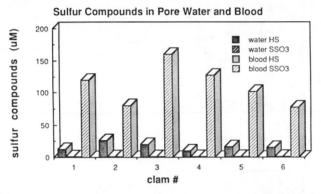

Figure 4. Reduced sulfur measurements in the blood and immediately surrounding pore water for 6 individuals of *Solemya reidi* collected from one square meter of sediment at the Los Angeles sewage outfall in Santa Monica Bay, CA. Water SSO_3^{2-} and blood HS⁻ levels were undetectable.

Since a portion of the hydrogen sulfide exists as the gaseous uncharged H_2S form at seawater pH it is difficult to believe that the organism can exclude it. Despite the low sulfide levels, the blood contains high levels of thiosulfate (Figure 4). This indicates that the animal has not adapted by being impermeable to sulfide but rather is actively detoxifying the sulfide diffusing into the animal. A similar detoxification mechanism, whereby sulfide is rapidly converted to the much less toxic metabolite thiosulfate, is present in the hydrothermal vent crab *Bythograea thermydron* and other decapod crustaceans (24).

Based on the results of Powell and Somero (22,25,48) there was reason to believe that much of this sulfide oxidizing activity was localized in the mitochondria of the animals. The addition of radiolabeled sulfide to isolated mitochondria has allowed us to show for the first time that the oxidation of sulfide by *Solemya* mitochondria is an organized biochemical process and not a non-specific oxidation caused by heme groups or other metal ions. As shown in Figure 5, the oxidation of radioactive sulfide in healthy mitochondria results in a stimulation, not an inhibition, of oxygen consumption and the production of thiosulfate as a final oxidation product. This corresponds to the appearance of thiosulfate in the blood. This activity was not present in rat mitochondria and was destroyed in *Solemya* mitochondria by boiling (Figure 5). The use of radiolabeled forms of sulfur is particularly powerful because the existing unlabeled pools can be measured, the stoichiometry of incorporation of a radiolabeled compound such as sulfide into polysulfurous compounds such as thiolsulfate or polysulfide can be monitored, and the metabolic intermediates can be traced.

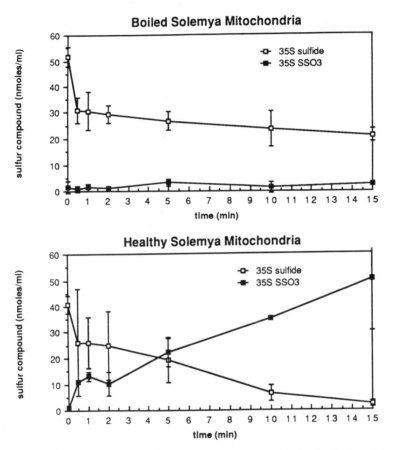

Figure 5. The metabolism of ^{35}-S radiolabeled sulfide by boiled and intact isolated *Solemya reidi* mitochondria. Time zero is prior to addition of mitochondria. When mitochondria are added there is a rapid decrease in sulfide concentration caused partially by dilution of the total volume and partially by binding of sulfide to protein. In the boiled preparation there is very little oxidation of sulfide and no appearance of oxidized products. In the healthy mitochondria the sulfide is rapidly oxidized to thiolsulfate. Sulfite and sulfate did not appear as oxidation products. The health of the isolated mitochondria was monitored by oxygen consumption rate in the chamber with succinate as substrate and by the ability to inhibit succinate stimulated respiration with respiratory inhibitors such as cyanide. Error bars represent plus and minus one standard deviation of the mean of three runs.

While sulfide is toxic to cytochrome (c) oxidase the most important enzyme of the respiratory chain, at concentrations in the tens of μM, thiosulfate is not. Thus, the animal has produced a soluble and excretable detoxification product and protected animal respiration. This thiosulfate still contains large amounts of energy and can serve as a substrate for bacterial chemoautotrophic metabolism. The bacterial endosymbionts can utilize the thiosulfate along with sulfide to fix CO_2 (50) and supply the host's nutritional needs.

In summary, it is the sensitivity, broad reactivity with reduced sulfur compounds, and the stability of the bimane-sulfur adducts that makes the bimane-HPLC method superior for the kinds of experiments which we are engaged in. The method should find increasing usage among marine scientists who wish to store and analyze at a later time large numbers of samples.

Literature Cited

1. Schindler, D. W. Science 1988, 239, 149-57.

2. Bates, T. S.; Charlson, R. J.; Gammon, R. H. Nature 1987, 329, 319-21.

3. Charlson, R. J.; Lovelock, J. E.; Andreae, M. O.; Warren, S. G. Nature 1987, 326, 655-61.

4. Lonsdale, P. F. Earth Planet. Sci. Let. 1977, 36, 92-110.

5. Spiess, F. N.; Macdonald, K. C.; Atwater, T.; Ballard, R.; Carranza, A.; Cordoba, D.; Cox, C.; Diaz Garcia, V. M.; Francheteau, J.; Guerrero, J.; Hawkins, J.; Haymon, R.; Hessler, R.; Juteau, T.; Kastner, M.; Larson, R.; Luyendyk, B.; Macdougall, J. D.; Miller, S.; Normak, W.; Orcutt, J.; Rangin, C. Science 1980, 207, 1421-33.

6. Cavanaugh, C. M.; Gardiner, S. L.; Jones, M. L.; Jannasch, H. W.; Waterbury, J. B. Science 1981, 213, 340-2.

7. Felbeck, H.; Childress, J. J.; Somero, G. N. Nature 1981, 293, 291-3.

8. Childress, J. J.; Felbeck, H.; Somero, G. N. Sci. Am. 1987, 255, 115-20.

9. Lake, J. A. Nature 1988, 331, 184-6.

10. Gilbert, H. F. Methods Enzymol. 1984, 107, 330-51.

11. Vairavamurthy, A.; Andreae, M. O.; Iverson, R. L. Limnol. Oceanogr. 1985, 30, 59-70.

12. Dacey, J. W. H.; Wakeham, S. G. Science 1986, 233, 1314-6.

13. Dickson, D. M.; Kirst, G. O. Planta 1986, 167, 536-43.

14. Turner, S. M.; Malin, G.; Liss, P. S.; Harbour, D. S.; Holligan, P. M. Limnol. Oceanogr. 1988, 33, 364-75.

15. Keller, M. D.; Bellows, W. K.; Guillard, R. R. L. In Biogenic Sulfur in the Environment; Saltzman, E. S.; Cooper, W. J., Eds.; American Chemical Society; this volume.

16. Andreae, M. O. In The role of air-sea exchange in geochemical cycling; Buat-Menard, P., Ed.; Reidel: 1986; pp 331-62.

17. Howarth, R. W.; Teal, J. M. Amer. Nat. 1980, 116, 862-72.

18. Vairavamurthy, A.; Mopper, K. Nature 1987, 329, 623-5.

19. Luther, G. W.; Church, T. M.; Scudlark, J. R.; Cosman, M. Science 1986, 232, 746-9.

20. Kelly, D. P. Phil. Trans. R. Soc. Lond. 1982, B298, 499-528.

21. Vetter, R. D. Mar. Biol. 1985, 88, 33-42.

22. Powell, M. A.; Somero, G. N. Biol. Bull. 1985, 169, 164-81.

23. Arp, A. J.; Childress, J. J.; Vetter, R. D. J. Exp. Biol. 1987, 128, 139-58.

24. Vetter, R. D.; Wells, M. E.; Kurtsman, A. L.; Somero, G. N. Physiol. Zool. 1987, 60, 121-37.

25. Powell, M. A.; Somero, G. N. Science 1986, 233, 563-6.

26. Cline, J. D. Limnol. Oceanogr. 1969, 14, 454-8.

27. Gilboa-Garber, N. Anal. Biochem. 1971, 43, 129-33.

28. Ellman, G. L. Arch. Biochem. Biophys. 1959, 82, 70-2.

29. Jocelyn, P. C. Methods Enzymol. 1987, 143, 44-66.

30. Luther, G. W.; Giblin, A. E.; Varsolona, R. Limnol. Oceanogr. 1985, 30, 727-36.

31. Dyrssen, D.; Wedborg, M. Anal. Chim. Acta 1986, 180, 473-9.

32. Boulègue, J. J. Geochem. Explor. 1981, 15, 21-36.

33. Mopper, K.; Delmas, D. Anal. Chem. 1984, 56, 2557-60.

34. Vairavamurthy, A.; Mopper, K. ES&T, accepted.

35. Fahey, R. C.; Newton, G. L. Methods Enzymol. 1987, 143, 85-95.

36. Matrai, P. A.; Vetter, R. D. Limnol. Oceanogr. 1988, 33, 624-31.

37. Kosower, N.; Kosower, E. Methods Enzymol. 1987, 143, 76-84.

38. Anderson, J. W. Sulphur in biology.; University Park Press, 1978.

39. Ackman, R. G.; Tocher, C. S.; McLachlan, J. J. Fish. Res. Bd. Canada 1966, 23, 357-64.

40. Barnard, W. R.; Andreae, M. O.; Iverson, R. L. Cont. Shelf Res. 1984, 3, 103-13.

41. Holligan, P. M.; Turner, S. M.; Liss, P. S. Cont. Shelf Res. 1987, 7, 213-24.

42. Matrai, P. A.; Eppley, R. W. Global Biogeochem. Cycles, subm.

43. Wakeham, S. G.; Howes, B. L.; Dacey, J. W. H. Nature 1984, 310, 770-2.

44. Cuhel, R. L.; Ortner, P. B.; Lean, D. R. S. Limnol. Oceanogr. 1984, 29, 731-44.

45. Anderson, J. W. 1980. In The Biochemistry of Plants; Stumpf, P. K.; Conn, E. E., Eds.; Academic: 1980; Vol. 5, pp 203-23.

46. Mullin, M. M. In Plankton Dynamics of the Southern California Bight; Eppley, R. W., Ed.; Verlag: 1986; pp 216-73.

47. Fahey, R. C.; Newton, G. L. J. Mol. Evol. 1987, 25, 81-8.

48. Powell, M. A.; Somero, G. N. Biol. Bull. 1986, 171, 274-90.

49. Powell, M. A.; Vetter, R. D.; Somero, G. N. In Comparative Physiology: Life in Water and on Land; DeJours, P.; Bolis, L.; Taylor, C. R.; Weibel, E. R., Eds.; Fidia Research Series; IX-Liviana Press: Padova, 1987; pp 241-50.

50. Anderson, A. E.; Childress, J. J.; Favuzzi, J. A. J. Exp. Biol. 1987, 133, 1-31.

RECEIVED October 13, 1988

Chapter 17

Enzymatic Steps and Dissimilatory Sulfur Metabolism by Whole Cells of Anoxyphotobacteria

Ulrich Fischer

Fachbereich Biologie, AG Gemikrobiologie, Universitaet Oldenberg, D−2900 Oldenberg, Federal Republic of Germany

Anoxyphotobacteria, comprising the purple and green phototrophic bacteria, perform an anoxygenic photosynthesis under anaerobic conditions and derive energy from light. Because they lack photosystem II they cannot use water as electron donor for their photosynthesis and thus never evolve oxygen in the light. Instead of water they are dependent on reduced sulfur compounds, such as sulfide, elemental sulfur or thiosulfate, hydrogen or simple organic carbon compounds as electron donors for photosynthesis and carbon dioxide reduction. In general, sulfide is oxidized via sulfite to sulfate, while thiosulfate can either be oxidized only to tetrathionate or may be split into sulfide and sulfite. Both compounds are then further oxidized to sulfate. During anaerobic sulfide or thiosulfate oxidation, elemental sulfur appears as sulfur globules inside or outside the cells. Sulfur metabolism in Anoxyphotobacteria has two main functions:

i) the assimilatory sulfate reduction is necessary for the biosynthesis of sulfur containing cell compounds, such as sulfolipids, amino acids (cysteine, methionine), vitamins (biotin), iron-sulfur proteins (ferredoxins) or coenzymes (acetyl CoA),

ii) the dissimilatory sulfur metabolism supplies these organisms with electrons necessary for their anaerobic photosynthesis. In most enzymatic reactions of anoxygenic sulfur oxidation cytochromes or iron-sulfur proteins serve as suitable electron carrier proteins.

The following presentation will give a general overview of our current knowledge of the in vivo and in vitro dissimilatory sulfur metabolism in Anoxyphotobacteria.

Dissimilatory Sulfur Metabolism by Whole Cells

Anoxyphotobacteria live in the anaerobic zones of waters, where they are sufficiently provided with sulfide and light. Typical habitats are salt marshes, lakes, sewage and coastal lagoons, hot springs, salt and soda lakes and the so-called "Farbstreifensandwatt" (versicolored, sandy tidal flat) (see 1-2a).

0097−6156/89/0393−0262$06.00/0
© 1989 American Chemical Society

According to our present knowledge the classical phototrophic bacteria comprise six families, shown in Figure 1 (see 3). The Anoxyphotobacteria perform an anoxygenic photosynthesis and cannot use water as photosynthetic electron donor and thus never produce oxygen in the light. Instead of water, most anoxygenic phototrophic bacteria use reduced sulfur compounds, at the oxidation level below that of sulfate, as electron donors for their photosynthesis and carbon dioxide reduction (1,4). In general, sulfur metabolism in phototrophic bacteria has four different functions:

1. In an aerobic or microaerobic dark metabolism a chemolithotrophic sulfur metabolism supplies the organisms with electrons necessary for growth. Most of the small cell Chromatiaceae, like Thiocapsa roseopersicina or Chromatium gracile, can grow chemolithotrophically or heterotrophically in the dark under microaerobic conditions. Under these conditions the organisms are able to assimilate simple organic carbon compounds, such as acetate, pyruvate or glycerol, but only in the presence of thiosulfate. On the other hand, large cell Chromatiaceae and Chlorobiaceae are strictly anaerobically living organisms and cannot live under these conditions. Purple nonsulfur bacteria, which can grow aerobically in the dark, change over from a phototrophic metabolism to respiration and stop oxidizing sulfur compounds (see 1,5-7).

2. In a fermentative dark metabolism, sulfur compounds serve as electron acceptors. In 1968 van Gemerden (8) could demonstrate that storage carbohydrates in form of polyglucose, which is synthesized in the light, disappeared in the dark. Simultaneously, poly-β-hydroxybutyric acid (PHB) is produced, CO_2 is released and the intracellularly stored elemental sulfur is reduced to sulfide. During this dark fermentative metabolism elemental sulfur serves as an electron sink for electrons of carbohydrate fermentation. The ratio between storage carbohydrate monomer consumed, sulfur reduced, PHB monomer produced and sulfide released is 1:3:1:3. The net gain per converted carbohydrate monomer is 3 ATP, enough to allow the organisms to survive dark periods (8).

3. The assimilatory sulfate reduction is necessary for the biosynthesis of sulfur containing cell compounds, such as sulfolipids, amino acids (cysteine, methionine), vitamins (biotin), iron-sulfur proteins (ferredoxins) or coenzymes. In all organisms sulfate reduction follows a unique scheme by activating the sulfate molecule by adenosine-5'-triphosphate (ATP) to perform adenosine-5'-phosphosulfate (APS) (Figure 2). Then the sulfate moiety in the APS molecule (oxidation level of the sulfur atom +6) is reduced to the oxidation level of sulfide (-2) by several enzymatic steps. In this state the sulfur atom is ready to be used for the biosynthesis of sulfur containing amino acids (see 9). In the individual organisms, sulfate reduction is initiated via APS or 3'-phosphoadenosine-5'-phosphosulfate (PAPS) (9-11). For Rhodospirillaceae, Imhoff (11) could recently demonstrate that phylogenetically related species also use the same sulfonucleotide (APS or PAPS).

4. The dissimilatory sulfur metabolism provides the organisms with electrons for the anaerobic photosynthesis and will be explained more precisely on the following pages.

Photosynthetic carbon fixation and oxidation of a reduced sulfur compound in Anoxyphotobacteria are stoichiometrically linked by van Niel's equations (12,1):

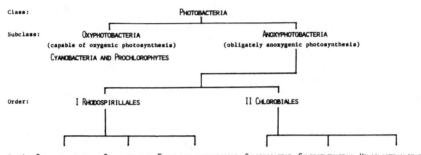

Class: PHOTOBACTERIA

Subclass: OXYPHOTOBACTERIA   ANOXYPHOTOBACTERIA
              (capable of oxygenic photosynthesis)   (obligately anoxygenic photosynthesis)
              CYANOBACTERIA AND PROCHLOROPHYTES

Order: I RHODOSPIRILLALES   II CHLOROBIALES

Family: RHODOSPIRILLACEAE CHROMATIACEAE ECTOTHIORHODOSPIRACEAE CHLOROBIACEAE CHLOROFLEXACEAE HELIOBACTERIACEAE

Figure 1. Higher taxa of phototrophic bacteria (modified after (3) and (9)).

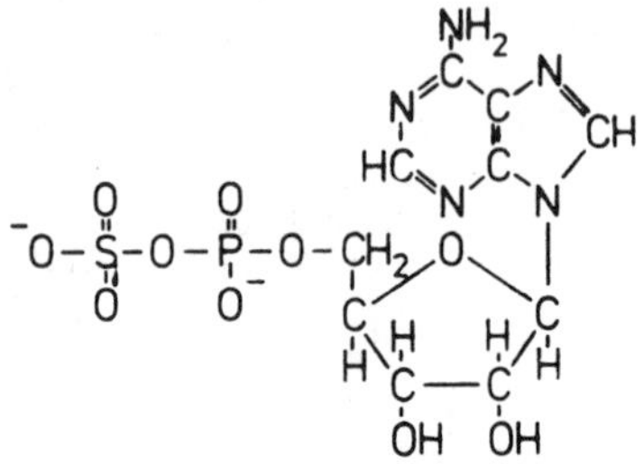

Figure 2. Structure of the adenosine-5'-phosphosulfate (APS) molecule
(Reproduced with permission from Ref. 9. Copyright 1986 U. Fischer).

For anaerobic sulfide oxidation

$$CO_2 + 2H_2S \text{ --- light ---} > CH_2O + H_2O + 2S^o$$

$$3CO_2 + 2S^o + 5H_2O \text{ --- light ---} > 3CH_2O + 2H_2SO_4$$

The summarizing equation of sulfide oxidation to sulfate is:

$$2CO_2 + H_2S + 2H_2O \text{ --- light ---} > 2CH_2O + H_2SO_4$$

and for anaerobic thiosulfate oxidation

$$CO_2 + 2Na_2S_2O_3 + H_2O \text{ --- light ---} > CH_2O + 2S^o + 2Na_2SO_4$$

$$3CO_2 + 2S^o + 5H_2O \text{ --- light ---} > 3CH_2O + 2H_2SO_4$$

The summarizing equation of thiosulfate oxidation to sulfate is:

$$3CO_2 + Na_2S_2O_3 + 3H_2O \text{ --- light ---} > 2CH_2O + H_2SO_4 + Na_2SO_4$$

(CH_2O symbolizes organic material)

The existence of a real stoichiometric relationship is confirmed by the fact that the organisms stop oxidizing sulfide or elemental sulfur when they have consumed all CO_2. But the further oxidation to sulfate immediately starts again, when CO_2 is added again to the culture ([13]).

Van Niel's equations clearly show that elemental sulfur is an intermediate during anaerobic sulfide or thiosulfate oxidation by whole cells. But it has turned out that elemental sulfur is not the only intermediate sulfur compound, as generally presumed. Other sulfur compounds formed during such a process are sulfite, thiosulfate, polysulfides or tetrathionate. These intermediates are also further oxidized to sulfate, if the organisms are able to do so ([14-17]).

During anaerobic sulfide oxidation only the members of Chromatiaceae store the elemental sulfur inside their cells, while the Ectothiorhodospiraceae, Chlorobiaceae and Chloroflexaceae deposit elemental sulfur outside their cells (see [4]). For a long time it was assumed that the sulfur globules contain the orthorhombic cycloocta sulfur consisting of S_8-rings ([1]).

Recently it was shown that the sulfur globules of Chromatiaceae (inside) and of Ectothiorhodospiraceae (outside) contain only very small amounts of S_8-rings but more over long-chained polysulfides or polythionates with S^- or SO_3^- groups at the end of the chains ([18]). But it seems that the structure of biological sulfur globules is not unique. The sulfur globules of Thiobacillus ferrooxidans consist of a hydrophobic core with mainly S_8-rings. On the surface of this core long-chained polythionate ions are attached which make the globule hydrophilic so that it is surrounded by water of hydration ([19]). On the basis of their findings, Steudel and coworkers ([19]) suggested the following formula for elemental sulfur:

$$xS_8 \cdot yH_2S_nO_6 \cdot zH_2O.$$

The above-mentioned findings on the structure of sulfur globules are more or less confirmed by Mas and van Gemerden ([20]) concerning the structure of the globules of Chromatium. The authors assumed that the sulfur globules could consist of a concentrical array of layers of sulfanes, arranged parallel to one another and that the space between the layers would allow the presence of water, due to the hydrophilic nature of the terminal groups of the sulfanes (20). Only the intracellular sulfur globules of Chromatiaceae proved to be surrounded by a protein layer consisting of a single protein with a molecular weight of 13,500 daltons (21). The existence of such a protein layer could not clearly be proved for the extracellularly stored sulfur globules of Chlorobiaceae and Ectothiorhodospiraceae (18).

Cinematographic studies have clearly shown that the formation of sulfur globules in Chromatiaceae begins all over the cytoplasm (H.G. Trueper, personal communication) and not by invagination of the cytoplasmic membrane, as assumed for a long time (21). On the other hand, nothing is known about the formation of extracellular sulfur globules by green sulfur bacteria and Ectothiorhodospiraceae. It is imaginable that the sulfide oxidation by these organisms occurs in the periplasmic space leading at first to the formation of polysulfides, which can pass the cell wall and outer membrane and then aggregate to sulfur globules which normally remain attached to the cells. Extracellular polysaccharide capsules, tubes or pili are responsible for such an attachment process (22).

Anoxyphotobacteria utilize mainly sulfide, thiosulfate or elemental sulfur as photosynthetic electron donors. The ability to use sulfite or tetrathionate is restricted to only a few species (4,23).

For a long time it was assumed that the Rhodospirillaceae were only capable of performing an assimilatory sulfate reduction and that sulfide was even toxic to these organisms. But it has turned out that the oxidation of reduced sulfur compounds by purple nonsulfur bacteria is more widespread than first of all expected (see 9,24) and that some species, such as Rhodobacter sulfidophilus, Rhodobacter veldkampii or Rhodopseudomonas sulfoviridis are even dependent on reduced sulfur compounds. These organisms perform a dissimilatory sulfur metabolism comparable to that of phototrophic red and green sulfur bacteria (14). On the other hand, Rhodopseudomonas blastica, Rhodopseudomonas rutila or Rhodospirillum salinarium are not able to utilize reduced sulfur compounds for growth (see 9). During anaerobic sulfide or thiosulfate oxidation by Rhodospirillaceae different final oxidation products are formed. Some species form elemental sulfur which is deposited outside the cells and cannot be further oxidized, other ones directly oxidize the sulfur compound offered to sulfate without the formation of elemental sulfur or to tetrathionate (see 9,24). The latter product is formed during thiosulfate oxidation by Rhodopila globiformis. It is of interest to know that this organism can utilize only thiosulfate and that 95 % of the thiosulfate offered are oxidized to tetrathionate which is not further metabolized, while only 5 % of the thiosulfate are assimilated (25). The course of anaerobic sulfide oxidation by Rhodobacter sulfidophilus is also remarkable. This organism is the first species among phototrophic bacteria which excretes sulfite during sulfide or thiosulfate oxidation. The excreted sulfite is also further oxidized to sulfate but sulfite as sole sulfur source is not utilized (14). A more detailed compilation of anaerobic sulfur metabolism by purple nonsulfur bacteria was recently given by several authors (1,4,9,14,23-24). As one can see from this short presentation, phototrophic sulfur oxidation by Rhodospirillaceae is much more diverse than expected. On the contrary, sulfur oxidation in red and green sulfur bacteria shows a higher degree of unity.

All species of Chromatiaceae examined so far begin to oxidize the intracellularly formed elemental sulfur to sulfate even before sulfide is consumed by the cells. This means, sulfate appears in the medium simultaneously with the elemental sulfur stored. The course of an anaerobic sulfide oxidation by Chromatium warmingii is shown in Figure 3 (26).

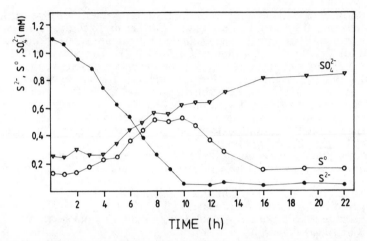

Figure 3.   Anaerobic sulfide oxidation by a growing culture of <u>Chromatium</u> <u>warmingii</u> (Reproduced with permission from Ref. 26. Copyright 1983 Verlag der Zeitschrift der Naturforschung).

All members of the Ectothiorhodospiraceae deposit the elemental sulfur outside their cells, which is formed via polysulfide. The average chain length of these polysulfides is 4. Only when sulfide is exhausted are these polysulfides further oxidized, and elemental sulfur droplets appear in the medium (1,16,18). Elemental sulfur is the final product during sulfide oxidation by the green coloured (containing bacteriochlorophyll b) extreme halophilic and alkaliphilic species E. abdelmalekii and E. halochloris (16,18), while red coloured species (containing bacteriochlorophyll a) of the Ectothiorhodospiraceae are able to further oxidize the elemental sulfur to sulfate, but only when sulfide is completely consumed by the cells (27). A different course during anaerobic sulfide oxidation has been described for thiosulfate and non-thiosulfate utilizing Chlorobiaceae.

Non-thiosulfate utilizing species of green sulfur bacteria oxidize sulfide directly to elemental sulfur which can only be further oxidized to sulfate when sulfide is completely consumed. An example for such an oxidation process is illustrated for Pelodictyon luteolum in Figure 4 (28). The constant level of thiosulfate in the medium is formed by a non-enzymatic reaction of sulfide and elemental sulfur (28). The first sulfur intermediate formed by thiosulfate-utilizing species of Chlorobiaceae during anaerobic sulfide oxidation is thiosulfate, which appears in the medium before elemental sulfur is formed. Depending on the species, thiosulfate can be converted via elemental sulfur or not (see 4). For the first case, the anaerobic sulfide oxidation by Chlorobium vibrioforme f. thiosulfatophilum is shown in Figure 5 (17).

If phototrophic bacteria can utilize externally offered elemental sulfur as sole electron source, they will oxidize it directly to sulfate without the formation of any other sulfur intermediate (27,29). If only thiosulfate is available, Anoxyphotobacteria have two possibilities to metabolize it:
   i) they oxidize it to form tetrathionate (25,30) or
   ii) thiosulfate is split into elemental sulfur and sulfite. Both sulfur moieties are then oxidized to sulfate (30).

For Chromatium vinosum it has been reported that the utilization of thiosulfate is influenced by the pH of the medium. At pH 7.3 thiosulfate is cleaved into elemental sulfur and sulfite, while at pH 6.25 it is oxidized to tetrathionate which cannot be further metabolized (30).

It seems worth mentioning that sulfur disproportionation of elemental sulfur is performed only by the thiosulfate-utilizing species of Chlorobiaceae (see 1,31). In the absence of CO_2 and under strictly anaerobic conditions in the light, these organisms form sulfide and thiosulfate simultaneously as long as sulfide is flushed out with an inert gas. The stoichiometry of this reaction is:

$$4S^0 + 3H_2O \text{ --- light ---> } 2H_2S + H_2S_2O_3$$

Thiosulfate was probably formed by a non-enzymatic condensation of sulfite with sulfur, because small amounts of sulfite were detected during the performance of the experiment. After the addition of CO_2, the disproportionation stops immediately and therefore it seems that elemental sulfur may replace CO_2 as an electron acceptor in this process (see 1).

Recently, Bak and Pfennig (32) described an anaerobic disproportionation of thiosulfate and sulfite, respectively, to sulfate and sulfide by the sulfate-reducing bacterium Desulfovibrio sulfodismutans, according to the following reactions:

$$S_2O_3^{2-} + H_2O \text{ ---> } SO_4^{2-} + HS^- + H^+$$

$$4SO_3^{2-} + H^+ \text{ ---> } 3SO_4^{2-} + HS^-$$

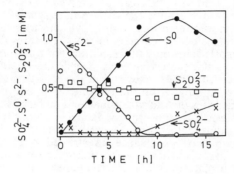

Figure 4. Anaerobic sulfide oxidation by a growing culture of Pelodictyon luteolum (Reproduced with permission from Ref. 28. Copyright 1982 Springer-Verlag New York Inc.).

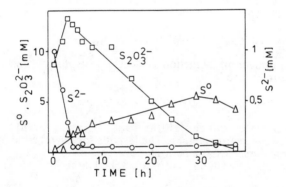

Figure 5. Anaerobic sulfide and thiosulfate oxidation by a growing culture of Chlorobium vibrioforme f. thiosulfatophilum (Reproduced with permission from Ref. 17. Copyright 1982 Springer-Verlag New York Inc.).

By the disproportionation of these inorganic sulfur compounds the organism is able to conserve energy for growth.

Enzymatic Steps in Dissimilatory Sulfur Metabolism

The oxidation or reduction of inorganic sulfur compounds takes for granted that the organisms are equipped with the corresponding enzymes and the necessary electron transfer proteins. About ten different enzymes are involved in phototrophic sulfur metabolism and mainly cytochromes and iron-sulfur proteins serve as potent electron carriers in most reactions (for review see 1,4,9). In most enzymatic steps only two electrons are transferred, except for the step from sulfite to sulfide and vice versa (see single equations below). In this reaction six electrons are transferred at once, without releasing any intermediate sulfur compound from the enzyme. The enzymes found and characterized in the families of Anoxyphotobacteria so far are summarized in Tables I-IV. The Tables clearly show that no family is uniformly equipped with the same enzymes of sulfur metabolism. Only the thiosulfate-utilizing enzymes occur in all families, while APS-reductase is restricted to the Chromatiaceae and Chlorobiaceae. Organisms lacking this enzyme have a sulfite: acceptor oxidoreductase instead. Subsequently, the single reactions and enzymes are described somewhat more precisely.

Thiosulfate Utilizing Enzymes

Three enzymes are responsible for thiosulfate utilization (4,9,33):
1) a thiosulfate:acceptor oxidoreductase

$$2 \, S_2O_3^{2-} + acceptor_{ox} \text{ --->} S_4O_6^{2-} + acceptor_{red}$$

Ferricyanide or cytochromes function as electron acceptors for the enzyme.
2) a thiosulfate reductase

$$S_2O_3^{2-} + e^--donor_{red} \text{ --->} SO_3^{2-} + S^{2-} + e\text{-}donor_{ox}$$

The enzyme can use reduced glutathione, reduced methyl viologen or dihydrolipic acid as electron donors. The last one seems to be the most efficient electron donor and may act as a natural reductant.
3) a thiosulfate sulfur transferase, also called rhodanese

$$S_2O_3^{2-} + CN^- \text{ --->} CNS^- + SO_3^{2-}$$

The enzyme needs thiophilic anions, such as CN^- or SO_3^{2-} as sulfur acceptor substrates.
    The first-named enzyme binds two thiosulfate molecules together to form tetrathionate, while the other two enzymes split thiosulfate into sulfide and sulfite.
    In Chlorobium limicola f. thiosulfatophilum cytochrome c-551 serves as electron acceptor of thiosulfate oxidation, before the electrons flow to oxidized bacteriochlorophyll via a soluble small cytochrome c-555 and a membrane-bound cytochrome (9). The absence of cytochrome c-551 in the non-thiosulfate-utilizing Chlorobiaceae might be one explanation why these organisms cannot use thiosulfate (4).
    For Chromatium vinosum three possible electron acceptors, a mobile cytochrome c-551 located in the periplasmic space, a membrane-bound

Table I. Enzymes of Dissimilatory Sulfur Metabolism of Rhodospirillaceae (data taken from 9,14,23,34)

	Rhodopseudomonas palustris	*Rhodopseudomonas sulfoviridis*	*Rhodobacter capsulatus*	*Rhodobacter sulfidophilus*	*Rhodopila globiformis*	*Rhodomicrobium vannielii*	*Rhodospirillum rubrum*	*Rhodobacter veldkampii*	*Rhodobacter adriaticus*
Rhodanese	+	+(2)	+	+	+	+	+	+	-
Thiosulfate-Reductase	+	+	+	+	+		+		-
Thiosulfate-Acceptor-Oxidoreductase	+	(+)	+	(+)	+				-
Reverse Sulfite-Reductase	-	-		-	+				
Sulfite-Acceptor-Oxidoreductase	+(m)	+(m)	-	+(m)	+			+(m)	+(m)72% / +(s)28%
APS-Reductase	-	-	-	-	-	-	-	-	-
ADP-Sulfurylase	-	-	-	-	-				-

(+) = low activity; - = not present; (m) = membrane-bound; s = soluble

Table II. Enzymes of Dissimilatory Sulfur Metabolism of Chromatiaceae
(data taken from 9,23,35)

	Thiocapsa roseopersicina	*Chromatium vinosum*	*Thiocystis violacea*	*Chromatium minutissimum*	*Chromatium gracile*	*Chromatium purpuratum*	*Chromatium minus*	*Chromatium warmingii*	*Thiocystis gelatinosa*
Rhodanese	+	+					+	+	+
Thiosulfate-Reductase	+	+					+	+	+
Siroheme-Sulfite-Reductase	-	+						+	
APS-Reductase	+	+	+	+	-	-		+	
ADP-Sulfurylase	+	+	+	+	+	-		+	
ATP-Sulfurylase	(+)	(+)		(+)	(+)	(+)			
Sulfite-Acceptor-Oxidoreductase	+	+	+	+	+	+	+	+	
Thiosulfate-Acceptor-Oxidoreductase	+	+							

(+) = very low activity; - = not present

Table III. Enzymes of Dissimilatory Sulfur Metabolism of
Ectothiorhodospiraceae (data taken from 9,23,27,36)

	E. mobilis	*E. shaposhnikovii*	*E. vacuolata*	*E. halophila*	*E. halochloris*	*E. abdelmalekii*
Rhodanese	+	+	+			
Thiosulfate-Reductase	+	+				
Siroheme-Sulfite-Reductase			-			
APS-Reductase	+	-	-	-	-	-
ADP-Sulfurylase	+	-	+	+		
ATP-Sulfurylase	-					
Sulfite-Acceptor-Oxidoreductase	+	+	+	+		

(+) = very low activity; - = not present; E. = Ectothiorhodospira

Table IV. Enzymes of Dissimilatory Sulfur Metabolism of Chlorobiaceae
(data taken from 9,23,33,37,38)

	Chlorobium limicola	*Chlorobium limicola f. thiosulfatophilum*	*Chlorobium vibrioforme f. thiosulfatophilum*
Rhodanese	-	+	+(2)
Thiosulfate-Reductase	-	+	+
Thiosulfate-Acceptor-Oxidoreductase		+	+
Sulfite-Acceptor-Oxidoreductase		(+)	-
APS-Reductase		+	+
ADP-Sulfurylase		-	+
ATP-Sulfurylase		+	-

(+) = very low activity; - = not present

flavocytochrome c-552 and a high-potential-iron sulfur protein (HIPIP) have been described (4). Even now there is no clear evidence whether these three components are part of a single electron transport chain or whether there exist three independent pathways (4,9). Studies of Kusai and Yamanaka (39) about the influence of sulfur compounds on the cytochrome pattern in Chlorobium limicola f. thiosulfatophilum have shown that the organisms contained more flavocytochrome when growing with sulfide than those which had had only thiosulfate available. These findings are in contrast to the results obtained with Chlorobium vibrioforme f. thiosulfatophilum (17). The ratio of cytochrome contents to each other (c-551:c-553:c-555) remains constant (3.5:1:2.7 with thiosulfate and 4:1:1.7 with sulfide) and is independent of the sulfur source offered. On the other hand, Kusche and Trueper (40) could clearly demonstrate that the absence or occurrence of cytochromes in Ectothiorhodospira shaposhnikovii is regulated by sulfur compounds. Without any reduced sulfur compound in the medium, this organism only contained a membrane-bound cytochrome c-553, while a soluble b-type cytochrome was synthesized, when sulfur compounds were available. A regulation of cytochrome concentration by sulfur compounds was also reported for Rhodobacter sulfidophilus (34).

The thiosulfate splitting enzymes rhodanese and thiosulfate reductase are widely distributed among Anoxyphotobacteria (see Tables I-IV). Chlorobium vibrioforme f. thiosulfatophilum, for example, contains two rhodanese enzymes, a basic and an acidic one, besides thiosulfate reductase and thiosulfate:acceptor oxidoreductase (33). According to its high activity it seems that rhodanese is the dominant enzyme, initiating thiosulfate cleavage. With cyanide as acceptor, rhodanese can only be regarded with respect to a cyanide-detoxifying function which may certainly be of minor importance (33). Another function for rhodanese is its reponsibility for the biosynthesis of the iron-sulfur cluster in ferredoxins by providing these proteins with the necessary "acid-labile" sulfur (41).

Sulfide Oxidation to Elemental Sulfur

Sulfide oxidation by phototrophic bacteria is catalyzed by c-type cytochromes, flavocytochromes and even cytochrome c complexes (see 4,9). A heat-labile cytochrome c-550 of Thiocapsa roseopersicina is responsible for the oxidation of sulfide. The end product is elemental sulfur and it is assumed that this cytochrome might also catalyze the reverse reaction by reducing the intracellularly stored elemental sulfur to sulfide (4,9).

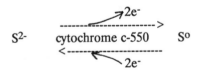

$$S^{2-} \quad \text{cytochrome c-550} \quad S^{o}$$

Elemental sulfur was also formed during sulfide oxidation by a cytochrome c-flavocytochrome c-552 complex in Chromatium vinosum (42). Flavocytochromes of different phototrophic bacteria act as sulfide: cytochrome c reductases and there was one report that a flavocytochrome possessed even elemental sulfur reductase activity (see 4,9). All flavocytochromes examined so far are heat-labile and are reduced by sulfide forming thiosulfate under strictly anaerobic conditions (4,9). The small acidic cytochromes c-551 of Ectothiorhodospira halochloris and Ectothiorhodospira abdelmalekii, both located on the outside of the cell membrane, stimulated the velocity of sulfide oxidation when they were added to spheroplast suspensions. The stimulation was light-dependent and was not observed with whole cells in the dark (16,18). The participation of cytochromes and iron-sulfur proteins in dissimilatory sulfur metabolism of phototrophic bacteria was recently reviewed by Fischer (4,9). But cytochromes are not always the first endogenous electron acceptors in sulfide oxidation in phototrophic bacteria. Brune and Trueper (43) could clearly demonstrate that sulfide at first reduced ubiquinone in the dark when added to an anaerobic chromatophore suspension of Rhodobacter sulfidophilus. From their studies the authors (43) suggest that the electron transport from sulfide to NAD in this organism is initiated by dark reduction of ubiquinone followed by a reverse electron transport from ubiquinol to NAD catalyzed by a NADH dehydrogenase.

Sulfide Oxidation to Sulfite

Sulfite reductases contain a new type of a non-covalently bound prosthetic group, designated as "siroheme". This enzyme catalyzes a six-electron reduction from sulfite to sulfide without releasing any sulfur intermediate from the enzyme and was found in plants, fungi and aerobic and anaerobic bacteria (44). A "reverse" siroheme containing sulfite reductase is responsible for the formation of sulfite directly from sulfide as well as from polysulfides and/or elemental sulfur. Until now, the only phototrophic sulfur bacterium containing such a siroheme sulfite reductase is Chromatium vinosum (45). All siroheme sulfite reductases examined so far contain iron-sulfur clusters and the siroheme prosthetic group exhibits two characteristic maxima at around 390 nm and 590 nm (29,45). The sulfite reductase of C. vinosum is only present in autotrophically grown cells with sulfide and bicarbonate and is fully repressed when the organism is grown on malate and sulfate (45).

Chlorobium limicola f. thiosulfatophilum (29) and Thiocapsa roseopersicina (46) do not contain siroheme sulfite reductases, although sulfite reductase activity was measured in the last organism.

Anoxyphotobacteria which do not possess a "reverse" sulfite reductase are nevertheless able to form sulfite from sulfide, if they contain a flavocytochrome and a thiosulfate reductase. Therefore Trueper and Fischer (23) assume that in these organisms the following reaction mechanism might be possible:

$$2H_2S \xrightarrow{\text{Flavocyt.}} S_2O_3^{2-} \xrightarrow[\text{reductase}]{\text{Thiosulfate}} SO_3^{2-} + H_2S.$$

Sulfite Oxidation to Sulfate

There are two possibilities to oxidize sulfite to sulfate in phototrophic bacteria:

1) via adenosine-5'-phosphosulfate (APS) by conserving energy in form of ATP
or
2) directly to sulfate without the gain of ATP.

The first-mentioned mechanism comprises several enzymatic steps and is initiated by the enzyme adenylylsulfate (APS) reductase catalyzing the following reaction:

$$5'AMP + SO_3^{2-} + acceptor_{ox} \longrightarrow APS + acceptor_{red}$$

All APS-reductases are nonheme iron-sulfur flavoproteins and are found either in the soluble fraction of the cells or tightly or weakly bound to the membranes (see 9). The APS-reductase from Thiocapsa roseopersicina also contains two heme groups of c-type cytochromes and may therefore be regarded as a flavocytochrome (47).

APS-reductase does not occur in Rhodospirillaceae, Ectothiorhodospiraceae, and was not found in two species of the Chromatiaceae (see Tables I-IV). In Chromatium vinosum, APS-reductase activity is very high at the end of the exponential growth phase (48), and in Thiocapsa roseopersicina the specific activity of this enzyme increases, when sulfide is completely consumed by the cells (U. Fischer, unpublished). Normally, ferricyanide is used as an artificial electron acceptor for the enzyme, but Ulbricht (35) could show for Chromatium gracile that the cell's own c-type cytochrome could replace ferricyanide as electron acceptor.

After the formation of APS, the enzymes ADP-sulfurylase or a dissimilatory ATP-sulfurylase now catalyze the exchange of sulfate with inorganic phosphate and inorganic pyrophosphate, respectively, on the APS molecule.

$$\text{APS} + P_a \xrightarrow{\text{ADP-sulfurylase}} \text{ADP} + SO_4^{2-}$$
$$\text{APS} + PP_a \xrightarrow{\text{ATP-sulfurylase}} \text{ATP} + SO_4^{2-}$$

Either or both of these enzymes can occur in the same organism (see Tables II-IV and 35,38). As far as known and examined, both enzymes do not occur in Rhodospirillaceae, while ADP-sulfurylase is widely distributed among Chromatiaceae (see Tables I and II). Both enzymes occur in Chromatium warmingii, where the dissimilatory ATP-sulfurylase competes with the enyzmes ADP-sulfurylase and pyrophosphatase for their substrates APS and pyrophosphate (35). But kinetic examinations have clearly shown that affinities for these substrates are much higher for the ATP-sulfurylase (km-values: 0.0639 mM for APS and 0.0083 mM for pyrophosphate) than for ADP-sulfurylase (km-value: 0.392 mM for APS) and for pyrophosphatase (km-value: 0.147 mM for PP_a) (35). Therefore, in this organism, energy conservation by substrate phosphorylation preferably sets off via the dissimilatory ATP-sulfurylase.

If phototrophic bacteria possess a dissimilatory ATP-sulfurylase, they convert APS with pyrophosphate directly to ATP and sulfate, without the help of an additional enzyme. Such an enzyme is necessary, if the organisms like Chlorobium vibrioforme f. thiosulfatophilum (Table IV) contain only the ADP-sulfurylase, because this enzyme liberates only ADP and sulfate from APS in the presence of inorganic phosphate. In this case, the organisms gain one ATP molecule from 2 molecules of ADP. This reaction is catalyzed by adenylate kinase which converts 2 ADP into 1 ATP and 1 AMP (38).

Microorganisms, which contain the enzymes APS-reductase, ADP-sulfurylase, adenylate kinase and/or ATP-sulfurylase have a certain advantage over those lacking these enzymes, because in addition to photophosphorylation they can also perform a substrate phosphorylation.

Rhodospirillaceae and Ectothiorhodospiraceae possess no APS-reductase and normally no ADP-sulfurylase (see Tables I,III). In these organisms, sulfite is directly oxidized to sulfate by an AMP-independent enzyme, called sulfite:acceptor oxidoreductase.

$$SO_3^{2-} + H_2 + acceptor_{ox} \longrightarrow SO_4^{2-} + 2H^+ + acceptor_{red}$$

Meanwhile, this enzyme was found in all families of Anoxyphotobacteria and appears mainly membrane-bound (see Tables I-IV; 27,34-36). Ferricyanide or c-type cytochromes serve as electron acceptors and Chromatium gracile can even use its own high-spin cytochrome c' (35). The activity of this enzyme is dependent on growth phase and growth conditions. High activities have been found in Chromatium vinosum and Chromatium gracile at the end of the exponential growth phase. In the stationary growth phase, the activities decrease rapidly (35). In Rhodopseudomonas sulfoviridis and Rhodobacter sulfidophilus the activity of the enzyme clearly increases, when the organisms are cultivated with sulfide and CO_2 (34). Therefore, it seems that sulfide stimulates an increased formation of this enzyme.

The visible absorption spectrum of a partially purified enzyme of Chromatium vinosum indicates a high-spin cytochrome c' (36,49). Whether cytochrome c' itself can be regarded as the real enzyme or whether c' is only attached to the enzyme during the purification procedure, still remains unclear. On the other hand, Ulbricht (35) could clearly demonstrate that the enzymes of Chromatium gracile and Chromatium minutissimum have no cytochrome character.

Summary and Conclusion

The dissimilatory sulfur metabolism of phototrophic bacteria is one aspect of sulfur metabolism. During anaerobic sulfide or thiosulfate oxidation to sulfate by whole cells, elemental sulfur - probably polysulfides or polythionates - is one intermediate or can even be the end product. Other products formed are sulfite, thiosulfate, polysulfides or tetrathionate. Rhodospirillaceae are not only capable of performing an assimilatory sulfur metabolism, there are some species now which are dependent on reduced sulfur compounds. Their sulfur metabolism is without doubt comparable to the dissimilatory sulfur metabolism of red and green sulfur bacteria. Even now little or nothing is known about enzymatic sulfur metabolism in purple nonsulfur bacteria and Chloroflexaceae, while there exists more information about purple and green phototrophic sulfur bacteria. Results from in vivo experiments with whole cells often differ from in vitro studies because expected enzymes' activities cannot be detected. On the other hand, there are some fundamental similarities in sulfur metabolism of the "classical" phototrophic bacteria. Cytochromes and other electron carrier proteins are involved in sulfide or thiosulfate oxidation and the thiosulfate-utilizing enzymes occur in all families of Anoxyphotobacteria. Our knowledge about sulfur metabolism in Anoxyphotobacteria is already rather extensive, but still incomplete. It is still unknown, which enzyme initiates the oxidation of the extracellularly stored elemental sulfur to sulfate in Chlorobiaceae and Ectothiorhodospiraceae, or which substance acts as natural sulfur acceptor molecule for the enzyme rhodanese.

In agreement with the statements of Trueper (1) one can say that principally different dissimilatory sulfur metabolic pathways exist in Anoxyphotobacteria for the oxidation of sulfite to sulfate (via APS or directly), the utilization of thiosulfate (splitting or formation of tetrathionate), and the oxidation of sulfide or elemental sulfur (by a "reverse" siroheme sulfite reductase or other mechanisms).

Literature Cited

1. Trueper, H. G. In Sulfur, Its Significance for Chemistry, for the Geo-, Bio- and Cosmosphere and Technology; Mueller, A.; Krebs, B., Eds.; Elsevier: Amsterdam, 1984; Vol. 5, pp 367-82.

2. Castenholz, R. W.; Pierson, B. K. In The Prokaryotes; Starr, M. P.; Stolp, H.; Trueper, H. G.; Balows, A.; Schlegel, H. G., Eds.; Springer: Berlin, 1981; pp 290-98.

2a Gerdes, G.; Krumbein, W. E.; Reineck, H. E. J. Sed. Petrology 1985, 55, 265-78.

3. Trueper, H. G. Microbiologia 1987, 3, 71-89.

4. Fischer, U. In Sulfur, Its Significance for Chemistry, for the Geo-, Bio- and Cosmosphere and Technology; Mueller, A.; Krebs, B., Eds.; Elsevier: Amsterdam, 1984; Vol. 5, pp 383-407.

5. Kaempf, C.; Pfennig, N. Arch. Microbiol. 1980, 127, 125-35.

6. Kaempf, C.; Pfennig, N. J. Basic Microbiol. 1986, 9, 507-15.

7. Kaempf, C.; Pfennig, N. J. Basic Microbiol. 1986, 9, 517-31.

8. van Gemerden, H. Arch. Mikrobiol. 1968, 64, 118-24.

9. Fischer, U. Habilitationsschrift, Oldenburg University, Fed. Rep. Germany, 1986.

10. Schmidt, A.; Trueper, H. G. Experientia 1977, 33, 1008-09.

11. Imhoff, J. F. Arch. Microbiol. 1982, 132, 197-203.

12. van Niel, C. B. Arch. Mikrobiol. 1931, 3, 1-112.

13. Trueper, H. G. In Environmental Regulation of Microbial Metabolism; Kulaev, I. S.; Dawes, E. A.; Tempest, D. W., Eds.; Academic Press: Orlando, 1985; pp 241-47.

14. Neutzling, O.; Pfleiderer, C.; Trueper, H. G. J. Gen. Microbiol. 1985, 131, 791-98.

15. Then, J.; Trueper, H. G. Arch. Microbiol. 1984, 139, 295-98.

16. Then, J.; Trueper, H. G. Arch. Microbiol. 1983, 135, 254-58.

17. Steinmetz, M. A.; Fischer, U. Arch. Microbiol. 1982, 131, 19-26.

18. Then, J. Ph.D. Thesis, Bonn University, Fed. Rep. Germany, 1984.

19. Steudel, R.; Holdt, G.; Goebel, T.; Hazen, W. Angew. Chem. 1987, 99, 143-46.

20. Mas, J.; van Gemerden, H. Arch. Microbiol. 1987, 146, 362-69.

21. Remsen, C. C. In The Photosynthetic Bacteria; Clayton, R. K.; Sistrom, W. R., Eds.; Plenum: New York, 1978; pp 31-60.

22. van Gemerden, H. Arch. Microbiol. 1986, 146, 52-56.

23. Trueper, H. G.; Fischer, U. Phil. Trans. R. Soc. Lond. B 1982, 298, 529-42.

24. Trueper, H. G. In Biology of Inorganic Nitrogen and Sulfur; Bothe, H.; Trebst, A., Eds.; Springer: Berlin, 1981; pp 199-211.

25. Then, J.; Trueper, H. G. Arch. Microbiol. 1981, 130, 143-46.

26. Wermter, U.; Fischer, U. Z. Naturforsch. 1983, 38c, 960-67.

27. Kusche, W. H. Ph.D. Thesis, Bonn University, Fed. Rep. Germany, 1985.

28. Steinmetz, M. A.; Fischer, U. Arch. Microbiol. 1982, 132, 204-10.

29. Schedel, M. Ph.D. Thesis, Bonn University, Fed. Rep. Germany, 1978.

30. Smith, A. J. J. Gen. Microbiol. 1966, 42, 371-80.

31. Paschinger, H.; Paschinger, J.; Gaffron, H. Arch. Microbiol. 1974, 96, 341-52.

32. Bak, F.; Pfennig, N. Arch. Microbiol. 1987, 147, 184-89.

33. Steinmetz, M. A.; Fischer, U. Arch. Microbiol. 1985, 142, 253-58.

34. Neutzling, O. Ph.D. Thesis, Bonn University, Fed. Rep. Germany, 1985.

35. Ulbricht, H. M. U. Ph.D. Thesis, Bonn University, Fed. Rep. Germany, 1984.

36. Brueckenhaus-Kruhl, I. Ph.D. Thesis, Bonn University, Fed. Rep. Germany, 1985.

37. Khanna, S.; Nicholas, D. J. D. J. Gen. Microbiol. 1983, 129, 1365-70.

38. Bias, U.; Trueper, H. G. Arch. Microbiol. 1987, 147, 406-10.

39. Kusai, A.; Yamanaka, T. Biochim. Biophys. Acta 1973, 325, 304-14.

40. Kusche, W. H.; Trueper, H. G. Z. Naturforsch. 1984, 39c, 894-901.

41. Cerletti, P. TIBS 1986, 11, 369-72.

42. Gray, G.O.; Knaff, D.B. Biochim. Biophys. Acta 1982, 680, 290-96.

43. Brune, D.C.; Trueper, H. G. Arch. Microbiol. 1986, 145, 295-301.

44. Siegel, L. M. In Developments in Biochemistry. Mechanisms of Oxidizing Enzymes; Singer, T. E.; Ondarza, R. N., Eds.; Elsevier: Amsterdam, 1978; Vol. 1, pp 201-14.

45. Schedel, M.; Vanselow, M.; Trueper, H. G. Arch. Microbiol. 1979, 121, 29-36.

46. Jorzig, E. B. Diploma Thesis, Bonn University, Fed. Rep. Germany, 1979.

47. Trueper, H. G.; Rogers, L. H. J. Bacteriol. 1971, 108, 1112-21.

48. Schwenn, J.D.; Biere, M. FEMS Microbiol. Lett. 1979, 6, 19-22.

49. Brueckenhaus, I. Diploma Thesis, Bonn University, Fed. Rep. Germany, 1977.

RECEIVED August 16, 1988

THE OCEANS:
CHEMICAL TRANSFORMATIONS

Chapter 18

Thermodynamics and Kinetics of Hydrogen Sulfide in Natural Waters

Frank J. Millero and J. Peter Hershey

**Rosenstiel School of Marine and Atmospheric Science,
University of Miami, Miami, FL 33149–1098**

The thermodynamic and kinetic measurements recently made on the H_2S system in natural waters have been critically reviewed. Thermodynamic equations are given for the solubility and ionization of H_2S

$$H_2S \text{ (g)} \text{-----} > \quad H_2S \text{ (aq)}$$

$$H_2S \quad \text{-----} > \quad H^+ + HS^-$$

in water, seawater and brines. Pitzer interaction coefficients are given so that the pK_1 for the ionization can be calculated from 0 to 50°C and I = 0 to 6m in natural waters containing the major sea salts (Na^+, Mg^{2+}, Ca^{2+}, K^+, Cl^-, SO_4^{2-}). The kinetics of oxidation of H_2S with oxygen and hydrogen peroxide

$$H_2S + O_2 \quad \text{-----} > \quad \text{Products}$$

$$H_2S + H_2O_2 \text{-----} > \quad \text{Products}$$

has also been examined as a function of pH, temperature and ionic strength. Equations are given for the second order rate constants for these oxidation reactions as a function of these variables. At the levels of O_2 (200 μM) and H_2O_2 (0.1 μM) in surface sea waters, the oxygen oxidation is 70 times faster than peroxide oxidation. In rain waters, however, the concentration of hydrogen peroxide (100 μM) is great enough so that it is the dominant oxidant for H_2S.

The formation of H_2S occurs in a variety of natural waters. It is formed by bacteria under anoxic conditions

$$SO_4^{2-} + 2H^+ \text{-----} > H_2S + 2O_2 \tag{1}$$

0097–6156/89/0393–0282$09.00/0

This bacterial production occurs in the pore fluids of sediments and in stagnant basins (seas, lakes, rivers and fjords). At the interface between anoxic and oxic waters the H_2S can be oxidized. This oxidation is frequently coupled to changes in the redox state of metals ([1,2]) and non-metals ([3]). Another major interest in the H_2S system comes from an attempt to understand the authigenic production of sulfide minerals as a result of biological or submarine hydrothermal activity and the transformation and disappearance of these minerals due to oxidation ([4]). For example, hydrothermally produced H_2S can react with iron to form pyrite, the overall reaction given by

$$2SO_4^{2-} + 4H^+ + 11Fe_2SiO_4 ----> FeS_2 + 7Fe_3O_4 + 11SiO_2 + 2H_2O \qquad (2)$$

The oxidation of pyrite (FeS_2) is thought to involve the dissolution of FeS_2 and subsequent oxidation of HS^- and Fe^{2+}.

Recently, workers ([5]) have been examining the equilibrium and kinetic factors that are important at the oxic-anoxic interface. The kinetic behavior is difficult to characterize completely due to varying rates of oxidation and absorption above the interface and varying rates of reduction, precipitation and dissolution below the interface ([2,5]). Bacterial catalysis may also complicate the system ([1]). Although one can question the importance of abiotic thermodynamic and kinetic processes at this interface, we feel it is useful to use simple inorganic models to approximate the real system. Recently, the thermodynamics and kinetics of the H_2S system in natural waters has been reviewed ([6]). From this review it became apparent that large discrepancies existed in rates of oxidation of H_2S and the thermodynamic data was limited to dilute solution. In the last few years we have made a number of thermodynamic ([7,8]) and kinetic ([9,10]) measurements on the H_2S system in natural waters. In the present paper we will review these recent studies. The results will be summarized by equations valid for most natural waters.

Thermodynamics of H₂S in Aqueous Solutions

The chemistry of H_2S in natural aqueous solutions is characterized by the following reactions

$$H_2S \text{ (g)} -----> H_2S \text{ (aq)} \qquad (3)$$

$$H_2S \quad -----> H^+ + HS^- \qquad (4)$$

$$HS^- \quad -----> H^+ + S^{2-} \qquad (5)$$

The concentration of H_2S gas in solution (C^*, mol kg⁻¹) at equilibrium with various gas fugacities (f_{H2S}) can be determined from

$$C^* = f_{H_2S}/H_S \qquad (6)$$

where H_S is the Henry's law constant. For pure water the value of H_S (atm - kg H_2O mol⁻¹) can be determined from ([11,12])

$$\log H_S = 103.70 - 4455.94/T - 37.1874 \log T + 0.01426T \qquad (7)$$

where T is the absolute temperature (°K) and the equation is valid from 25 to 260°C (σ = 0.001).

The most extensive measurements of the solubility of H_2S in seawater were made by Douabel and Riley (13). Their results from 2 to 30°C and 0 to 40 salinity [$I = 19.92S/(10^3 + 1.005\ S)$] have been refitted to

$$\ln K_S\ (mol\ l^{-1}\ atm^{-1}) = -44.6931 + 71.4632/T + 16.8818\ \ln\ (T)\ +$$

$$S[0.87627 - 0.40081/T - 0.13037\ \ln\ (T)] \tag{8}$$

where $K_S = 1/H_S$ and the results are valid for $f_{H2S} = 1$ atm of H_2S ($\sigma = 0.5\%$).

The decrease in the solubility of H_2S with the addition of salt or the salting out can be represented by the Setchenow equation

$$\ln(C_0/C) = \ln\ \gamma_g = k\ I \tag{9}$$

where C_0 and C are the solubilities in water and salt solution, γ_g is the activity coefficient of the gas, k is the salting coefficient and I is the ionic strength. The value of k for seawater can be determined from

$$10^3k = 87.43 - 2.5317t + 3.1982\ x\ 10^{-2}\ t^2 \tag{10}$$

where t is in °C and the $\sigma = 0.4\ x\ 10^{-3}$ in k. The value of γ_g is 1.03 for average seawater (S = 35 or I = 0.723) at 25°C. This is similar to the value found for other acids ($\gamma_g = 1.0$ for HF (14)).

The value of γ_g for other ionic media can be estimated from the solubility of H_2S in NaCl solutions (15). Results at 25°C are given by (8)

$$k_{25} = 0.1554 - 0.00806\ I \tag{11}$$

Values of γ_{H2S} at other temperatures can be determined from

$$\log\ \gamma_{H_2S} = [k_{25} - 2.5321\ x\ 10^{-3}\ (t-25) + 3.1984\ x\ 10^{-5}\ (t^2-25^2)]\ I \tag{12}$$

which uses the temperature dependence of k in seawater (8). New measurements of the solubility of H_2S in the major sea salts as a function of ionic strength and temperature would be useful, but since NaCl is the major salt of most natural brines, this is not a serious limitation. Since γ_g is near 1.0 for seawater, the interactions of H_2S and the major sea salts are quite small.

Since H_2S dissociates in aqueous solutions (equations (4) and (5)), it is necessary to have reliable equilibrium constants for the ionization. These constants are defined by

$$K_1 = ([H^+][HS^-]/[H_2S])(\gamma_H\ \gamma_{HS}/\gamma_{H_2S}) \tag{13}$$

$$K_2 = ([H^+][S^{2-}]/[HS^-])(\gamma_H\ \gamma_S/\gamma_{HS}) \tag{14}$$

where [i] and γ_i are the concentrations and activity coefficients of species i. As discussed elsewhere (6), experimental results for the K_2 of H_2S cover a wide range of values (pK_2 = 12.4 to 17.1 near 25°C). Recent work of Meyer et al. (16) supports the higher value of pK_2= 17.1 measured spectroscopically by Giggenbach (17). The lower values are apparently in error due to the oxidative formation of polysulfides that interfere with the determination of pK_2. Since we feel this higher value is correct and in the pH range of most natural waters the concentration of S^{2-} is quite small, we will not consider the second dissociation of H_2S any further.

The thermodynamic first dissociation constant for H_2S at infinite dilution can be determined from ([8])

$$pK_1 = -98.080 + 5765.4/T + 15.0455 \ln T \qquad (15)$$

which is valid from 0 to 300°C. The equation has a $\sigma = 0.04$ from 0 to 100°C and is thought to be valid to ± 0.1 above 100°C ([18]).
Measurements of the first dissociation constant for H_2S in seawater have been made by a number of workers. The measurements of Savenko ([19]) and Goldhaber and Kaplan ([20]) were made using the National Bureau of Standards (N.B.S.) pH scale ([21])

$$K_1' = a_H' [HS^-]/[H_2S]_T \qquad (16)$$

where a_H' is the apparent activity of the proton obtained with N.B.S. dilute solution buffers. The measurements of Almgren et al. ([22]) and Millero et al. ([7]) were made using the total proton scale ([23])

$$K_1^* = [H^+]_T [HS^-]_T/[H_2S]_T \qquad (17)$$

where the subscript T is used to denote the total concentration, i.e., $[H^+]_T = [H^+]_F + [HSO_4^-] + [HF]$. The two scales are related by ([24])

$$a_H' = f_H[H^+]_T \qquad (18)$$

where f_H is the apparent activity coefficient of the proton. This value includes effects of liquid junctions, the definition of the N.B.S. scale and the activity coefficient of the proton ([24]).
The apparent constants of Savenko ([19]) and Goldhaber and Kaplan ([20]) have been fitted to

$$pK_1' = pK_1 + A' S^{1/2} + B'S \qquad (19)$$

where

$$A' = 0.0057 - 19.98/T \qquad (20)$$

$$B' = 0.0028 \qquad (21)$$

with a $\sigma = 0.019$ in pK_1' (mol kg-SW^{-1}). An examination of the residuals between the measured and calculated results as a function of T and S is shown in Figure 1, demonstrating that the two studies are in good agreement and that the electrode systems used give similar values of f_H. The results of Millero et al. ([9]) were combined with the adjusted pK_1' values of Savenko ([19]) and Goldhaber and Kaplan ([20]) to yield the consensus equation

$$pK_1^* = pK_1 + A^* S^{1/2} + B^*S \qquad (22)$$

where

$$A^* = -0.1498 \qquad (23)$$

$$B^* = 0.0119 \qquad (24)$$

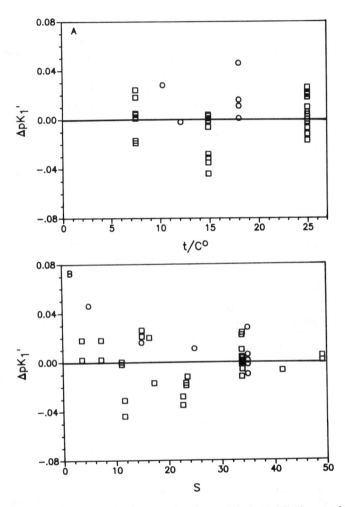

Figure 1. Values of $\Delta pK_1'$ using the data of (□) Goldhaber and Kaplan ([20]) and (○) Savenko ([19]) fitted to Equation 19 as a function of (A) T and (B) S.

with a σ = 0.028 in pK^*_1 (mol kg-SW^{-1}). The values of pK^*_1 at 25°C from the two studies are shown in Figure 2. The experimentally determined values of f_H of Mehrbach et al. (25), Culberson and Pytkowicz (26) and Millero (27) were fitted to

$$f_H = 0.739 + 3.07 \times 10^{-3} + 7.94 \times 10^{-5} S^2 +$$

$$6.443 \times 10^{-5} T - 1.17 \times 10^{-4} TS \tag{25}$$

which has a σ = 0.006. The values of f_H from this equation were used to calculate the pK^*_1 from the pK_1' using

$$pK^*_1 = pK'_1 + \log f_H \tag{26}$$

It should be pointed out that the pK^*_1 results of Almgren et al. (22) were 0.07 ± 0.02 higher than other workers and were not used to obtain the consensus equation.

The residuals of errors for the pK^*_1 fit for various workers are shown in Figure 3. These residuals demonstrate that the three studies are in good agreement when adjusted to the same pH scale.

In our earlier work (28,6) we demonstrated how Pitzer parameters can be generated from pK^*_1 measurements in various ionic media. Recently we have extended these calculations to higher ionic strengths from 0 to 50°C (8).

The activity coefficients for γ_H and γ_{HS} using the Pitzer (29) equations are given (28,30,31)

$$\ln \gamma_H = f^\gamma + \Sigma 2m_X (B_{HX} + E\, C_{HX} + \Sigma\Sigma m_M m_X (B'_{MX} + C_{MX})$$

$$+ \Sigma m_M (2\theta_{MH} + m_X \psi_{HMX}) \tag{27}$$

$$\ln \gamma_{HS} = f^\gamma + \Sigma 2m_M (B_{MHS} + E\, C_{MHS}) + \Sigma\Sigma m_M m_X (B'_{MX} + C_{MX}) \tag{28}$$

where X = Cl$^-$, SO_4^{2-}, etc. and M = Na$^+$, Mg^{2+}, Ca^{2+}, etc. The definition of the Debye-Hückel slope, f^γ, and the interaction parameters, B_{MX}, C_{MX} etc., are given in Appendix I. The parameters needed to calculate the values of γ_H and γ_{HS} for various natural waters are given in Table I. These parameters yield calculated values of pK^*_1 in NaCl, KCl, $MgCl_2$ and $CaCl_2$ solutions from 5 to 45°C and I = 0 to 6 to ± 0.02 in pK^*_1.

The reliability of these parameters can be demonstrated by comparing the measured and calculated values of pK^*_1 for seawater. The differences, ΔpK^*_1, are shown in Figure 4. The agreement is quite good and well within the standard error of the experimental data (σ = 0.026). Measurements of pK^*_1 in artificial Dead Sea brines (32) gave pK^*_1 = 7.25 ± 0.03 at 25°C compared to a calculated value of pK^*_1 = 7.30. The agreement is quite good and indicates that the parameters are valid to I = 6.0.

Since the γ_{HS} is approximately equal to γ_{Cl} (28), a reasonable approximation of pK^*_1 can be made to high temperatures using the temperature coefficients for Cl$^-$ salts. Future measurements using spectrophotometric techniques should be made on the pK^*_1 in the major sea salts at high temperatures.

The effect of pressure on the K_1 can be estimated from

$$\ln(K^P_1/K_1^0) = -(\Delta V_1/RT)P + 0.5\,(\Delta K_1/RT)P^2 \tag{29}$$

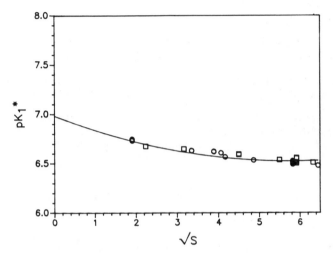

Figure 2. Values of pK^*_1 versus $S^{1/2}$ at 25°C from the present work and adjusted values of (■) Savenko ([19]) and (○) Goldhaber and Kaplan ([20]) and (□) the present work. Fitted curve determined from the values of the present work.

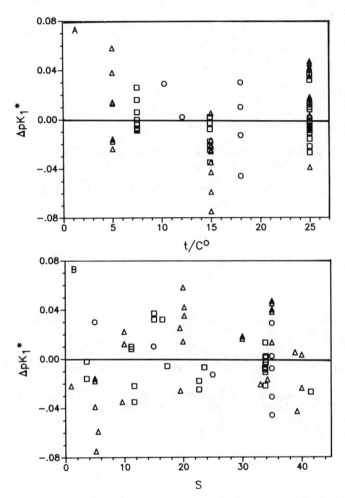

Figure 3.   Values of △ pK*₁ using the data of (△) this work, (▣) Goldhaber and Kaplan (20), and (◯) Savenko (19) as a function of (A) T and (B) S.

Table I. Pitzer Coefficients used to Calculate Ion
Activity Coefficients for H and HS[a]

	HCl[b]	KCl[b]	NaCl[c]	NaHS	KHS	MgHS[d]	CaHS[d]
q_1	0.32779	0.22518	25.7819	0.3662	0.6371	0.170	-0.105
q_2	3.88370	-	-777.03	-67.54	-138.5	-	-
q_4	-5.478×10^{-4}	-5.9309×10^{-3}	8.946×10^{-3}	-	-	-	-
q_6	-0.14553	-0.22796	0.1528	-	-	2.78	3.43
q_7	-1.4759×10^{-3}	1.4763×10^{-3}	6.161×10^{-5}	-	-	-	-
q_9	1.523×10^{-4}	-1.5852×10^{-3}	-0.63337	-0.01272	-0.1935	-	-
q_{10}	2.1725×10^{-5}	2.4993×10^{-6}	33.317	-	-	-	-

a. $\beta^{(0)} = q_1 + q_2/T + q_3 \ln T + q_4 + q_5 T^2$
   $\beta^{(1)} = q_6 + q_7 T + q_8 T^2$
   $C^{\phi} = q_9 + q_{10}/T + q_{11}^{\ln T} + q_{12} T$
b. Calculated from the mean activity coefficients tabulated in Harned and Owen ([61]).
c. Values of $q_3 = -4.4706$, $q_5 = -3.3158 \times 10^{-6}$, $q_8 = 1.0715 \times 10^{-6}$, $q_{11} = 9.421 \times 10^{-2}$, $q_{12} = -4.655 \times 10^{-5}$ are also from Silvester and Pitzer ([62]).
d. Calculated at 25°C.

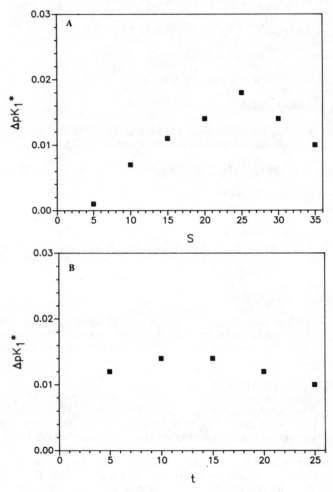

Figure 4. Values of ΔpK^*_1 versus (A) S at T = 25°C and (B) T at S = 35.

where $\Delta V_1 = \overline{V}_H + \overline{V}_{HS} - \overline{V}_{H2S}$ is the change in the partial molal volumes and $\Delta K_1 = -(\partial \Delta V_1/\partial P) = \overline{K}_H + \overline{K}_{HS} - \overline{K}_{H2S}$ is the change in the partial molal compressibilities. For water (0 to 200°C)

$$-\Delta V^0_1 = 16.33 - 0.0573t + 7.9048 \times 10^{-4}t^2 \tag{30}$$

$$-10^3 \Delta K^0_1 = 2.18 - 7.6507 \times 10^{-3}t + 1.054 \times 10^{-4}t^2 \tag{31}$$

and for seawater (S = 35 and 0 to 30°C)

$$-\Delta V^*_1 = 11.07 + 0.009t + 0.942 \times 10^{-3}t^2 \tag{32}$$

$$-10^3 \Delta K^*_1 = 2.89 - 0.054t \tag{33}$$

Values of ΔV_1 and ΔK_1 for other ionic media can be estimated using various models for ionic interactions. Details are given elsewhere (33).

The Oxidation of H₂S by O₂ in Natural Waters

The oxidation of H_2S by oxygen

$$H_2S + O_2 -----> \text{Products} \tag{34}$$

has been studied by a number of workers (9,34-40). The oxidation kinetics are complicated and the results of various workers are not in good agreement (6). The overall rate equation is given by

$$d[H_2S]_T/dt = -k [H_2S]^a[O_2]^b \tag{35}$$

where n = a + b is the overall rate of the reaction. Although Chen and Morris (41) found a = 1.34 and b = 0.56, most workers have found the reaction to be second order (first order with respect to H_2S and O_2. When $[O_2] > [H_2S]$ the reaction is pseudo-first-order

$$d[H_2S]_T/dt = -k_1' [H_2S] \tag{36}$$

where $k_1' = k[O_2]$ and the half-time, $t_{1/2} = (\ln 2)/k_1'$. As discussed in an earlier review (6), the values of $t_{1/2}$ vary from 0.4 to 65h (see Table II). We have made a number of measurements on the oxidation of H_2S by O_2 as a function of temperature, ionic strength and pH to resolve these discrepancies.
    In our first series of measurements we determined the pseudo-first-order rate constant (k_1') for the oxidation of H_2S in water and seawater (S = 35) as a function of temperature at a pH = 8.0. Our results for the overall rate constant $k = k_1'/[O_2]$ were calculated from these results using the values of $[O_2]$ determined from the equations of Benson and co-workers (44,45). The results are shown plotted versus 1/T (°K) in Figures 5 and 6. The energies of activation, E_a, for the oxidation calculated from

$$E_a = (d\ln k/dT)RT^2 \tag{37}$$

were found to be $E_a = 56 \pm 4$ kJ mol⁻¹ and $E_a = 66 \pm 5$ kJ mol⁻¹, respectively, for water and seawater. Within the experimental error (0.18 in log k) a value of $E_a = 57 \pm 4$ kJ mol⁻¹ can be used to represent the effect of temperature on the oxidation of H_2S in water and seawater. This energy of activation is slightly higher than the value of $E_a = 46$ kJ mol⁻¹ at a pH = 12 in water (36). The

Table II. Comparisons of the Half-Times for the Oxidation of H_2S
in Air Saturated Solutions at 25°C and pH = 8.0

Media	$t_{1/2}$	References
Water	50^h	Our results
	37^a	O'Brien and Birkner(42)
	50	Chen and Morris(41)
	18^b	Avrahami and Golding(36)
Seawater	26	Our results
	27^a	O'Brien and Birkner(42)
	10-40	Sorokin(38)
	65	Skopintsev et al.(35)
	$6-28^c$	Cline and Richards(37)
	2-5	Almgren and Hagstrom(43)
	0.4	Ostlund and Alexander(34)

a) Water results extrapolated to I = 0, seawater results at I = 0.7.
b) At a pH = 12.
c) At 9.8°C.

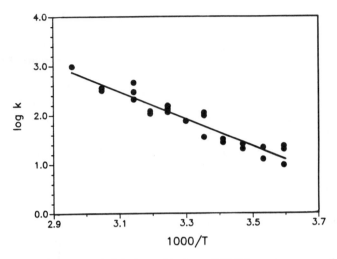

Figure 5. Values of log k for the oxidation of H$_2$S in water versus 1/T at pH = 8.0.

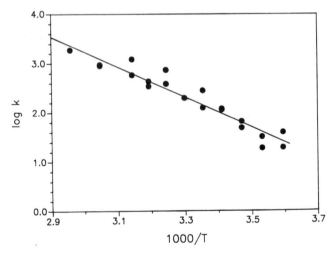

Figure 6. Values of log k for the oxidation of H$_2$S in seawater (S = 35) versus 1/T at pH = 8.0.

difference could be related to the changes in E_a as a function of pH and possibly due to the strange pH behavior of the oxidation of H_2S above a pH = 10 found by earlier workers (41).

In our next series of measurements we determined the effect of pH on the oxidation of H_2S in buffered dilute solutions. The measurements were made at 55°C to speed up the acquisition of data. The results are shown in Figure 7. The results from pH = 2 to 8 are similar to the earlier measurements of Chen and Morris (41). Above a pH = 8 we find the rate to be independent of pH unlike the results of Chen and Morris (41) who find a complicated pH dependence. This could be related to trace metal impurities in the buffers used by Chen and Morris (41). Hoffmann and Lim (46) have examined these trace metal effects and the base catalysis of the oxidation of H_2S.

As discussed elsewhere (6,9,46), the effect of pH on the oxidation of H_2S can be related to the reactions

$$H_2S + O_2 \xrightarrow{\ k_0\ } \text{Products} \tag{38}$$

$$HS^- + O_2 \xrightarrow{\ k_1\ } \text{Products} \tag{39}$$

where k_0 and k_1 are, respectively, the rate constants for the oxidation of H_2S and HS^-. The observed rate constant is related to these values (6) by

$$k = k_0 \alpha_{H_2S} + k_1 \alpha_{HS} \tag{40}$$

The fraction of H_2S and HS^- are given by

$$\alpha_{H_2S} = 1/(1 + K_1/[H^+]) \tag{41}$$

$$\alpha_{HS} = 1/(1 + [H^+]/K_1) \tag{42}$$

where K_1 is the thermodynamic dissociation constant for the ionization of H_2S. The rearrangement of equation 40 yields the linear equation

$$k/\alpha_{H_2S} = k_0 + k_1 K_1/[H^+] \tag{43}$$

Values of k/α_{H2S} plotted versus $1/[H^+]$ are shown in Figure 8. The least squares straight line gives $k_0 = 80 \pm 17$ (kg H_2O) mol^{-1} hr^{-1} and $k_1 = 344 \pm 7$ (kg H_2O) mol^{-1} hr^{-1} using $pK_1 = 6.68$ (6). The values of k determined from

$$k = (k_0 + k_1 K_1/[H^+])/(1 + K_1/[H^+]) \tag{44}$$

are shown as the smooth curve in Figure 7. The standard deviation between the measured and calculated values of log k is 0.10.

We have also determined the effect of ionic strength on the oxidation of H_2S in NaCl solutions. The values of log k measured at 45°C are shown as a function of $I^{1/2}$ in Figure 9. They can be represented by

$$\log k = 2.33 + 0.50\, I^{1/2} \tag{45}$$

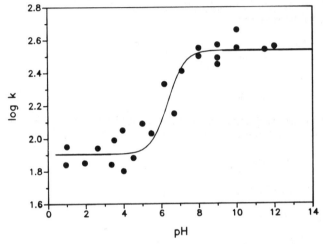

Figure 7. Values of log k for the oxidation of H_2S in water versus pH in water at 55°C. Fitted curve determined from Equation 44.

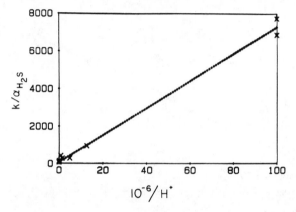

Figure 8. Values of k/α_{H2S} versus $1/[H^+]$ at 25°C.

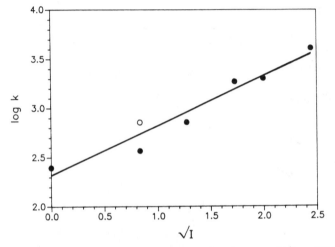

Figure 9. Values of log k for the oxidation of H_2S versus $I^{1/2}$ at $45^\circ C$ and pH = 8 in NaCl (●) and seawater, S = 35 (○).

with a σ = 0.12 in log k. The slope of 0.50 ± 0.08 can be compared to a value of 0.49 ± 0.2 obtained at 25°C by O'Brien and Birkner (42) from 0.16 to 1.78 in NaCl solution. The agreement is good and within the combined experimental error of the measurements.

All of our experimental measurements at a pH = 8.0 can be combined to yield the equation

$$\log k = 11.78 - 3.0 \times 10^3/T + 0.44 \, I^{1/2} \tag{46}$$

which has an overall σ = 0.18 in log k. The slope of log k (0.44 ± 0.06) versus $I^{1/2}$ for seawater found over the entire temperature range is the same within experimental error as found for NaCl (0.50 ± 0.08) to saturation at 45°C. A comparison of our results expressed as half-times with other workers is given in Table II. Our results for water are in good agreement with the work of O'Brien and Birkner (42) and Chen and Morris (41). The lower half times obtained by Avrahami and Golding (36) may be related to the high pH = 12 used for the measurements. Over a pH range of 11 to 14, they found large changes in the half-times (an increase in pH of one unit decreased the half time by two). Thus, the extrapolation of their results to a pH = 8 would increase the half time eight times. We have used the U.V. spectra method at a pH = 8.0 and obtained half-times that are in reasonable agreement with the values obtained with the methylene blue technique (see Figure 10).

Our seawater results are in good agreement with the work of O'Brien and Birkner (42) in NaCl solutions at 0.7 m and in fair agreement with the seawater work of Sorokin (38) and Skopintsev et al. (35). The seawater results of Cline and Richards (37) show a wide range depending upon the experimental conditions used (47,48). An average value of $t_{1/2}$ = 12 ± 7 h is obtained at 9.8°C and [O₂] = 225 μM from all the experiments except for the addition of iron (48). At 25°C this would give a $t_{1/2}$ = 4 h which is much lower than our results. We have no logical explanation for this difference.

The lower half-times determined using the emf technique (34,43) requires some discussion. In our earlier studies of the H₂S oxidation, we made numerous seawater measurements using this emf technique at a pH = 8.0 and 25°C. As found by Ostlund and Alexander (34) and Almgren and Hagstrom (43) in low concentrations of [H₂S]$_T$ = 20 μM, reasonably linear plots were obtained (see Figure 11). We obtained half-times of 2 to 3 hours at 25°C that were only slightly dependent upon salinity. The results were quite variable and we frequently obtained a non-linear behavior of the emf and dramatic changes if the experiments were carried out over long time periods. A methylene blue analysis of the solutions after a run ($\approx$ 6 hours) showed only small changes in the [H₂S]$_T$ giving $t_{1/2} \approx$ 24 h compared to 2.5 h with the emf technique. This led us to believe that the emf technique was sensing the disappearance of electrochemically active sulfur (S^{2-}) due to the formation of polysulfides in the solution or on the surface of the electrode. This was demonstrated by the changes in the emf after the addition of 0.01 M tributylphosphine (which breaks S-S bonds) to the solutions. Although we cannot state with certainty the exact cause of the observed effects, we feel that the emf method does not give reliable results for the oxidation of H₂S with O₂. It is interesting to note that similar emf measurements of the oxidation of H₂S with 1000 μ M H₂O₂ (Figure 12) gave half-times (14 to 30 min) for seawater that were in good agreement with the U.V. spectra technique (see Figure 13). These half-times are sufficiently fast to avoid the problems of deactivation of the electrode or formation of polysulfides. More will be said about the oxidation of H₂S by H₂O₂ in the next section.

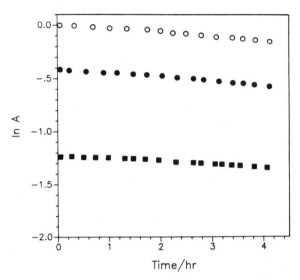

Figure 10. Values of ln absorbance at 229 nm versus time for the oxidation of H_2S with O_2 in water at 25°C and pH = 8.0; (◯) $[H_2S]^o$ = 161 μM, $t_{1/2}$ = 30 hr.; (●) $[H_2S]^o$ = 107 μM, $t_{1/2}$ = 30 hr.; (■) $[H_2S]^o$ = 54 μM, $t_{1/2}$ = 28 hr.

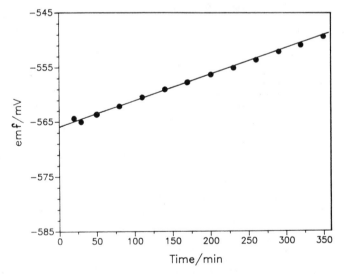

Figure 11. Values of emf versus time for air oxidation of H_2S with O_2 at pH = 8.0 in seawater (S = 35) where $[H_2S]^o$ = 20 μM.

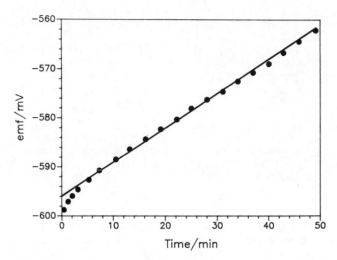

Figure 12. Values of emf versus time for the oxidation of H_2S with 1000 μM H_2O_2 at 25°C at pH = 8.0.

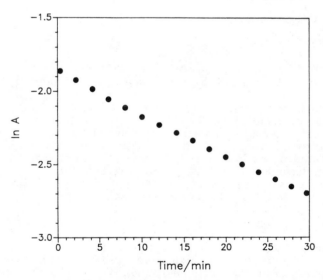

Figure 13. Values of ln absorbance at 229nm versus time for the oxidation of H_2S with H_2O_2 in water at 25°C at pH = 8.

Since $\alpha_{HS} \approx 0.9$ at a pH = 8.0, the effect of temperature on the overall k can be equated to the effect of temperature on k_1 for the oxidation of HS^-. To determine the temperature dependence of the k_0 for the oxidation of H_2S, we have measured the oxidation of H_2S at 25°C and pH = 4.0. We found a value of log k = 1.17 ± 0.03. Since α_{H2S} = 0.999 at pH = 4.0 at 25°C, this value can be equated to log k_0 for the oxidation of H_2S. Combining this value with the k_0 determined at 55°C yields the equation

$$\log k_0 = 9.22 - 2.4 \times 10^3/T \tag{47}$$

The energy of activation for the oxidation of H_2S is 46 ± 8 kJ mol^{-1}. If the value of k_0 is assumed to be independent of ionic strength, the overall rate as a function of pH, temperature and ionic strength can be determined from Equations (40), (46) and (47). Over a pH range from 4 to 8, from 5 to 65°C and 0 to 6 m, the equation can be simplified to

$$\log k = 10.50 + 0.16pH - 3.0 \times 10^3/T + 0.44 \, I^{1/2} \tag{48}$$

This equation should be valid for most natural waters. In our future studies we plan to examine the oxidation products formed as a function of temperature and ionic strength. These measurements will hopefully lead to a better understanding of the mechanism of the oxidation.

The Oxidation of H_2S by H_2O_2 in Natural Waters

The oxidation of H_2S by hydrogen peroxide (H_2O_2)

$$H_2S + H_2O_2 -----> \text{Products} \tag{49}$$

has been studied by a number of workers (49-53). The most extensive measurements were made at 25°C by Hoffmann (47) as a function of pH (2 to 8.1).

We became interested in the oxidation of H_2S by H_2O_2 due to the findings of Zika and co-workers (55-56) of 0.1 μM concentrations of H_2O_2 in surface waters. The concentrations in rainwater can be higher (100 μM); thus, peroxide may be the preferred oxidant in rainwaters and marine aerosols. To elucidate the kinetics of oxidation of H_2S by H_2O_2, we have made measurements on the effect of temperature, ionic strength and pH on the reaction (9).

In our first series of measurements we determined the pseudo-first-order rate constant (k'_1) for the oxidation of H_2S by H_2O_2 in water at a pH = 8.0 (0.01M borax) and 25°C as a function of $[H_2S]^o_T$ = 25 to 200 μM at $[H_2O_2]^o$ = 5000 μM and as a function of $[H_2O_2]^o$ = 500 μM to 60 mM at $[H_2S]^o_T$ = 25 μM, where the superscript (^{o}) indicates initial concentration. The pseudo-first-order behavior as a function of $[H_2S]^o$ is shown in Figure 14. The value of k'_1 was found to be 0.13 ± 0.02 min^{-1}. This first order behavior relative to $[H_2S]^o_T$ agrees with the earlier findings of Hoffmann (46). The order with respect to H_2O_2 was examined by plotting log k'_1 versus the log $[H_2O_2]^o$ as shown in Figure 15. The slope is 0.94 ± 0.04 which is essentially first order. If the reaction is assumed to be first order the value of k = $k_1'/[H_2O_2]^o$ = 30 ± 5 min^{-1} M^{-1} is found for the 27 measurements made at pH = 8 and 25°C. These findings are also in reasonable agreement with the results of Hoffmann (46) who found k = 10 to 87 min^{-1} M^{-1} between pH = 6.8 to 8.1 and 25°C. In all our subsequent discussions we will examine the overall rate constant k for

$$d[H_2S]_T/dt = -k[H_2S]_T[H_2O_2] \tag{50}$$

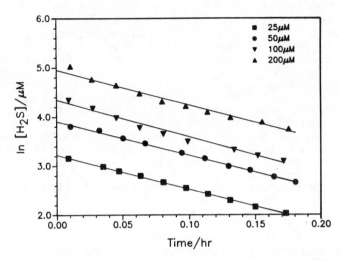

Figure 14. Values of ln $[H_2S]_T$ versus time for the oxidation of H_2S with H_2O_2 for different values of $[H_2S]_T^o$ (pH = 8.0, in 0.01 M Borax at 25°C).

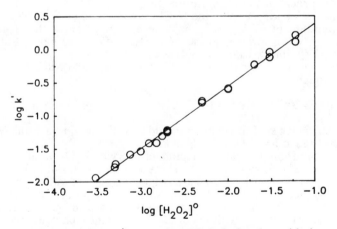

Figure 15. Values of log k_1' versus log $[H_2O_2]^o$ for the oxidation of H_2S with H_2O_2 (k_1' is the pseudo first order rate constant) (pH = 8, in 0.01M borax at 25°C).

In our second series of measurements we examined the effect of ionic strength on the rate of oxidation at a pH = 8 and at 5, 25 and 45°C. These results are shown plotted versus the square root of ionic strength in Figure 16. Over most of the ionic strength range the values of log k were found to be linear functions of $I^{1/2}$ independent of whether the measurements were made in seawater or NaCl. The slopes were almost independent of temperature and ranged between 0.04 to 0.12 (average of 0.08 ± 0.04 from 5 to 45°C). The slopes were smaller than the values of 0.44 ± 0.06 found in our measurements for the oxidation of H_2S by O_2 (9). As mentioned earlier, extensive measurements made in seawater, using the emf and spectrophotometric technique as function of salinity shown in Figures 12 and 13, give results that agree with the more recent results (10) and also demonstrate that the rate constants are nearly independent of ionic strength.

We also made a few measurements as a function of ionic strength at pH = 3 and 13. The results at pH = 13 gave log k = 1.33 ± 0.01 min^{-1} M^{-1} for four measurements between I = 0 to 3m. At a pH = 3 in dilute solutions below 0.04M, no ionic strength dependence was found; however, at I = 3.0m, the rate was ten times faster that at I = 0. We attribute this increase in rate to the presence of trace metals. All of our runs at pH = 8 to 13 were made with enough borax to complex these trace metals and suppress the catalytic effect. An experiment at pH = 11 without borax was completed within 5 minutes compared to 1.5 hours with 0.01 M borax. These results support our contention that the effect of ionic strength on the rates of oxidation are independent of pH if the catalytic effects of trace impurities are avoided.

In our next series of measurements we examined the effect of temperature on the rate of oxidation of H_2S by H_2O_2. These results are shown plotted versus the reciprocal of the absolute temperature (1/T) in Figure 17. The energies of activation for seawater and NaCl were found to be E_a = 39 ± 1.2 kJ mol^{-1} independent of ionic strength. The energy of activation for H_2S oxidation with H_2O_2 is lower than the value (E_a = 57 ± 4 kJ mol^{-1}) found in our earlier work (9) for the oxidation of H_2S with O_2.

All of our measurements at pH = 8.0 in seawater and NaCl have been fitted to the equation

$$\log k = 8.60 - 2052/T + 0.084\ I^{1/2} \tag{51}$$

(with a σ = 0.07 in log k. This equation should be valid for most natural waters from 0 to 50°C and to I = 6.0 near a pH of 8.0.

The effect of pH on the rate of oxidation of H_2S with H_2O_2 was determined from pH = 2 to 13 at 5, 25 and 45°C. These results are shown in Figure 18. Our results at 25°C from pH = 5 to 8 are in good agreement with the results of Hoffmann (46) (See Figure 19). At lower values of pH, his results are faster than ours. This may be due to problems with the emf technique he used. For the slower reactions of H_2S with O_2 or H_2O_2, the emf technique may yield unreliable results due to problems with the electrode response.

The effect of pH on the oxidation of H_2S at various temperatures can be divided into two linear portions: from pH = 2 to 7.5 and from pH = 7.5 to 13. The increase between 2 to 7.5 has been fitted to (σ = 0.18)

$$\log k = 6.38 - 3420/T - 0.902\ pH \tag{52}$$

and the decrease between 7.5 to 13 has been fitted to (σ = 0.13)

$$\log k = 12.04 - 2641/T - 0.186\ pH \tag{53}$$

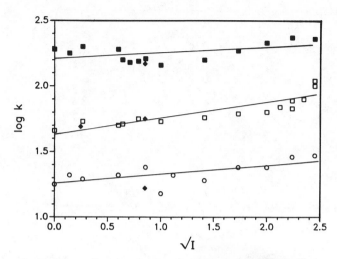

Figure 16. Values of log k versus $I^{1/2}$ for the oxidation of H_2S with H_2O_2 in seawater (◆) and NaCl at 5, (O) 25 (□) and 45°C (■) (pH = 8.0, in 0.01M borax).

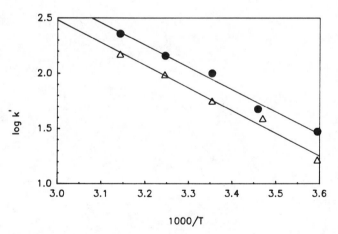

Figure 17. Values of log k' versus 1/T (T°K) for the oxidation of H_2S with H_2O_2 for seawater (△) and 6m NaCl (●) solution.

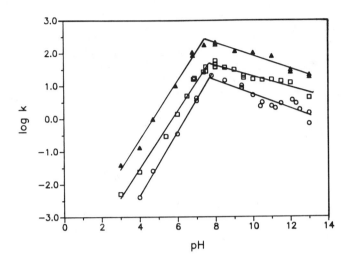

Figure 18. The effect of pH on the log k for the oxidation of H_2S with H_2O_2 at 5 (O), 25 (□) and 45°C (▲).

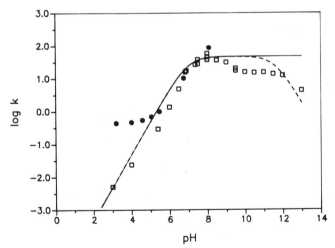

Figure 19. Values of log k versus pH for the oxidation of H_2S with H_2O_2 compared to the experimental values at 25°C: (□) present work; (●) Hoffmann; (- - -) fitted curve accounting for α_{HS^-} and α_{H2O2}; (-) fitted curve accounting for α_{HS}.

These linear fits of log k as a function of pH are given in Figure 18. The effect of temperature on the results above pH = 7.5 is the same as found for the results in NaCl at pH = 8 (equation 51). The effect of temperature on the results below pH = 8 is different because of the effect of temperature on the ionization of H_2S that occurs in this pH range.

From a pH = 2 to 8, the rate increases in a near linear manner with increasing pH. This is related to the ionization of H_2S and indicates that the HS^- species is more reactive than H_2S. It is noteworthy that log k determined in this study does not appear to level off at low pH values as was found by Hoffmann (46) for the oxidation of H_2S by H_2O_2. The leveling off at low pH was also observed for the oxidation of H_2S by oxygen (9). The leveling off at low values of pH can be related to the difference in the rates of oxidation of H_2S and HS^- according to equation 40. A plot of k/α_{H_2S} versus $k_1/[H^+]$ is shown in Figure 20. The intercept $k_0 = 0$ at all temperatures is within the experimental error. The slopes give $k_1 = 12.0 \pm 0.5$, 36.2 ± 0.4, and 211 ± 5 min^{-1} M^{-1}, at 5, 25 and 45°C, respectively. The values of $\ln k_1$ at the three temperatures have been plotted versus $1/T$ in Figure 21 and fitted to

$$\ln k_1 = 25.0 - 6306/T \tag{54}$$

which gives an energy of activation of $E_a = 51 \pm 3$ kJ mol^{-1} for the oxidation of H_2S. This is larger than the value given earlier at a pH = 8 ($E_a = 39 \pm 2$ kJ) because it does not contain terms due to the effect of temperature on the dissociation of H_2S. The overall rate constant, k, is related to k_1 by $k \approx \alpha_{HS}$ k_1. Thus, the differential of k with respect to T contains terms due to the effect of T on α_{HS} and k_1. The effect of temperature on α_{HS} is related to the $\Delta H°$ for the ionization of H_2S.

The simplest mechanism suggested by the pH dependence between 2 and 8, is a rapid pre-equilibrium of the dissociation of H_2S followed by a nucleophilic attack of HS^- on H_2O_2 with heterolytic cleavage of OH^- in the rate determining step

$$H_2S \ \text{-----} > H^+ + HS^- \tag{55}$$

$$HS^- + H_2O_2 \ \text{-----} > HSOH + OH^- \tag{56}$$

where

$$rate = -k_1[HS^-][H_2O_2] \tag{57}$$

Additional steps are the formation of polysulfides HS_n^- and their subsequent oxidation by HSOH (46).

By substituting the expression for $[HS^-]$, the following rate equation can be derived

$$rate = -k_1 K_1 [H_2S]_T [H_2O_2]/(K_1 + [H^+]) \tag{58}$$

where the second order rate constant should be equal to $k_1 K_1/(K_1 + [H^+])$. Using a value for $k_1 = 36$ min M^{-1} and $pK_1 = 6.98$ at 25°C (6), Figure 19 illustrates the pH dependence based on this simple mechanism. This mechanism predicts that when $[H^+] >> K_1$, that is, at the lower values of pH, the slope of log k vs pH should be a straight line with a slope of one. The

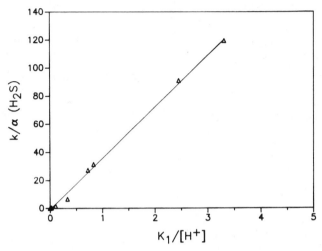

Figure 20. Values of k/α_{H2S} for the oxidation of H_2S with H_2O_2 versus $K_1/[H^+]$ at 25°C.

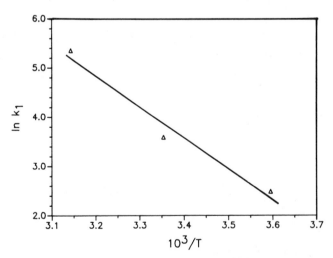

Figure 21. Values of $\ln k_1$ versus $1/T$ for the oxidation of H_2S with H_2O_2.

experimental values of log k are within experimental error of this line at all three temperatures.

The experimental points are lower than the calculated curve at pH values above 8.0 and falling off in a near linear manner. At $pH \geq 11$ the $[H_2O_2]$ is reduced due to its dissociation

$$H_2O_2 = H^+ + HO_2^-$$ (59)

where $pK_{H2O2} = 11.6$ (58). The formation of HO_2^- causes the rates to decrease above a $pH \approx 8.0$ apparently due to the slow reaction of HS^- with HO_2^-.

If we assume that the reaction between HS^- and HO_2^- is small, the decrease in k can be attributed to

$$k = k_1 \alpha_{HS} \alpha_{H_2O_2}$$ (60)

where

$$\alpha_{H_2O_2} = [H^+]/([H^+] + K_{H_2O_2})$$ (61)

The 25°C results (dotted line in Figure 19) show that the addition of the correction for H_2O_2 ionization does improve the fit above a pH = 10, but does not explain the nearly linear decrease above pH = 8. The 5 and 45°C results shown in Figure 22 look slightly better. Obviously, this simple explanation does not completely explain the near linear dependence above pH = 8. Other factors such as the formation of polysulfide ions (HS^-_n) and S^{2-} may be important.

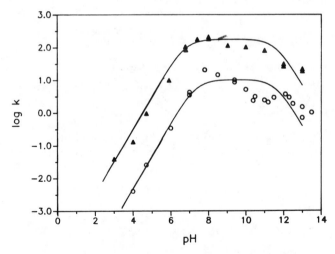

Figure 22. Values of log k versus pH for the oxidation of H_2S with H_2O_2 at (▲) 45°C; (○) 5°C; fitted curve accounting for α_{HS^-} and α_{H2O2} (Equation 60).

Over the entire range of pH and temperature studied there was little or no dependence of the rate on ionic strength, that is, no salt effect. This result is consistent with the simple mechanism in that the slow step does not involve two ions. With at least one neutral molecule in the slow step, transition state theory predicts that the rate of the reactions will be independent of ionic strength.

The half-life for the oxidation of sulfide in seawater with oxygen was 30 hours at 25°C. If one uses the $[H_2O_2] = 1.0 \times 10^{-7}M$ found (54,55) for surface seawater then the half-life for the oxidation of sulfide by peroxide in seawater would be 2800 hours. At concentrations of $H_2O_2 > 10^{-5}$ M, the oxidation of H_2O_2 with H_2S becomes competitive with O_2. Such concentrations of H_2O_2 are found in rain waters (58); thus, peroxide oxidation of H_2S may be more important than oxygen oxidation in aerosols or rainwaters.

APPENDIX I

The activity coefficients for γ_H and γ_{HS} using the Pitzer (29,59) equations are given (30,31,28) by

$$\ln \gamma_H = f^\gamma + \Sigma 2m_X (B_{HX} + E\,C_{HX} + \Sigma\Sigma m_M m_X (B'_{MX} + C_{MX})$$

$$+ \Sigma m_M (2\theta_{MH} + m_X \psi_{HMX}) \tag{A1}$$

$$\ln \gamma_{HS} = f^\gamma + \Sigma 2m_M (B_{MHS} + E\,C_{MHS}) + \Sigma\Sigma m_M m_X (B'_{MX} + C_{MX})$$

$$+ \Sigma m_X (2\theta_{XHS} + m_M \psi_{MXHS}) \tag{A2}$$

where M and X are cations and anions, respectively, and

$$f^\gamma = -A^\phi [I^{1/2}/(1+1.2I^{1/2}) + (2/1.2)\ln(1+1.2I^{1/2})] \tag{A3}$$

$$B = \beta^{(0)} + (2\beta^{(1)}/\alpha_1^2 I)[1 - (1 + \alpha_1 I^{1/2})\exp(-\alpha_1 I^{1/2}) \tag{A4}$$

$$B' = (2\beta^{(1)}/\alpha_1^2 I^2)[-1 + (1 + \alpha_1 I^{1/2} + \alpha_1^2 I)\exp(-\alpha_1 I^{1/2})] \tag{A5}$$

$$C = C^\phi/2|Z_M Z_X|^{1/2} \tag{A6}$$

and $E = 1/2\ \Sigma|m_i Z_i|$ where $\alpha_1 = 1.2$ and the Debye Hückel limiting law $A^\phi = 1/3(2\pi N_o d_w/1000)^{1/2}(e^2/DkT)^{3/2}$ has been previously defined (59). The parameters θ and ψ represent interactions between ions of the same charge and triple interactions, respectively.

The Pitzer parameters have been fitted as a function of temperature to the following equations

$$\beta^{(0)} = q_1 + q_2/T + q_3\ln T + q_4 + q_5 T^2 \tag{A7}$$

$$\beta^{(1)} = q_6 + q_7 T + q_8 T^2 \tag{A8}$$

$$C^\phi = q_9\ q_{10}/T + q_{11}\ln T + q_{12}T \tag{A9}$$

The values of q_i are given in Table I.

Further details for the method used for determining the Pitzer parameters are given by Pitzer (59), Millero (28,31), Harvie and co-workers (30,60).

Acknowledgement

The authors wish to acknowledge the support of the Office of Naval Research (N00014-87-G-0116) and the Oceanographic Section (OCE-8600284) of the National Science Foundation for this study.

Literature Cited

1. Jacobs, L.; Emerson, S. Earth Planet Sci. Lett. 1982, 60, 237-52.

2. Boulegue, J.; Lord, III, C. J.; Church, T. M. Geochim. Cosmochim. Acta 1982, 46, 453-64.

3. Zhang, J.; Whitfield, M. Mar. Chem. 1986, 19, 121-37.

4. Morse, J. W.; Millero, F. J.; Cornwell, J. C.; Rickard, D. Earth Science Rev. 1987, 24, 1-42.

5. Emerson, S.; Jacobs, L.; Tebo, B. In Trace Metals in Seawater; Wong, S.; Burton, J. D.; Bruland, K.; Goldberg, E., Eds.; Plenum: New York, 1983; pp 579-608.

6. Millero, F. J. Mar. Chem. 1986, 18, 121-47.

7. Millero, F. J.; Plese, T.; Fernandez, M. Limnol. Oceanogr. 1987, in press.

8. Hershey, J. P.; Plese, T.; Millero, F. J. Geochim. Cosmochim. Acta 1987, submitted.

9. Millero, F. J.; Hubinger, S.; Fernandez, M.; Garnett, S. Environ. Sci. Technol. 1987, 21, 439-43.

10. Millero, F. J.; LeFerriere, A.; Fernandez, M.; Hubinger, J. P.; Hershey, S. Envir. Sci. Technol. 1987, submitted.

11. Mason, D. M.; Kao, R. In Thermodynamics of Aqueous Systems with Industrial Applications; Newman, S. A ., Ed.; American Chemical Society: Washington, DC, 1980; pp 107-38.

12. Clarke, E. C. W.; Glew, D. N. Can. J. Chem. 1971, 49, 691-98.

13. Douabul, A. A.; Riley, J. P. Deep-Sea Res. 1979, 26A, 259-68.

14. Millero, F. J.; Schreiber, D. R. Amer. J. Sci. 1982, 282, 1508-40.

15. Gamsjager, von H.; Schindler, P. Helv. Chim. Acta 1969, 52, 1395-1402.

16. Meyer, B.; Ward, K.; Koshlap, K.; Peter, L. Inorg. Chem. 1983, 22, 2345-6.

17. Giggenbach, W. Inorg. Chem. 1971, 10, 1333-8.

18. Barbero, J. A.; McCurdy, K. G.; Tremain, P. R. Can. J. Chem. 1982, 60, 1872-80.

19. Savenko, V. S. Oceanology 1977, 16, 346-50.

20. Goldhaber, M. B.; Kaplan, I. R. Mar. Chem. 1975, 3, 83-104.

21. Bates, R. Determination of pH, Theory and Practice; Wiley: New York, 1973; pp. 479.

22. Almgren, T.; Dyrssen, D.; Elgquist, B.; Johansson, O. Mar. Chem. 1976, 4, 289-97.

23. Hansson, I. Deep-Sea Res. 1973, 20, 479-91.

24. Dickson, A. Geochim. Cosmochim. Acta 1983, 48, 2299-308.

25. Mehrbach, C.; Culberson, C. H.; Hawley, J. E.; Pytkowicz, R. M. Limnol. Oceanogr. 1973, 18, 897-907.

26. Culberson, C.; Pytkowicz, R. M. Mar. Chem. 1973, 1, 304-16.

27. Millero, F. J. Limnol. Oceanogr. 1986, 31, 839-47.

28. Millero, F. J. Geochim. Cosmochim. Acta 1983, 47, 2121-9.

29. Pitzer, K. S. J. Phys. Chem. 1973, 77, 268-77.

30. Harvie, C. E.; Moeller, N.; Weare, J. H. Geochim. Cosmcohim. Acta 1984, 48, 723-51.

31. Millero, F. J. Thalis. Jugoslav. 1982, 18, 253-91.

32. Dyrssen, D.; Johansson, O.; Wedborg, M. Mar. Chem. 1978, 6, 275-9.

33. Millero, F. J. In Chemical Oceanography; Riley, J. P.; Chester, R., Eds.; Academic Press: London, England, 1983; Vol. 8, pp 1-88.

34. Ostlund, G. H.; Alexander, J. J. Geophys. Res. 1963, 68, 3995-7.

35. Skopintsev, B. A.; Karpov, A. V.; Vershinina, O. A. Sov. Oceanogr. 1964, 4, 55-73.

36. Avrahami, M.; Golding, R. M. J. Chem. Soc. A 1968, 647-51.

37. Cline, J. D.; Richards, F. A. Environ. Sci. Technol. 1969, 3, 838-43.

38. Sorokin, Y. L. Okeanologiya (Moscow) 1971, 11, 423-31.

39. Sorokin, Y. L. Arch. Hydrobiol. 1970, 66, 391-446.

40. Boulegue, J.; Ciabrini, J. P.; Fouillac, C.; Michard, G.; Ouzonian, G. Chem. Geol. 1979, 25, 19-29.

41. Chen, K. Y.; Morris, J. C. Environ. Sci. Technol. 1972, 6, 529-37.

42. O'Brien, D. J.; Birkner, F. G. Envrion. Sci. Technol. 1977, 11, 1114-20.

43. Almgren, T.; Hagstrom, L. Water Res. 1974, 8, 395.

44. Benson, B. B.; Krause, D., Jr.; Peterson, M. A. J. Solution Chem. 1979, 8, 655-90.

45. Benson, B. B.; Krause, D., Jr. Limnol. Oceanogr. 1984, 29, 620-32.

46. Hoffmann, M. R.; Lim, B. C. Environ. Sci. Technol. 1979, 13, 1406-14.

47. Hoffmann, M. R. Environ. Sci. Technol. 1977, 11, 61-6.

48. Cline, J. D. Limnol. Oceanogr. 1969, 14, 454-58.

49. Cline, J. D. Ph.D. Thesis, University of Washington, Seattle, WA, 1969.
50. Classen, A.; Bauer, O. Ber. 1883, 16, 1061.
51. Wasserman, A. Liebigs Ann. 1933, 503, 249.
52. Feher, F.; Heuer, E. Angew. Chem. A 1949, 59, 237.
53. Satterfield, C. N.; Reid, R. C.; Briggs, D. R. J. Amer. Chem. Soc. 1954, 76, 922.
54. Moffett, J. M.; Zika, R. G. Mar. Chem. 1983, 13, 239.
55. Zika, R. G.; Saltzman, E.; Cooper, W. J. Mar. Chem. 1985, 17, 265.
56. Zika, R. G.; Moffett, J. M.; Cooper, W. J.; Petasne. R. G.; Saltzman, E. Geochim. Cosmochim. Acta 1985, 49, 1173.
57. Gray, R. J. Am. Chem. Soc. 1969, 91, 56.
58. Kok, G. L. Atmos. Environ. 1980, 14, 653.
59. Pitzer, K. S. In Activity Coefficients in Electrolyte Solutions; Pytkowicz, R. M., Ed.; CRC: Boca Raton, Florida, 1982; Vol. 1, pp. 157-208.
60. Harvie, C. E.; Weare, J. H. Geochim. Cosmochim. Acta, 1980, 44, 981.
61. Harned, H. S.; Owen, B. B. The Physical Chemistry of Electrolyte Solutions; Reinhold Publishing Co.: New York; p 803.
62. Silvester, L. F.; Pitzer, K. S. J. Phys. Chem. 1977, 81, 1822.

RECEIVED October 18, 1988

Chapter 19

Hydrogen Sulfides in Oxic Seawater

Scott Elliott, Eric Lu, and F. Sherwood Rowland

Department of Chemistry, University of California, Irvine, CA 92717

Carbonyl sulfide hydrolysis has been identified as a quantifiable source capable of maintaining the hydrogen sulfides at picomolar or greater in mixed layer seawater. Such levels are thermo-dynamically sufficient for sulfide complexes to dominate inorganic fractions of several trace metals which also have concentrations in the pico- to nanomolar range. Preliminary measurements now confirm total sulfide at 0.1-1 nM in the western Atlantic, but speciation remains unclear. Field evidence indicates that copper, an abundant and strong soft ligand interactant, is organically bound, and competition from sulfide is difficult to evaluate because of uncertainties in equilibrium constants. The extent of metal complexation will be a key in determining the direction and magnitude of H_2S fluxes at the sea surface. Diurnal variations in the total sulfide signal require input additional to hydrolysis and removal additional to direct oxidation by O_2.

The hydrogen sulfides (H_2S, SH^-, S^{2-} and their metal complexes) are well known in restricted reducing regions of the world ocean such as anoxic basins (1), but they have traditionally been dismissed as unimportant for, or even non-existent in, most oxic seawaters (2-4). Several lines of reasoning are now beginning to suggest that sulfides actually do exist in the surface ocean, and enter into a rich metal chemistry there. Extensive measurements of carbonyl sulfide (OCS) in seawater (5,6) permit the quantification of a mixed layer source, the hydrolysis reactions (7-11)

$$OCS + H_2O \rightarrow CO_2 + H_2S, \tag{1}$$

$$OCS + OH^- \rightarrow CO_2 + SH^-. \tag{2}$$

Based on laboratory oxidation rates for

$$SH^- + O_2 \rightarrow \text{oxidized products} \tag{3}$$

(12,13), hydrolysis alone is capable of supporting 10^1 to 10^3 picomolar of the free species SH^-, and so also, of converting the inorganic fraction to sulfide complexes for several trace metals with total concentrations ranging upward

0097–6156/89/0393–0314$06.00/0

from pM (8-11). The OCS arguments have stimulated measurement activity, and preliminary determinations indicate 0.1-1 nM of total dissolved sulfide in the western Atlantic (14-16).

In the present work, we review these early developments in the study of open ocean hydrogen sulfides, and address some matters arising from them. Perhaps most importantly, speciation of the measured totals appears difficult to establish. In equilibrium calculations, a conserved injection of sulfide tends to compete successfully for, and be titrated by, soft or transition metals (17-19). However, the most concentrated tight sulfide binder (copper) has been associated almost exclusively with organic complexes in seawater (e.g. 20-22). Relative amounts of the sulfide forms are sensitive to several uncertain parameters, including the degree of organic interactions, and linear free energy estimates of the many unknown sulfide constants. The non-metal complexed species are of particular relevance to sulfur cycling investigations because H_2S can cross the sea air interface. Atmospheric models had interpreted hydrogen sulfide measurements made over the open ocean as an indication that the gas passes out of some surface seawater (23,24). An interference akin to reactions 1 and 2 now casts doubt on all marine H_2S vapor data (25), substantially weakening any tropospheric hinge for flux calculations.

The preliminary total sulfide values exhibit diurnal variations and depth profiles which require sources and sinks beyond those considered in the laboratory (14-16). We list some of the likely candidates.

Historical Marine H_2S Chemistry

The members of the hydrogen sulfide family are both thermodynamically and kinetically unstable in oxic seawater. The thermodynamic bias against them can be appreciated in a simple redox couple to sulfate, the predominant form of oceanic sulfur.

$$SH^- + 2O_2 \rightarrow H^+ + SO_4^{2-} \tag{4}$$

Sulfide is here represented by the bisulfide ion, the major free species at the oceanographic pH of 8.2. Relative to sulfate (0.01 M), equilibrium sulfide concentrations are on the order of 10^{-140} M in this system (26). It can be calculated that an imaginary search through 10^{100} terrestrial oceans would be needed to yield even a single molecule of the bisulfide if it were under equilibrium control. A number of kinetic studies have shown that the sulfides are also short lived in oxic seawater. Oxidation experiments are complicated by catalytic and wall effects so that lifetimes have varied from minutes to days, but the latest values fall towards the upper end of the range (12,13).

Several types of anoxic environment harbor H_2S along the periphery of the bulk ocean, but in order to exist in oxic waters more than a few hours of mixing removed from these isolated localities, the sulfides must be kinetically supported. They have generally been ignored in the open sea, because no internal sources have been known for them. In an early paper on trace metals, for example, Krauskopf (2) restricted metal sulfide interactions to anoxic basins. Ostlund and Alexander (3) later determined an oxidation lifetime of about half an hour in O_2 saturated seawater, and noted that sulfide generated in sediments was expected to reach the atmosphere only from very shallow waters (< a few meters). With no sources between the sea bottom and surface, the sulfides were presumed absent from most of the sea. This sentiment is echoed almost to the present (4).

The atmospheric conception of H_2S has been at odds with these conclusions for several decades. Early atmospheric sulfur balances required input of some uncharacterized reduced species, and a sea to air H_2S flux was assigned arbitrarily, because hydrogen sulfide is the simplest divalent sulfur compound (27). It was eventually learned that biological activity supports a wide variety of reduced sulfur volatiles in seawater. Carbonyl and dimethyl sulfide, for example, actually dominate marine tropospheric fluxes of divalent S (24). Early measurements of remote H_2S vapor, however, were still consistent with sea to air transport (23). Although an upward flux of H_2S implies the presence of considerable amounts of the sulfide anions in seawater, their potential for open ocean chemistry went completely unexplored during this period.

A Quantifiable Source: OCS Hydrolysis

It became apparent several years ago in the course of atmospheric sulfur budget work that carbonyl sulfide is very evenly distributed through surface seawater, at 10-100 pM (5,6). Johnson (7) noted that rate constants were available for OCS hydrolysis, and reactions 1 and 2 were soon recognized as a quantifiable and ubiquitous source of the hydrogen sulfides to the surface ocean (8-11). One dimensional modeling had also shown that carbonyl sulfide was long enough lived relative to vertical transport to penetrate many tens of meters below the surface (7). Reactions 1 and 2, then, must produce sulfide throughout the oceanic mixed layer. Open ocean depth profiles have recently become available to confirm the 1-D calculations (personal communication with J. Johnson and T. Bates, 1987).

Although it was clear that alternate modes of cycling were also likely to be operative (8,11), 1, 2 and 3 were at the time, and remain at present, the only quantifiable input and output. This simple reaction set provides a means for establishing preliminary estimates for kinetically supported mixed layer sulfide levels (8-11). Rate constants for 1 and 2 have recently been remeasured (8,11), verifying several earlier experiments (28,29). The Ahrrenius expressions are k_1 = 1.0 x 10^{11} exp(-10770/T) s^{-1} and k_2 = 8.1 x 10^9 exp(-6040/T) $(Ms)^{-1}$, for both loss of OCS and H_2S production, with no observed salt effect. Millero (12) has argued that some oxidation rate constants for reaction 3 have been overestimated, and gives the empirical expression Log k_3 $(Ms)^{-1}$ = 8.59 - 3000/T for pH = 8 in seawater. The effective activation energy for OCS hydrolysis by the combination of 1 and 2, and for reaction 3 are coincidentally both close to 14 kcal/mole. Coupled in a hypothetical kinetic steady state, 1 through 3 give the ratio $(SH^-)/(OCS)$ = 6-7, independent of temperature, and the bisulfide ion concentrations listed in Table I for various water types. The rate of production in the hydrolyses suggests that it is reasonable to expect non-thermodynamic levels of the hydrogen sulfides between pico- and nanomolar in mixed layer seawater (8-11).

By analogy with the carbonic acid system, OCS hydrates to form $H_2O.OCS$ in aqueous solution, and the bicarbonate and carbonate like anions $OCS.OH^-$ and $OCS.O^{2-}$, known collectively as monothiocarbonates (MTC). The entire carbonate analog system is shown in Figure 1, and it should be clear that MTC species are likely to serve as intermediates in both overall hydrolysis channels (7). Several pieces of information point to a lack of reversibility to OCS, or, schematically, to

$$OCS \rightarrow MTC \rightarrow sulfide.   (5)$$

In one study (29), a fleeting anion which was probably $OCS.O^{2-}$ could be stabilized at high pH. The intermediates seemed to form at +B, and decay through a unimolecular reaction F. Indirect evidence for irreversibility has also come from our own laboratories. As shown in Figure 2, over the pH range

Table I. Average Surface Carbonyl Sulfide Concentrations for Various Productivity Zones (6), and Bisulfide Ion Levels Corresponding to them in the Hydrolysis-Oxidation Steady State (Equations 1, 2, and 3). Units are Picomolar

	OCS	SH^-
Oligotrophic	10	65
Transition	20	130
Coastal/Shelf	110	720

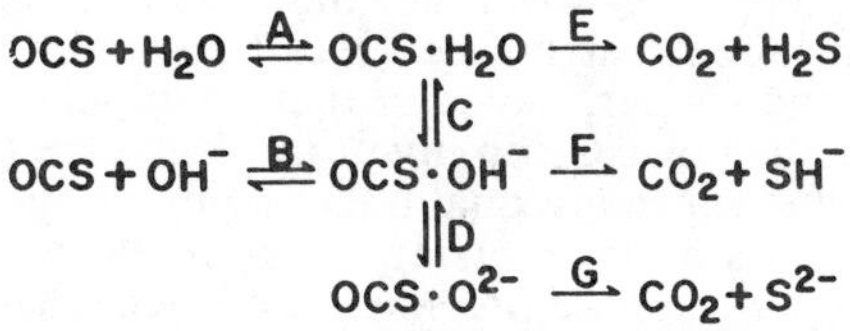

OCS + H2O ⇌A OCS·H2O →E CO2 + H2S
C
OCS + OH⁻ ⇌B OCS·OH⁻ →F CO2 + SH⁻
D
OCS·O2- →G CO2 + S2-

Figure 1. The carbonyl sulfide hydrolyses, including monothiocarbonate intermediates.

SOLUBILITY / SOLUBILITY (pH = 4)
10^5
10^0
CO2
OCS
5
10
pH

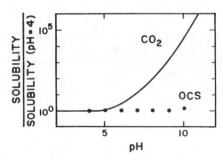

Figure 2. Solubility of carbonyl sulfide in aqueous solution as a function of pH, with the analogous values for carbon dioxide and the carbonate system shown in comparison.

where CO_2 solubilities rise because of equilibration with carbonates, values for carbonyl sulfide are invariant and correspond to the low pH Henry's Law constant of around 2.0. By pH 10 there is some hint at a solubility change. Measurements are uncertain in this regime, however, because hydrolysis and dissolution time scales become comparable. The two sets of observations are nicely unified by assigning rate control to the hydration steps, or in other words, attaching the Ahrrenius expressions k_1 and k_2 to +A and +B. All kinetic results can then be explained (30) by assuming that MTC dehydration rates (-A,-B) and acid base properties are similar to those of the carbonates, with $k_F > 10$ s^{-1} from Phillip and Dautzenberg (29). Over the pH range of natural waters, conversion to sulfide outstrips dehydration for the monothiocarbonates, and their concentrations lie orders of magnitude below those of the parent OCS.

Sulfide Metal Chemistry

Several groups have now evaluated mixed layer metal interactions by including sulfide in equilibrium speciation calculations (8-11,15-18). The models have been constructed at very different levels of sophistication, but some consistencies have emerged. Elliott and coworkers (8-11) compared the stabilities of sulfide complexes with other inorganics in light of the hydrolytic steady state, and concluded that sulfide speciation could be significant for at least mercury, silver and copper. Some of their manipulations are outlined in this section, with further discussion given in reference (19). Dyrssen and coworkers obtain results which are qualitatively similar by distributing western Atlantic sulfide levels among a variety of species in the HALTAFALL computer program (17,18). These previous assessments have not incorporated measurements of conditional organic binding reported for copper (e.g. 20-22). We take the opportunity to add some organic considerations here.

The generalized reaction

$$M^{m+} + sS^{2-} + hH^+ = MS_sH_h^{m-2s+h}; K_{sh} \qquad (6)$$

encompasses many of the sulfide complexes which have been studied in the laboratory. Because bisulfide is the major unbound form at pH = 8, it is often appropriate for oceanographic purposes to build from the subset s = h, and the bisulfide formation reactions

$$M^{m+} + nSH^- = M(SH)_n^{m-n}; K_n \qquad (7)$$

where n = s = h. Species with s and h nonequivalent are then handled as acid base conjugates.

Determinations of K_n are most numerous for the 2$^+$ metals, and so we focus on them here, but even within this set, data are scarce. The non-parenthetical entries in Table II represent a fairly comprehensive collection of the published information (19). Linear free energy correlation can be performed against sulfide solubility products (8,10,19) or the model ligand dithizone (1,17) in order to estimate unknown constants for n > 1. Agreement between the two methods suggests bands of uncertainty an order of magnitude or so in breadth on either side of the preferred values in the Table. Larger error bars surround some of the n = 1 estimates, because only one measurement is available (cadmium) and interpolation is therefore impossible. Dyrssen (17) arrives at the higher figures for copper and mercury by extrapolating parallel to the n = 2 correlations. Elliott (19) bases the lower set on free energies for hydroxide complexation. It is not certain that the actual equilibria are bracketed by these approaches (19). Obviously, experimental work is sorely needed.

At the trace sulfide levels in open ocean seawater, mass action dictates that smaller complexes are more important (8). For purposes here, it is sufficient to limit ourselves to the single and double sulfides (n = 1 and 2). Where their acidity is known, pK_a lies near or well below the oceanographic pH, so that conjugate bases will be dominant sulfides in the sea. A pair of established dissociation constants are pK_{a1} and pK_{a2} for $Hg(SH)_2$, at 6.2 and 8.3 (31). Elliott (8) argued on a simple electrostatic basis that the n = 1 complexes should be several orders of magnitude more acidic. Dyrssen (17) has recently realized that a pH independent minimum solubility of 10^{-7} M observed for solid cadmium sulfide is probably best explained by the neutral single sulfide complex CdS^o, which can be thought of as an n = 1 base. Coupled with the solubility product $Log_{10} K_{sp}$ = -25 (17), and the Table II K_1, the pK_a for $CdSH^+$ is about 2. Acidities for the other 2+ metals must currently be estimated by setting them equal to mercury n = 2 or cadmium n = 1 values (1,8,11,17). We stress again, however, that orders of magnitude of uncertainty may be involved in such approximations (19).

The stability of a complex in solution can be conveniently expressed by normalizing to the concentration of the aquo ion, M^{m+}, as in

$$M(SH)_n^{m-n}/M^{m+} = K_n(SH^-)^n \qquad (8)$$

for the sulfides. Deprotonation adds factors of about 10^6 and 200 for n = 1 and 2 given the cadmium and mercury acidities, and uncertainty contributions enter from both K_n and pK_a. In seawater, metals simultaneously complex other anions to form better understood species, and ratios analogous to 8 can be derived relating them to the aquo as well. Chlorine and carbonate are typical inorganic ligands, and stabilities are listed in Table III for their complexes with the sulfide interactive metals of interest here. Organic binding sites have been identified for oceanic copper, and conditional concentrations and equilibrium constants measured (20-22). We take an upper limit of 10^5 for the ratio of organically held metal to Cu^{2+} consistent with a recent study by Moffett and Zika (21). Actual values range downward several log units. Some of the variation may be artificial (21), but spatial effects have also been documented (20-22).

Bisulfide ion levels at which n = 1 and 2 sulfides have stabilities similar to the better known species are listed in the latter two columns of Table III, with the single sulfide K_n set at our lower values. The relationship with concentrations in Table I indicates that it is reasonable to expect the hydrogen sulfides to be competitive in the complexation of several trace metals in the mixed layer. Even at the more cautious estimates for bisulfide binding, the conclusion is strong for n = 1. It would be insensitive to adjustments of a factor of ten or more for n = 2. Equilibrium modelling, then, has supported the possibility of a rich metal sulfide chemistry since the recognition of the OCS hydrolysis source (8). This is demonstrated in Figure 3 for copper by plotting species stability versus free sulfide. Total dissolved concentrations of the Tables II and III metals are above picomolar in the mixed layer (32), so that in the hydrolysis situation, complexes could bear a sulfide load near steady state free levels.

Measured Total Sulfide

The technique recently applied by Cutter and Oatts (14) and Krahforst and Cutter (15) to preliminary measurements of mixed layer sulfide was developed

Table II. Bisulfide Complexation Constants (K_n in Equation 7) for Selected
Oceanic Trace Metals, all as Base 10 Logarithms. Parenthetical Entries
have been Estimated Through Linear Free Energy Comparisons;
Others are Taken from References (19) or (31), for 25°C,
and an Ionic Strength of 1.0 in Most Cases

n =	1	2	3	4
Cu^{2+}	(10-15)	(22.5)	(23.8)	23.7
Zn^{2+}	(7)	(14)	14.2	
Cd^{2+}	6.4	13.8	16.0	18.4
Hg^{2+}	(15-30)	37.7	40.3	

Table III. The Ratio of Concentrations for Selected Complexes to those of the
Aquo Ion M^{2+} (17,31,32), and Free Bisulfide Ion Concentrations (Molar)
Required to Sustain Like Stability, all as Base 10 Logarithms.
In Each Case K_1 is a Rough Lower Limit from Elliott (19)

	Complex	Complex/M^{2+}	(SH-),n=1	(SH-),n=2
Cu^{2+}	$CuCO_3^0$	1.3	-15	-12
Cu^{2+}	Organic	< 5	<-11	<-10
Zn^{2+}	Aquo	0	-13	8
Cd^{2+}	$CdCl_c^{2-c}$	1.6	-11	-7
Hg^{2+}	$HgCl_c^{2-c}$	14.2	-7	-13

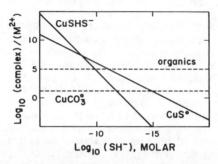

Figure 3. Rough ratios of equilibrium concentrations for inorganic (31,32) and organic (20-22) copper species to those of the aquo ion Cu^{2+}, with sulfide complexes plotted as functions of free bisulfide ion level and K_1 set at the lower limit in Table II.

for sedimentary pore seawater, and is designed to detect the sum of metal bound and free forms. Samples are acidified to low pH for conversion of complexes and anions to H_2S, which is then stripped into a gas stream. In view of the formation constants in Table II, it is not entirely clear that all of the complexes can be volatilized easily, but until tests are available, we take this to be the case. The first reports for the open ocean averaged 0.5 nM (14). Krahforst and Cutter (15,16) have presented one particularly interesting data set, obtained 100 km from the mouth of Chesapeake bay in the western Atlantic. At both the surface and 50 meters, total sulfide builds to 0.5 nM overnight, then decays by more than a factor of 3 during daylight. A lower mixed layer maximum was observed on occasion.

Measured total sulfide levels confirm the major features of the hypotheses of Elliott and coworkers (8-11), in that sulfide seems to exist in remote oxic waters at concentrations approaching nanomolar, sufficient for meaningful metal interactions. The amplitude of the diurnal variation is too strong, however, for carbonyl sulfide hydrolysis to be the primary input, or for direct oxidation by O_2 to be the sole sink, and alternate cycling processes are indicated.

Some abiotic, or more precisely, indirectly biotic options include metal catalysis of the hydrolyses and oxidation, or photolysis. Several lines of evidence suggest that reactions 1 and 2 are subject to acceleration in the presence of trace metals in natural water systems. OCS and H_2S corrode sulfide attracting metal surfaces at the same rate (33) and the explanation may lie in rapid hydrolysis in a thin moisture coating (personal communication with T. Graedel, 1986). Cooper and Saltzman (25) have proposed silver catalyzed versions of 1 and 2 as the interference in attempts to measure remote marine H_2S vapor on $AgNO_3$ impregnated filters. Trace metal catalysis of sulfide oxidation has often been noted, but only at reactant concentrations many orders of magnitude higher than ambient (12-16). If the effect depends on relative amounts of bound and unbound sulfide, it could be enhanced for the trace levels in seawater. An anonymous referee for this work notes that the association of sulfide with redox reactive metals such as copper (20) may influence oxidation rates. Dissolved organic sulfur compounds (DOS) constitute a reduced sulfur pool additional to OCS. Photolysis of DOS has produced carbonyl sulfide in the laboratory (6), although the wavelengths employed did not match the mixed layer spectrum closely. H_2S might be sought in similar experiments.

More direct biological channels also seem promising as sources. Land plants release H_2S, but the process has not been considered for marine algae (6). Intermittent deep sulfide maxima could be connected with anoxic microenvironments recently located in marine snow. These organic particulates accumulate in the pycnocline and offer potential sites for contrary redox reactions such as dissimilatory sulfate reduction (34).

The overall diurnal signal requires daylight removal, generally suggesting photochemistry or biology, and/or nighttime input. With regard to the latter, it is interesting that the lysing of cells by grazing zooplankton has been implicated in the release of DMS precursors in laboratory cultures (35). Zooplankton move upward through the water column at night to feed, and there has been speculation that this could be a reason for nighttime DMS peaks. The potential for simultaneous injection of H_2S might also be investigated.

Speciation of the Measured Totals

Distribution of measured concentrations into the various bound and free forms can in principle be examined by inserting sulfide into equilibrium models as a conservative property. In one early calculation, for example, Dyrssen and coworkers ($\underline{17}$) adopted K_n values near the upper limits for the Table II elements, and added the sulfide interactants Co, Ni, Pb and Ag. With dissolved metal concentrations set at rough mixed layer averages ($\underline{32}$), the n = 1 complex CuS⁰ dominated all of 0.5 nM sulfide by titrating SH⁻ to 10^{-20} molar, and embraced a major portion of total copper as well. In practice, complications arising from uncertainties in equilibrium relationships enter at several levels. Copper is often associated with organics in seawater samples (e.g. $\underline{20\text{-}22}$). This in itself may place restrictions on any theoretical sulfide fraction. Conditional stability constants obtained in such studies were not incorporated in the early model, but would be expected to moderate sulfide complexation somewhat. Effects of the large errors inherent in sulfide constant estimates also deserve attention.

In order to illustrate these points, we have simulated more elaborate computations by focusing a simple sensitivity test on the reactions of copper. In a conceptual experiment, sulfide is added to water containing 1 nM Cu(II), an amount normally attached to the total dissolved metal in the western Atlantic ($\underline{32}$), accounting also for 1-10% photochemically maintained as Cu(I) ($\underline{20}$; personal communication with J. Moffett, 1987). The standard seawater ligands are included, and organic complexing agents at unspecified levels well above those of sulfide, with the stability of their copper species fixed at the rough upper limit 10^5 against Cu^{2+} (Table III; $\underline{21}$). Equilibration with CuS⁰ for K_1 = 10^{10} gives the concentration curves in Figure 4a. Titration of free sulfide is relaxed to above picomolar, but sulfide complexes still take precedence when the total added exceeds half of labile copper. As noted in the metal chemistry section, it is possible that real binding constants lie below the Table II range, or that the CuSH⁺ acidity is not comparable to the sole literature value (CdSH⁺; $\underline{17}$). In Figure 4b we have replaced CuS⁰ with the n = 2 species, and lowered K_2 to 10^{20}. Most of copper is now tied to organics over the entire sulfide range, but CuSHS⁻ makes a sizable contribution, and free sulfide goes untitrated.

If copper interactions were minimized in real seawater, abundant metals of lesser sulfide affinity would take up some of the slack. This is partially evident from analyses of the type in Table III. For example, nickel has mixed layer concentrations on the order of nanomolar ($\underline{32}$), and its sulfide equilibria and inorganic seawater speciations may resemble those of zinc ($\underline{17\text{-}19,31,32}$). Titration, however, should only lower free sulfide to a Table III SH⁻ equivalence point, or, to roughly picomolar. In a follow up to $\underline{17}$, Dyrssen and coworkers treat Cu(II) as a variable parameter, and find that in its absence, nickel, zinc and lead can all become sulfides while the bisulfide ion still hovers well above pM ($\underline{18}$). Again, it must be emphasized that error margins in the various equilibria remain to be investigated.

The divergence between earlier work ($\underline{17}$) and the simple calculation presented here underscores critical sensitivities to both sulfide and non-sulfide binding. Uncertainty in bisulfide ion concentrations, or equivalently, in aqueous phase H_2S, should be of particular importance in atmospheric sulfur cycling studies. Marine tropospheric models constructed over the last decade or so have translated open ocean H_2S vapor measurements into a sea to air flux ($\underline{23,24,11}$). The argument has been that the gas is too short lived relative to hydroxyl radical attack to survive an ocean crossing, and so must emanate from the sea surface. A central round figure for reports of remote marine tropospheric H_2S is 10 pptv, and the Henry's law constant is close to 0.3 at oceanographic temperatures ($\underline{13}$), requiring > > 25 pM bisulfide ion in surface waters to support oxidation above the ocean ($\underline{11}$). If titration of sulfide by copper were general over the sea surface ($\underline{17,18}$), a conflict would be created. However, it has recently become problematic to judge dissolved free sulfide

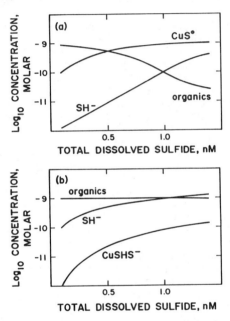

Figure 4. Concentrations in a model of seawater chemistry including the copper sulfide reactions. a) $n = 1$ with $K_1 = 10^{10}$ and a value $R = 10^5$ for the ratio of organic copper complexes to Cu^{2+}. b) $n = 2$ with $K_2 = 10^{20}$, $R = 10^5$.

from gas phase measurements. Cooper and Saltzman (25) have demonstrated that OCS hydrolyzes on filter surfaces to interfere with the standard Natusch technique, some version of which has been used in all marine H_2S vapor determinations. Appropriate corrections have been applied only in the latest study, and while an open ocean signal remains, it is near the lower limit of detection (personal communication with E. Saltzman, 1987). In retrospect, the Natusch method seems non-ideal, and trace H_2S determinations made with it should be treated with discretion.

Conclusion

Carbonyl sulfide hydrolysis comprises a source of the hydrogen sulfides through much of the oceanic mixed layer (8-11). Hydrolysis and oxidation by O_2 remain the only two pathways involving open ocean sulfide which can be quantified. Based on a hydrolytic steady state, it is reasonable to anticipate non-thermodynamic sulfide levels above pM, sufficient for sulfide complexes to compete with traditional speciations for several trace metals with labile concentrations between pico and nanomolar. OCS has therefore implied a significant sulfide carrying capacity and rich metal chemistry in the mixed layer (8-11). Preliminary measurements confirm the presence of total sulfide at 0.1-1 nM (14-16). The observed diurnal variation and depth dependence require alternate production, perhaps trace metal catalysis of the hydrolysis, photolysis of dissolved organic sulfur, or direct release by biology, and also alternate sinks. In equilibrium models including conservative sulfide, speciation remains uncertain because the central binding constants are unknown and difficult to estimate. Measurements of remote marine tropospheric H_2S are currently taken as an indication of sea to air flux from some open waters (24), but an interference now clouds all data acquired prior to the work of Cooper and Saltzman (25). Aqueous sulfide complexation must be considered in any re-examination of direction and magnitude for the transport.

A general theme of this body of work has been to introduce new, kinetically supported ligands into the ocean, and a logical extension of our ideas is worth noting for future reference. Just as OCS implies H_2S, the presence of the hydrogen sulfides forecasts an entire array of reduced sulfur intermediates leading along the oxidation sequence to the eventual sink at sulfate. Polysulfides, elemental sulfur, thiosulfate and sulfite have all been observed as products in sulfide oxidation experiments (12,13). Insofar as open ocean degradation is abiotic, some of these valence states should exist. Their anions bind metals tightly (31), and are completely unscrutinized as constituents of oxic seawater. If kinetic and thermodynamic stability correspond along this series, there may be further unexpected ligands to add to speciation models.

Acknowledgements

We thank J. Moffett, D. Dyrssen, F. Millero, E. Saltzman, G. Cutter, C. Krahforst, T. Church and G. Luther for preprints and helpful discussions. This work has been supported by United States Department of Energy Grant number ATO3-73ER-70126.

Literature Cited

1. Dyrssen, D. Mar. Chem. 1985, 15, 285-93.

2. Krauskopf, K. B. Geochim. Cosmochim. Acta 1956, 9, 1-32.

3. Ostlund, H. G.; Alexander, J. A. J. Geophys. Res. 1963, 68(13), 3995-4000.

4. Liss, P. S. In Air Sea Exchange of Gases and Particles; Liss, P. S.; Slinn, W. G. N., Eds.; Reidel: Dordrecht, 1983; p 285.

5. Johnson, J. E.; Harrison, H. J. Geophys. Res. 1986, 91(D7), 7883-8.

6. Andreae, M. O. In The Role of Air Sea Exchange in Geochemical Cycling; Buat-Menard, P.; Liss, P. S.; Merlivat, L., Eds.; Reidel: Dordrecht, 1986; pp 331-56.

7. Johnson, J. E. Geophys. Res. Letts. 1981, 8(8), 938-40.

8. Elliott, S. Ph.D. Thesis, University of California, Irvine, 1984.

9. Elliott, S.; Lu, E.; Rowland, F. S. EOS Trans. AGU 1985, 66(51), 1309.

10. Elliott, S.; Lu, E.; Rowland, F. S. EOS Trans. AGU 1986, 67(16), 295.

11. Elliott, S.; Lu, E.; Rowland, F. S. Geophys. Res. Letts. 1987, 14(2), 131-4.

12. Millero, F. J.; Hubinger, S.; Fernandez, M.; Garnett, S. Environ. Sci. Technol. 1987, 21(5), 439-43.

13. Millero, F. J. Mar. Chem 1986, 18, 121-47.

14. Cutter, G. A.; Oatts, T. J. Anal. Chem. 1987, 59, 717-21.

15. Krahforst, C. F.; Cutter, G. A. EOS Trans. AGU 1987, 68(16), 339.

16. Cutter, G. A.; Krahforst, C. F. Geophys Res. Letts. 1988, submitted.

17. Dyrssen, D. Mar. Chem. 1988, 24, 143-53.

18. Dyrssen, D.; Wedborg, M. Anal. Chem. 1988, submitted.

19. Elliott, S. Mar. Chem. 1988, in press.

20. Moffett, J. W.; Zika, R. In Photochemistry of Environmental Aquatic Systems; Zika, R. G.; Cooper, W. J., Eds.; American Chemical Society: Washington, D.C., 1987; pp 116-30.

21. Moffett, J. W.; Zika, R. Mar. Chem. 1987, 301-13.

22. Buckley, P. J. M; Van Den Berg, C. M. G. Mar. Chem. 1986, 19, 281-96.

23. Graedel, T. E. Geophys. Res. Letts. 1979, 6(4), 329-31.

24. Toon, O. B.; Kasting, J. F.; Turco, R. P.; Liu, M. S. J. Geophys. Res. 1987, 92(D1), 943-63.

25. Cooper, D. J.; Saltzman, E. S. Geophys. Res. Letts. 1987, 14(3), 206-9.

26. Latimer, W. M. Oxidation Potentials; Prentice-Hall: Englewood Cliffs, New Jersey, 1952; 392 pp.

27. Erikson, E. J. Geophys. Res. 1963, 68(13), 4001-8.

28. Thompson, H. W.; Kearton, L. F.; Lamb, S. A. J. Chem. Soc. 1935, 1033-7.

29. Phillip, B.; Dautzenberg, H. Z. Phys. Chem. 1965, 229, 210-24.

30. Elliott, S; Lu, E.; Rowland, F. S. Env. Sci. Technol. 1988, submitted.

31. Martell, A. E.; Smith, R. W. Critical Stability Constants, Vol. 4; Plenum Press: New York, 1976; 257 pp.

32. Bruland, K. W. In Chemical Oceanography; Riley, J. P.; Chester, R., Eds.; Academic Press: London, 1983; Vol. 8, pp 157-215.
33. Graedel, T. E.; Kammlott, G. W.; Franey, J. P. Science 1981, 212, 663-4.
34. Alldredge, A. L.; Cohen, Y. Science 1987, 235, 689-91.
35. Dacey, J. W. H. In Biogenic Sulfur in the Environment; Saltzman, E.; Cooper, W. J. Eds.; American Chemical Society: Washington, D.C., 1988; this volume.

RECEIVED September 15, 1988

THE ATMOSPHERE: DISTRIBUTION

Chapter 20

Dimethyl Sulfide and Hydrogen Sulfide in Marine Air

Eric S. Saltzman and David J. Cooper

Division of Marine and Atmospheric Chemistry,
Rosenstiel School of Marine and Atmospheric Science,
University of Miami, Miami, FL 33149-1098

In this Chapter we discuss the distribution of DMS and H_2S in marine air. The discussion focuses on: 1) analytical techniques used to obtain the existing data base, 2) the measurements of DMS and H_2S over the oceans, and 3) modelling efforts to test current concepts of tropospheric cycling of these compounds. Results from simple box model of the marine boundary layer are presented for comparison of estimated rates of sea/air exchange and photochemical oxidation with atmospheric concentrations of DMS and H_2S in the marine boundary layer.

Our understanding of the sulfur cycling in marine air has recently undergone a number of important developments relating to the speciation of reduced sulfur in the surface oceans, the biological factors controlling the distribution of these compounds, and the chemical and physical transformations of these sulfur compounds in the atmosphere over the oceans. In this Chapter we focus on atmospheric measurements of reduced sulfur gases in marine air and on our current understanding of the atmospheric chemistry of these compounds. The discussion includes an brief overview of the existing data base of measurements of DMS and H_2S in marine air and of the analytical techniques used to obtain those data. Specifically, we discuss the data with regard to two points: 1) the relative importance of DMS and H_2S as fluxes of reactive sulfur to the marine atmosphere, and 2) the interpretation of observed diurnal variations of these compounds in terms of atmospheric oxidation pathways.

Global sulfur budgets constructed during the 1960's and early 1970's recognized that a natural, and presumably biogenic, source of volatile sulfur to the atmosphere was needed to provide a precursor for non-sea-salt sulfate in the aerosol over the remote oceans. The magnitude of the flux needed to balance the sulfur budget was substantial, 35-270 x 10^{12} g S per year ([1-3]), which is of the same order of magnitude as the estimated anthropogenic emissions of SO_2. The oceanic source was presumed to be H_2S which was the primary natural emission known at the time. This proposition has been problematic in view of the fact that bacterial sulfate reduction to sulfide occurs only in anoxic environments. The oxidation of H_2S in the oxygenated surface ocean is relatively rapid, with a 50 hour half life ([4]), keeping sulfide levels in the oceans

0097–6156/89/0393–0330$06.50/0

low. Support for a substantial oceanic sulfide flux was obtained primarily from shipboard and terrestrial measurements of H_2S in marine air.

Lovelock et al. (5) made the first quantitative measurement of DMS in the surface ocean and suggested that it, rather than H_2S, made up the principle oceanic sulfur source. Since that time, a number of measurements have been made of both water column and atmospheric DMS and flux calculations support the view that the emissions of this organic sulfur compound constitute a major global flux. However, the relative importance of the fluxes of DMS and H_2S as precursors of the background sulfate aerosol have not been well established, at least in part because the two compounds have not, until recently, been measured simultaneously in the same air or water mass. In this paper we compare our own simultaneous measurements of DMS and H_2S in marine air (6) with existing data and discuss analytical considerations which alter the interpretation of earlier data. These results clearly support the Lovelock et al. hypothesis.

Current research on the atmospheric cycling of sulfur compounds involves the experimental determination of reaction rates and pathways (see Plane review, this volume) and the field measurement of ambient concentrations of oceanic emissions and their oxidation products. Photochemical models of tropospheric chemistry can predict the lifetime of DMS and H_2S in marine air; however there is considerable uncertainty in both the concentrations and perhaps in the identity of the oxidants involved. The ability of such models to simulate observed variations in ambient concentrations of sulfur gases is thus a valuable test of our assumptions regarding the rates and mechanisms of sulfur cycling through the marine atmosphere.

Analytical Methods

Dimethylsulfide. In this section we discuss the analytical techniques which have been used to determine the concentration of DMS in marine air. Unlike the determination of DMS in seawater, the analysis of DMS in air is complicated by the presence of atmospheric oxidants which can cause variable and often severe sampling losses of DMS. For this reason, a number of different techniques have been used to determine atmospheric DMS and the accuracy of data reported in the literature is often difficult to assess.

The first reported attempt to determine atmospheric DMS was that of Lovelock et al. (5) in the Atlantic ocean. They attempted to cryotrap DMS directly from air using liquid nitrogen, and to analyze the gas using a gas chromatograph with electron capture detection. Although DMS was present in the underlying water, it was not observed in air samples. It was concluded that the DMS was destroyed during sampling by atmospheric oxidants.

All subsequent studies have utilized the flame photometric detector (FPD), operating in the sulfur specific mode. Briefly, this detector operates by detecting the light emitted from excited state S_2 (394 nm) formed in a reducing flame. The FPD has a highly selective response to sulfur compounds, but can be susceptible to both negative and positive interferences from non-sulfur compounds under certain conditions. A detailed discussion of the principles of operation and response of this detector has recently been published by Farwell et al. (7). No significant interferences relating to the FPD have been noted with regard to the analysis of DMS in air, providing that adequate chromatographic separation from CS_2 is achieved. The detector has been used with and without the addition of a sulfur gas to the flame hydrogen supply to provide a constant background of sulfur in the detector. This practice, often called "doping" the flame, results in improved sensitivity and linearity and reveals the presence of non-sulfur interferents (e.g. CO_2, hydrocarbons) in samples as negative peaks.

Maroulis and Bandy (8) determined DMS in air at Wallops Island by cryotrapping air drawn through a teflon tube packed with glass beads immersed in liquid argon. The DMS was thermally desorbed onto a GC column with polyphenyl ether/H_3PO_4 liquid phase coated onto a teflon substrate and was detected via doped FPD. No mention was made in that paper of sampling losses, however it has been suggested that the presence of the glass beads may have served to remove co-trapped oxidants (D. Thornton, personal communication).

Braman et al. (9) and Ammons (10) developed a preconcentration method for DMS in air using adsorption onto gold coated sand or gold wire. The DMS was recovered by thermal desorption in an inert gas stream into a liquid nitrogen cooled cryotrap. Analysis was done by GC/FPD. Both trapping and desorption were quantitative using standards in inert gases, however losses were observed while sampling from ambient air. Similar losses were observed in synthetic air mixtures when ozone or NO_2 were present. While using gold foil for preconcentration of atmospheric H_2S, Ammons (10) evaluated a number of scrubbers for oxidant removal including Teflon and Tygon shavings, and various substrates (glass fiber filters, Chromasorb, Anakrom, and glass beads) coated with Na_2CO_3 or MnO_2. The Na_2CO_3 based scrubbers appeared to give the best results and were used in the field study of Braman et al. (9). The gold method was also subject to positive interference from sulfur dioxide, however this was removed from the sample stream by the carbonate scrubber.

The gold wire method with carbonate based scrubbers has subsequently been used extensively in the field by numerous workers (11-17). Various materials have been used as supports for the carbonate scrubbers, probably with varying degrees of effectiveness. Early work (Barnard et al. (11); Andreae et al. (12); Andreae and Raemdonck (14)) used packed tubes containing Na_2CO_3/chromasorb, which were referred to in the literature as sulfur dioxide scrubbers. In 1985, Andreae et al. (13) reported the failure of the chromasorb scrubbers in high oxidant air masses, and attributed this to a limited capacity of the tubes for oxidant removal. At that time they began using a carbonate scrubber with Anakrom C22 as a support, which appeared to have a greater capacity than other scrubbers.

Kuster et al. (18) conducted a laboratory comparison of the efficiency of various oxidant scrubbers at removing ozone from an air stream, and concluded that KOH impregnated filters were more effective than carbonate based scrubbers. An intercomparison between the carbonate based anakrom scrubber and the KOH scrubber in ambient air revealed rapid loss of efficiency in the latter (P. Goldan and R. Ferek, personal communication), once more raising questions concerning possible sampling losses in the existing reports of DMS measurements. To our knowledge, no atmospheric data from marine air have been published using the KOH scrubbers.

Nguyen et al. (19) collected DMS by drawing air through a teflon tube packed with tenax and held at -100°C. The samples were thermally injected onto a Chromosil 310 column. No discussion of possible sampling losses is given.

Luria et al. (20) collected air samples in flasks aboard aircraft for subsequent analysis. No details of the collection or analysis procedures have been published.

In our own studies we have collected DMS by drawing air through a teflon tube packed with teflon wool and immersed in liquid oxygen. In that work, water was allowed to collect in the trap along with the sulfur gases; this resulted in blockage of the traps and ultimately limited the sampling flow rate and volume. The samples were thermally desorbed onto a Chromosil 330 column and detected using an undoped FPD. Two carbonate coated glass fiber filters in series were used as oxidant scrubbers on cruises to the Bahamas and Gulf of Mexico, and a tube packed with carbonate coated chromasorb was used on a cruise across the Caribbean. During the Gulf of Mexico cruise, episodes of extremely low DMS concentrations were observed over waters with normal (1-2 nM) DMS concentrations. Although the scrubbers appeared to work effectively in the field during those studies, as evidenced by quantitative recovery of periodic DMS standard additions, the question of the effectiveness of the carbonate scrubbers in polluted air was raised. Andreae et al. (12) had observed a similar phenomenon previously in the Gulf of Mexico, which was later attributed to the failure of scrubbers (13).

Although much work has been done on the analysis of DMS in air, the nature of the oxidant interference remains largely unknown. The gas phase reaction of DMS with atmospheric oxidants such as ozone or NO_2 is extremely slow. In the trapping procedure it is possible that through cryogenic condensation or surface adsorption the concentrations of both oxidant and sulfur compound are increased such that the reaction proceeds more rapidly than in the gas phase. Alternatively, the disappearance of DMS may be caused by the production of reactive free radical species as a result of the decomposition of the oxidant. Ozone (and perhaps NOx) may thus be precursor rather than the active oxidant. Alkyl sulfides are known to react rapidly with radicals such as OH and NO_3 in both the gas phase (see Plane review, this volume) and the aqueous phase (21), and such radicals are likely to be generated in condensed air. This is of course speculative, and it is certainly possible that the oxidation process on gold tubes is entirely different from that in cryogenic traps.

In an effort to resolve some of the questions relating to the effectiveness of the oxidant scrubbers in various air masses, we have recently developed a new oxidant scrubber based on the aqueous chemistry of iodide. At neutral pH, excess aqueous iodide reacts with a variety of oxidants according to the net reaction:

$$3I^-_{aq} + 2O_3 \text{---}> I_3^- + 3O_2$$

This reaction has been utilized for more than a century for the determination of gaseous oxidants such as ozone, with detection based on thiosulfate titration (22) or the visible absorption of I_3^- in solution (23). The kinetics of the iodide reaction are also rapid for NO_2. At neutral pH the reaction of I^- with oxygen is slow, and the uncatalyzed reduction of hydrogen peroxide occurs on the order of several minutes. For the measurement of DMS, we used a 200 ml glass bubbler containing 60 ml of KI solution at the inlet of the DMS sampling train (24). The bubbler was immersed in an ice water bath (0°C) to lower the water vapor concentration of the air being sampled. The bubbler was refilled with fresh solution once per day during use. The oxidation state of the solution is easily monitored visually from the gradual yellowing due to the accumulation of the I_3^- ion, although no deterioration in performance has ever been found. In this system, residual water vapor in the sample stream was removed prior to cryotrapping by passage through a -30°C teflon trap.

Without understanding the actual mechanism of the artifact testing of scrubbers must be an empirical process. We have tested a variety of scrubbers in Miami air by using standard additions to ambient air samples. Two air samples were collected simultaneously, and a liquid standard containing DMS was added to the inlet of one channel. The standard additions were done at levels equal to or less than the amount of DMS collected in the ambient samples. Complete recovery of standards was achieved with all scrubbers when sampling marine air masses, i.e. in the case of onshore (easterly) winds. Losses of DMS from standards became evident as the degree of pollution increased, as evidenced by offshore winds and elevated ozone concentrations. In agreement with the earlier field data, the Chromasorb/Na_2CO_3 scrubbers were found to fail under these conditions (Figure 1). The Anakrom/Na_2CO_3 and aqueous KI solutions gave successful recovery of DMS standards under all conditions sampled. We also performed similar successful tests on the aqueous KI scrubbers in more severely polluted conditions on a recent cruise from Miami to the Gulf of Maine (24). We did not, unfortunately, test the Anakrom/Na_2CO_3 scrubbers at that time.

Hydrogen Sulfide. To date the successful analysis of hydrogen sulfide in marine air has been achieved using only a single method. This is the silver nitrate impregnated filter method introduced in its current form by Natusch et al. (25), and used subsequently in marine air by Slatt et al. (26), Delmas and Servant (27), Hermann and Jaeschke (28) and Saltzman and Cooper (6). In this method, H_2S is collected by drawing air through a cellulose filter impregnated with acidic (.01N HNO_3) silver nitrate. H_2S is trapped on the filter as Ag_2S which is highly insoluble and resistant to oxidation. Sampling rates are typically less than 10 SLPM and collection efficiencies are better than 99% (29). In our own shipboard measurements (6) we have utilized 90 minute samples at 8 SLPM to achieve a detection limit of 3 pptv.

After sample collection the filter is extracted with basic NaOH/NaCN solution (0.1 M), which dissolves the Ag_2S. An aliquot of the fluorescent complex fluorescein mercuric acetate (FMA) is added to the solution and the fluorescence (500 nm excitation/520 nm emission) of the solution is determined. The presence of sulfide in the solution results in quenching of the fluorescence of the FMA solution, presumably as a result of the interaction of sulfide with the mercuric ion. The sulfide concentration is determined by measuring the decrease in fluorescence relative to that of a blank filter extract. Calibration is done by adding successive aliquots of a fresh aqueous Na_2S standard (prepared in 0.1 M NaOH) to a blank filter extract.

The sensitivity of the method is limited by the reproducibility of the filter blanks. These blanks originate both from sulfide absorbed during filter impregnation and drying, and from unknown matrix contamination which is not removed by washing the filters. In our own work, we have found that under ideal conditions variability in the filter blank yields a practical detection limit of 1×10^{-10} moles $S^=$ per filter.

The method has been tested for artifacts from a wide variety of both sulfur containing and non-sulfur containing compounds (30); however, until recently, no interferences were identified which would affect measurements in marine air. Cooper and Saltzman (30) observed the apparent breakthrough of H_2S through a silver nitrate filter onto the backup filter during sampling large volumes of remote marine air. Laboratory investigation revealed that a similar artifact could be induced by addition of OCS to the sampled air stream. It was determined that a small but significant fraction (1-2%) of the OCS in air was adsorbed onto the filter. The magnitude of the artifact becomes significant in the sampling of remote marine air where H_2S levels are low, while OCS levels remain constant. The artifact was found to increase with increasing temperature. This resulted in the production of a spurious midday maximum in field data. Because the artifact remained constant over several filters placed in series, it is reasonable to assume that the artifact production on the front filter was equal to that on subsequent filters. The sulfide on the front filter can therefore be corrected for the artifact by subtraction of the amount collected on a backup filter. This effect is illustrated in Figure 2.

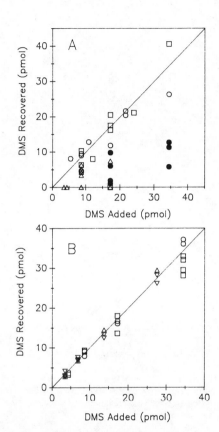

Figure 1. Recovery of DMS standards added to atmospheric samples using various oxidant scrubbers. (A) 0.4g Na$_2$CO$_3$/Chromasorb in oceanic air with low ozone concentrations (○); 0.4g Na$_2$CO$_3$/Chromasorb in oceanic air with intermediate ozone levels (●); 0.4g (△) and 1.5g (□) Na$_2$CO$_3$/Chromasorb in continental air with ozone concentrations of 55-65 ppb. (B) 0.4g Anakrom/ Na$_2$CO$_3$ in oceanic (▽) and continental (△) air; neutral aqueous KI in oceanic (□) and continental (○) air.

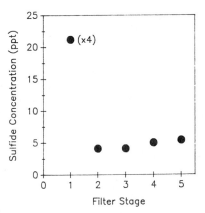

Figure 2. Sulfide concentrations measured on silver nitrate impregnated filters in series. Data shown from sample collected in Miami on 10/8/86 for 170 min at 7.8 l/min. The front filter extract was diluted by a factor of four.

Atmospheric Concentrations of DMS and H₂S Over the Oceans

Dimethylsulfide. The literature data base concerning the atmospheric concentration of DMS in the marine atmosphere can be divided into two subsets, 1) shipboard studies and 2) aircraft studies. The former yield information concerning the variability within the marine boundary layer, while the latter yield information concerning the vertical structure and mixing processes affecting the DMS concentrations. The boundary layer data are summarized in Table I.

The data from remote marine air are of particular interest because both the sulfur sources and the atmospheric oxidant fields should be fairly homogeneous. Andreae and Raemdonck ($\underline{14}$) presented the first data showing the diurnal variations of DMS over the open ocean, from the equatorial Pacific ocean. They obtained a mean concentration of 128 ppt with a standard deviation of 53 ppt. By averaging the data into 4 hour intervals they showed a daytime minimum and nighttime maximum with a diurnal range of 90 to 149 ppt. Dividing the maximum by the minimum gives a factor of 1.65 diurnal variation for this data set.

The range of this data set is slightly higher than that of Nguyen et al. ($\underline{19}$) which was measured at mid latitudes in the South Pacific. Their data had a mean of 79 ppt with a range of 43 to 153 ppt. Insufficient data was available to discern any diurnal cycle. They attempted to measure the gradient of DMS in the atmospheric surface layer just above the sea surface. The observed gradients were substantially greater than predicted using estimated sea surface DMS emissions from a gas exchange model and vertical mixing rates calculated from the horizontal wind speed. This was interpreted as indicating that DMS has a short and variable chemical lifetime in the marine boundary layer. This data is used by Becker et al. ($\underline{31}$) to argue for the importance of IO as an atmospheric sink for DMS. The conclusion that DMS is oxidized on the order of minutes in marine air is in conflict with the observation of consistent diurnal cycles as discussed later in this paper. Further work is clearly needed to reconcile these differences.

Saltzman and Cooper ($\underline{6}$) measured the DMS concentration in the trade winds on two north-south transect across the Caribbean Sea. The mean of these data is 57 ppt. Averaging of the data into 4 hour intervals yielded a diurnal amplitude of 39 to 58 ppt, a factor of 1.49 variation in the maximum/minimum concentration. The data show an early morning maximum and a broad afternoon minimum, similar to that found in the Pacific by Andreae and Raemdonck ($\underline{14}$). Interestingly, the diurnal variation found between the minimum and maximum concentration is similar for the two data sets.

Similar diurnal features were observed by Berresheim ($\underline{17}$) in pristine air over the subantarctic and Antarctic oceans during fall. The concentrations found in that study had a mean of 106 ppt, with highest levels occurring at night during times of low wind speed. The data yield a diurnal variation of a factor of approximately 1.4, although it was suggested that these may at least in part reflect diurnal variations in boundary layer height. This effect is due to nighttime ground level cooling of the nearby land surface and consequent increased vertical stability which traps surface emissions near the ground. Vertical mixing increases during the daytime due to ground level heating, resulting in dilution of the surface later concentrations. This cycle complicates the interpretation of diurnal cycles in coastal regions.

Attempts to measure DMS in air masses influenced by polluted continental regions have yielded quite different results, with a higher degree of variability and less well defined diurnal cycles. In the North Atlantic, Andreae et al. ($\underline{13}$) used trajectory analysis to separate their data set into air masses of continental

Table I. Summary of DMS Measurements in the Marine Boundary Layer
and Related Surface Waters

Location	Mean (ppt)	sd	Range	Diurnal Range	DMSaq (nM)	Ref.
Temperate Pacific	79	25	43-153	n/a	4.4	27
Eq. Pacific	128	53	52-317	90-150	3.8	14
Coastal Tasmania	128	n/a	26-368	115-180	n/a	13
Bahamas	93	n/a	4-513	n/a	2.6	13
N. Atlantic (Marine)	138	n/a	n/a	n/a	4.4	13
N. Atlantic (Cont.)	45	n/a	n/a	n/a	4.4	13
Sargasso Sea	178	n/a	38-283	n/a	n/a	13
G. of Mexico (Marine)	27	30	n/a	n/a	n/a	30
G. of Mexico (Cont.)	7	3	n/a	n/a	n/a	30
Tropical Atlantic	73	19	40-102	n/a	n/a	15
Atlantic Pollut. Plume	n/a	n/a	0.5-18	n/a	n/a	16
Bermuda	n/a	n/a	35-195	n/a	n/a	16
Southern Ocean	106	55	17-236	90-140	1.8	17
NE Pacific	n/a	n/a	6-30	n/a	0.8	31
Bahamas	154	93	25-300	n/a	n/a	6
Gulf of Mexico	25	17	5-70	n/a	1.9	6
Remote Caribbean	57	29	16-104	39-58	1.4	6
W. Atlantic	30	26	1-110	14-51	1.5	21
Gulf of Maine	101	67	11-335	65-185	3.6	21

and marine origin. The continentally influenced air masses had consistently lower DMS concentrations and lacked significant diurnal variations. Similarly, low DMS levels have been observed in the Gulf of Mexico in air originating over the continental US (6) after passage of a cold front. Sea surface DMS levels in both of these studies were within the range of typical marine concentrations, so the phenomenon must be atmospheric rather than marine in origin. Andreae et al. (13) argued that the lower DMS levels were due to the increased concentration of atmospheric oxidants in polluted air masses. They also proposed that the lack of diurnal variation was due to the nighttime oxidation of DMS by the nitrate radical. These observations also raise the question of the effectiveness of various oxidant scrubbers in polluted air which was discussed earlier in this paper.

In order to further investigate this question, we conducted a cruise in the Western Atlantic ocean from Miami, FL to the Gulf of Maine using the KI oxidant scrubber (24). The results from this cruise showed significant differences from the previous data sets. On the north-south transects along the east coast of the U.S., strong diurnal cycles were observed. The mean DMS concentration was 30 ppt and a day/night variation of 14-51 ppt, a factor of 3.6, was observed. Routine collection of simultaneous standard additions demonstrated that the sampling procedures used in that study were quantitative for DMS. The averaged diurnal data from this cruise are shown in Figure 3. Examination of meteorological data suggests that local boundary layer variations were not important during this cruise (24). This would suggest that the large diurnal variation in DMS reflects the concentration of photochemical oxidants during the daytime. This is the first data set from polluted air in which strong diurnal cycles have been observed. Further experiments with the KI scrubber in polluted air are underway.

The vertical profile of DMS in marine air was first determined by Ferek et al. (15), over the tropical Atlantic ocean. They found that under stable meteorological conditions, the mixing depth of DMS was about 1 km, with a rapid decline in concentration above this altitude. This distribution was considered consistent with the chemical lifetime of a few days predicted by photochemical models. Under convective conditions DMS concentrations at higher altitudes (4-6 km) were considerably greater than under stable conditions. This data set was interpreted as confirmation of earlier models of vertical transport of chemicals by convective clouds (32).

Similar vertical profiles have been reported in subsequent aircraft studies. Luria et al. (20) measured DMS concentrations over the Gulf of Mexico averaging 27 ppt in the marine boundary layer, 7 ppt in boundary layer of continental origin, and <3 ppt in the free troposphere. Van Valin et al. (16) reported highly variable DMS levels over the north Atlantic Ocean in the pollutant plume from the Northeastern U.S., measured in the vicinity of Boston. Those workers also reported a vertical profile from Bermuda which was virtually identical to the stable case of Ferek et al. (15) off Barbados.

Andreae et al. (33) found a similar profile over the Northeast Pacific ocean, but with lower boundary layer concentrations of less than 30 ppt. These low values were apparently related to low sea surface DMS concentrations in the open ocean (0.8 nM) during that study.

Taking the data set as a whole, we observe that fairly consistent results have been obtained from remote or clean marine air. A higher degree of variability is associated with measurements in continentally influenced air masses. This variability may reflect the distribution of free radical oxidant precursors, such as NOx and O_3. However, it may also reflect analytical difficulties associated with the sampling of air masses containing high levels of oxidants as discussed in the

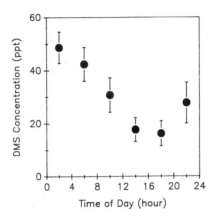

Figure 3. Atmospheric DMS data from two north-south transects in the western Atlantic Ocean. Data are averaged into four-hour time intervals. Compiled from Cooper and Saltzman (24).

previous section, particularly in the case of earlier data where low capacity oxidant scrubbers may have been used.

Hydrogen Sulfide. There have been several studies of the concentration of H_2S in marine air using variations of the Natusch technique described earlier (26-29,6). The results of the earlier studies, summarized in Table II, indicated that H_2S concentrations over the oceans were substantial, suggesting that the flux of H_2S from the surface ocean is significant relative to other sulfur sources. Anomalous diurnal behavior of hydrogen sulfide was noted by Delmas and Servant (27) in which a midday maxima and nighttime minima were observed. As discussed earlier, both the elevated H_2S levels and the diurnal variations are likely to be the result of artifact sulfide production from atmospheric OCS collected on the filters.

The magnitude of the artifact is such that coastal and wetland data in the literature is not likely to be seriously affected. An example of this is our Bahamas data (6) which illustrates the dominance of local island sources; elevated concentrations were preceded by wind shifts bringing air from the island of Andros which has extensive tidal flats. Similarly, the H_2S concentrations measured on the Gulf of Mexico cruise in the same paper most likely reflect the advection of continental and tidally derived H_2S. Brief periods of low H_2S concentrations (< 10 ppt) were obtained in the Gulf Stream near the end of both of these cruises during brief exposure to easterly open ocean air.

The Caribbean transects data shown by Saltzman and Cooper (6) was obtained in oceanic air with easterly winds which should be representative of the subtropical/tropical Atlantic Ocean. On this cruise two filters were used in series, and correction was made for the OCS artifact (30). The resultant data showed H_2S concentrations averaging less than 10 ppt with a slight daytime minimum, as expected for a species such as H_2S which should be removed predominantly by photochemically produced OH. The maximum/minimum concentration ratio is 1.7, which is similar to that observed for DMS on the same cruise. However, it should be noted that this diurnal variation is only a few ppt, which is close to the precision of the method. We wish to stress that considerably more data is needed to verify this observation, as the data used are compiled from measurements over a considerable geographic range and because the variations involved are small.

These measurements comprise the only set of simultaneous measurements of DMS and H_2S in remote marine air, and caution should be used in extrapolation of these results to the world oceans. However, the data certainly support the hypothesis of Lovelock (5) who proposed that DMS rather than H_2S as the major sea-to-air flux of sulfur. We have estimated from this data that hydrogen sulfide contributes on the order of 10% of the biogenic excess sulfate in marine air.

Recently, the first direct measurements of sulfide in seawater have been made (34,35). Interestingly, the concentrations of sulfide in oxic seawater appear to be of the order of 1 nM, which is similar to those of DMS. However, the measurements detect total sulfide, rather than free sulfide ion concentration or activity. The extent of sequestration of sulfide in strong, presumably metal, complexes is unknown. Luther et al. (35) present some indirect evidence suggesting that the sulfide may be largely complexed; this effect could certainly reduce the sea-to-air flux to levels consistent with our atmospheric observations and with modelled results presented later in this chapter. More information is needed about the chemical state of sulfide in seawater before a detailed comparison can be made.

Table II. Summary of Atmospheric H_2S Measurements in the
Marine Boundary Layer

Location	Mean (ppt)	sd	Range	Diurnal Range	Ref.
S. Florida, marine	34-82	10-54	20-140	n/a	23
S. Florida, cont.	336	258	121-829	n/a	23
Barbados	7	8	0-45	n/a	23
Sal Island	4 & 10	4 & 6	0-18	n/a	23
Gulf of Guinea	15	10	6-47	reversed	24
Coastal Ireland	n/a	n/a	11*-144	11*-144	25
N-S Atlantic transect	29	23	11*-111	n/a	25
Bahamas	82	56	10-260	n/a	6
Gulf of Mexico	45	30	5-90	n/a	6
Remote Caribbean	8.5	5.3	0-20	7-11	6

*Indicates lower limit of detection

Cycling of DMS and H$_2$S Through the Marine Boundary Layer

In this section, we address the question of whether or not the atmospheric concentrations of DMS and H$_2$S concur with current models of sulfur cycling in marine air. The discussion focuses primarily on DMS, as little is currently known about the concentrations of H$_2$S in seawater. Several simple box models of the marine boundary layer have been used generate hypothetical atmospheric diurnal cycles for comparison with field data. The first such model was that of Graedel (36), who examined the behavior of DMS and H$_2$S, assuming that oxidation via OH was the sole chemical loss mechanism for both compounds. In that study, the oceanic source term was adjusted to provide agreement between the average calculated atmospheric concentration and those reported by Maroulis and Bandy (8) for DMS, and Slatt et al. (26) for H$_2$S. The model shows an early morning maximum and mid-afternoon minimum for both sulfur compounds, with an amplitude of nearly 100% of the mean. While this model predicted much larger diurnal amplitudes than have been subsequently observed, it did provide both impetus and a conceptual framework for much of the field and modelling work that followed.

Andreae et al. (13) presented a series of model calculations using a similar box model to examine the diurnal behavior of DMS. In their model the flux of DMS into the boundary layer was fixed at 0.1 g S (DMS) m^{-2} yr^{-1}, the global average flux estimated by Andreae and Raemdonck (14). The loss of DMS was via chemical oxidation by reaction with OH and NO$_3$ radicals, and the concentrations of these were specified on the basis of the CO-O$_3$-NO$_x$ chemistry described by Chatfield and Crutzen (32). Three model cases were run. In a model run with OH oxidation only, ie. no nighttime oxidation, a very large diurnal amplitude was obtained. Relatively high OH concentrations were used in this model (10^7 molec/cc daytime maximum) which gave average DMS levels of about 120 ppt. A calculation with the same OH level and adding nighttime oxidation via NO$_3$ yielded slightly lower average DMS levels (about 75 ppt), and significantly reduced the diurnal amplitude to a factor of approximately 2 (maximum/minimum). The final case also allowed nighttime NO$_3$ reaction, this time with lower OH concentrations. This simulation gave higher average DMS levels of about 200 ppt with a greatly reduced diurnal amplitude of only a factor of 1.3. They also noted the nonlinearity which appears in the model when NO$_3$ chemistry is introduced. This effect arises because DMS is, itself, a major sink for NO$_3$. For high DMS levels the nighttime loss rate becomes simply the production rate of NO$_3$, resulting in a zero order loss of DMS for a given NO$_x$ concentration. These authors point out that there is considerable latitude in the values of the parameters, and that reasonable agreement can be obtained between the model simulations and the atmospheric DMS data.

A critical parameter in all models of DMS chemistry in the marine atmosphere is the sea-to-air flux. Although the sea surface concentrations of DMS have been measured in a wide variety of environments (14), the flux has never been measured directly. Instead, it has been calculated using observed concentrations and various models of gas transfer across the air sea interface. All of the models parameterize the transfer as a first order loss, as follows

$$\text{Flux} = k \cdot [\text{DMS}_{sw} - \alpha P_{DMS}]_g$$

where α is the solubility of DMS in seawater. Since seawater is greatly supersaturated with DMS relative to the atmosphere, i.e. $\alpha P_{DMSg} << [\text{DMS}]_{sw}$, the expression becomes simply

$$\text{Flux} = k \cdot [\text{DMS}]_{sw}$$

where the exchange constant (k) has the units of velocity (cm/sec or m/day) and is often referred to as a piston velocity (V_p). This simple expression belies the physical complexity of the interface and the gas transfer process. Numerous approaches have been taken to the estimation of piston velocities for natural waters (see Liss (37) for review). Here, we discuss only those used in the literature to calculate fluxes of sulfur gases.

Liss and Slater (38) made an early estimate of the flux using the thin film model. In this model the flux of a gas across the sea surface is assumed to be controlled by diffusive transport through surface layers on either side of the interface. In the case of DMS, Liss and Slater (38) demonstrated that the resistance to exchange is essentially confined to the liquid phase. In their calculations a piston velocity of 4.8 m/day was assumed; this was obtained from global average ^{14}C data which reflects the average air sea exchange of CO_2 on a world wide, long-term basis. In order to extrapolate to DMS and other gases, a small correction factor of $(M.W._{gas}/M.W.CO_2)^{1/2}$, or $(62/44)^{1/2} = 1.18$, was applied to the piston velocity to allow for the differences in their diffusivities. The theoretical basis for this correction was not given, and it has not been used subsequently in such calculations. At the time this estimate was made, only the sea surface measurements of Lovelock et al. (5) were available and these low concentrations 3.7 ng S/l resulted in a calculated global air/sea flux of 3.7 Tg S/yr.

Andreae and Raemdonck (14) applied the thin film parameterization described above in a slightly different way. The thin film model predicts that the exchange coefficient is linearly related to the diffusivity of the gas as follows,

$V_p = D/z$, where z is the film thickness.

These authors used the global average piston velocity determined by Broecker and Peng (39) by the radon deficit method, 2.8 m/day. The Othmer-Thakar relationship was used to calculate the diffusivity of DMS, which has never been determined experimentally. Since the calculated diffusivity for DMS (1.2×10^{-5} cm^2/s) is similar to that calculated for radon, the radon deficit piston velocity was assumed to apply to DMS without correction. The DMS concentrations used in this study were based on more than 600 surface ocean samples from a variety of environments. The global area weighted concentration used for the calculation was 102.4 ng S/l, resulting in a flux of 39×10^{12} g S/yr.

Bates et al. (40) attempted to use regional estimates of piston velocities in conjunction with regional variations in DMS concentrations to make flux estimates. They adopted the "surface renewal" model of gas exchange, which predicts a square root dependence of the piston velocity on diffusivity. The premise of this model is that some turbulent exchange occurs within the thin film, and some experimental support for this model has been published (41-43). Piston velocities as a function of wind speed were obtained from the radon deficit measurements of Smethie et al. (44) and the diffusivity of DMS as a function of temperature and water viscosity were estimated from the empirical equation of Wilke and Chang (45) (also Hayduk and Laudie (46)). The resultant calculated piston velocities ranged from 1.5 m/day in the equatorial regions during summer to a maximum of 3.35 m/day in temperate regions during winter and yield a global area weighted mean of 2.3 and 2.7 m/day for summer and winter, respectively. Sea surface DMS concentrations were obtained by combining the authors' own extensive data set from the Pacific ocean with that of Andreae and Raemdonck (14). The global oceanic flux calculated was 16×10^{12} g S/yr. The marked decrease compared to Andreae and Raemdonck's estimate primarily reflects the use of new data providing lower estimates of DMS concentrations in non-tropical upwelling and coastal areas. This estimate is in good agreement with an estimate of 24 x 10^{12} g S/yr for the total (wet + dry) depositional flux of biogenic sulfate to the oceans (47). A more sensitive test of the assumptions inherent in our concept of sea/air cycling of DMS is to attempt to model data from an actual field experiment. Here we present a series of calculations to based on the data from the Pacific and Caribbean Sea data discussed above. The model used is essentially the same box model of the boundary layer used by Andreae et al. (13) except that it is somewhat simpler in that it includes specified, rather than calculated, oxidant concentrations. We assume that these two cruises sampled clean background marine air, hence reaction with OH is assumed to be the only loss pathway for DMS. NO_3 is likely to be quantitatively unimportant in clean remote marine air masses because of the low NOx concentrations (48,13). The height of the boundary layer is fixed at 1 km and no venting of air into the free troposphere is allowed. The DMS flux into the box is calculated using the average wind speed (corrected to 20 m above the sea surface) and sea surface DMS levels measured on board the ships. The piston velocity for each cruise was obtained in two ways: (1) from the lake SF_6 measurements of Wanninkhof et al. (49) and (2) from the Rn data of Smethie et al. (44). The piston velocities were corrected for the diffusivity of the various gases (SF_6, Rn, DMS, H_2S) assuming a $D^{1/2}$ relationship and using their calculated diffusivities (46).

We first describe a series of model runs using our data from the Caribbean transect. The sea surface concentrations of DMS during the transect averaged 1.4 nM and the average wind speed was 7.6 m/s. In each run, we adjusted the OH concentration to bring the mean atmospheric DMS level into agreement with the field data. The results, Figure 4a, show that the amplitude of the diurnal variations are overestimated using either the SF_6 or the Rn based piston velocities. The large diurnal amplitude in the model curves result from, 1) the fact that only daytime oxidation is specified, and 2) that relatively high OH levels are required to bring the mean DMS concentrations down to the observed level. Better agreement between the model and data can be obtained by assuming a lower piston velocity. Figure 4b shows that better fits to the data in terms of both the diurnal amplitude and mean levels can be obtained by allowing both the OH concentration and piston velocity to vary. The ranges of both the piston velocity and the OH concentration shown are probably within the uncertainty of the independent estimates of these parameters.

We can attempt to apply the same type of model to the H_2S data, however there are two additional unknown factors involved. First, we do not have a measurement of the sea surface concentrations of H_2S. Second, the piston velocity of H_2S is enhanced by a chemical enrichment factor which, in laboratory studies, increases the transfer rate over that expected for the unionized species alone. Balls and Liss (50) demonstrated that at seawater pH the HS^- present in solution contributes significantly to the total transport of H_2S across the interface. Since the degree of enrichment is not known under field conditions, we have assumed (as an upper limit) that the transfer occurs as if all of the labile sulfide (including HS^- and weakly complexed sulfide) was present as H_2S. In this case, the piston velocity of H_2S would be the same as that of Radon for a given wind velocity, with a small correction (a factor of 1.14) for the estimated diffusivity difference. If we then specify the piston velocity and OH concentration we could calculate the concentration of H_2S in the surface waters. Using the input conditions from model run B from Figure 4a (OH = 5 x 10^6 molecules/cm^3, Vp = 3.1 m/day) yields a sea surface sulfide concentration of approximately 0.1 nM. Figure 5 illustrates the diurnal profile of atmospheric H_2S which results from these calculations.

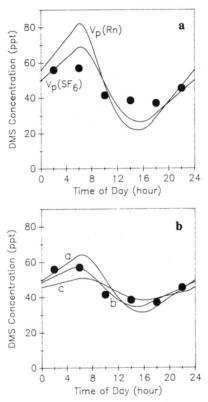

Figure 4. Box model results compared to Caribbean transect data ((6), solid symbols). Units are m/day for V_p and 10^6 molecule/cm^3 for midday maximum OH. (a) Runs using piston velocities obtained from the radon deficit (V_pRn = 3.1) and SF$_6$ lake study (V_pSF_6 = 2.2) wind speed relationships. Midday maximum OH concentrations (shown on plot) were adjusted to give mean DMS levels in agreement with the shipboard data. (b) Model runs with lower piston velocities and lower OH showing less diurnal variation. Conditions used were (a) V_p = 1.7, OH = 8.0; (b) V_p = 1.1, OH = 5.0; (c) V_p = 0.6, OH = 3.0.

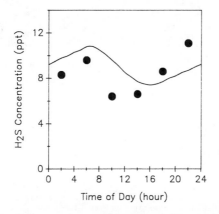

Figure 5. Box model results for H_2S compared to Caribbean transects data (solid symbols), using the conditions specified in Figure 4b run (b) after correction for H_2S diffusivity, as discussed in the text. V_p = 3.5 m/day and midday OH maximum = 5.0×10^6 molecule/cm³.

Next, we apply the same model to the equatorial Pacific DMS data of Andreae and Raemdonck (14). Here the average wind speed was slightly less than the last case, 6 m/s, giving considerably smaller piston velocities. The Rn piston velocities converge with the SF_6 estimates at low wind speeds, so in this case they yield similar results. The average sea surface DMS concentration was greater than the previous case, at 3.8 nM. Figure 6a shows that in this case better agreement is obtained between the model and data, although the diurnal variation is still slightly overestimated by the model. As with the Caribbean data, a somewhat better fit is obtained by lowering the sea-to-air flux and the OH levels. Figure 6b shows a range of parameters which provide a good fit to the data.

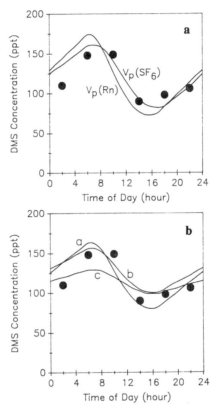

Figure 6. Box model results compared to the equatorial Pacific DMS data of Andreae and Raemdonck (14, solid symbols). Units are m/day for V_p and 10^6 molecule/cm^3 for midday maximum OH. (a) Runs using piston velocities obtained from the radon deficit ($V_pRn = 1.95$) and SF_6 lake ($V_pSF_6 = 1.5$) wind speed relationships. Midday maximum OH concentrations (shown on plot) were adjusted to give mean DMS levels in agreement with the shipboard data. (b) Model runs with lower piston velocities and lower OH showing less diurnal variation. Conditions used were (a) $V_p = 1.6$, OH = 8.0; (b) $V_p = 1.1$, OH = 5.0; (c) $V_p = 0.6$, OH = 3.0.

These calculations suggest that the variations in DMS observed in two remote oceanic regions can be explained reasonably well with current models of tropospheric cycling and air/sea exchange. Interestingly, the model results show that given the piston velocities estimated on the basis of the gas exchange field studies, fairly high OH levels are needed to obtain agreement with the field data. These levels are higher than predicted on the basis of photochemical models of the remote marine troposphere. Two limitations of the data used should be stressed, however: (1) the averaging of chemical and meteorological data collected over numerous days from a moving vessel may bias the data and obscure the natural geographic variability, and (2) little or no supporting chemical data is available for these data sets with which to constrain the oxidant fields. It is clear that observations of diurnal variations are needed from many other areas of the oceans and an effort should be made to characterize the oxidation state of these air masses with respect to OH, NO_3, and other potential oxidants.

Summary and Conclusions

1) Analytical considerations complicate the interpretation of the existing data base of atmospheric measurements of DMS and H_2S. In the case of DMS the observed reduction in polluted air masses may have, at least in part, resulted from to oxidative losses during sampling. For H_2S, only one analytical method has been used, and much of the existing data may be biased by an OCS artifact. New methods, preferably chromatographic or spectroscopic, are needed for H_2S.

2) The relative abundance of H_2S and DMS suggest that the former is a minor component of the total sea-to-air flux of reduced sulfur.

3) Measurements of DMS from remote oceanic air exhibit diurnal cycles, with a pronounced daytime minimum and nighttime maximum. These variations are generally consistent with the idea that photochemically generated OH is the major sink of DMS. Detailed comparison of the data with current models of air/sea exchange suggest that either OH levels are greater than predicted or, more likely, that the piston velocities have been overestimated in the past

Acknowledgments

We wish to thank John Plane, Kim Holmen, Dennis Savoie, and Rana Fine for helpful discussions on the subject of kinetic modelling and gas exchange. This work was supported by grant ATM 87-09802 from the National Science Foundation.

Literature Cited

1. Granat, L.; Rodhe, H.; Hallberg, R. O. In: Nitrogen, Phosphorus and Sulfur-Global Cycles, SCOPE Report 7, Ecol. Bull.: Stockholm, 1976; 22, 89.

2. Friend, J. P. In: Chemistry of the Lower Atmosphere; Rasool, S. I., Ed.; Plenum, New York, 1973; p 177.

3. Eriksson, E. J. Geophys. Res. 1963, 68, 4,001.

4. Millero, F. J.; Hubinger, S.; Fernandez, M.; Garnett, S. Env. Sci. Technol. 1986, 21, 439.

5. Lovelock, J. E.; Maggs, R. J., Rasmussen, R.A., Nature (London) 1972, 237, 452.

6. Saltzman, E. S.; Cooper, D. J. J. Atmos. Chem. 1988, 7, 191.

7. Farwell, S. O.; Barinaga, C. J. J. Chrom. Sci. 1986, 24, 483.

8. Maroulis, P. J.; Bandy, A. R. Science 1977, 196, 647.

9. Braman, R. S.; Ammons, J. M.; Bricker, J. L. Anal. Chem. 1978, 50, 992.

10. Ammons, J. M. PhD Dissertation, University of South Florida, 1980.

11. Barnard, W. R.; Andreae, M. O.; Watkins, W. E., Bingemer, H.,; Georgii, H.-W. J. Geophys. Res. 1982, 87, 8,787-94.

12. Andreae, M. O.; Barnard, W. R.; Ammons, J. M. Ecol. Bull. (Stockholm) 1983, 35, 167.

13. Andreae, M. O.; Ferek, R. J.; Bermond, F.; Byrd, K. P., Engstrom, R. T.; Hardin, S.; Houmere, P. D.; LeMarrec, F.; Raemdonck, H.; Chatfield, R. B. J. Geophys. Res. 1985, 90, 12891.

14. Andreae, M. O.; Raemdonck, H. Science 1983, 221, 744.

15. Ferek, R. J.; Chatfield, R. B.; Andreae, M. O. Nature 1986, 320, 514.

16. Van Valin, C. C.; Berresheim, H.; Andreae, M. O.; Luria, M. Geophys. Res. Lett. 1987, 7, 715.

17. Berresheim, H. J. Geophys. Res. 1987, 92, 13,245.

18. Kuster, W. C.; Goldan, P. D.; Albritton, D. L. EOS 1986, 67, 887.

19. Nguyen, B. C.; Bergeret, C.; Lambert, G. In Gas Transfer at Water Surfaces; Brutsaert and Jirka, Ed.; D. Reidel, 1984, pp 539-45.

20. Luria, M.; Van Valin, C. C.; Wellman, D. L., Pueschel, R. F. Environ. Sci. Technol. 1986, 20, 91.

21. Farhataziz; Ross, A. B., NSRDS-NBS 59, U.S. Dept. of Commerce: Washington, D.C., 1977.

22. Brodie, B. C. Phil. Trans. 1874, 162, 435.

23. Schecter, H. Water Res. 1973, 7, 729.

24. Cooper, D. J.; Saltzman, E. S. J. Geophys. Res., in preparation.

25. Natusch, D. F. S; Klonis, H. B.; Axelrod, H. D.; Teck, R. J.; Lodge, J. P. Jr., Anal. Chem. 1972, 44, 2067.

26. Slatt, B. J.; Natusch, D. F. S.; Prospero, J. P.; Savoie, D. L. Atmos. Environ. 1978, 12, 981-91.

27. Delmas, R.; Servant, J. J. Geophys. Res. 1982, 87, 11,019.

28. Herrmann, J.; Jaeschke, W. J. Atmos. Chem. 1984, 1, 111.

29. Cooper, D. J. M.S. Thesis, University of Miami, Florida, 1986

30. Cooper, D. J.; Saltzman, E. S. Geophys. Res. Letts. 1987, 14, 206.

31. Barnes, I.; Becker, K. H.; Martin, D.; Carlier, P.; Mouvier, J.; Jourdain, J. L.; Laverdet, G.; Le Bras, G. In Biogenic Sulfur in the Environment, Saltzman, E. S.; Cooper, W. J., Eds.; Amer. Chem. Soc., this volume.

32. Chatfield, R. B.; Crutzen, P. J. J. Geophys. Res. 1984, 89, 7,111.

33. Andreae, M. O.; Berresheim, H.; Andreae, T. W.; Kritz, M. A.; Bates, T. S.; Merrill J. T. J. Atmos. Chem. 1988, 6, 149.

34. Cutter, G. A.; Krahforst, C. F. Geophys. Res. Lett. 1988, 15, 1393-7.

35. Luther, G. W. III; Tsamakis, E., Preprint, (1988).

36. Graedel, T. E. Geophys. Res. Lett. 1979, 6, 329.

37. Liss, P. S. In Air-Sea Exchange of Gases and Particles; Liss, P. S. and Slinn, W. G. N., Eds.; 1983, pp 241-98.

38. Liss, P. S.; Slater, P. G. Nature 1974, 247, 181.

39. Broecker, W. S.; Peng, T. H. Tellus 1974, 26, 21.

40. Bates, T. S.; Cline, J. D.; Gammon, R. H.; Kelly-Hansen, S. R. J. Geophys. Res. 92, 2,930.

41. Holmen, K.; Liss, P. S. Tellus 1984, 36B, 92.

42. Ledwell, J.R. In Gas Transfer at Water Surfaces; Brutsaert, W.; Jirka, G. H., Eds.; D. Reidel: Mass., 1984; pp. 293-302.

43. Jahne, B.; Munnich, K. O.; Bosinger, R.; Dutzi, A.; Huber, W.; Libner, P. J. Geophys. Res. 1987, 92, 1,937.

44. Smethie , W. M. Jr.; Takahashi, T.; Chipman, D. W. J. Geophys. Res. 1985, 90, 7,005.

45. Wilke, C. R.; Chang, P. Am. Inst. Chem. Eng. J. 1955, 1, 264.

46. Hayduk, W.; Laudie, H. Am. Inst. Chem. Eng. J. 1974, 20, 611.

47. Savoie, D. L. Ph.D. Dissertation, University of Miami, Florida, 1984.

48. Winer, A. M.; Atkinson, R.; Pitts, J. N., Jr. Science 1984, 224, 156-9.

49. Wanninkhof, R.; Ledwell, J. R.; Broecker, W. S. Science 1985, 227, 1,224.

50. Balls, P. W.; Liss, P. S. Atmos. Environ. 1983, 17, 735.

RECEIVED December 2, 1988

Chapter 21

Distribution of Biogenic Sulfur Compounds in the Remote Southern Hemisphere

H. Berresheim[1], M. O. Andreae[2], G. P. Ayers[3], and R. W. Gillett[3]

[1]School of Geophysical Sciences,
Georgia Institute of Technology, Atlanta, GA 30332
[2]Max-Planck Institute for Chemistry, Mainz, Federal Republic of Germany
[3]Division of Atmospheric Research, Commonwealth Scientific
and Industrial Research Organization, Mordialloc, Australia

The Southern Ocean south of 40°S has been considered to be a potentially major source of biogenic sulfur compounds to the remote atmosphere due to reported high primary productivity rates and intensive winds in these latitudes. Most of this area is remote from man-inhabited continents. Therefore it offers the possibility to study the natural atmospheric sulfur cycle without major interferences by continental air masses. In this paper we discuss measurements of atmospheric and seawater concentrations of dimethylsulfide (DMS) which were made on a cruise across the Drake Passage and in inshore waters of Antarctica. The annual DMS emission from the Southern Ocean to the atmosphere is estimated to be 0.2 Tmol yr^{-1}. Some data for atmospheric methylmercaptan (MeSH) concentrations are also reported. We further report measurements of vertical distributions of atmospheric DMS which were made during flights over the west coast of Tasmania. During both field expeditions we also determined the atmospheric concentrations of the DMS oxidation products sulfur dioxide (SO_2), methanesulfonic acid (MSA), and non-sea-salt sulfate (nss-SO_4^{2-}). The results suggest a higher yield of MSA and a lower yield of SO_2 from DMS oxidation compared to other world ocean areas.

The global natural flux of sulfur compounds to the atmosphere has recently been estimated to be about 2.5 Tmol yr^{-1} ([1]) which is comparable to the emissions of sulfur dioxide (SO_2) from anthropogenic sources ([2]). A substantial amount of the natural sulfur contribution (0.5-1.2 Tmol yr^{-1}) is attributed to the emission of dimethylsulfide (DMS) from the world's oceans to the atmosphere ([3,4]). One of the major uncertainties in this estimate is due to a scarcity of DMS and other sulfur data from the Southern Hemisphere, particularly the Southern Ocean region between about 40°S and the Antarctic continent, which represents about one fifth of the total world ocean area.

Very high primary productivity rates have been reported for the coastal and inshore areas around the Antarctic continent ([5,6]). Recently, high densities of phytoplankton species like Phaeocystis poucheti, which is known to be an important source for marine DMS ([7,8]) have been observed in the same areas ([9]). On the other hand, strong winds which are associated with the intensive

0097–6156/89/0393–0352$06.00/0

circum-Antarctic weather systems prevailing between 40°-60°S are very common in these latitudes. One may therefore expect high DMS emission rates from the Southern Ocean, with even a temporary depletion of DMS levels in surface seawater due to strong ventilation, which may significantly contribute to the atmospheric sulfur cycle of the remote Southern Hemisphere.

Once DMS is emitted into the atmosphere it will eventually be oxidized by OH or NO_3 radicals to sulfur dioxide (SO_2), methanesulfonic acid (MSA), and, via SO_2 oxidation, to non-sea-salt sulfate (nss-SO_4^{2-}) as major reaction products (e.g. 10,11). The Southern Ocean represents a relatively unpolluted marine environment. It offers a unique possibility to study the natural sulfur cycle in an atmosphere far remote from man-inhabited continents.

In this work we compare some of the major results from two field expeditions to the Southern Ocean region. The first expedition was a ship cruise between Punta Arenas/Chile and the west coast of the Antarctic Peninsula, which was conducted during the austral fall season (March 20 - April 28, 1986) as part of the United States Antarctic Research Program (USARP). Details of the ship cruise and a complete discussion of the results have been published elsewhere (12). The second expedition consisted of a series of flight measurements off the west and south coast of Tasmania during the austral summer season (December 3-18, 1986). We discuss preliminary results from 2 of 8 flights which were conducted in marine air masses behind cold fronts moving over the open ocean.

Experimental

The determination of DMS in seawater and in the atmosphere has been described in detail elsewhere (12-14). Methods used to determine atmospheric concentrations of MeSH, SO_2, MSA, and nss-SO_4^{2-} have also been discussed in previous publications (12,15). A brief summary is presented here.

DMS dissolved in seawater was purged from 10-15 ml samples by a helium carrier gas stream, passed through a K_2CO_3 drying column, and then trapped cryogenically on a chromatography column (15% OV3 on Chromosorb W AW-DCMS 60/80 mesh) which was immersed in liquid nitrogen. After removal of the liquid nitrogen the DMS was separated from interfering compounds by controlled heating of the GC column and was then detected by flame photometric detection.

An atmospheric DMS sample was obtained by preconcentration on ca. 2g gold wool enclosed in a quartz glass sampling tube. Before entering the sampling tube the sample air first passed through a preconditioned scrubber material (5% Na_2CO_3 on Anakrom C22, 40/50 mesh) to remove SO_2 and strong oxidants like ozone. Samples were taken in duplicate. In the laboratory the DMS was transferred to the GC column after heat desorption from the gold wool under a helium or hydrogen carrier gas flow and then analyzed as described previously. For the analysis of MeSH 1 ml propyliodide vapor was added to the carrier gas prior to the heat desorption step to convert MeSH into propylmethylsulfide which is more stable and easier to detect. All DMS and MeSH analyses were calibrated using permeation devices (Metronics, Dynacal).

Particle-bound MSA and nss-SO_4^{2-} were collected in a two-stage aerosol filter sampler using a Nuclepore filter (diameter: 47 mm, nominal pore size: 8.0 μm) on the first stage and a Zefluor filter (Gelman, diameter: 47 mm, nominal pore size: 2.0 μm) on the second filter stage. The aerosol particles were thereby separated into coarse and fine-mode particles (particle cutoff diameter: 1.5 μm). SO_2 was trapped on a third stage consisting of a K_2CO_3/glycerol impregnated membrane filter (Schleicher and Schuell, FF#2, diameter: 25 mm). The aerosol filters were leached out with 10 ml of 18 megohm deionized water. The impregnated filters were treated with 10 ml of 0.06% H_2O_2 to convert sulfur (IV) to sulfate. All solutions were analyzed on a Dionex ion chromatograph (model 2020i). The overall precision of the filter measurements was between 10-20%, with the uncertainty largely due to the filter blank corrections.

In this work all concentrations are reported in units of nanomole per cubic meter (nmol m^{-3}) at 20°C and 1013 hPa, and/or as parts per trillion by volume (pptv) with 1 nmol m^{-3} corresponding to 24 pptv.

Results and Discussion

Ship Cruise Measurements

Figure 1 gives an overview of the cruise tracks and the DMS sampling stations in the Drake Passage and in Bransfield Strait. The land base in Antarctica was Palmer Station on Anvers Island. Most of the measurements were made in open ocean areas (Drake Passage) and in the inshore areas of Gerlache Strait.

Ten-day isobaric back-trajectories were calculated for each day of the cruise (Harris, J., personal communication, 1986) indicating that air masses were predominantly advected from the open South Pacific Ocean. Wind velocities over the Drake Passage ranged between 3-30 m s^{-1}, whereas in the inshore areas they never exceeded 10 m s^{-1}. During about 70% of the cruise period the weather was characterized by overcast conditions with occasional snowfall and fog episodes. A high pressure weather situation occurred between April 8-14 which was responsible for sustained cloud-free conditions with intense daylight during this short period. Surface seawater temperatures were typically around 8°C in the Subantarctic and 2°C in the Antarctic and inshore waters, whereas air temperatures were around -0.5°C and showed much less latitudinal difference.

Table I summarizes the results of the DMS and MeSH measurements in the major study areas. No significant differences in the concentrations were found between the individual areas. The average surface seawater DMS concentration was 1.80 nmol l^{-1} which is comparable to values observed in temperate latitudes and in oligotrophic areas of the world's oceans (16). This is consistent with lower primary productivity rates in the study areas during austral fall relative to spring and summer conditions (7,17). Exceptionally high DMS concentrations in seawater (6-9 nmol l^{-1}) were measured in Gerlache Strait during periods with intense daylight conditions between April 9-14 and April 18-19, which likely caused a strong increase in primary productivity and hence, DMS production rates. Generally the DMS distribution in surface seawater was very patchy, particularly in the inshore areas. No regular diurnal variations were observed. High concentrations were often found in the vicinity of icebergs, which is consistent with reports that sea ice may be an ideal growth medium for certain algal species (18). Measurements in areas with high krill populations, or in krill-rich water basins at Palmer Station did not show a significant production of DMS by these animals, as one may have suspected from reports about high DMS concentrations in frozen body tissue of krill (19). Only decomposition processes of dead krill produced significant levels of DMS and dimethyldisulfide (DMDS). Both MeSH and DMDS were observed in some seawater samples but could not be quantified due to the lack of a reliable calibration procedure. However, from the obtained chromatograms we estimate the concentrations did not exceed a few percent of those of DMS.

The average atmospheric DMS concentration was 4.4 nmol m^{-3} (106 pptv), similar to the global mean value over the world's oceans (4.7 nmol m^{-3}, 14). MeSH concentrations were mostly below the detection limit (0.03 nmol m^{-3}). Measurable MeSH values were frequently observed in coincidence with high

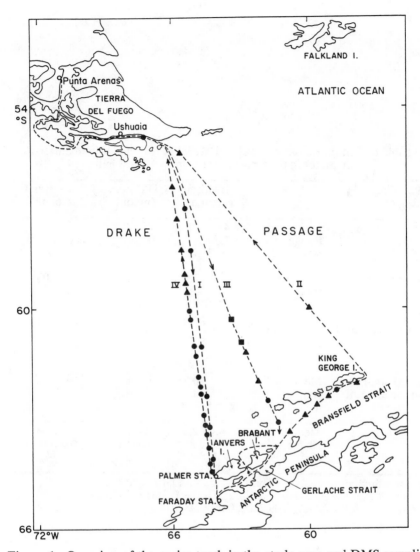

Figure 1. Overview of the cruise track in the study area and DMS sampling locations in the Drake Passage and in Bransfield Strait. Cruise legs in the Drake Passage (1986): I, March 21-24, II, March 28-29, III, March 30-31, IV, April 24-27; Circles indicate stations including both air and seawater samples, triangles seawater samples only, squares air samples only, and hatched areas represent multiple adjacent stations where air and/or seawater samples were taken. (Reprinted with permission from Ref. 12. Copyright 1987 by the American Geophysical Union).

Table I. Concentrations of DMS and MeSH in Air and DMS in Surface
Seawater in the Cruise Area During Austral Fall

Sampling Region		DMS(air) (nmol m^{-3})	DMS(seaw.) (nmol L^{-1})	MeSH(air)[*] (nmol m^{-3})
Drake Passage	Mean±s.d	4.8±1.9	1.9±0.5	---
	Range	1.7-8.3	0.7-3.2	<0.04-0.15
	n	23	31	9
Gerlache Strait	Mean±s.d.	4.4±2.5	1.8±1.2	<0.03
	Range	0.7-9.8	0.6-8.6	<0.03-0.09
	n	80	104	64
Bransfield Strait	Mean±s.d.	5.2	1.5±0.8	<0.04
	Range	4.6-5.7	0.8-2.8	---
	n	2	7	1
Coastal Shelf/ Gerlache Strait	Mean±s.d.	2.3±0.5	1.7±0.8	---
	Range	1.5-2.8	0.7-4.7	---
	n	5	7	---
Total Area	Mean±s.d.	4.4±2.3	1.8±1.1	<0.04
	Range	0.7-9.8	0.6-8.6	<0.03-0.15
	n	110	149	74

Here n is number of analyzed samples; s.d., standard deviation.
[*]Values of MeSH above detection limit: Drake Passage, 0.15 (n=1);
Gerlache Strait, 0.07, 0.07, 0.09, 0.05, 0.05, 0.08, 0.08 (n=7).

atmospheric DMS levels. Because of technical difficulties it was not possible to determine MeSH concentrations in all of the gold tube samples. The average diurnal variation of the DMS atmospheric concentration during the cruise period was rather low, possibly because of frequent overcast conditions which diminished the photochemical production of OH radicals. Moreover, the differences between daytime and nighttime levels were probably not caused by photochemical processes alone but may have also been influenced by diurnal fluctuations of the atmospheric mixed layer height. The 4-hour average values ranged from 3.9 nmol m^{-3} at 1800 hours local time to 5.6 nmol m^{-3} at 0600 hours, which represents a difference of a factor of 1.4 between the nighttime maximum and the daytime minimum. The average daytime OH concentration was estimated to be about 3×10^5 cm^{-3} (12). The nighttime NO_3 concentration was assumed to be negligible, based on low NO_x and O_3 concentrations reported by Sheppard et al. (20). Using the presently most reliable kinetic data for the DMS oxidation by OH (21) the atmospheric residence time of DMS is calculated to be about 6.5 days, which is somewhat longer than values reported for lower latitudes (1), and by far longer than recent estimates of only a few hours or less given by Nguyen et al. (22) and Martin et al. (this book), who suggest a major importance of the IO radical for DMS oxidation.

A conspicuous decrease in atmospheric DMS concentrations was observed between April 8-11 in coincidence with the clear-weather conditions described above. The diurnal DMS variation showed its highest amplitudes during this period, with a maximum/minimum ratio of 3.8 between April 8-9 (Figure 2). The corresponding average 12-hour daytime OH concentration has been estimated to 8.5 x 10^5 cm^{-3} (12), which suggests a mean atmospheric residence time of DMS of about 2-3 days. It must be emphasized, however, that strong katabatic winds occurred during the night between April 8-9 which may have contributed significantly to the particular diurnal variation shown in Figure 2.

The sea-to-air flux of DMS in the study area was calculated to be about 4.4 μmol $m^{-2}d^{-1}$ over the open ocean (Drake Passage) and about 1.2 μmol $m^{-2}d^{-1}$ from the inshore waters of Gerlache and Bransfield Strait (12). The calculations were based on a simple parameterization of the stagnant film model (23). The results are estimated to be uncertain by a factor of 2 (1). The difference between the open ocean and inshore area values can be attributed mainly to differences in wind velocities rather than sea surface temperatures or aqueous DMS concentrations between both regions.

Berresheim (12) extrapolated the DMS emission fluxes obtained for the cruise area to estimate the total annual DMS emission from the Southern Ocean area south of 40°S. He calculated a total DMS emission flux of 0.2 Tmol yr^{-1}. For this estimate he used data of the seasonal variation of primary productivity in the Southern Ocean reported by Holm-Hansen et al. (24) and assumed a linear relationship between primary productivity and DMS emission rates. The seasonal pattern of primary productivity is also consistent with the seasonal variation of atmospheric MSA concentrations measured by Ayers et al. (25) at Cape Grim, Tasmania. The summer DMS emission rates in the inshore areas were estimated to reach several hundreds of μmol $m^{-2}d^{-1}$ during phytoplankton blooms, which suggests surface seawater concentrations of DMS of 50 nmol l^{-1} and higher. This is consistent with the measurements of Deprez et al. (7) who observed similar seawater DMS concentrations at Davis Bay during the austral summer. It is likely that high DMS emissions from these areas could have an important impact on the sulfur budget of the Antarctic troposphere during the summer season. On the other hand, the yearly average sea-to-air flux of DMS from these areas and from the whole Southern Ocean is similar compared to the DMS fluxes from other world ocean areas (12).

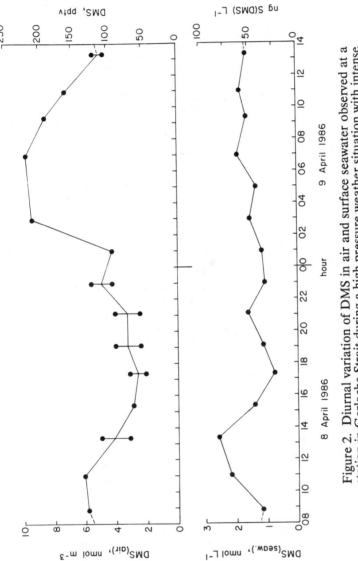

Figure 2. Diurnal variation of DMS in air and surface seawater observed at a station in Gerlache Strait during a high pressure weather situation with intense daylight and strong katabatic winds at night. (Reprinted with permission from Ref. 12. Copyright 1987 by the American Geophysical Union).

Table II shows the results of the shipboard measurements of atmospheric SO_2, MSA, and nss-SO_4^{2-} concentrations. Sampling integration times were typically 48 hours unless sampling had to be discontinued intermittently during high sea-spray, fog, or precipitation events. As in the case of DMS the values obtained over the open ocean and inshore areas were not significantly different. The mean MSA concentrations were similar to the monthly mean values for March/April obtained by Ayers et al. ([25]) at Cape Grim, and to values reported for other world ocean areas ([26,27]). The SO_2 concentrations were very low, but were consistent with the lowest values previously obtained by Nguyen et al. ([22,28]) over the Southern Indian Ocean. In agreement with the SO_2 concentrations the nss-SO_4^{2-} levels were also extremely low and only comparable to the lowest sulfate concentrations reported for the South Pole station ([29]), and for Punta Arenas ([30]). The nss-SO_4^{2-} fraction relative to total sulfate was only about 15- 20%. Based on these results the MSA/nss-SO_4^{2-} ratios were surprisingly high compared to average values of 0.07 over other ocean areas ([15,26,27]). However, they agree very well with similar ratios observed in Antarctic ice core samples ([31]).

The concentrations of the individual sulfur compounds appeared to be most strongly correlated during the sunny period between April 8-14. While DMS concentrations decreased sharply from about 8 nmol m^{-3} to 2 nmol m^{-3}, both MSA and nss-SO_4^{2-} levels increased by a factor of 2-3 relative to the preceding sampling period (Table II). An increase in atmospheric SO_2 levels was also observed. However, it was relatively modest (factor of 1.3) compared to the changes in MSA and nss-SO_4^{2-} concentrations and was limited to only the first half of the clear weather period. During the second half (April 11-14) the average SO_2 concentration decreased sharply by a factor of 2.6. These observations indicate the importance of two major chemical processes: 1. A significant increase of the photochemical oxidation rate of DMS and of the production rate of both MSA and SO_2, and 2. A significant removal of atmospheric SO_2 by oxidation leading to elevated nss-SO_4^{2-} concentrations. SO_2 oxidation may have been substantially promoted by photochemical processes during this period. However, as discussed by Berresheim ([12]), absorption and aqueous phase oxidation of SO_2 in sea-salt particles was more likely the dominant SO_2 removal process, since sea-salt particle concentrations were extremely high during the same period. A more detailed discussion of the atmospheric sulfur budget over the study areas is given in the paper by Berresheim ([12]).

Flight Measurements

Previous measurements of the vertical profiles of DMS, SO_2, MSA, and nss-SO_4^{2-} over the temperate North Pacific Ocean ([15]) have shown the important influence of meteorological processes on the vertical distributions of the individual compounds. Those measurements were made in air masses behind cold fronts which were moving over the open ocean towards the coastline of the northwestern United States. The boundary between the atmospheric mixed layer and the free troposphere was well defined by the subsidence of the postfrontal air. One of the most striking results from these measurements was the observation of a strong increase of the nss-SO_4^{2-} concentration in the free troposphere in contrast to the profiles of all other sulfur compounds. Isentropic trajectory calculations supported the suggestion that this increase in nss-SO_4^{2-} levels could only be explained by the advection of continental air masses from Asia with an average transport time of about 10 days to reach the study area.

Table II. Results Obtained from Shipboard Measurements of MSA, SO$_2$, and nss-SO$_4^{2-}$ Concentrations During Austral Fall

Sampling Region	Date (1986)	MSA (nmol m^{-3})	SO$_2$ (nmol m^{-3})	nss-SO$_4^{2-}$ (nmol m^{-3})	MSA/nss-SO$_4^{2-}$ (mol/mol)	MSA/SO$_2$ (mol/mol)
Drake Passage	March 22-23	0.39	---	0.35	1.11	---
	March 23-24	0.66	0.56	0.50	1.32	1.18
	March 30-31	0.23	0.17	---	---	1.35
	March 24-27	0.03	0.41	0.05	0.60	0.07
Mean ± s.d.		0.33±0.27	0.38±0.19	0.30±0.23	1.01±0.37	0.87±0.69
Gerlache Strait	April 02-06	0.13	0.54	0.36	0.36	0.24
	April 07-11	0.22	0.70	0.20	1.10	0.31
	April 11-14	0.34	0.27	0.63	0.54	1.26
	April 22-23	0.07	0.51	0.19	0.37	0.14
Mean ± s.d.		0.19±0.12	0.50±0.18	0.34±0.20	0.60±0.35	0.49±0.52
Coastal shelf/ Gerlache Strait	April 17-21	0.39	0.47	0.21	1.86	0.83

In view of these results we concluded that in the Northern Hemisphere very few occasions are available to study the vertical distribution of biogenic sulfur in the marine atmosphere without a significant interference by continental air masses. Therefore we decided to conduct a similar flight program in the Southern Hemisphere, west of Tasmania, with the Australian continent as the only possible source for contaminated air masses. The flights were made with a Fokker F-27 aircraft owned by CSIRO. Samples were taken offshore from the Tasmanian coast behind cold fronts which were advected by strong westerly or southerly winds from the open sea. Figure 3 shows the data from flights #7 and #8, which were conducted during the same day, on December 14. Flight #7 was in the morning, and flight #8 was a one-level flight at 5.3 km altitude during mid-afternoon. The atmospheric mixed layer height was between 1-1.5 km. Thermodynamic conditions were neutral in the mixed layer and stable in the free troposphere.

All sulfur compounds showed relatively constant profiles throughout the lower flight levels which is consistent with the neutral conditions in the mixed layer. As compared to the data obtained during the ship cruise (Tables I and II) the concentrations in the lowest flight level (30 m) were about 2-3 times lower for DMS, nearly the same for SO_2, and about 3 times higher for MSA and nss-SO_4^{2-}. The fraction of nss-SO_4^{2-} to total sulfate at this level was 18%, similar to the results from the ship cruise. The relatively higher concentrations observed for MSA and nss-SO_4^{2-} indicate a significant accumulation of aerosol particles in the mixed layer due to the postfrontal inversion between the mixed layer and the free troposphere.

High-volume MSA samples were taken at Cape Grim before and after the flight measurements. Weekly average concentrations showed an increase from 1.1 to 12.7 pptv in November and a decrease from 16.7 to 10.5 pptv in February 1987, bracketing maximum MSA values which are usually observed around the end of each year at Cape Grim (25). These data are in good agreement with the MSA concentrations measured at the lowest flight level.

Our results for the low flight level DMS concentrations are consistent with previous ground-level measurements by Andreae et al. (14) at Cape Grim during the summer season of 1984. They reported concentrations ranging from 25-360 pptv. In this study we measured ground-level concentrations at Kingston and Cape Grim in the range of 39-82 pptv. The relatively low productivity of the waters off the west and south coast of Tasmania (32) may explain the observed DMS concentrations which were somewhat lower than expected for the summer season. We also assume that primary production had not yet developed to its full extent because the biological peak of the summer season is more likely reached in January rather than in December. This is also consistent with the annual maximum of the aerosol MSA concentration at Cape Grim, which is usually observed in early January (25).

As shown in Figure 3 the DMS concentration dropped sharply above the mixed layer to values around the detection limit (1 pptv) at 3 km. This can be explained by the subsidence of the postfrontal air masses which largely suppressed an upward diffusion of DMS to the free troposphere. On the other hand, we also think that DMS had been largely oxidized in free tropospheric air masses which were advected from greater distances. This is supported by the observed increase of SO_2 and nss-SO_4^{2-} concentrations above the mixed layer, and of MSA concentrations above 2.3 km. The concentration profiles of these compounds in the free troposphere can also be attributed to the postfrontal meteorological conditions during the flights. The relative humidity dropped sharply from 90% to about 40% between the 1 km and 2 km levels. The dryness of the air around the 2 km level may have caused a large scale evaporation of clouds over the study area. Relatively high aerosol sulfate concentrations at this

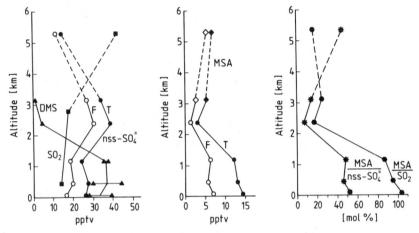

Figure 3. Vertical distributions of sulfur compounds over the Tasman sea measured during flights #7 and #8 on December 14, 1986. Dashed lines connect the data from flight #7 with the high altitude data obtained during flight #8. Symbols F and T denote fine and total particle fractions, respectively.

level may have been produced from in-cloud oxidation of SO_2 and subsequent evaporation of the droplets. This would explain the observed maximum in the nss-SO_4^{2-} profile shown in Figure 3. Although no air mass trajectories have yet been calculated we assume that advection of continental air masses was negligible and therefore not responsible for these elevated concentrations. This can be concluded from synoptic weather charts and satellite pictures. It is also consistent with the sharp decrease of the nss-SO_4^{2-} concentrations at higher altitudes, which was in contrast to our previous observations over the North Pacific Ocean (15). Figure 3 also shows that nss-SO_4^{2-} was predominantly present in fine particles and that the highest values of the fine particle fraction of nss-SO_4^{2-} were observed above the 2 km level. This indicates negligible concentrations of soil-derived and/or sea-salt sulfate particles in the free troposphere.

The fine particle fraction of MSA did not significantly exceed 50% of the total MSA, except at the highest flight level. A remarkably parallel behavior of the DMS and MSA profiles was observed in the mixed layer and the lower free troposphere. It is likely that both DMS and MSA concentrations were predominantly of local origin at these altitudes. The relatively strong correlation between both profiles, as compared to the nss-SO_4^{2-} profile, suggests that DMS was mainly oxidized to MSA and/or that it was converted much more rapidly to MSA than to nss-SO_4^{2-} at lower flight levels.

SO_2 samples had to be integrated over several flight levels during flight #7 and over 1.5 hours of sampling time at the 5.3 km level during flight #8 because of the low SO_2 concentrations in these altitudes. SO_2 concentrations were similar in the mixed layer and the lower free troposphere and showed no significant correlation with the DMS concentration profile, in contrast to MSA. A strong increase of the SO_2 concentration, which is in clear contrast to the nss-SO_4^{2-} profile, was observed in the highest flight levels. It was also much more pronounced than the vertical increase of the MSA concentration at these levels. Either one or several of the following explanations for this observation are possible: 1. Changes in the DMS (and possibly SO_2) oxidation kinetics as a function of altitude, 2. A longer atmospheric residence time of SO_2 at higher, cloud-free levels, 3. Additional chemical sources for SO_2 at these levels, and/or 4. Long-range transport and advection of SO_2 from distant sources.

Vertical profiles of the molar MSA/nss-SO_4^{2-} and the MSA/SO_2 ratio have been derived from the flight data and are also shown in Figure 3. The MSA/nss-SO_4^{2-} ratio was at all altitudes significantly higher than the average value of 7% reported by Saltzman et al. (26,27) for other ocean areas in lower latitudes. The lower altitude data compare reasonably well with the results obtained from the ship cruise measurements (Table II). It was observed during all flights that both the MSA/nss-SO_4^{2-} and the MSA/SO_2 profiles follow the vertical DMS distribution closely in the mixed layer and the lower free troposphere. However, at higher levels they show a substantial increase with altitude which may indicate a DMS oxidation chemistry different from that in lower altitudes, as was suggested above.

One possible explanation is offered by the results of a recent laboratory study (21) which showed that the mechanism of the DMS oxidation by OH is a function of temperature and that the OH addition mechanism dominates over H abstraction at temperatures below 285 K. Hynes et al. (21) speculated that OH addition to DMS may lead primarily to the formation of MSA and that the yield of MSA may increase inversely with temperature. Temperatures during our flights were around +12°C at ground level and decreased to about -23°C at the highest flight altitude. If MSA is indeed the dominant product of the OH addition reaction we may calculate a corresponding increase of the MSA yield from 50% to 76%, respectively, based on the data of Hynes et al. (21). The same authors speculate that SO_2 may be formed via both OH addition and H abstraction mechanisms, which show a contrary dependence on temperature to each other. This hypothesis is consistent with the relatively neutral SO_2 profiles which we observed in the mixed layer and the lower free troposphere. More recent laboratory studies (Tyndall and Ravishankara, this book; Hatakeyama, this book; Barnes et al. Int. J. Chem. Kinet., in press) suggest that the MSA/SO_2 yield ratio may also strongly depend on ambient NO_x levels. This needs to be verified in future field studies which should include measurements of both NO_x and biogenic sulfur compounds in the atmosphere. Also, more laboratory studies of the DMS oxidation by OH are needed which involve specifically the direct identification of individual reaction products.

Our data obtained from both the ship cruise (Table II) and the flight measurements (Figure 3) suggest that photooxidation of DMS in the Southern Ocean atmosphere leads to a higher yield of MSA and a lower yield of SO_2 compared to other world ocean areas. We conclude that these yields are possibly a function of latitude and altitude and, basically, of temperature as suggested by the recent laboratory studies of Hynes et al. (21).

Summary of Major Results

1. Average DMS concentrations in surface waters and in the atmosphere over the Southern Ocean were similar to values reported for other world ocean areas.

2. Atmospheric residence times of DMS were estimated to be between 2-3 days during cloud-free conditions and 6-7 days during overcast conditions in austral fall.

3. DMS emission fluxes from Antarctic inshore waters may be important for the tropospheric sulfur budget of Antarctica during summer. The contribution of the Southern Ocean to the global atmospheric sulfur budget (ca. 0.2 Tmol yr^{-1}) is consistent with present estimates of the total global DMS emission from the world's oceans (0.5-1.2 Tmol yr^{-1}).

4. SO_2 and nss-SO_4^{2-} concentrations were extremely low compared to MSA concentrations. Atmospheric MSA levels were strongly correlated with horizontal and vertical DMS distributions. The $MSA/nss-SO_4^{2-}$ and MSA/SO_2 concentration ratios were significantly higher than previously reported values for other oceanic regions of the world. These results suggest a higher yield of MSA and a lower yield of SO_2 from DMS oxidation over the Southern Ocean compared to other world ocean areas.

Acknowledgments

The authors wish to thank the following people for their valuable contributions to this work: R. Duce, J. Galloway, A. Pszenny, A. Castelle, the staff of ITT Antarctic Services, the captain and crew of the R/V Polar Duke, W. Hamner, J. Harris, W. Keene, the pilots of the Fokker F-27, C. Maher, J. Smith, and M. Dancy. This work was financially supported by the U.S. National Science Foundation through grant ATM-8407137 and by the Australian CSIRO Office of Space Science and Applications.

Literature Cited

1. Andreae, M. O. In The Role of Air-Sea Exchange in Geochemical Cycling; Buat-Menard, P., Ed.; Reidel: Dordrecht, 1986; pp 331-62.

2. Cullis, C. F.; Hirschler, M. M. Atmos. Environ. 1980, 14, 1263-78.

3. Andreae, M. O.; Raemdonck, H. Science 1983, 221, 744-7.

4. Bates, T. S.; Cline, J. D.; Gammon, R. H.; Kelly-Hansen, S. R. J. Geophys. Res. 1987, 92, 2930-8.

5. El-Sayed, S. Z. In Marine Phytoplankton and Productivity; Holm-Hansen, O; Bolis, L; Gilles, R., Eds.; Springer: New York, 1984; pp. 19-34.

6. Platt, T.; Subba Rao, D. V. In Photosynthesis and Productivity in Different Environments; Cooper, J. P., Ed.; Cambridge University Press: Cambridge, 1975; pp 249-80.

7. Deprez, P. P.; Franzmann, P. D.; Burton, H. R. J. Chromatogr. 1986, 362, 9-21.

8. Barnard, W. R.; Andreae, M. O.; Iverson, R. L. Cont. Shelf Res. 1984, 3, 103-13.

9. von Bodungen, B.; Smetacek, V. S.; Tilzer, M. M.; Zeitschel, B. Deep-Sea Res. 1986, 33, 177-94.

10. Yin, F.; Grosjean, D.; Seinfeld, J. H. J. Geophys. Res. 1986, 91, 14,417-38.

11. Atkinson, R. Chem. Rev. 1986, 86, 69-201.

12. Berresheim, H. J. Geophys. Res. 1987, 92, 13,245-62.

13. Andreae, M. O.; Barnard, W. R. Anal. Chem. 1983, 55, 608-12.

14. Andreae, M. O.; Ferek, R. J.; Bermond, F.; Byrd, K. P.; Engstrom, R. T.; Hardin, S.; Houmere, P. D.; LeMarrec, F.; Raemdonck, H.; Chatfield, R. B. J. Geophys. Res. 1985, 90, 12,891-900.

15. Andreae, M. O.; Berresheim, H.; Andreae, T. W.; Kritz, M. A.; Bates, T. S.; Merrill, J. T. J. Atmos. Chem. 1988, 6, 149.

16. Andreae, M. O. In The Biogeochemical Cycling of Sulfur and Nitrogen in the Remote Atmosphere, Galloway, J. N.; Charlson, R. J.; Andreae, M. O.; Rodhe, H., Eds.; Reidel: Dordrecht, 1985; pp 5-25..

17. El-Sayed, S. Z. Antarctic Res. Ser. 1967, 11, 15-47.

18. Bunt, J. S.; Wood, E. J. Nature 1963, 199, 1254-55.

19. Tokunaga, T.; Iida, H.; Nakamura, K. Bull. Japan. Soc. Sci. Fish. 1977, 43, 1209-17.

20. Sheppard, J. C.; Hardy, R. J.; Hopper, J. F. Antarct. J. U.S. 1983, 18, 243-4.

21. Hynes, A. J.; Wine, P. H.; Semmes, D. H. J. Phys. Chem. 1986, 90, 4148-56.

22. Nguyen, B. C.; Bergeret, C.; Lambert, G. In Gas Transfer at Water Surfaces; Brutsaert W.; Jirka, G. H., Eds.; Reidel: Dordrecht, 1984, pp. 539-45.

23. Liss, P. S. Deep-Sea Res. 1973, 20, 221-38.

24. Holm-Hansen, O.; El-Sayed, S. Z.; Franceschini, G. A.; Cuhel, R. L. In Adaptations within Antarctic Ecosystems; Llano, G. A., Ed.; Proceedings of the third SCAR Symposium on Antarctic Biology, Gulf Publ. Co.: Houston, 1977; pp 11-50.

25. Ayers, G. P.; Ivey, J. P.; Goodman, H. S. J. Atmos. Chem. 1986, 4, 173-85.

26. Saltzman, E. S.; Savoie, D. L.; Zika, R. G.; Prospero, J. M. J. Geophys. Res. 1983, 88, 10,897-902.

27. Saltzman, E. S.; Savoie, D. L.; Prospero, J. M.; Zika, R. G. J. Atmos. Chem. 1986, 4, 227-40.

28. Nguyen, B. C.; Bonsang, B.; Lambert, G. Tellus 1974, 26, 241-9.

29. Cunningham, W. C.; Zoller, W. H. J. Aerosol Sci. 1981, 12, 367-84.

30. Lawson, D. R.; Winchester, J. W. Science 1979, 205, 1267-9.

31. Ivey, J. P.; Davies, D. M.; Morgan, V.; Ayers, G. P. Tellus 1986, 38B, 375-379.

32. Jitts, H. R. Austral. J. Mar. Freshwater Res. 1965, 16, 151-62.

RECEIVED November 14, 1988

Chapter 22

Sulfur Isotope Ratios

Tracers of Non-Sea Salt Sulfate in the Remote Atmosphere

Julie A. Calhoun[1] and Timothy S. Bates[2]

[1]Department of Chemistry, University of Washington, Seattle, WA 98195
[2]Pacific Marine Environmental Laboratory, National Oceanic and Atmospheric Administration, Seattle, WA 98115

The atmospheric biogeochemical sulfur cycle is being significantly impacted by increasing anthropogenic sulfur emissions. The effect of these emissions on the concentration of sulfate aerosol particles in the remote marine atmosphere is difficult to assess due to uncertainties surrounding the relative contributions of natural and anthropogenic sources. Sulfur isotope ratios can be used to determine the relative magnitude of these sources in the remote atmosphere provided 1) the isotopic ratios of the potential sulfur sources are known, 2) the isotopic compositions of the various sources are different from one another, and 3) the isotopic changes that occur during transformations are thoroughly documented. In the text which follows, these aspects of sulfur isotope chemistry are addressed. Isotopic interpretation of sulfur sources to the remote atmosphere is severely limited by the absence of critical isotopic measurements, yet it appears that continental sulfur sources are isotopically distinguishable from seasalt or marine biogenic sulfur sources. Improved analytical techniques will soon provide the means to obtain the necessary data.

Concerns about the environment have drawn considerable attention to anthropogenic sulfur emissions, particularly their affects in remote areas (1). On a global scale, anthropogenic sulfur emissions from the combustion of coal and oil are now equivalent to those from natural sources (2). However, the relative importance of anthropogenic sulfur in the remote atmosphere is unclear. Biogenic reduced sulfur compounds from the ocean, particularly dimethylsulfide (DMS) from phytoplankton (3-6), are the major natural source of sulfur to the remote marine atmosphere. Other natural sources include biogenic sulfur emissions from the continents, sea spray and volcanic emissions (2). Understanding the relative importance of these sources to the remote atmosphere is crucial for determining the environmental significance of anthropogenic sulfur emissions.

The presence of non-seasalt sulfate particles in the remote marine atmosphere has important environmental consequences. As a result of their size and hydrophylicity, sulfate particles make good cloud condensation nuclei

0097–6156/89/0393–0367$06.00/0
© 1989 American Chemical Society

(CCN). Changes in the concentrations of CCN may alter the cloud droplet concentration, the droplet surface reflectivity, the radiative properties of clouds (cloud albedo) (7), and hence, the earth's climate (8-10). This mechanism has been proposed for the remote atmosphere, where the radiative properties of clouds are theoretically predicted to be extremely sensitive to the number of CCN present (10). Additionally, these sulfate particles enhance the acidity of precipitation due to the formation of sulfuric acid after cloud water dissolution (11). The importance of sulfate aerosol particles to both radiative climate and rainwater acidity illustrates the need to document the sources of sulfur to the remote atmosphere.

One means to assess the relative contributions of the various sources of sulfur to the atmosphere is through the use of sulfur isotope ratios (12-16). Isotopic ratios may be used as source tracers if 1) the isotopic composition of the sources as they enter the atmosphere are known, 2) the isotopic compositions of the various sources are different from one another, and 3) the isotopic changes that occur during biological, physical and/or chemical transformations are understood. Presently, isotopic data for sulfur compounds in the remote atmosphere (Table I) are limited. However, collection and analytical techniques are now available to make isotopic measurements of the critical species. In the text that follows, various aspects of sulfur isotope chemistry will be discussed.

Background

Review of Isotope Concepts. The average relative concentration of the two most abundant stable isotopes of sulfur, ^{32}S and ^{34}S, are 95.0% and 4.2%, respectively. Ratios of these two isotopes are measured with specialized isotope ratio mass spectrometers, normalized to standard Canyon Diablo meteoritic sulfur, and expressed as a delta value ($\delta^{34}S$) according to the relation:

$$\delta^{34}S = \left[\frac{(^{34}S/^{32}S)\text{sample}}{(^{34}S/^{32}S)\text{standard}} - 1 \right] (1000) \tag{1}$$

The unit for the calculated $\delta^{34}S$ value is o/oo, which is expressed "per mil".

The mass difference between the two isotopes of sulfur can alter the isotope ratio of a system during chemical and physical transformations (17-21) a process referred to as "fractionation." Quite simply, the zero point energy of a molecule, and hence its rotational and vibrational energy, is mass dependent (22). Therefore, molecules containing different isotopic masses will have different reactivities and fractionation will occur during mass dependent transformations.

The degree to which the isotope ratio is fractionated differs for individual processes. The per mil isotope difference between the $\delta^{34}S$ value for the reactant, R, and the product, P, at any instant in time is termed discrimination, D, where $D = \delta^{34}S(R) - \delta^{34}S(P)$. By convention, if the lighter isotope reacts faster resulting in an isotopically lighter product, D is positive (14).

Reactions may exhibit kinetic isotope effects as a result of the lighter isotope reacting faster. To illustrate this type of isotope effect, consider the oxidation of methane in the atmosphere (23). The oxidation is initiated by a reaction with the hydroxyl free radical (OH) in which OH irreversibly abstracts a hydrogen atom from the carbon. Methane molecules containing the lighter carbon isotope react faster, and as a consequence, the $\delta^{13}C$ value of the product is lower.

Although kinetic effects typically lead to positive discriminations, the discrimination from equilibrium interactions may be either positive or negative

Table I. $\delta^{34}S$ Values for Compounds Potentially Involved in
the Production of Remote Atmospheric Sulfate

Compound	$\delta^{34}S$ (in o/oo)	Reference
Seasalt Sulfate		
Seawater	+21	Rees et al. (35)
Atmospheric	+21	Luecke et al. (51)
Dimethylsulfonio-Proprionate (DMSP)	+19.8	Fry and Andreae (pers. commun.)*
Dimethylsulfide		
Seawater	none	
Atmospheric	none	
Hydrogen Sulfide		
Terrestrial Biogenic	-32 to -6	Krouse et al. (61)
Geothermal	+5	Robinson et al. (56)
	+8.5	Spedding et al. (55)
Sulfur Dioxide		
Volcanic	-10 to +10	Nielsen (54)
Anthropogenic	{ -1 to +6 }	
Power Plant Plume	0 to +6	Holt et al. (52)
Non-urban N.E. US	-1.1 to +2.3	Saltzman et al. (24)
Coal fired Power Plant	+5	Newman et al. (53)
Oil fired Power Plant	+1.3 to +3.6	Newman et al. (50)
Terrestrial Biogenic	-34 to -35	Krouse et al. (61)
Atmospheric Aerosol Non-Seasalt Sulfate		
Non-Remote Marine	{ -10 to +13 }	
San Francisco Bay	-10 to +12	Ludwig (63)**
N.W. Atlantic	+7 to +13	Gravenhorst (62)***
HBEF, non-urban US	+1 to +4	Saltzman et al. (24)
Miami, Florida	+1 to +2	Saltzman (pers. commun.)
Mauna Loa Observatory	+4 to +6	Zoller & Kelly (pers. commun.)
Rainwater Sulfate		
Non-remote Marine		
Atlantic Ocean	+12 to +15	Chukhrov et al. (60)
Pacific Ocean	+9 to +16	Chukhrov et al. (60)
New Zealand Coast	+16	Spedding et al. (55)
New Zealand Coast	-1.5 to +19	Mizutani et al. (66)
Continental		
Non-industrial Japan	+14	Jensen et al. (64)
Industrial Japan	+6	Jensen et al. (64)
Pisa, Italy	-3 to +7	Cortecci et al. (65)
N. America/Scandinavia	+2	Ostland (67)
Great Salt Lakes	+2 to +8	Nriagu et al. (68)

* $\delta^{34}S$ value for seawater sulfate was 20.1 o/oo]
** Unusually low $\delta^{34}S$ values were reported for seawater sulfate, ranging from +9.7 to 17 o/oo]
*** Neglecting value of -12 o/oo]

and they are often larger than those from kinetic reactions (23). A general rule for determining the sign of the discrimination is that heavier isotopes tend to concentrate in the more strongly bonded molecules (22). It should be noted, however, that exceptions to this rule exist for some species and that fractionation may vary with temperature.

Consider, the equilibrium governed dissolution of sulfur dioxide (SO_2) in cloud droplets or on the surface of wet aerosol particles:

$$SO_2 \text{ (gas)} <\text{---}> SO_2 \cdot H_2O \text{ (dissolved)} \qquad (2A)$$

$$SO_2 \cdot H_2O \text{ (dissolved)} <\text{---}> HSO_3^-\text{(aq)} + H^+\text{(aq)} \qquad (2B)$$

which is followed by a fairly rapid aqueous phase (aq) oxidation,

$$HSO_3^-\text{(aq)} \text{----}> SO_4^{-2}\text{(aq)} + H^+ \qquad (2C)$$

In this case, both equilibrium (reactions 2A,2B) and kinetic isotope effects (reaction 2C) are expected. A large negative discrimination has been observed for reaction (2B), with the heavier isotope concentrating in the HSO_3^- (24). Kinetic isotope effects for reaction (2C), however, would produce sulfate which is isotopically lighter than the HSO_3^-. The equilibrium effects are probably larger than the kinetic effects (24), making the overall discrimination for the aqueous phase oxidation of SO_2 negative.

Mass Spectral Techniques. Samples for isotope ratio analysis are typically converted to sulfates or sulfides, then to SO_2(g) for analysis on a mass spectrometer (MS). The precision of the SO_2 measurement is commonly reported as 0.1 to 0.2 o/oo (16,24), yet systematic errors of 1 o/oo or larger may result from 1) memory effects due to adsorption of SO_2 on the walls of the MS, and 2) secondary isotope effects due to the existence of two stable isotopes of oxygen, ^{16}O and ^{18}O (25). Both of these errors can be eliminated by using SF_6 rather than SO_2 as the analyte in the MS (25,26). However, existing sulfur fluorination procedures are relatively dangerous and tedious, making the SF_6 method less desirable as a routine environmental technique (26).

Another procedure for sulfur isotope measurements has been developed where samples are converted to solid arsenic sulfide, As_2S_3 (s), and measured by thermal ionization mass spectrometry (TIMS) (27). This technique offers several advantages over the gaseous methods in that both memory and isotope effects are eliminated, and the chemical procedure is simpler. A precision of 1 o/oo, and the capability of making measurements on small samples, makes the TIMS technique competitive with gas phase MS techniques.

Sources of Sulfur to the Remote Atmosphere

Biogenic Sulfur Emissions from the Ocean. The ocean is a source of many reduced sulfur compounds to the atmosphere. These include dimethylsulfide (DMS) (2,4,5), carbon disulfide (CS_2) (28), hydrogen sulfide (H_2S) (29), carbonyl sulfide (OCS) (30,31), and methyl mercaptan (CH_3SH) (5). The oxidation of DMS leads to sulfate formation. CS_2 and OCS are relatively unreactive in the troposphere and are transported to the stratosphere where they undergo photochemical oxidation (32). Marine H_2S and CH_3SH probably contribute to sulfate formation over the remote oceans, yet the sea-air transfer of these compounds is only a few percent that of DMS (2).

DMS in the ocean is produced during assimilatory sulfate reduction (ASR) by phytoplankton (Figure 1) (3,33). This involves the uptake of oceanic sulfate

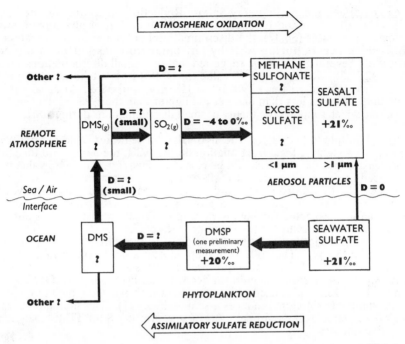

Figure 1. Transformations in the ocean and overlying atmosphere which lead to the production of sulfate from a marine biogenic source (dark arrows). DMS is produced in the ocean after the uptake of seawater sulfate by phytoplankton and the production and breakdown of DMSP. Sulfate formation occurs after DMS is transferred across the sea-air interface and undergoes atmospheric oxidation. The δ^{34}S values for the individual sulfur pools are indicated in the boxes and measured or estimated discriminations (D) are indicated above the arrows. Clearly, data for the remote atmosphere are limited.

by phytoplankton, followed by biological production of methionine and dimethylsulfonioproprionate (DMSP) (3,33). Enzymatic breakdown of DMSP leads to the production of DMS and acrylic acid (34). The transfer of DMS to the atmosphere accounts for the major portion of the reduced sulfur being emitted from the ocean to the atmosphere (2).

Oceanic Sulfate and Assimilatory Sulfate Reduction. Seawater sulfate has a $\delta^{34}S$ value of +21 o/oo (35). Although fractionation occurs during most biological processes (36,37), the discrimination between seawater sulfate and DMS which occurs during ASR by phytoplankton has not been directly measured. Several groups have suggested that only small discriminations would result (12,37,38). Recently, a $\delta^{34}S$ value for DMSP from a laboratory plankton culture was found to be +19.8 o/oo [B. Fry, and M.O. Andreae; pers. commun.]. This was less than one per mil lighter than the $\delta^{34}S$ value measured for sulfate which supports the theory of small discriminations during ASR.

Dimethyl Sulfide in Seawater. The $\delta^{34}S$ value for seawater DMS has not been measured. Assuming little fractionation during ASR, the $\delta^{34}S$ value for DMS upon release into the ocean should be close to +20 o/oo. This $\delta^{34}S$ value may be altered during the transfer of DMS across the sea-air interface, or during photochemical oxidation (33), and/or biological degradation of DMS in seawater. Until the relative importance of these processes and the discriminations expected for each removal pathway are measured, it is difficult to estimate the $\delta^{34}S$ value for the resulting DMS pool.

Recently, a method has been developed to measure the $\delta^{34}S$ value for seawater DMS (39). The collection method uses a gold wool system (40) to preconcentrate DMS from air equilibrated with surface seawater. DMS is thermally desorbed from the gold surface with hot hydrogen gas. DMS reacts with the hydrogen to produce hydrogen sulfide which, when bubbled through an arsenic ammonia solution, produces arsenic sulfide (27). This method makes it possible to collect sufficient quantities of seawater DMS for TIMS analysis as arsenic sulfide.

Transfer of DMS Across the Sea-Air Interface into the Atmosphere. At this time there is no empirical evidence for isotope discrimination during the sea-air transfer of DMS. Theoretically, the transfer of DMS is estimated from gas exchange models (Equation 3)

$$F = k \, ([DMS]_{sw} - b[DMS]_{air}) \qquad (3)$$

where the flux (F) across the sea-air interface is dependent on a first order exchange coefficient (k), the difference in concentrations of DMS in seawater and in the atmospheric mixed layer, $[DMS]_{sw}$ and $[DMS]_{air}$ respectively, and the molecular solubility, which is represented by the Bunsen coefficient (b) (4,41).

The molecular solubility term (b) depends on the chemical interaction between the solute (DMS) and the solvent (water). Because the internal energy of molecules is mass dependent, these chemical interactions should differ for the isotopically substituted DMS molecules and result in isotopic fractionation. However, in the case of DMS, the concentration in seawater is much larger than that in the atmospheric mixed layer, and Equation (3) reduces to (3a)

$$F = k \, [DMS]_{sw} \qquad (3a)$$

which depends solely on the DMS concentration in seawater, and the transfer coefficient (k) ($\underline{4}$).

The transfer coefficient (k) is a function of wind speed and temperature, and varies from molecule to molecule with the square root of molecular diffusivity ($\underline{42}$). Since molecular diffusivity does not vary with molecular mass, but rather with molecular volume or shape ($\underline{43,44}$), k too is independent of molecular mass and is unimportant in terms of isotopic effects. Based on the above theoretical discussion, little or no fractionation would be expected for the sea-air transfer of DMS.

<u>Atmospheric Oxidation of DMS.</u> DMS reacts fairly rapidly in the atmosphere to produce either SO_2, which is further oxidized to sulfate, or methane sulfonate ($CH_3SO_3^-$) (Figure 1). The relative abundances of the products in the remote troposphere suggest that sulfate is the major atmospheric end product ($\underline{45}$), although methane sulfonate has been identified under laboratory conditions as a dominant end product ($\underline{46,47}$).

The oxidation of DMS is initiated by the hydroxyl free radical which either abstracts a hydrogen from DMS or undergoes an addition reaction ($\underline{46-48}$). Kinetic isotope effects would be associated with these types of reactions. However, the remaining oxidation steps may involve larger equilibrium isotope effects resulting in heavier end products. These effects have not been measured. The $\delta^{34}S$ values for sulfate and methane sulfonate should be related if they are derived from the same DMS pool ($\underline{49}$). At this time there are no $\delta^{34}S$ measurements for methane sulfonate or sulfate which are unambiguously from the oxidation of DMS.

<u>Atmospheric Oxidation of SO_2 to Sulfate.</u> Regardless of the source, sulfur dioxide is oxidized under atmospheric conditions in the gas phase, cloud droplets or on the surface of wet aerosol particles. The gas phase reaction (Equation 4)

$$SO_{2(gas)} + \text{Oxidants} \text{ ---> } SO_{3(gas)} \text{ ---> } SO_4^{2-}{}_{(aq)} \tag{4}$$

is irreversible so isotopic equilibrium with sulfate is not obtained ($\underline{24}$). Instead kinetic isotope effects are expected. Sulfate formed from gas phase oxidation should have a positive discrimination and be isotopically lighter than SO_2 ($\underline{24}$). On the other hand, sulfate formed from the aqueous phase oxidation of SO_2 should exhibit equilibrium isotope effects and be isotopically heavier than SO_2.

Isotope ratios have been used with some success in the past to determine the importance of gas phase (Equation 4) verses aqueous phase (Equations 2A,2B,2C) oxidation of SO_2. Saltzman et al. ($\underline{24}$) compared the $\delta^{34}S$ values for SO_2 and sulfate from samples collected from Hubbard Brook Experimental Forest (HBEF) in the non-urban northeastern US. They found discriminations which were intermediate to those expected for the individual oxidation mechanisms and suggested that both gas and aqueous phase oxidation were important. Newman ($\underline{50}$) found that $\delta^{34}S$ values for SO_2 in the plume of an oil fired power plant decreased with distance (and time) from the stack which they attributed to equilibrium isotope effects.

<u>Summary of Biogenic Sulfur Emissions from the Ocean.</u> To summarize the source fractionation patterns for biogenic sulfur emissions from the ocean, a simple isotopic model has been constructed (Figure 1). From this, the $\delta^{34}S$ value of sulfate from a marine biogenic source of atmospheric sulfur can be estimated.

Although limited, the data suggest that DMS from marine phytoplankton would most likely be enriched in the heavier isotope and would have a $\delta^{34}S$ value slightly less than seawater sulfate as a result of ASR. Until the $\delta^{34}S$ values for seawater DMS, remote SO_2 and sulfate are actually measured or until the uncertainties which surround the removal pathways for DMS and its atmospheric oxidation are addressed, questions remain as to the $\delta^{34}S$ value of atmospheric sulfate from this source.

Sea Spray Sulfate. A non-biogenic marine source of atmospheric sulfur results from sulfate in seawater being directly injected into the atmosphere. With no fractionation during this sea- air transfer, atmospheric seasalt sulfate particles have a $\delta^{34}S$ value of 21 o/oo (51). The amount of atmospheric sulfate derived from seawater is generally estimated from the sodium to sulfate ratio in seawater, and sulfate which exceeds this amount is termed "non-seasalt" sulfate. Seasalt sulfate can also be distinguished from the sulfate which forms from gas-to-particle conversions on the basis of particle size. Seasalt sulfate generally forms large particles, with diameters greater than one micrometer (um), while the sulfate from gas phase reactions is found on sub- micrometer particles.

Non-Marine Sulfur Sources. Non-marine sources of atmospheric sulfur to the remote atmosphere require long distance transport from continental areas. The following section summarizes the isotopic data from anthropogenic, volcanic, and terrestrial biogenic sulfur sources.

Anthropogenic Sulfur Emissions. Anthropogenic sulfur emissions are the largest source of sulfur dioxide to the atmosphere (2). Many studies have reported $\delta^{34}S$ values for sulfur dioxide in non-remote air masses. Delta values for SO_2 ranged from -1.1 to +2.3 o/oo at HBEF in the non-urban northeastern US (24), from 0 to +6 near Chicago, Illinois (52) and from -1 to -2 o/oo in Miami, Florida [E. Saltzman, pers. commun.]. Delta values for SO_2 from oil fired power plants averaged +5 o/oo, while those from coal fired plants ranged from +1.3 to +3.6 o/oo (47,50). These data suggest that sulfur from anthropogenic emissions has $\delta^{34}S$ values between -2 and +6 o/oo prior to atmospheric oxidation.

Volcanic Sulfur Emissions. During quiescent periods, the majority of the volcanic sulfur which enters the atmosphere comes from deep crustal and mantle outgassing and has $\delta^{34}S$ values in the range of -10 to +10 o/oo (54). Isotopic studies from inland geothermal areas in New Zealand reported the $\delta^{34}S$ value for hydrogen sulfide at +8.5 o/oo (55) and +5 o/oo (56). The $\delta^{34}S$ values for sulfur emissions during volcanic eruptions have not been measured. However, Castleman et al. (57) reported the isotopic composition of sulfate in stratospheric aerosols increased from + 2.6 o/oo to as much as +13 o/oo subsequent to eruptions which inject sulfur directly into the stratosphere. These $\delta^{34}S$ values decreased over time, which they attributed to the oxidation of SO_2 to sulfate.

Terrestrial Biogenic Sulfur Emissions. Soils and vegetation are sources of hydrogen sulfide and DMS to the atmosphere (58,59). The isotopic composition of these sulfur compounds upon release into the atmosphere varies, depending on the original source of sulfate, and the degree of bacterial activity in the soil. Krouse et al. (61) found that H_2S released from soil and vegetation samples at various sites had $\delta^{34}S$ values between -32 and -6 o/oo. SO_2 from samples collected within the same site had $\delta^{34}S$ values which were 2 to 3 o/oo lighter than the H_2S, which they attributed to kinetic isotope effects from the atmospheric oxidation of H_2S.

Little or no fractionation accompanies the uptake of sulfate in soils by plants during ASR (60,61). Chukhrov et al. (60) showed that in cases where atmospheric sulfate is not subject to bacterial reduction in the soil, the $\delta^{34}S$ value of the plant sulfur was identical to rainfall sulfur. In soils subject to dissimilatory sulfate reduction, the $\delta^{34}S$ value of plant sulfur differed from that of local rainfall. Additionally, Chukhrov et al. (60) found that plants from oceanic islands had a sulfur content with higher $\delta^{34}S$ values than those from continental areas, which they attribute to the relative influence of marine sulfate to these areas.

Summary of Non-Marine Sulfur Emissions. $\delta^{34}S$ values for continental sources of atmospheric sulfur dioxide vary, ranging between -32 and +10 o/oo. This makes it difficult to use sulfur isotope ratios to distinguish sulfate from these individual sources. It appears that the $\delta^{34}S$ value for marine biogenic sulfur is much more enriched in the heavier isotope than sulfur from continental origins. Therfore, it should be possible to isotopically distinguish between marine biogenic and continentally-derived sulfur.

Non-Seasalt Sulfate

Atmospheric Aerosol Sulfate. Isotope measurements of non-seasalt sulfate in marine aerosols (24,52,63) require that sulfate from sea spray be either physically or mathematically removed from the sample medium. Mathematically, mass balance relationships are used to correct the $\delta^{34}S$ value for the presence of seasalt sulfate in the sample. Physical means employ impactors or cyclone separators to segregate particles based on size so that $\delta^{34}S$ value for non-seasalt sulfate can be directly measured.

Several studies have reported $\delta^{34}S$ values for non-seasalt sulfate in aerosols collected in the marine atmosphere, which have varying degrees of continental influence (Table I). Gravenhorst (62) reported $\delta^{34}S$ values for sulfate aerosols collected over the N. Atlantic Ocean ranging from +7 to +13 o/oo (neglecting one value of -12 o/oo). Ludwig (60) reported $\delta^{34}S$ values for sulfate aerosols from the San Francisco Bay area ranging from -10 to +12 o/oo, however, unusually low $\delta^{34}S$ values (between +9.7 to +17 o/oo) were reported for seawater sulfate. Aerosol sulfate from the Mauna Loa Observatory in Hawaii had a $\delta^{34}S$ values of +5 o/oo [W.H. Zoller and W.R. Kelly; pers. commun.]. And finally, aerosol sulfate collected in Miami, Florida ranged from +1 to +2 o/oo [E. Saltzman, pers. commun.].

Rainwater Sulfate. Rain is the main pathway for removal of sulfate aerosol particles from the atmosphere (2). The isotopic composition of sulfate in rain has been well studied in several different regions (Table I). Jensen et al. (64) reported $\delta^{34}S$ values for rain of +14 o/oo in non-industrial regions of Japan and + 6 o/oo in industrial regions. Cortecci et al. (65) reported $\delta^{34}S$ values for sulfate in rain from Pisa, Italy ranging between -3 and +7 o/oo. Ostland (63) found $\delta^{34}S$ values near +2 o/oo for sulfate from rain in non-remote areas of North America and Scandinavia. Rainwater from New Zealand coastal regions collected during onshore flows had $\delta^{34}S$ values averaging +16.3 o/oo (55). Mizutani et al. (64) measured $\delta^{34}S$ values for sulfate in rain from coastal areas which ranged from -1.5 to +19.4 o/oo and found a direct correlation between the $\delta^{34}S$ values of sulfate and the seasalt content of the rainwater.

Several studies have demonstrated the relative contributions of natural versus anthropogenic sulfur to a given area based on $\delta^{34}S$ values for rainwater sulfate. Grey and Jensen (14) compared the $\delta^{34}S$ values of SO_2 to those in rainwater sulfate before and after a strike by smelter plant workers. The data showed that non-smelter sources were almost as important as industrial sources to the study area on a seasonal basis.

Recently, Nriagu and co-workers (15) used the isotopic composition of sulfur in weekly bulk precipitation samples to show that biogenic sulfur sources accounted for over 30% of the acidifying sulfur burden at a remote location in northern Ontario. Their data suggested that significant re-emission of anthropogenic sulfur to the atmosphere was occurring. Re-emission entails sulfate deposition in bogs, marshes and wetlands, the biological reduction of sulfate to H_2S (under anaerobic systems) or DMS (under aerobic conditions) followed by transfer of the reduced sulfur to the atmosphere. The atmospheric oxidation of this sulfur to sulfate continues the cycle, prolonging the acidification of even remote environments.

The effects of anthropogenic sulfur dioxide on the remote marine atmosphere may be evident from rainwater studies by Chukhrov et al. (60) in which the isotopic composition of sulfur in rain was studied systematically at great distances from the continent. Rainwater sulfate ranged from +12.1 to +15.0 o/oo over the Atlantic and from +9.5 to +16.2 o/oo over the Pacific, with a one month average $\delta^{34}S$ value of +13.3 o/oo for the two oceans. Their study included measurements of rainwater sulfate from a wide variety of continental areas and found that most inland $\delta^{34}S$ values were significantly lower than those over the oceans. The oceanic rainfall sulfate was most likely a mixture of the isotopically lower continental sulfate and the more enriched marine sulfate.

Summary and Future Research

Existing data suggest that the isotopic signature from continental sulfur emissions is significantly lower than seasalt and marine biogenic sulfate. However, the isotopic interpretation of the remote atmospheric sulfur cycle is seriously limited by the absence of $\delta^{34}S$ measurements for several critical compounds. Both laboratory and field experiments are needed to understand the isotopic signature for marine biogenic sulfate. The questions to be addressed include:

a) What is the $\delta^{34}S$ value for DMSP and DMS from laboratory cultures? What is the isotopic discrimination that occurs during assimilatory sulfate reduction by phytoplankton?

b) What is the $\delta^{34}S$ value for seawater DMS and how does this compare with laboratory $\delta^{34}S$ values? Does this $\delta^{34}S$ value suggest fractionation as a result of photochemical oxidation or bacterial degradation?

c) What is the $\delta^{34}S$ value of DMS in the remote atmosphere? Is there fractionation during the sea-air transfer?

d) What are the $\delta^{34}S$ values for SO_2, methane sulfonate, and non-seasalt sulfate from the oxidation of DMS? Do $\delta^{34}S$ values for non-seasalt sulfate from this natural source exhibit variations, spatially and/or seasonally, in the remote atmosphere?

It is already possible to measure the $\delta^{34}S$ values for seawater sulfate, DMSP, and seawater DMS, but only minimal data exist. Coordinated measurements of these compounds, along with simultaneous $\delta^{34}S$ measurements of atmospheric DMS, SO_2, methane sulfonate and non-seasalt sulfate, from atmospheres free of continental influence, are needed. These data will help in the isotopic interpretation of sulfur sources so that their relative contributions to the remote atmosphere can be assessed.

Acknowledgments

The authors are grateful for the reviews and communications of R. Charlson, D. Covert, B. Fry, J. Johnson, R. Kelly, P. Quinn, E. Saltzman, and B. Zoller. This project was funded in part by the Washington Sea Grant Program. Contribution number 1055 from the Pacific Marine Environmental Laboratory.

Literature Cited

1. Ryaboshapko, A.G. In The Global Biogeochemical Sulphur Cycle; Ivanov, M.V.; Freney, J. R., Eds.; John Wiley & Sons Pub.: N.Y., 1983; pp 278-96.

2. Andreae, M.O. In The Biogeochemical Cycling of Sulfur and Nitrogen in the Remote Atmosphere; Galloway, J.; Charlson, R.; Andreae, M.; Rhode, H., Eds.; Reidel Pub. Co., 1985; pp 5-25.

3. Andreae, M. O. In The Role of Air-Sea Exchange in Geochemical Cycling; Buat-Menard, Ed.; Reidel: Dordrecht, 1987; pp 331-62.

4. Bates, T.S.; Cline, J. D.; Gammon, R. H.; Kelly-Hansen, S. J. Geophys. Res. 1987, 92, 2930-8.

5. Berresheim, H. J. Geophys. Res. 1987, 92, 13245-62.

6. Lovelock, J.; Maggs, R.; Rasmussen, R. Nature 1972, 237, 452-3.

7. Coakley, J. A.; Bernstein, R. L; Durkee, P. A. Science 1987, 237, 1020-2.

8. Bates, T. S.; Charlson, R. J.; Gammon, R. H. Nature 1987, 329, 319-21.

9. Shaw, G. E. Atmos. Environ. 1987, 21, 985-6.

10. Charlson, R. J.; Lovelock, J. E.; Andreae, M. O.; Warren, S. G. Nature 1987, 326, 655-61.

11. Charlson, R. J.; Rhode, H. Nature 1982, 295, 683-95.

12. Fry, B.; Krouse, H. R. Food Web Studies with Sulfur Stable Isotopes; 1987; in press.

13. Fuller, R. D.; Mitchell, J. J.; Krouse, H. R.; Wyskowski, B. J. Water, Air, and Soil Pollution 1986, 28, 163-71.

14. Grey, D. C.; Jensen, M. L. Science 1972, 177, 1099-100.

15. Nriagu, J. O.; Holdway, D. A.; Coker, R. D. Science 1987, 237, 1189-92.

16. Peterson, B. J.; Fry, B. Ann. Rev. Ecol. Syst. 1987, 18, 293-320.

17. Kaye, J. Rev. Geophys. 1987, 25, 1626-8.

18. Sakai, H. Geochim. Cosmochim. Acta. 1957, 12, 150-60.

19. Thorton, E. K.; Thorton, E. R. ACS Monograph 167; Collins, C. J.; Bowman, N. S., Eds.; 1970; pp 213-80.

21. Krouse, H. R. (1980) In Handbook of Environmental Isotope Geochemistry; Fritz, P.; Fontes, J. Ch., Eds.; Elseverier: Amsterdam, 1980; Vol. 1.

22. Urey, H. C. J. Chem. Soc. 1947, 562-81.

23. Davidson, J.; Cantrell, C.; Tyler, S.; Shetter, R.; Cicerone, R.; Calvert, J. J. Geophys. Res. 1987, 92, 2195-9.

24. Saltzman, E. S.; Brass, G.; Price, D. Geophys. Res. Let. 1983, 10, 513-6.

25. Rees, C. E. Geochim. Cosmochim. Acta. 1978, 42, 383-9.

26. Puchelt, H.; Sabels, B. R.; Hoering, T. C. Geochim. Cosmochim. Acta. 1971, 35, 625-8.

27. Paulsen, P. J.; Kelly, W. Anal. Chem., 1984, 56, 708-713.

28. Kim, K; Andreae, M. O. J. Geophys. Res. 1987, 92, 14733-8.

29. Saltzman, E.; Cooper, D. J. Atmos. Chem. 1988, in press.

30. Ferek, R. J.; Andreae, M. O. Nature 1984, 307, 148-50.

31. Johnson, J.; Harrison, H. J. Geophys. Res. 1986, 91, 7883-8.

32. Sze, N.; Ko, M. Nature 1979, 280, 308-10.

33. Andreae, M. O. Limnol. Oceanogr. 1980, 25(6), 1054-63.

34. Cantoni, G.; Anderson, D. J. Biol. Chem. 1956, 222, 171-7.

35. Rees C. E.; Jenkins, W. J.; Monster, J. Geochim. Cosmochim. Acta. 1978, 42, 377-81.

36. Kaplan, I. R. (1963) In Biogeochemistry of Sulfur Isotopes; Proc. Nat. Sci. Found. Sym., 1962: Yale Univ., 1963; pp 80-93.

37. Kaplan, I. R.; Rittenberg, S. J. Gen. Microbiol. 1964, 34, 195-212.

38. Chambers, L. A.; Trudinger, P. A. Geomicrobiol. J. 1979, 1, 249-93.

39. Calhoun, J. C.; Zoller, W. H.; Charlson, R. J.; Kelly, W. R. Isotope Analysis for Understanding the Importance of DMS in the Production of Excess Sulfate over the Remote Oceans. Solid Source Mass Spectrometry: The Technique and its Application. 194th National Meeting of the American Chemical Society; Biogenic Sulfur in the Environment Symposium, New Orleans; 1987, 261-2.

40. Ammons, J. Ph.D. Thesis, Univ. of S. Florida, Ann Arbor, MI, 1980.

41. Liss, P. (1973) Deep Sea Res. 1973, 20, 221-38.

42. Broecker, W. S.; Peng, T. H. Tellus 1974, 29, 21-35.

43. Peng, T. H.; Takahashi, T.; Broecker, W. S. J. Geophys. Res. 1974, 79, 1772-80.

44. Wilke, C. R.; Chang, P. A.I.Ch.E. J. 1955, 1, 264-70.

45. Saltzman, E.; Savoie, D.; Prospero, J.; Zika, R. J. Geophys. Res. 88, 10,897-902.

46. Grosjean, D.; Lewis, R. Geophys. Res. Let. 1982, 9, 1203-6.

47. Hatakeyama, S.; Akimoto, H. J. Phys. Chem. 1983, 87, 2387-95.

48. Yin, F.; Grosjean, D.; Seinfeld, J. J. Geophys. Res. 1986, 91, 14,417-38.
49. Monson, K. D.; Hayes, J. M. J. Biol. Chem. 1980, 255, 11,435-41.
50. Newman, L.; Forrest, J.; Manowitz, B. Atmos. Environ. 1975, 9, 959-68.
51. Luecke, W.; Nielsen, H. Fortschr. Min. 1972, 50 (Beiheft3), 36-7.
52. Holt, B.; Engelkemeir, A.; Venters, A. Environ. Sci. and Technol. 1972, 6, 338-41.
53. Newman, L.; Forrest, J.; Manowitz, B. Atmos. Environ. 1975, 9, 969-77
54. Nielsen, H. Tellus 1974, 26, 213-21.
55. Spedding, D.; Cope, D. Atmos. Environ. 1984, 18, 2703- 6.
56. Robinson, B.; Sheppard, D. New Zealand. J. Volcanol. Geotherm. Res. 1986, 27, 135-51.
57. Castleman, A., Jr.; Munkelwitz, H.; Manowitz, B. Tellus 1974, 26, 222-34.
58. Goldan, P.; Kuster, W.; Albritton, D.; Fehsenfeld, F. J. Atmos. Chem. 1987, 5, 439-62.
59. Lamb, B.; Westberg, H.; Allwine, G.; Banesberger, L.; Guenther, A. J. Atmos. Chem. 1987, 5, 469-91.
60. Chukhrov, F. V.; Ermilova, L. P.; Churikov, V.; Nosik, L. Org. Geochem. 1980, 2, 69-75.
61. Krouse, H. R.; Van Everdingen, R. Water, Air and Soil Pollution 1984, 23, 61-7.
62. Gravenhorst, G. Atmos. Environ. 1978, 12, 707-13.
63. Ludwig, F. L. Tellus 1976, 28, 427-33.
64. Jensen, M.; Nakai, N. Science 1961, 134, 2102-4.
65. Cortecci, G.; Longinelli, A. Earth Planet. Sci. Lett. 1970, 8, 36-40.
66. Mizutani, T.; Rafter, T. New Zealand J. Sci. 1969, 12, 69-80.
67. Ostlund, G. Tellus 1959, 11, 478-80.
68. Nriagu, J.; Coker, D. Tellus 1978, 30, 365-75.

RECEIVED September 15, 1988

Chapter 23

Two Automated Methods for Measuring Trace Levels of Sulfur Dioxide Using Translation Reactions

Purnendu K. Dasgupta, Liu Ping, Genfa Zhang, and Hoon Hwang

Department of Chemistry and Biochemistry, Texas Tech University, Lubbock, TX 79409-1061

Two disparate translation methods are investigated for the measurement of sulfur dioxide. Both involve interaction with an aqueous solution. In the first, collected S(IV) is translated by the enzyme sulfite oxidase to H_2O_2 which is then measured by an enzymatic fluorometric method. The method is susceptible to interference from $H_2O_2(g)$; efforts to minimize this interference is discussed. The second method involves the translation of SO_2 into elemental Hg by reaction with aqueous mercurous nitrate at an air/liquid interface held in the pores of hydrophobic membrane tubes. The liberated mercury is measured by a conductometric gold film sensor. Both methods exhibit detection limits of ~ 100 pptv with response times under two minutes. Ambient air measurements for air parcels containing sub-ppbv levels of SO_2 show good correlation between the two methods.

At the present time, there is a lack of simple, affordable methods capable of measuring background concentrations of sulfur dioxide in an automated fashion. As a minimum, such methods should have a limit of detection (LOD) of 200 pptv, with a temporal resolution not exceeding a few minutes. Among the more recently developed methods, cryogenic preconcentration followed by GC/MS analysis has a claimed LOD of 1 pptv, with a temporal resolution of 3 minutes, if isotopically labeled SO_2 is used for calibration ([1]). Aside from the cost and complexity of such instrumentation, considerable efforts must be directed towards removing water vapor from the sample air prior to cryogenic collection and such predryers exhibit memory effects far in excess of the claimed LOD. In another recently developed approach, Cvijn et al ([2]) have reported an LOD of 200 pptv using pulsed UV laser photoacoustic spectroscopy; attainable temporal resolution was not specified. The stated detection limit corresponds to an S/N ratio of 1 and aside from affordability, the instrumentation is not well-suited for field deployment. (In accordance with the guidelines of the American Chemical Society ([3]), the limit of detection throughout the rest of this paper, unless otherwise specifically stated, connotes a signal value equal to three times the standard deviation of the blank, over the blank.) Most recently, modified pulsed-fluorescence SO_2 instruments have become available (Model 43S, Thermo Environmental Instruments, Inc); with

0097–6156/89/0393–0380$06.50/0

an integration time of 3 minutes, an LOD of < 100 pptv can be obtained (4). However, no intercomparison study has thus far been conducted to determine the applicability in real sampling situations; the instrument responds to NO to an extent 0·022 times its sensitivity for SO_2.

Chemical derivation methods designed to determine an analyte A by reaction with some reagent R and resulting in a product P are generally such that the product contains one or more of the original atoms or functional groups belonging to, and generally characteristic of, the analyte A. It is product P which is actually measured. Translation methods constitute a special subclass in which more than one product is formed in the derivatization reaction and the product measured does not contain any atoms of the original analyte. Thus, the analyte translates the reagent to the measured product. In order for such a method to be attractive, the reaction must exhibit the necessary specificity and the product should be detectable with excellent sensitivity. Translation methods are quite common in clinical analysis. Indeed, assay of urinary or blood glucose, in the form of test-strips, is based on such a translation reaction. The product actually detected is H_2O_2. Enzymatic mediation endows the measurement with the necessary specificity. There are a variety of oxidase enzymes which function by oxidizing the specified substrate with molecular oxygen and simultaneously producing a stoichiometric amount of H_2O_2

$$\text{glucose Oxidase}$$
$$(\text{e.g., Glucose} + O_2 \text{ -----------------> D-Gluconic acid} + H_2O_2).$$

Sulfite oxidase (SOD) is a hepatic enzyme responsible for detoxifying S(IV) produced endogenously in mammalian systems due to the catabolism of sulfur containing amino acids. SOD is one of four known molybdenum-Center mammalian enzymes. SOD functions by oxidizing S(IV) with oxygen to produce sulfate and H_2O_2. Commercially available preparations of SOD are derived from chicken liver (5). There are at least two reports in the literature concerning the determination of S(IV) with SOD. The first involves an immobilized enzyme reactor used in a flow injection configuration with amperometric detection of the H_2O_2 produced (6). Concentrations of S(IV) down to 10 μM could be measured. In another approach, SOD was immobilized on a dissolved oxygen electrode and the depletion of oxygen in the solution, as measured by the electrode, was taken as a measure of S(IV) concentration (7).

The basic chemistry of this process was attractive to us because in recent years a number of excellent approaches have been developed to achieve very high quotients (gas phase flow rate/liquid phase flow rate) in continuous gas-liquid scrubber designs; these lead to excellent liquid phase concentrations (8-11). In particular, the diffusion scrubber (D.S.), a membrane based diffusion denuder (12-19), was developed in this laboratory and with admitted bias, we believe it to be superior to available alternatives (19). Further, we have been interested in the measurement of trace levels of aqueous H_2O_2 and through a series of developments (20-23), inexpensive automated analytical systems capable of quantitating nanomolar levels of H_2O_2 have evolved. The chemistry involves the oxidation of p-hydroxyphenylacetate by H_2O_2 at pH 5.5, mediated by the enzyme peroxidase, to 5,5'-dihydroxybiphenyl-2,2'-diacetic acid and then raising the pH sufficiently to ionize the phenolic protons. The ionized product fluoresces intensely (λ_{ex} 329 nm, λ_{em} 412 nm). This liquid phase chemistry was coupled to a diffusion scrubber for the measurement of atmospheric H_2O_2 (16,18,19). Such an instrument is essentially directly applicable for SO_2 measurement, provided that an additional step, involving the SOD mediated translation of collected S(IV) to H_2O_2, is incorporated.

The second translation method involves the translation of SO_2 into elemental Hg. The strategy utilizes the disproportionation reaction of mercurous ion into mercuric ion and elemental mercury, driven thermodynamically by the high formation constant (24) of the sulfonato complex of mercury (II): $Hg_2^{2+} + H_2O + SO_2 \longrightarrow HgSO_3H^+ + Hg$. The reaction was first used by Marshall and Midgley (25,26) and subsequently commercialized by Severn Science Limited (U.K.). The commercial version claims an LOD of 200 pptv (S/N = 2) with a response time of ~ 2 minutes. The instrument utilizes a modified bubbler through which a solution of mercurous nitrate is made to continuously flow, while sample air bubbles through the liquid and also carries the liberated mercury to the detector. The problem stems from the fact that an optical detector (cold vapor atomic absorption) is used to measure the absorption due to elemental Hg at 253.7 nm. Such a detector is particularly susceptible to interferences from water vapor. Consequently, pains must be taken to remove the water vapor, resulting in much the same shortcomings involved in the predrying step in cryogenic SO_2 collection. In fact, the problem is in some ways worse than the cryogenic collection situation because the stream to be dried in the present case is obligatorily saturated with water vapor.

Additionally, generation/liberation of mercury from the fairly large mass of bubbler solution (at least 5 mL) is a slow process and the equilibrium values of elemental Hg liberated from any given solution of mercurous nitrate are both substantial and acutely temperature dependent.

Drying remains a necessity with atomic fluorescence detection for Hg (27). Drying also had to be used with a gold-coated piezoelectric mass-sensitive crystal detector (28) and this scheme does not provide adequate sensitivity (estimated LOD ~ 500 pptv SO_2 based on stated LOD of 80 pg SO_2 and stated air flow rate of 55 mL/min) and temporal resolution (10 minute zero gas purging of bubbler solution required between samples). Our strategy to use this translation reaction for the determination of SO_2 centered on a commercially available humidity-insensitive conductometric gold film sensor (Arizona Instruments, Inc., Jerome Division). Mercury is adsorbed on gold with extremely high specificity and results in a marked change in electrical resistance. Two identical gold film patterns are deposited on a ceramic substrate and constitutes two arms of a Wheatstone bridge. In some proprietary manner, one of the film patterns is made insensitive to mercury and this serves as the reference arm. The use of the reference arm compensates for temperature variations and permits an LOD of ~ 300 pg Hg. Initially, substantial efforts were made to carry out the translation reaction on an impregnated filter, e.g., a glass fiber mat impregnated with a mercurous salt formulation and a humectant (29). Although detectabilities well below 100 pptv were obtainable, batch to batch filter performance showed marked variability and impregnated filters could not be stored over a month-long period without loss of performance. We therefore carried out the desired reaction using a diffusion scrubber: air was sampled through a porous hydrophobic membrane tube while a dilute mercurous nitrate solution was circulated on the outside of the membrane tube. Mercury liberated by reaction at the gas-liquid interface in the pores is carried by the air stream to the detector.

Experimental details, performance characteristics and the results of an ambient air intercomparison study are presented in this paper for these two disparate translation methods for measuring low levels of gaseous SO_2.

Experimental

Figure 1 shows the diffussion scrubbers used in these studies. The D.S.

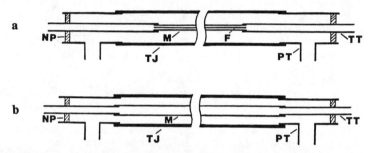

Figure 1.   (a) Diffusion scrubber used for the H$_2$O$_2$-translation system.
(b) Diffusion scrubber used for the Hg-translation system.

schematically shown in (a) is our present design of choice where the intent is to concentrate a gas-phase analyte into the liquid phase with subsequent analysis of the liquid phase. Briefly, the heart of the device is the microporous polypropylene tube M (Celgard X-20, 400 μm i.d., 25 μm wall, 40% surface porosity, 0.02 μm mean pore size, Hoechst-Celanese Corporation, Charlotte, NC) filled with a 300 μm dia. nylon monofilament (F) and connected to end-flared 300 μm i.d. in/out PTFE tubes (TT) through which the scrubber liquid (in the present case, water) is pumped. The membrane tube is suspended concentrically within a PTFE jacket tube TJ (5 mm i.d. and maintained in a linear configuration by an external rigid tube, not shown). Air is sampled through the tee-ports PT and end-seals are provided by a series of nested plastic tubes (NP). Greater details of construction are available elsewhere (19). This D.S. was used for the H_2O_2 translation system (hereinafter called system 1) with an active membrane length of 40 cm.

The second D.S., used in the Hg translation system (hereinafter called system 2) and shown in Figure 1(b), utilizes a reverse flow configuration. Air flows through the membrane tube M (type TA002 Gore-Tex porous PTFE tube, 2 mm i.d., 0 4 mm wall, 50% surface porosity, 2·5 μm pores; W. L. Gore and Associates, Elkton, MD) maintained concentrically within a linear PTFE Jacket tube TJ. Air is sampled through the connecting PTFE tubes TT (16 SW, Zeus Industrial Products, Raritan, NJ) while liquid (dilute acidic mercurous nitrate solution) is pumped through the tee-ports PT.

The gas phase pre-D.S. setup for both systems is shown schematically in Figure 2. Zero air, governed by a mass flow controller MFC (FC 280, Tylan Corp., Carson, CA) flows through a temperature controlled enclosure (30 ± 0·5°C) CT in which a SO_2 permeation wafer PW (VICI Metronics, Santa Clara, CA; two different devices were used in this work and were gravimetrically calibrated to be emitting 63 and 20 ng SO_2/min respectively) is maintained. Valve V1 is a 3-way PFA-Teflon Solenoid Valve (Galtek Corp., Chaska, MN) used to select the calibrant stream or ambient air for sampling. Similarly, Valve V2, identical to V1, either allows the V1-selected stream to proceed directly or removes the SO_2 from it by passing the stream through a soda-lime filter F. The valve ports are so chosen that the unenergized positions of V1 and V2 correspond respectively to ambient air and direct passage to the scrubber. For further precautions that should be taken with these valves to prevent sample degradation, see ref. 19. The valves are controlled by a microprocessor-governed timer (Chrontrol model CD-4S, Lindberg Enterprises, San Diego, CA).

The flow schematic for system 1 is shown in Figure 3. A household aquarium pump P, modified to aspirate air, provides an air sampling rate of 1 L/min in conjunction with mass flow controller MFC. Water is pumped through the D.S. at 84 μL/min and is merged with a 16 μL/min SOD stream (composition described later in this section) immediately after the D.S. and proceeds through a 1 m long line L1 (0.3 mm i.d., 30 SW PTFE tube, $t_R \sim 42$ s) used in a knotted configuration to promote better mixing (20). All reactions occured at room temperature (22°C). Following the SOD mediated translation of collected S(IV) to H_2O_2, p-hydroxyphenylacetic acid, PHPA (Eastman Kodak, recrystallized, 0·08% w/v filtered solution) and peroxidase (Sigma Chemical, Type II, 0·1 mg/mL in 0·1 M KH_2PO_4 buffer adjusted to pH 5·0, filtered) were added immediately following each other. A second reaction line L2 (25 cm, 0·3 mm i.d.) provided a reaction time of ~ 8.5 s and then the stream passed through a cation exchange membrane reactor R containing concentrated NH_3 (23) to raise the pH. The resulting fluorescence was detected by a filter fluorometer FD (Fluoromonitor III, Laboratory Data Control, Riviera Beach, FL) equipped with a Cd-excitation lamp (325 nm interference filter for

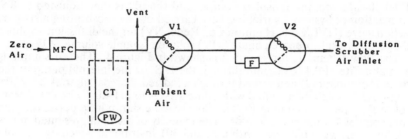

Figure 2. Pre-diffusion scrubber experimental arrangement.

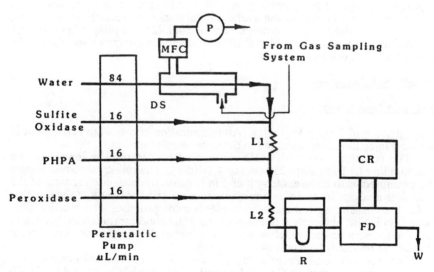

Figure 3. Analytical flow schematic, H$_2$O$_2$-translation system.

excitation, 370 nm high-pass emission filter) and recorded by a strip-chart recorder CR. A peristaltic pump (Minipuls 2, Gilson Medical Electronics, Middleton, WI) equipped with a 4-channel head was used for pumping the liquids. The internal diameters of the pump tubing were 0·0075 in. for all channels except for the water channel for which a 0 025 in. i.d. tube was used. The SOD reagent was prepared by diluting 250 units of the enzyme derived from chicken liver (Sigma Chemical) to 10 mL (stock solution, stored at 4°C). Immediately before using, 2 mL of the stock solution is diluted to 10 mL with pH 9·0 10 mM sodium phosphate buffer and the pH is then adjusted to 8·5. Optimization of System 1 variables were carried out by substituting a six-port injection valve (HVLX 6-6, Hamiton Co., Reno, NV) for the diffussion scrubber and injecting freshly prepared dilute S(IV) solution (sample volume ~ 17 μL).

The flow schematic for system 2 is shown in Figure 4. The air sample is drawn at a rate of 0.4-0.6 L/min through the D.S. by the internal pump of the Jerome Hg monitor (J) and vented through a iodized charcoal filter. A 2.5 x 10^{-6} M Hg$_2$(NO$_3$)$_2$ solution, acidified with dilute HNO$_3$ to pH 4.0 is circulated from a bottle B by peristaltic pump P. The flow scheme shown here pertains to the Jerome model 421 mercury monitor. The majority of the data presented in this work however, were obtained with a model 301 instrument which uses a gold coil for initial collection of the Hg vapor followed by flash desorption to deliver the collected mercury to the sensor proper. The flow scheme, especially in regards to temporal programming, is slightly different. This instrument is no longer available. However, the overall performance characteristics attainable with either model are comparable. The model 421 sampling sequence was controlled by a Radio-Shack Model 100 portable microcomputer and the results printed out by a dot-matrix printer.

Results and Discussion

Peroxide Translation System.

Optimization of System Variables. All optimization efforts were carried out in the flow injection mode, as stated in the experimental section.

The reaction time with SOD was varied by varying the length of the reaction line L1 from 10 to 250 cm (t$_R$ 4.2-105 s). The results, shown in Figure 5a, prompted us to choose a length of 1 m because the response lag time of the complete analyzer increases in direct relationship with the length of the reaction line. The chosen length therefore represents a compromise. The reaction time necessary is also obviously a function of the SOD concentration used. It would be nice to use a shorter reaction line and a higher SOD concentration, were it not for the relatively high cost of the enzyme.

There is some differences in literature reports regarding the optimum pH for the enzymatic action of SOD. MacLeod et al (30) and Cohen and Fridovich (31) report optimum pH levels of 8.5-8.6 while Smith (7) found best results at pH ~ 7 for enzyme-coated electrodes. For the immobilized enzyme, Masoom and Townshend (6) reported optimum performance between pH 8.5 and 9.0. Our results, shown in Figure 5b, largely confirms the original reports (30,31); in these experiments, the buffer composition in the peroxidase reagent were altered to maintain the same pH for the second enzymatic reaction step. We have subsequently found however, that the quality and the behavior of the enzyme purchased in different batches can vary significantly. At least one enzyme batch performed optimally at a pH substantially lower than pH 8.5, lending credence to the report of Smith (7).

While operation at an elevated temperature increases the rate of the translation reaction mediated by SOD (6,7), the enzyme itself is rapidly

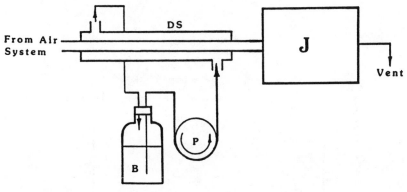

Figure 4. Analytical flow schematic, Hg-translation system.

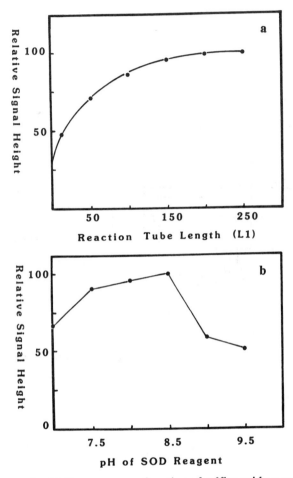

Figure 5. (a) Response as a function of sulfite oxidase reaction time.
(b) Performance of sulfite oxidase as a function of pH.

deactivated. Even at room temperature (N 22°C) we found the rate of loss of activity to be ~ 5%/h and therefore found it necessary to keep the SOD solution in ice or refrigerated during use; this reduced the decay rate to ~ 1%/h.

Utilization as a Liquid Phase S(IV) Analysis System. Under the optimum experimental conditions as described in the experimental section, the liquid phase analytical system provides a simple, convenient way to measure S(IV) in precipitation samples. Typical system performance is shown in Figure 6 for a 20 μM S(IV) sample repeatedly injected over a 4-hour period. Although the detection limit of the system is ~ 0.1 μM, the performance data are shown for this relatively high concentration level because of the difficulty of maintaining a stable concentration of S(IV) at much lower levels. Note that formaldehyde, which often accompanies S(IV) in precipitation samples (32), does interfere in the S(IV) determination because of the formation of the stable adduct, hydroxymethanesulfonate (33). However, the alkaline condition of the SOD reaction is capable of at least partly dissociating hydroxymethanesulfonate. At a S(IV) level of 10 μM, the the formaldehyde interference is negligible at concentrations $\leq$ 5 μM and increases nonlinearly at higher concentrations. Thus, a simple expedient to determine the presence of any interference from HCHO is to dilute the sample and redetermine the S(IV) concentration.

For the described system, the sample throughput rate is ~ 25 samples/h with < 1% carryover. The response plot is best described in terms of a second order equation, as described later for gaseous SO₂.

Performance as a Gas Phase Analysis System. With the stated liquid flow rates and an air sampling rate of 1 L/min, the attainable LOD ranged from 78 to ~ 170 pptv (based on the noise level with zero air and signal height with 680-2000 pptv calibrant) during the course of the study. This variation in S/N performance was in large part due to the differences in activity of the SOD enzyme from lot to lot, even though the vendor supplied each lot as ostensibly containing the same units of activity. The signal rise time (10% to 90%) was 0.96 min and the lag time (time from switching the sample from zero air to SO₂ to the first onset of the signal peak) was 1.82 min. Purely liquid phase experiments in the flow injection mode show the response time and lag time to be 0.43 and 1.37 min respectively in that system. The response speed is adequate for most purposes; although an improvement in the LOD will be desirable. Albeit not explicitly attempted during the present experiments, a number of simple changes can substantially improve the observed LOD. Since the efficiency of collection of an analyte gas by a D.S. is independent of the scrubber liquid flow, the attainable LOD improves in a direct relationship with the total liquid flow rate. For low blank reaction systems such as the present one, scrubber liquid flow rates down to 30 μL/min are easily permissible before one perceives any significant dependence of response behavior upon relative humidity (19). Merely scaling down the liquid flow rates and appropriately decreasing the length of reaction lines (so as not to unduly increase the lag and rise times) can be used to improve the LOD by a factor of at least 2.5. Further for this particular D.S. system, doubling the air sampling rate to 2 L/min is known to increase the amount of SO₂ collected by a factor of 1.2 (19). Thus, a three-fold improvement of LOD should be attainable without major modifications.

The use of water as a scrubber liquid deserves some discussion. We previously found (19) that the extent of oxidative loss occuring with such a scrubber liquid when collecting SO₂ is relatively irreproducible and therefore used very dilute formaldehyde solutions for the purpose (albeit the use of the

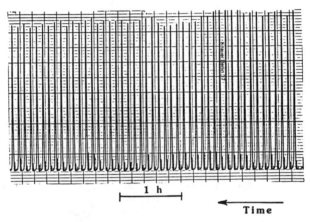

Figure 6. System performance: Repetitive analysis of 20 μM S(IV).

latter scrubber solution is not without problems). More recent experiments reveal that the use of very high purity water essentially eliminates oxidative loss; specifically, a particularly convenient and effective measure is to use a small column of a mixed bed ion exchange resin immediately ahead of the D.S.

The response of the system to SO$_2$ is linear at high levels (ca. > 1 ppbv) but nonlinear at lower levels. This is a characteristic of the reaction system since the same behavior is exhibited by the liquid phase analysis system. Figure 7 shows this nonlinearity at low SO$_2$ levels for the gas phase analyzer. A second-degree equation (e.g., for the data shown, $Y = aX^2 + bX + c$ produces excellent fit, correlation coefficient > 0.999) and may be used for calculations. It should be noted that both the measures suggested above for improving the LOD actually result in an increase of the analyte concentration and thus do not involve an increased need for manipulation in the nonlinear response region.

Whenever any analytical instrument is operated near the limits of its capabilities, considerations of baseline drift (zero drift) becomes important. With D.S. based instrumentation, our preferred mode of operation is to utilize alternate sample/zero periods. Instrument performance at the 2 ppbv SO$_2$ level is shown in Figure 8 (protocol: 2 min sample, 8 min Zero).

Interferences. The strategy of enzymatic translation of SO$_2$ into H$_2$O$_2$ results in a highly detectable product and the enzymatic specificity of the translation reaction constitute the principal merits of the approach. Therein lies its vulnerability as well: interference from H$_2$O$_2$, the translation product, concurrently present in the sample gas.

In order to test the effects from H$_2$O$_2$, a membrane-based H$_2$O$_2$(g) generation system (19,34) was used. In the absence of SO$_2$, the response of the instrument to H$_2$O$_2$ ranged from 1.1 to 1.2 times compared to SO$_2$, on an equivalent mixing ratio basis. This could be solely attributed to the higher diffusion coefficient of H$_2$O$_2$ and the consequently better collection efficiency for H$_2$O$_2$ as compared to SO$_2$. For the present type of D.S., (membrane length 40 cm) we have previously reported the collection efficiency of H$_2$O$_2$ to be 1.1 times better than that for SO$_2$ at a sampling rate of 1 L/min (19). Experiments were also conducted in which 0 to 3 ppbv H$_2$O$_2$ (g) were added to 4 and 7 ppbv SO$_2$. The extent of the observed interference was lower than that expected on the basis of the results obtained in absence of SO$_2$. For example, the response due to 4 ppbv SO$_2$ increased 33%, 67% and 100% upon addition of 1, 2, and 3 ppbv H$_2$O$_2$ respectively (this translates to an interference equivalent 0.75 ppbv H$_2$O$_2$ = 1 ppbv SO$_2$) while the response due to 7 ppbv SO$_2$ increased 4.2% and 9.7% upon addition of 1 and 2 ppbv H$_2$O$_2$ respectively (which translates to an average interference equivalent 3 ppbv H$_2$O$_2$ = 1 ppbv SO$_2$). The reason for this decreased interference level may be due to a loss by mutual reaction either within the mixing manifold or after collection in the aqueous phase; however, available rate data for the H$_2$O$_2$-S(IV) reaction (35) do not suggest that the aqueous phase reaction can be important. Another possibility is that the prior presence of H$_2$O$_2$ itself affects the rate of the SOD mediated translation reaction. In any case, these results would suggest that serious errors may be encountered when determining low levels of SO$_2$ ($\leq$ 5 ppbv) in the presence of significant amounts of H$_2$O$_2$ unless appropriate measures are taken.

Attempts were therefore made to catalytically destroy H$_2$O$_2$ by incorporating a small packed bed reactor immediately after the D.S. and before the introduction of SOD. We have previously used microsized reactors packed with granular MnO$_2$ (18,21,23) for the catalytic decomposition of H$_2$O$_2$. Experiments in the flow injection mode showed that using a large amount of MnO$_2$ (ca. 25 mg) in the reactor resulted in not only decomposition of H$_2$O$_2$ but also the oxidation of more than half of the collected S(IV). Reducing the

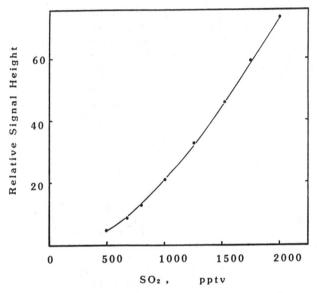

Figure 7. Response to low SO_2 levels: H_2O_2-translation system.

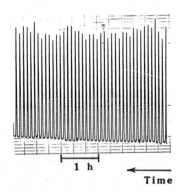

Figure 8. Performance of H_2O_2-translation system: 2 ppbv SO_2.

amount of MnO$_2$ in the reactor was found to improve the selectivity to a degreee but was not judged good enough to be worthy of further pursuit. Metallic silver, also known to catalytically destroy H$_2$O$_2$ (36), provided even less selectivity when used as a catalytic reactor in the liquid phase.

Substantially better results were obtained with attempts to remove H$_2$O$_2$ in the gas phase. The use of an internally silvered glass tube ahead of the D.S. inlet was particularly convenient. At the stated sampling rate of 1 L/min, such a denuder (6 mm i.d.) reduced the signal by 85.4, 95.5 and 97.6% for respective denuder lengths of 15, 25 and 40 cm when sampling a 2.6 ppbv H$_2$O$_2$ source. The same 15 and 25 cm silver denuders respectively reduced the signal by only 7.0 and 10.1% when sampling a 2.2 ppbv SO$_2$ source. A mixture of 2.0 ppbv SO$_2$ with 2.5 ppbv H$_2$O$_2$ showed a 108% increase in signal compared to 2.0 ppbv SO$_2$ alone, but incorporation of the 15 cm silver denuder prior to the D.S. when sampling the same mixture resulted in a signal equal to 1.9 ppbv SO$_2$. It is thus felt that deploying a silvered denuder in the gas phase can be used to largely ameliorate the H$_2$O$_2$ interference problem.

Mercury Translation System

Optimization of System Variables. The dependence of the blank level and the total signal (blank + analyte response) on the liquid flow rate is shown in Figure 10. The conditions are the same as those for Figure 9 except that 10^{-6} M Hg$_2$(NO$_3$)$_2$ at pH 4 is used. Down to the lowest flow rate studied (1500 μL/min), the net response to ~ 5 ppbv SO$_2$ is essentially constant. Unfortunately, this flow rate dependence was examined fairly late in the study and the other data reported here were obtained with a liquid flow rate of 2600 μL/min. It is clear, however, that down to at least 1500 μL/min, the response/blank ratio improves; it may be advantageous to use a lower flow rate. This behavior also strongly suggests that the transport of mercury from the bulk solution (liberated due to the intrinsic disproprotionation equilibrium) to the carrier air stream is controlled by liquid phase mass transfer.

The dependence of the blank value and the total signal for a ~ 7 ppbv SO$_2$ sample (other conditions as for Figure 10) on the air sampling rate is shown in Figure 11. The blank increases linearly with the air flow rate, suggesting its direct influence on interfacial mass transfer of mercury at the liquid-air interface in the pores. The net signal (the difference between the solid and the dashed line) reaches a maximum around a flow rate of ~ 0.6 L/min and slowly decreases thereafter. This behavior is solely due to the kinetics of the translation reaction; the collection efficiency for SO$_2$ for such a D.S. should be > 98% up to a flow rate of 1 L/min based on previously published data (13). Based on these results, we estimate that the translation efficiency of SO$_2$ into Hg at an air flow rate of 0.6 L/min is ~ 30% or ~ 60% depending on whether the monosulfonato or disulfonato complex of Hg(II) is being formed (24). An air flow rate of 0.57 L/min was used for the other data reported here. The flow rate of the Jerome 421 instrument is preadjusted by the manufacturer to 0.4 L/min; this can be altered by replacing an internal voltage regulator that governs the pump.

The response of the analytical system to various concentrations of SO$_2$ as a function of the Hg$_2$(NO$_3$)$_2$ concentration is shown in Figure 12 (other conditions as in Figure 11). It should be noted that the blank values are not directly proportional to the Hg$_2$(NO$_3$)$_2$ concentration; we find that a 2.5 μM solution displays a blank 1.7 times higher than a 1 μM solution and the latter is only marginally higher than the blank from a 0.5 μM solution. As may be intuitively apparent, the primary effect of increasing the Hg$_2$(NO$_3$)$_2$ concentration is to extend the applicable dynamic range to higher values; the

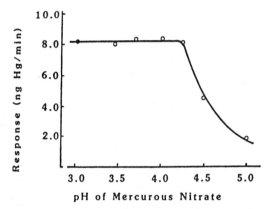

Figure 9. Response of Hg-translation system as a function of the pH of Hg₂(NO₃)₂ solution.

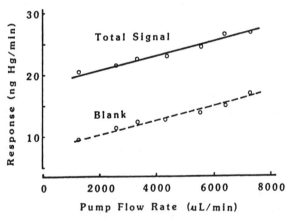

Figure 10. The dependence of the Hg-system response upon liquid flow rate.

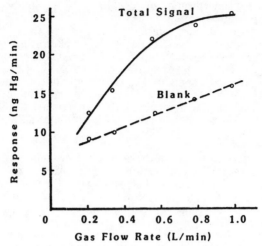

Figure 11. The dependence of blank and response values upon air sampling rate.

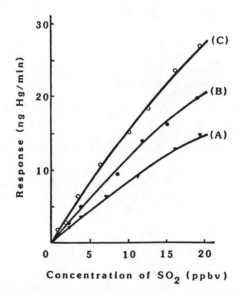

Figure 12. The response of the Hg-translation system to various SO_2 levels as a function of $Hg_2(NO_3)_2$ concentration. A,B,C:0.5, 1, 2.5 μM $Hg_2(NO_3)_2$.

calibration slope also increases with increasing $Hg_2(NO_3)_2$ concentration. While we have chosen to use a concentration of 2.5 μM $Hg_2(NO_3)_2$ solution to be able to routinely measure high ambient values of SO_2 (ca. up to 25 ppbv) it is obvious that for measurement of background levels of SO_2 ($\leq$ 10 ppbv), a 1 μM solution may be more suitable because of decreased blank values.

The effect of the membrane length upon the blank and the calibration slope are shown for a 1 μM $Hg_2(NO_3)_2$ solution in Figure 13. Both blank and response increase with increasing membrane length but plateaus after about 0.5 m. The response/blank ratios for the 0.1, 0.2, 0.3, 0.5 and 1.0 m lengths are 0.44, 0.53, 0.61, 0.73 and 0.73 in arbitrary units respectively. Since the blank noise was found to be approximately related to the absolute value of the blank, we chose a length of 0.5 m.

Interferences. The effect of temperature was studied by equilibrating the inlet sample air through a thermal equilibration coil (25 - 35°C). The results are shown in Figure 14. The blank remained constant while the response increased linearly at a rate of 2.5%/°C. This temperature dependence is far smaller than that observed with a bubbler. Since air temperature is easily measured, routine corrections can also be made.

Interference testing for specific chemical species have been made for this reaction by Marshall and Midgley for the liquid phase reaction (25) and by Suleiman and Guilbault (30) for the gas phase reaction. The first set of investigators found no significant interferences. The latter study, which looked at NH_3, HCl, NO_2, CO, CO_2, $COCl_2$, C_2H_3Cl, H_2S, O_3 and HNO_3 at very high levels (2-320 ppmv) concluded that only H_2S at levels of 20 ppmv or higher may pose a problem. We therefore limited our study to H_2S and CH_3SH. No measurable effects on the blank value could be detected with 14 ppbv H_2S and 12 ppbv CH_3SH. There was no measurable effect of either H_2S or CH_3SH, up to a concentration of 5 ppbv each on the response due to 10 ppbv SO_2. However, increasing the concentration to 8.5 ppbv H_2S depressed the response of 10 ppbv SO_2 by 4.5%. Similarly, a 1.3% decrease of response was observed for 10 ppbv SO_2 when 6.5 ppbv CH_3SH was added. The precision of the determination at the 10 ppbv SO_2 level is sufficiently good to accurately determine the stated change.

It does not appear that the method is likely to suffer from any serious potential interference.

Performance. In routine operation, the LOD varied from 60 to 150 pptv on a day to day basis for a 1 minute sample and was slightly better for the model 301 instrument than the model 421 instrument. Longer sampling times improve the LOD; typically the LOD improves in proportion to the square root of the sampling time. The total time the instrument can operate continuously before the gold film is saturated (and must be rejuvenated) depends on the blank levels and the levels of SO_2 measured. Under the typical operating conditions with SO_2 levels $\leq$ 10 ppbv, the model 421 instrument requires film regeneration (10 min) every 2 hours. While the model 301 instrument could operate for more than 8 hours under the same conditions, the regeneration time is longer (26 min).

Intercomparison Studies

Ambient air experiments were conducted in Lubbock, TX in the winter of 1987; the two instruments shared a common manifold. No attempts were made to remove or correct for interference due to H_2O_2 in system 1. A set of representative results are shown in Figure 15 and a correlation plot is shown in

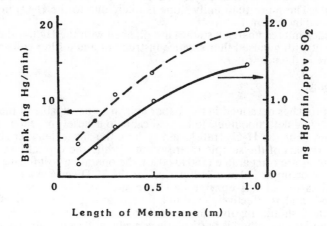

Figure 13. The effect of membrane length upon the response and blank values.

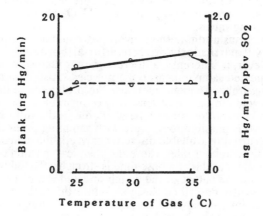

Figure 14. The effect of sample air temperature for the response and blank values.

Figure 16. The data shown in Figure 15 are hourly averages and the bars surrounding each point represent the variation (± 1 s.d.) of the measured values during the hour; the standard deviation of individual measurements are far smaller. The correlation, as shown in Figure 16, is good (correlation coefficient 0.88; System 1 SO_2 = 1.162 x System 2 SO_2 + 59 pptv) with a relatively small intercept. The more than unity slope is likely due to the H_2O_2 interference experienced by System 1.

Further intercomparison studies are planned with the Hg translation based instrument with a pulsed fluorescence instrument and a filter collection based procedure in the near future.

Conclusions

Both approaches described in this paper appear to be capable of the necessary specificity and detection limits to be applicable to a variety of real measurement situations. For the H_2O_2-translation system, the silver denuder should be a mandatory part of the sample inlet system. While the capital investment for both systems are comparable ($5000-6000), the operating cost for the enzymatic system can be up to $50 in a 24 h period for the SOD enzyme alone. Since the immobilization of this enzyme has been successfully carried out and the supported catalyst effectively used in a flow system (6), future efforts to utilize this method should involve an immobilized enzyme reactor. Pursuing the H_2O_2-translation method is particularly convenient if one is already using a gas phase H_2O_2 measuring instrument such as that described in (19) because only minor modifications will be necessary to measure SO_2. In principle, a time-shared approach to measure both SO_2 and H_2O_2 should be feasible with appropriate switching valves.

The mercury translation method involves a compact instrument package and thus makes it particularly attractive for airborne applications. The instrument, however, is inherently a discrete, rather than continuous analyzer, operating on repeated sample-measurement cycles. Also, film regeneration is necessary at undesirably frequent intervals. According to the manufacturer, a truly continuous gold film sensor of sensitivity comparable to presently available instrumentation is possible and may be available in the future. It should be noted that while it is possible to operate with sampling times as short as one minute with the currently available discrete analyzers, the actual response time is controlled by the scrubber inlet where the chemical reaction occurs. Present experiments indicate that this response time is strongly affected by the details of scrubber design and the nature of the membrane and can be easily as long as 5 min. for the presently described design. While this does not appear to be prohibitively large for most applications involving measurement of background levels of SO_2, improved scrubber designs utilizing membranes of smaller wall thickness may be necessary for the reduction of response time, of potential benefit to airborne sampling applications.

For the user interested in choosing between the two methods (when utilization of instrument components already in the laboratory is not a consideration), we favor the Hg-translation method. We further suggest that for airborne applications, the total costs involved may well justify the use of two gold film sensors such that one detector is operative when the other is undergoing film regeneration. It is not necessary that two completely independent detectors be used, the instrument electronics can be modified to perform film regeneration on one sensor with a separate film heating power supply while the other sensor is being used.

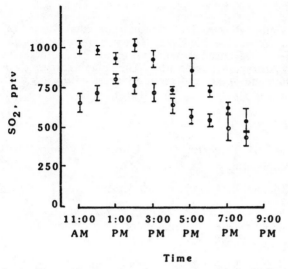

Figure 15. Ambient Air data; hourly averages with variations indicated as ± 1 s.d. Closed circles: H_2O_2-translation system; open circles: Hg-translation system.

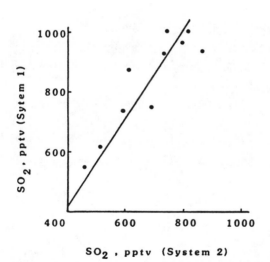

Figure 16. Correlation plot: H_2O_2 (system 1) vs Hg (system 2) translation methods.

Acknowledgments

This research was supported by the U.S. Environmental Protection Agency through CR812366-010, the Electric Power Research Institute through RP1630-28, the State of Texas Advanced Research Program on Synthetic membrane Technology and by Arizona Instruments Inc. The opinions expressed herein are those of the authors and do not represent endorsement by any of the above agencies.

Literature Cited

1. Driedger, A. R., III; Thornton, D. C.; Lalevic, M.; Bandy, A. R. Anal. Chem. 1987, 59, 1196-1200.

2. Cvijn, P. V.; Gilmore, D. A.; Leugers, M. A.; Atkinson, G. H. Anal. Chem. 1987, 59, 300-4.

3. American Chemical Society Committee on Environmental Improvement. Anal. Chem. 1980, 52, 2242-9.

4. Appel, D. Thermo Environmental Instruments, Inc., Franklin, MA. Personal Communication, 1988.

5. Kessler, D. L.; Rajagopalan, K. V. J. Biol. Chem. 1972, 247, 6566-73.

6. Masoom, M.; Townshend, A. Anal. Chim. Acta, 1986, 179, 399-405.

7. Smith, V. J. Anal. Chem. 1987, 59, 2256-9.

8. Abbas, R. and Tanner, R. L. Atmos. Environ. 1981, 15, 277-81.

9. Lazrus, A. L.; Kok, G. L.; Lind, J. A.; Gitlin, S. N.; Heikes, B. G.; Shetter, R. E. Anal. Chem. 1986, 58, 594-7.

10. Cofer, W. R., III; Edahl, R. A., Jr. Atmos. Environ. 1986, 20, 979-84.

11. Vecera, Z.; Janák, J. Anal. Chem. 1987, 59, 1494-8.

12. Dasgupta, P. K. Atmos. Environ. 1984, 18, 1593-9.

13. Dasgupta, P. K.; McDowell, W. L.; Rhee, J.-S. Analyst (London) 1986, 111, 87-90.

14. Dasgupta, P. K.; Phillips, D. A. Sep. Sci. Technol. 1987, 22, 1255-67.

15. Tanner, R. L.; Markovitz, G. Y.; Ferreri, E. M.; Kelly, T. J. Anal. Chem. 1986, 58, 1857-65.

16. Kleindienst, T. E.; Shepson, P. B.; Hodges, D. N.; Nero, C. M.; Arnts, R.R.; Dasgupta, P. K.; Hwang, H.; Kok, G. L.; Lind, J. A.; Lazrus, A. L.; Mackay, G. I.; Mayne, L. K.; Schiff, H. I. Environ. Sci. Technol. 1988, 22, 53-61.

17. Kleindienst, T. E.; Shepson, P. B.; Nero, C. M.; Arnts, R. R.; Tejada, S.B.; Mackay, G. I.; Mayne, L. K.; Schiff, H. I.; Lind, J. A.; Kok, G. L.; Lazrus, A. L.; Dasgupta, P. K.; Dong, S. Atmos. Environ. (in press).

18. Dasgupta, P. K.; Dong, S.; Hwang, H. J. Aerosol Sci. Technol. (in press).

19. Dasgupta, P. K.; Dong, S.; Hwang, H.; Yang, H.-C.; Genfa, Z. Atmos. Environ., 1988, 22, 949-63.

20. Hwang. H.; Dasgupta, P. K. Anal. Chim. Acta, 1985, 170, 347-52.

21. Dasgupta, P. K.; Hwang, H. Anal. Chem. 1985, 57, 1009-12.

22. Hwang, H.; Dasgupta, P. K. Mikrochim. Acta, 1985, III, 77-87.

23. Hwang, H.; Dasgupta, P. K. Anal. Chem. 1986, 58, 1521-4.

24. Dasgupta, P. K.; DeCesare, K. B. Atmos. Environ. 1982, 16, 2927-34.

25. Marshall, G.; Midgley, D. Anal. Chem. 1981, 53, 1760-5.

26. Marshall, G.; Midgley, D. Anal. Chem. 1982, 54, 1490-4.

27. Lang, H.; Jiang, L.; Ren, D.; Xie, Z. Huanjing Kexue, 1984, 5(4), 56-8.

28. Suleiman, A. A.; Guilbault, G. G. Anal. Chem. 1984, 56, 2964-6.

29. Hisamatsu, Y.; Dasgupta, P. K. Unpublished Studies, Texas Tech University, 1987.

30. MacLeod, R. M.; Farkas, W.; Fridovich, I.; Handler, P. J. Biol. Chem. 1961, 236, 1841-6.

31. Cohen, H. J.; Fridovich, I. J. Biol. Chem. 1971, 246, 359-66.

32. Dong, S.; Dasgupta, P. K. Environ. Sci. Technol. 1987, 21, 581-8.

33. Dasgupta, P. K.; Dong, S. Atmos. Environ. 1986, 20, 1635-7.

34. Hwang, H.; Dasgupta, P. K. Environ. Sci. Technol. 1985, 19, 255-8.

35. Martin, L. R.; Damschen, D. E. Atmos. Environ. 1981, 15, 1615-21.

36. Quagliano, J. V. Chemistry, 2nd ed.; Prentice-Hall Inc.: Englewood Cliffs, NJ. 1963; p 110.

RECEIVED July 30, 1988

THE ATMOSPHERE:
GAS-PHASE TRANSFORMATIONS

Chapter 24

Gas-Phase Atmospheric Oxidation of Biogenic Sulfur Compounds

A Review

John M. C. Plane

Division of Marine and Atmospheric Chemistry,
Rosenstiel School of Marine and Atmospheric Science,
University of Miami, Miami, FL 33149-1098

This Chapter is a review of the recent gas-phase kinetic studies
that have been carried out to elucidate the oxidation chemistry of
reduced sulfur species in the troposphere. It also serves as an
introduction to the following chapters in this section of the
Symposium Series. Each of these chapters is concerned either
with investigations of specific elementary reactions in the
oxidation mechanisms, or with measurements of the overall
product yields of these mechanisms. The purpose here is to
provide an overview which links these investigations together at a
more general level; the reader is referred to the relevant chapters
for detailed background and discussion. At the same time,
material outside the papers presented at the symposium has been
included here to produce a broader picture of the oxidation
chemistry of these species. Finally, in reviewing this subject the
discrepancies between the results of different studies and
experimental techniques are discussed and evaluated; in the case
of multiple reported determinations of the rate coefficient and
Arrhenius parameters of a particular elementary reaction, a
recommended value is given.

The oxidation of reduced sulfur compounds of biogenic origin such as the thiols
(RSH, R = alkyl group), sulfides (RSR) and disulfides (RSSR) is an important
source of SO_2, sulfate and methanesulfonic acid (MSA) in the troposphere.
The contribution from this biogenic source to the global budget of SO_2 and
sulfate has been estimated to be comparable to anthropogenic emissions of
SO_2, and may therefore provide a strong background source of acid deposition
(1). In particular, there is the need to understand the regional impact of strong
localized biogenic sources of these compounds, through the use of quantitative
models. Thus, a complete elucidation of the kinetics of the pathways by which
the sulfur in these compounds is oxidised from the S(-II) to S(+IV and +VI)
states is crucial to the full understanding of the tropospheric chemistry of sulfur.
In this chapter the significant progress that has been made towards this goal will
be reviewed, and this will serve as an introduction to the detailed accounts of
experimental measurements which occur in the following chapters of this
section.

0097–6156/89/0393–0404$06.00/0
© 1989 American Chemical Society

Here we will consider the oxidation chemistry of the five most abundant sulfur compounds which are of biogenic origin and undergo substantial oxidation in the troposphere: dimethyl sulfide (CH_3SCH_3, DMS), methanethiol (CH_3SH), dimethyl disulfide (CH_3SSCH_3, DMDS), carbon disulfide (CS_2) and hydrogen sulfide (H_2S). First, the primary reactions between these species and their most important tropospheric oxidants, namely, the hydroxyl radical (OH), the nitrate radical (NO_3), and various halogen species, will be discussed. This will lead on to the chemistry of intermediate radical species such as the methyl thiyl radical (CH_3S), and stable intermediates and products such as dimethyl sulfoxide (CH_3SOCH_3, DMSO), dimethyl sulfone ($CH_3SO_2CH_3$, DMSO$_2$), SO_2, sulfate and MSA (CH_3SO_3H). Figure 1 is a schematic of the reaction pathways which are discussed below, and is provided as a reference point for the reader.

Where appropriate, recommended rate coefficients and Arrhenius parameters will be listed for the elementary reactions under discussion. The following procedure has been used for obtaining these recommendations. If more than two determinations of the rate coefficient of a particular reaction at 298 K are reported, the majority of those that agree within experimental error are used to obtain an unweighted mean value; this yields the recommended rate constant at 298 K. The recommended activation energy is obtained from the unweighted mean of the activation energies reported in the studies used to obtain the recommended rate constant at 298 K. Where an exception is made, usually where the temperature range used to determine the activation energy in a study is relatively small, thus leading to greater inaccuracy, this will be pointed out in the text. The pre-exponential Arrhenius A-factor is then computed from the recommended activation energy and the recommended rate constant at 298 K.

Oxidation Chemistry of Reduced Sulfur Species

Dimethyl Sulfide. DMS is one of the major reduced sulfur compounds produced by biological activity, mainly in the oceans (2,3). For this reason its oxidation reactions have been intensively studied, especially the reaction

$$OH + DMS \rightarrow products \tag{1}$$

The eventual products in reaction (1) have been identified as SO_2 and MSA from experiments involving the steady photolysis of mixtures of DMS and a photolytic precursor of OH (4-9). Absolute measurements of k_1 have been obtained using the discharge-flow method with resonance fluorescence or electron paramagnetic resonance (EPR) detection of OH (10-14), and the flash photolysis method with resonance fluorescence or laser induced fluorescence (LIF) detection of OH (14-18). Competitive rate techniques where k_1 is measured relative to the known rate constant for a reaction between OH and a reference organic compound (18-21) have also been employed to determine k_1 at atmospheric pressure of air.

In the absence of O_2, the reported measurements of k_1 (see Table I) have shown significant discrepancies, varying by a factor of three from 3.2 to 9.8 x 10^{-12} cm^{3}molecule^{-1}s^{-1}. The earlier measurements using flash photolysis (14,15) and discharge-flow (10,11) fell into a group with k_1(298 K) $\approx$ (8 - 9.8) x 10^{-12} cm^{3}molecule^{-1}s^{-1} and a small negative temperature dependence. More recent work, again using flash photolysis (16-18) and discharge-flow (12,13), has been grouped with $k_1 \approx 4$ x 10^{-12} cm^{3}molecule^{-1}s^{-1} and a zero or slightly positive temperature dependence of k_1. Wine et al. (16) established that, in the flash photolysis studies, the photolysis of mixtures containing DMS at $\lambda < 165$ nm led

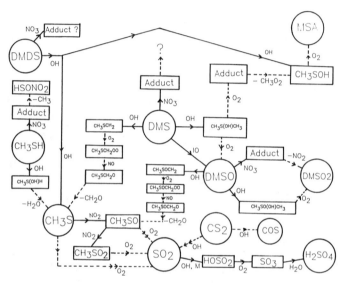

Figure 1. A schematic diagram of the oxidation chemistry of the reduced sulfur compounds DMS, CH_3SH, DMDS and CS_2, illustrating the chemistry discussed in this review. The solid lines indicate reaction pathways for which absolute rate constants have been measured, the broken lines indicate reation pathways for which there is circumstantial evidence mainly derived from product study analyses.

Table I. Reactions of Dimethyl Sulfide

Reaction/ Technique[*]	Temp. Range K	10^{12} k(298 K)/ $cm^3 molecule^{-1}s^{-1}$	10^{12} A/ $cm^3 molecule^{-1}s^{-1}$	$E_{act}/R(K)$	Reference
OH + DMS	(Absence of O_2)				
DF-EPR	373-573	9.2 ± 0.6 (373K)	5.1	-176 ± 151	MacLeod et al.[10,11]
DF-EPR	273-318	3.22 ± 0.16	-	-	Martin et al.[12]
FP-RF	300-427	9.9 ± 1.2	5.47	-179 ± 151	Atkinson et al.[14]
FP-RF	273-400	8.28 ± 0.87	6.25 ± 4.19	-131 ± 215	Kurylo[15]
FP-RF	297-400	3.9 ± 0.2	2.5 ± 0.9	-130 ± 102	Wallington et al.[18]
FP-RF	248-363	4.26 ± 0.56	6.8 ± 1.1	274 ± 91	Wine et al.[16]
CK-GC	300	5.3 (ref.: OH + cyclohexane)	-		Wallington et al.[18]
FP-RF	276-397	4.46 ± 0.66	13.6 ± 4.0	332 ± 96	Hynes et al.[17]
PLP-LIF	298	4.8	-	-	Hynes et al.[17]
CK-GC/FTIR	300	4.4 ± 0.4 (ref.: OH + ethene)		-	Barnes et al.[20]
DF-RF	260-393	5.5 ± 1.0	11.8 ± 2.2	236 ± 150	Hsu et al.[13]
Recommended Values:		4.50	11.6	281	

OH + DMS	(In air and approximate atmospheric pressure)				
CK-GC	297	9.78 ± 1.54 (ref.: OH + ethene)		-	Cox and Sheppard[19]
CK-GC	296±2	9.7 ± 1.0 (ref.: OH + n-hexane)		-	Atkinson et al.[21]
CK-GC	298	9.3 ± 0.7 (ref.: OH + n-hexane)		-	Wallington et al.[18]
CK-GC	298	8.5 ± 0.3 (ref.: OH + cyclohexane)		-	Wallington et al.[18]
CK-GC/FTIR	298	8.0 ± 0.5 (ref.: OH + ethene)		-	Barnes et al.[20]
PLP-PLIF	298	6.3 ± 0.3 -		-	Hynes et al.[17]

Recommended Value: 6.3
Recommended temperature dependence of k_1(760 torr air):[17]

$$k_1 = \frac{Texp(-234)/T) + 8.46x10^{-10}exp(7230/T) + 2.68x10^{-10}exp(7810/T)}{1.04x10^{11} + 88.1exp(7460/T)}$$

Recommended branching ratio for reaction 1:[17]

$$k_{1a}/k_1 = 9.6x10^{-12}exp(-234/T)/k_1$$

NO_3 + DMS					
CK-FTIR	298	0.96 ± 0.13	-	-	Atkinson et al.[21,24]
FP-A	280-350	0.75 ± 0.05	0.47 ± 0.26	-170 ± 130	Wallington et al.[26]
FP-A	298	0.81 ± 0.13	-	-	Wallington et al.[27]
MP-A	278-315	1.0 ± 0.2	1.0 ± 0.2	0	Tyndall et al.[25]
FT-LIF	257-376	1.06 ± 0.13	0.179 ± 0.022	-530 ± 40	Dlugokencky and Howard[24]
Recommended Values:		0.96	0.30	-350	

IO + DMS					
SP-FTIR	296	30 ± 15	-	-	Barnes et al.[31]
DF-MS	296	15 ± 5	-	-	Martin et al.[32]
Recommended Value:		22.5			

to the higher rate constants in the earlier studies (14,15). They repeated their study of reaction (1) using pulsed laser photolysis at λ = 248 nm of H_2O_2 to create OH, followed by LIF detection, and obtained essentially the same result (17). Martin and co-workers (12) demonstrated by coating their flow tube with a halocarbon wax that their earlier higher estimate of k_1 (10,11) was probably due to heterogeneous reactions on the quartz walls of the tube. The lower estimate for k_1 was recommended by Atkinson in an extensive review (22), and has since been reaffirmed in a study by Hsu et al. (13) who used discharge-flow with resonance fluorescence detection of OH, and by a competitive rate study by Barnes et al. (20). The recommended k_1 is the mean of the results from these later studies (12,13,16-18,20). The majority of these studies indicate that E_{act} is positive; the recommended value is taken from References (13), (16) and (17).

In the presence of an atmosphere of air, reaction (1) has been studied by several groups using relative rate techniques under conditions of steady photolysis (18-21), and by Hynes et al. (17) who employed the pulsed laser photolysis of H_2O_2 with LIF detection of OH. Inspection of Table I reveals that the direct time-resolved study (17) yielded a significantly lower result for k_1. Hynes et al. (17) discuss the likelihood of there being a secondary reaction channel for DMS in the competitive kinetics experiments which leads to the high rate constants. Indeed, this discrepancy is repeated for the cases of OH + CH_3SH and OH + CS_2 (see Table II). For this reason the rate constants under atmospheric conditions obtained by the direct technique are recommended for all three of these reactions.

Reaction (1) has been shown by Wine et al. (17) to proceed by two pathways:

Hydrogen-atom abstraction:

$$OH + CH_3SCH_3 \dashrightarrow H_2O + CH_3SCH_2 \qquad (1a)$$

Reversible addition:

$$OH + CH_3SCH_3 <==> CH_3S(OH)CH_3 \qquad (1b)$$

The addition complex formed in (1b) has also been shown to react rapidly with O_2 (17),

$$CH_3S(OH)CH_3 + O_2 \dashrightarrow products \qquad (2)$$

The rate constant for the addition pathway increases with decreasing temperature. Thus, at 300 K reaction (1) proceeds in air 25% via the addition pathway, which increases to about 70% at 250 K (see Table I) (17).

A number of product studies have been carried out on reaction (1) under atmospheric conditions (4-9,20). These studies used NOx containing precursors of OH in their steady photolysis systems, which Barnes et al. (20) have now demonstrated leads to complications in the subsequent kinetics. Barnes et al. (20) have utilized H_2O_2 as an alternative photolytic precursor of OH, and observed 70% production of SO_2 at room temperature. However, they also observed about a 30% yield of $DMSO_2$, which is inconsistent with a recent study by Hynes and Wine (this volume). The major products that have been observed in these product studies (4-9,20) have included SO_2, MSA, CH_2O and DMSO. It is still not clear whether each of the initial reaction pathways (1a) or (1b) leads to a single product or a mixture of these products. Hynes and Wine (this volume) discuss the likely products arising from the addition mechanism (1b)

Table II. Reactions of Methanethiol

Reaction/ Technique[*]	Temp. Range K	10^{12} k(298 K)/ cm^{3}molecule^{-1}s^{-1}	10^{12} A/ cm^{3}molecule^{-1}s^{-1}	E_{act}/R(K)	Reference
OH + CH$_3$SH	(absence of O$_2$)				
FP-RF	300-423	33.9 ± 3.4	8.89	-397 ± 151	Atkinson et al.[33]
FP-RF	244-366	33.7 ± 4.1	11.5 ± 3.9	-338 ± 102	Wine et al.[16]
FP-RF	254-430	31.8 ± 1.0	10.1 ± 1.9	-346 ± 59	Wine et al.[34]
PLP-LIF	270-300	33.0 ± 3.0	-	-	Hynes and Wine[35]
DF-EPR	293	21 ± 2	-	-	MacLeod et al.[10,11]
DF-RF	296	25.6 ± 4.4	-	-	Lee and Tang[36]
CK-GC	296	36.0 ± 4.0 (ref.: OH + propene)	-		Barnes et al.[20,37]
OH + CH$_3$SH	(In air and approximate atmospheric pressure)				
CK-GC	297±2	90.4 ± 8.5 (ref.:OH + ethene)	-		Cox and Sheppard[19]
CK-GC	298	124 - 140 (ref.:OH + propene, precursor: RONO)			Barnes et al.[37]
CK-GC	298	41 (ref.: OH + propene, precursor: H$_2$O$_2$)			Barnes et al.[20]
PLP-LIF	298	31.3			Hynes and Wine[35]
Recommended Values:		33.6	10.1	-360	
NO$_3$ + CH$_3$SH					
SP-FTIR	300	1.0 ± 0.3	-	-	MacLeod et al.[39]
FP-A	280-350	0.81 ± 0.06	0.1 $^{+.27}_{-.07}$	-600 ± 400	Wallington et al.[27]
FT-LIF	256-376	1.09 ± 0.13	1.09 ± 0.17	0 ± 50	Dlugokencky and Howard[24]
Recommended Values:		0.97	0.35	-300	

and subsequent reaction (2) in air; these include DMSO and $DMSO_2$ which may themselves be oxidized further (see Barnes et al., this volume), and possibly the direct production of MSA through the reaction sequence

$$CH_3S(OH)CH_3 + O_2 --> CH_3SOH + CH_3O_2 \qquad (2)$$

$$CH_3SOH + O_2 --> MSA \qquad (3)$$

Atkinson (22) suggested that the H-abstraction reaction (1b) would proceed in air as follows:

$$CH_3SCH_2 + O_2 --> CH_3SCH_2OO \qquad (4)$$

$$CH_3SCH_2OO + NO --> CH_3SCH_2O + NO_2 \qquad (5)$$

$$CH_3SCH_2O --> CH_3S + CH_2O \qquad (6)$$

The atmospheric chemistry of the alkyl thiyl (CH_3S) radical, probably leading to the production of SO_2, is discussed below and in the chapters by Tyndall and Ravishankara and by Hatakeyama.

Thus, the eventual products of the two pathways (1a) and (1b) by which OH reacts with DMS under atmospheric conditions have not yet been fully established. However, circumstantial evidence, namely the 70% yield of SO_2 in the study of Barnes et al. (21) being roughly equal to the branching ratio for the H-atom abstraction (reaction (1a)) at 298 K determined by Hynes et al. (17), indicates that SO_2 is a major product of the abstraction reaction and that DMSO, $DMSO_2$ and MSA probably result from the primary addition of OH to the sulfur atom (reaction (1b)).

Whilst OH is almost certainly the primary oxidant of sulfur species during the daytime in the troposphere, at nighttime its concentration becomes insignificant. Instead, the nitrate radical, which is photolyzed during the day but has been observed at concentrations in the range 0.3 - 450 pptv (1 pptv = 2.4 x 10^9 molecule cm^{-3}) at night, is a major nighttime oxidant (23). The reaction between DMS and the nitrate radical,

$$NO_3 + DMS --> products \qquad (7)$$

has been studied by a number of investigators (Table I), who have employed several different techniques including a fast flow tube with LIF detection of NO_3 (24), modulated and flash photolysis with resonance absorption detection at λ = 623, 662 nm (25-27), and a relative rate study in a smog chamber with detection of NO_3 by FTIR (21). The most recent study of reaction (7) is that of Dlugokencky and Howard (24), which also reviews the earlier work. There is good agreement among all of the studies except that of Wallington et al. (26). The recommended rate constant at T = 298 K in Table I is the mean of the other four studies. (Note: The relative rate constant from Atkinson et al. (21) is corrected from the original following a revised measurement of NO_3 + trans-2-butene (24).) The activation energy is computed from the studies of Dlugokencky and Howard (24) and Wallington et al. (27) which were over significantly larger temperature ranges than the study of Tyndall et al. (25).

Dlugokencky and Howard (24) concluded that the small negative temperature dependence that they observed (see Table I) suggests that the mechanism for this reaction involves addition of NO_3 to the sulfur atom

$$NO_3 + DMS --> [adduct]^{\neq} --> products \qquad (7a)$$

$$M \downarrow$$

$$adduct \qquad (7b)$$

where an excited adduct ($\neq$) is formed as in the case of $O(^3P)$ reactions with DMS, methanethiol and DMDS (28,29). Furthermore, since their low pressure (1 torr = 133 Nm^{-2}) measurements (24) of k_7 agree very well with the measurements of Wallington et al. (27) made at atmospheric pressure, the absence of any pressure effect indicates that, if the formation of a stable adduct in reaction (7b) is important, then the recombination reaction is close to its high pressure, second-order limit; this is quite likely given the number of vibrational degrees of freedom in the excited adduct.

Other possible pathways for the reaction between NO_3 and DMS include:

$$NO_3 + DMS --> HNO_3 + CH_3SCH_2 \qquad (7c)$$

$$--> DMSO + NO_2 \qquad (7d)$$

However, pathway (7c) involves the abstraction of a hydrogen atom from a methyl group by NO_3; such reactions are generally described by rate constants about three orders of magnitude smaller than k_7, and so this pathway is probably minor. In the case of reaction (7d), Tyndall et al. (25) failed to observe either NO_2 or DMSO as a reaction product; Dlugokencky and Howard (24) also did not observe the formation of NO_2 in this reaction and were able to impose an upper limit of 5% for this pathway.

Finally, the oxidation of DMS by halogen atoms and halogen oxides is considered. Following a model of iodine chemistry in the marine boundary layer by Chameides and Davis (30) which indicated concentrations of the IO radical in the 0.1 - 10 pptv range, the localized oxidation of DMS in the marine environment by IO could be significant (see the chapters by Barnes et al.). This became apparent after recent laboratory studies of the reaction

$$IO + DMS --> DMSO + I \qquad (8)$$

by Barnes et al. (31) and Martin et al. (32) showed that the reaction is fast (Table I) and leads quantitatively to the products indicated. These studies were performed using quite different techniques, namely, by steady-state photolysis of NO_2 in an excess of I_2 as a source of IO radicals (31), and by discharge-flow and mass-spectrometric detection (32), and are in very good agreement. Barnes et al. (this volume) also report the determination of the rate constant for the very rapid reaction between Cl atoms and DMS.

Methanethiol (Methyl Mercaptan). The reaction between OH and CH_3SH can proceed by three pathways:

$$OH + CH_3SH --> CH_2SH + H_2O \qquad (9a)$$

$$--> CH_3S + H_2O \qquad (9b)$$

$$--> CH_3S(OH)H \qquad (9c)$$

The rate constant k_9 has been determined by the flash photolysis technique with resonance fluorescence (16,33,34) or LIF (35) detection of OH; by the discharge flow method with detection of OH by EPR (10,11) or resonance fluorescence (36); and by three relative rate techniques (19,20,37). As in the case of OH + DMS, the relative rate results in the presence of O_2 are all significantly higher than those from the absolute studies (Table II). These higher values for k_9 cannot be ascribed to the reactions being carried out in air at 1 atmosphere, since Hynes and Wine (35) have demonstrated no pressure dependence of k_9 or enhancement in the presence of O_2. The four flash photolysis studies (16,33-35) and the relative rate study of Barnes et al. (20) (in the absence of O_2) are in excellent agreement and are somewhat higher than the discharge flow studies (10,11,36). Their mean yields the recommended value at 298 K and the negative temperature dependence of k_9 given in Table II.

Atkinson (22) has observed that the room-temperature rate constants and the temperature dependences of the reactions between OH and the thiols are essentially invariant of the alkyl group, including t-butanethiol, and that this demonstrates that H-atom abstraction from the C-H bonds must be a minor pathway. Hynes and Wine (this volume) conclude from the absence of a significant isotope effect for CH_3SH and CH_3SD (34), that reaction (9b), whilst possibly having a smaller activation energy than reaction (9a) because of the weaker S-H bond (38), is also a minor pathway. This evidence and the small negative temperature dependence of reaction (9) indicates that the addition pathway, reaction (9c), is probably the major channel.

Hatakeyama and Akimoto (6) have concluded from a product study of the irradiation of CH_3SH-RONO-NO-air mixtures that the adduct formed in reaction (9c), $CH_3S(OH)H$, either reacts with NO_x containing species to form CH_3SNO or can thermally decompose at room temperature:

$$CH_3S(OH)H \text{ --> } CH_3S + H_2O \qquad (10)$$

Since CH_3SNO will rapidly photolyze to yield CH_3S and NO, and the alkyl thiyl radical is thought to form SO_2 in the atmosphere (see below), the eventual product of reaction (9) is probably SO_2.

The reaction

$$NO_3 + CH_3SH \text{ --> } \text{ products} \qquad (11)$$

has been studied by three of the groups that also studied reaction (7) (24,27,39). Referring to Table II, there is excellent agreement between the work of Dlugokencky and Howard (24) and MacLeod et al. (39), whose results are ≈ 20% higher than those of Wallington et al. (27). The recommended value at 298 K is the mean of these three studies. Dlugokencky and Howard (24) have discussed the mechanisms of reaction (11) and observe that abstraction of either the methyl or thiol hydrogen atoms is very unlikely because reactions involving such abstractions are typically three orders of magnitude slower than k_{11} (see reaction (17), $NO_3 + H_2S$). Those authors (24) also did not observe NO_2 as a product of reaction (11), indicating that, as in the case of the analogous reaction of DMS, a simple O-atom transfer to the sulfur atom is not a major reaction pathway.

The only remaining mechanism is that of adduct formation (analogous to reaction (7c) for NO_3 + DMS), which is supported by the zero or small negative activation energy (Table II). Indeed, the study of MacLeod et al. (27), which employed the steady photolysis of a mixture of $N_2O_5/NO_2/CH_3SH/N_2/O_2$, found two of the products to be H_2CO and CH_3ONO_2, which indicates that CH_3 or CH_3O radicals are probable reaction intermediates. Dlugokencky and Howard (24) suggest that the CH_3 radicals might result from the decomposition of such an adduct formed between NO_3 and CH_3SH, since the C-S bond is not thought to be very strong ($\approx$ 309 kJmol^{-1} (38)). Their study shows that the $HSONO_2$ which would be formed does not decompose immediately to yield NO_2.

<u>Dimethyl Disulfide</u>. The reaction

$$OH + DMDS \text{ --> products} \tag{12}$$

has been studied by Wine et al. (16) using the flash photolysis/resonance fluorescence technique, and by Cox and Sheppard (19) who measured k_{12} relative to the reaction OH + ethene. The studies are in good agreement (Table III), and the recommended $k_{12}(T = 298$ K) and Arrhenius parameters are taken from the review of Atkinson (22). The initial reaction is thought to involve the formation of an adduct, which then rapidly decomposes to CH_3S and CH_3SOH radicals (6,16,19,22):

$$OH + CH_3SSCH_3 \text{ --> } [CH_3S(OH)SCH_3]^{\#} \text{ --> } CH_3SOH + CH_3S$$

Product studies (6) indicate that these radicals are then involved in further reactions leading eventually to SO_2, HCHO, and MSA. The reaction

$$NO_3 + DMDS \text{ --> products} \tag{13}$$

has been studied by three groups (24,27,39). Unfortunately, the results listed in Table III reveal a wide discrepancy in $k_{13}(T = 298$ K). The value from the smog chamber study of MacLeod et al. (39) is more than a factor of ten smaller than the two direct studies (24,27). Wallington et al. (27) suggested that this small value of k_{13} could be due to the regeneration of DMDS in the smog chamber. In fact, DMDS was also formed in the smog chamber study (39) of reaction (11), where Dlugencky and Howard (24) observe it could not be formed from the recombination of alkyl thiyl radicals, and so there appears to be a serious problem with the smog chamber technique where DMDS is regenerated by an unknown mechanism. The two direct studies (24,27) do not agree within their quoted experimental errors but are within about 30%: the recommended values $k_{13}(T = 298$ K) and E_{act} in Table III are the means from these two studies.

<u>Carbon Disulfide</u>. The reaction

$$OH + CS_2 \text{ --> products} \tag{14}$$

is discussed in detail in the chapter by Hynes and Wine (this volume). Those authors demostrate that the reaction only proceeds rapidly in the presence of O_2, through the formation of an adduct:

$$OH + CS_2 + M(=\text{third body}) <==> CS_2OH + M \tag{14a}$$

$$CS_2OH + O_2 \text{ --> products} \tag{15}$$

By means of the sub-microsecond time resolution achieved by employing a pulsed laser photolysis/pulsed laser induced fluorescence technique, Hynes and Wine are able to obtain k_{14a}, k_{-14a} and k_{15} and obtain an overall expression for k_{14} as a function of temperature and $[O_2]$. Their result in 700 torr of air is again substantially lower than the result from three competitive studies (Table IV)

Table III. Reactions of Dimethyl Disulfide

Reaction/ Technique*	Temp. Range K	10^{12} k(298 K)/ cm^3molecule^{-1}s^{-1}	10^{12} A/ cm^3molecule^{-1}s^{-1}	E_{act}/R(K)	Reference
OH + DMDS					
CK-GC	297 ± 2	240 ± 86 (ref.: OH + ethene)	-	-	Cox and Sheppard[19]
FP-RF	249-367	198 ± 1	59 ± 33	-380 ± 160	Wine et al.[16]
Recommended Values:[22]		205	51.2	-414	
NO$_3$ + DMDS					
SP-FTIR	300	0.042 ± 0.009	-	-	MacLeod et al.[39]
FP-A	280-350	0.49 ± 0.08	0.19 ± 0.03	-290 ± 50	Wallington et al.[27]
FT-LIF	334-381	0.739 ± 0.15	0.74 ± 0.15	0 ± 200	Dlugokencky and Howard[24]
Recommended Values:		0.61	0.38	-145	

Table IV. Reactions of Carbon Disulfide

Reaction/ Technique*	Temp. Range K	10^{12} k(298 K)/ cm^3molecule^{-1}s^{-1}	10^{12} A/ cm^3molecule^{-1}s^{-1}	E_{act}/R(K)	Reference
OH + CS$_2$	(Atmospheric pressure of Air)				
SP-GC	300	2.0 ± 1.0	-	-	Jones et al.[40,41]
SP-FTIR	300	2.7 ± 0.6	-	-	Barnes et al.[42]
PLP-LIF	247-298	1.1			(see Hynes and Wine, this volume)
Recommended Value:		1.1			

Recommended Temperature Dependence: $k_{14} = \dfrac{1.25 \times 10^{-16} \exp(4550/T)}{T + 1.81 \times 10^{-3} \exp(3400/T)}$

(40-42). This study provides the recommended k_{14} and the complex temperature dependence. The product studies of Barnes et al. (42), Jones et al. (40,41) and Yang et al. (43) indicate that COS is a major product of reaction (15), along with SO_2 and CO.

Hydrogen Sulfide. The reaction

$$OH + H_2S \text{--}> H_2O + HS \tag{16}$$

is discussed in a recent NASA compilation (44). The recommended k_{16} (T = 298 K) and Arrhenius parameters (Table V) are now well established from a number of studies (16,19,20,45-48).

The reaction

$$NO_3 + H_2S \text{ --}> \text{ products} \tag{17}$$

is extremely slow and k_{17} has only been measured as an upper limit (24,27) (Table V). Dlugokencky and Howard (24) have suggested that this reaction involves a hydrogen atom abstraction to form HNO_3. This is good evidence that H-atom abstraction from an S-H bond can only form a very minor pathway in the much more rapid reactions between the thiols and NO_3 (e.g., reaction 11).

Reactions of Thiyl Radicals

As is evident from the discussion in the previous section, thiyl radicals are considered to be very important intermediates in the oxidation of reduced sulfur compounds after the primary reaction with OH or NO_3. Here we will limit the discussion to the atmospheric chemistry of the methyl thiyl radical, CH_3S, which will be much more abundant than the higher carbon thiyls, but make reference to the work of Black and Jusinski. These investigators demonstrated the use of laser induced fluorescence to monitor the thiyls CH_3S, C_2H_5S and i-C_3H_7S (49), and have determined the rate coefficients for the reactions between C_2H_5S and i-C_3H_7S with NO_2, NO and O_2 (50,51).

The atmospheric reactions of CH_3S are discussed in detail in the chapters by Tyndall and Ravishankara and by Hatakeyama (this volume). Briefly, the reaction between CH_3S and NO_2 has been considered to proceed by the following pathways (39,52-54):

$$CH_3S + NO_2 \text{ --}> CH_3SO + NO \tag{18a}$$

$$\text{--}> CH_2S + HONO \tag{18b}$$

$$\text{--}> CH_3SNO_2 \tag{18c}$$

The recombination pathway (reaction (18c)) is reported to be minor (52,53). Both studies in this volume indicate that reaction (18a) is the dominant pathway. Hatakeyama used labelled $NO^{18}O$ in a steady photolysis experiment, which demonstrated that CH_3SO and NO were the major products and that $SO^{18}O$ was an eventual product in this reaction.

Tyndall and Ravishankara employed the pulsed laser photolysis of DMDS at λ = 248 nm to generate CH_3S which was then detected by LIF at λ > 371 nm. NO was observed as a product of reaction (18) by LIF at λ = 226 nm. Furthermore, these authors observed the formation of a second NO molecule on a longer time-scale than reaction (18), which they attribute to

$$CH_3SO + NO_2 \text{-->} CH_3SO_2 + NO \tag{19}$$

Hatakeyama (this volume) postulates further reactions between CH_3SO and CH_3SO_2 with O_2 to yield the observed products of SO_2 and CH_2O in his study. The determination of k_{18} by Tyndall and Ravishankara (this volume) is somewhat lower than a previous measurement by Balla et al. (54) (Table VI). The recommended value is the mean of these studies.

The recombination reaction

$$CH_3S + NO + M(=\text{third body}) \text{-->} CH_3SNO + M \tag{20}$$

has also been studied by Balla et al. (54) and k_{20} is listed in Table VI.

The reaction between CH_3S and O_2 is of particular interest because several studies have indicated that under very low NO_x conditions the overwhelming product is SO_2, without forming CH_3 radicals (6,8,55,56):

$$CH_3S + O_2 \text{-->} \text{-->} SO_2 \tag{21}$$

However, the two direct studies of k_{21} (Tyndall and Ravishankara, this volume, and (54)) demonstrate that this reaction is extremely slow; this is supported by the relative rate measurements of Grosjean (8) and Hatakeyama and Akimoto (6) (Table VI). Furthermore, Barnes et al. (57) have demonstrated that O_2 can add to the CH_3SO radical, which may then lead to the formation of MSA. The interested reader is directed to the discussion in the chapter by Tyndall and Ravishankara.

Stable Intermediates in the Oxidation of Reduced Sulfur

Dimethyl Sulfoxide. The first reported direct measurements on the kinetics of DMSO with atmospheric oxidants are reported by Barnes et al. (this volume). This study was undertaken after the recent laboratory measurements on the reaction between IO and DMS (reaction (8)) showed that this reaction could be an efficient source of DMSO in the marine troposphere (31,32). The rate coefficients for the reactions between DMSO and OH, Cl, and NO_3 are listed in Table VII. With the exception of the NO_3 reaction, they are faster than the rate coefficients for the DMS analogues (cf. Table I).

Barnes et al. (this volume) propose the following mechanism for the OH reaction based on a product analysis:

$$OH + CH_3SOCH_3 \text{-->} CH_3SOCH_2 + H_2O \tag{22a}$$

$$OH + CH_3SOCH_3 \text{-->} CH_3S(OH)OCH_3 \tag{22b}$$

These pathways are analogous to those postulated for reaction (1), OH + DMS. By further analogy with reactions (4-6), the CH_3SOCH_2 produced in reaction (22a) is expected to form the CH_3SO radical: Barnes et al. (this volume) suggest that $DMSO_2$ may be formed by reaction of O_2 with the adduct produced in reaction (22b):

$$O_2 + CH_3S(OH)OCH_3 \text{-->} CH_3SO_2CH_3 + HO_2 \tag{23}$$

In the case of the reaction between DMSO and NO_3, Barnes et al. (this volume) consider that the H-abstraction channel is minor and that the major channel leads to formation of $DMSO_2$ via an unstable adduct:

Table V. Reactions of Hydrogen Sulfide

Reaction/ Technique[*]	Temp. Range K	10^{12} k(298 K)/ cm^3molecule^{-1}s^{-1}	10^{12} A/ cm^3molecule^{-1}s^{-1}	E_{act}/R(K)	Reference
OH + H$_2$S NASA/JPL Compilation[44]		4.7	5.9	65 ± 65	20,44,16,19,45-48
NO$_3$ + H$_2$S FP-A	300	≤0.03	-	-	Wallington et al.[27]
FT-LIF	300	≤0.0008	-	-	Dlugokencky and Howard[24]

Recommended Upper Limit: k_{17} ≤ 8 x 10^{-16} cm^{3}molecule^{-1}s^{-1}

Table VI. Reactions of Methyl Thiyl Radicals

Reaction/ Technique[*]	Temp. Range K	10^{12} k(298 K)/ cm^3molecule^{-1}s^{-1}	10^{12} A/ cm^3molecule^{-1}s^{-1}	E_{act}/R(K)	Reference
NO$_2$ + CH$_3$S PLP-LIF	298	61 ± 3	-	-	Tyndall and Ravishankara (this volume)
PLP-LIF	298	100 ± 10	83 ± 14	80 ± 60	Balla et al.[54]
Recommended Value:		81	83	80	
NO + CH$_3$S + M PLP-LIF		k_∞ = (1.81 ± 0.84) x 10^{-12} exp[(905 ± 160)/T] cm^{3}molecule^{-1}s^{-1}			Balla et al.[54]
O$_2$ + CH$_3$S CK-GC	298	≤ 3x10^{-5} (ref.:CH$_3$S + NO$_2$)	-		Grosjean[8]
CK-GC	298	≤ .02 (ref.: CH$_3$S + NO)	-		Hatakeyama and Akimoto[6]
PLP-LIF	298	≤ 1.0x10^{-4}	-	-	Balla et al.[54]
PLP-LIF	298	≤ 3.0x10^{-6}	-	-	Tyndall and Ravishankara (this volume)

Recommended Upper Limit: k_{21} ≤ 3.0x10^{-18} cm^{3}molecule^{-1}s^{-1}

Table VII. Reactions of Dimethyl Sulfoxide

Reaction/ Technique[*]	Temp. Range K	10^{12} k(298 K)/ cm^3molecule^{-1}s^{-1}	10^{12} A/ cm^3molecule^{-1}s^{-1}	E_{act}/R(K)	Reference
OH + DMSO SP-FTIR	298	71 ± 25	-	-	Barnes et al. (this volume)
NO$_3$ + DMSO SP-FTIR	298	0.17 ± 0.03	-	-	Barnes et al. (this volume)
Cl + DMSO SP-FTIR	298	74 ± 18	-	-	Barnes et al. (this volume)

$$CH_3SOCH_3 + NO_3 \longrightarrow CH_3SO(ONO_2)CH_3 \longrightarrow CH_3SO_2CH_3 + NO_2 \quad (24)$$

Dimethyl Sulfone and Methane Sulfonic Acid. No studies of the gas-phase kinetics of these species appear to have been reported.

Sulfur Dioxide. The formation of SO_2 from CH_3S and CH_3SO was previously discussed. SO_2 is currently understood to be oxidized in the gas phase by OH radicals in a series of steps leading to sulfuric acid (H_2SO_4) and is thus of major importance to acid deposition processes. The chapter by Anderson et al. (this volume) describes new experimental studies of the mechanism of the oxidation of SO_2.
The initial oxidation step is considered to be the recombination reaction between OH and SO_2,

$$OH + SO_2 \longrightarrow HOSO_2 \qquad\qquad (25)$$

which has been studied by several groups at atmospheric pressure and room temperature (20,58-63). The measurements of k_{25} by a variety of techniques are in good agreement, and their mean yields the recommeded value of k_{25} in Table VIII.

The $HOSO_2$ radical may then react with O_2:

$$HOSO_2 + O_2 \longrightarrow HO_2 + SO_3 \qquad\qquad (26)$$

k_{26} has been measured in four different studies, discussed in the chapter by Andersen et al. (this volume). Three of the studies have been carried out by observing reactions (25-26) in the presence of NO and O_2. The HO_2 radical which is then produced reacts rapidly with NO to reform OH and NO_2. Thus the overall rate of disappearance of OH is slowed down when NO is present. The first of these studies was a flash photolysis/resonance fluorescence study by Margitan et al. (64). Martin et al. (65) and Anderson et al. (this volume) both utilized the discharge-flow method with EPR and resonance fluorescence detection of OH, respectively; these systems require chemical modeling to interpret the results. Gleason et al. (66) monitored both $HOSO_2$ and SO_3 by chemical ionization mass spectrometry; Anderson et al. (this volume) discuss the problems which may be encountered with using this detection technique unambiguously. Finally, Bandow and Howard (67) studied reaction (26) by monitoring the HO_2 product using laser magnetic resonance spectroscopy in a flow-tube experiment. These studies are all in good agreement (Table VIII) and the recommended value is their weighted mean.
The SO_3 formed in reaction (26) is then considered to react in the gas phase with H_2O (68):

$$SO_3 + H_2O \longrightarrow H_2SO_4 \qquad\qquad (27)$$

The reaction

$$SO_2 + NO_3 \longrightarrow \text{products} \qquad\qquad (28)$$

has been investigated by four groups: studies by Wallington et al. (27) and Burrows et al. (69) who used flash photolysis with detection of NO_3 by absorption; a study by Daubendiek and Calvert (70) who investigated the N_2O_5-SO_2 system in a smog chamber with IR detection of N_2O, SO_3 and SO_2; and a study by Duglokencky and Howard (24) who studied the reaction in a flow tube

Table VIII. Reactions of SO_2 and $HOSO_2$

Reaction/ Technique*	Temp. Range K	10^{12} k(298 K)/ $cm^3molecule^{-1}s^{-1}$	10^{12} A/ $cm^3molecule^{-1}s^{-1}$	$E_{act}/R(K)$	Reference
OH + SO$_2$	(Atmospheric Pressure of air)				
SP-GC	300	1.2 ± 0.2			Cox[58]
SP-GC	300	1.14			Castleman and Tang[59]
FP-RF	300	0.9 ± 0.25			Davis et al.[60]
CK-GC	300	0.72 ± 0.16			Cox and Sheppard[19]
FP-RA	300	0.95 ± 0.13			Paraskevopoulos et al.[61]
SP-GC/UVF	300	1.22 ± 0.13			Izumi et al.[62]
FP-RF	298	0.8 ± 0.08			Wine et al.[63]
SP-GC	300	1.1 ± 0.2			Barnes et al.[20]
Recommended Value:		1.00			
O$_2$ + HOSO$_2$					
FP-RF	298	0.4 ± 0.2	-	-	Margitan[64]
DF-LMR	298	0.49 ± 0.29	-	-	Bandow and Howard[67]
DF-EPR	298	0.35 ± 0.1	-	-	Martin et al.[65]
DF-CIMS	298	0.437 ± 0.066	-	-	Gleason et al.[66]
DF-RF	298	0.4	-	-	Andersen et al. (this volume)
Recommended Value:		0.415			
NO$_3$ + SO$_2$					
FT-LIF	298	< 1x10^{-3}	-	-	Dlugokencky and Howard[24]
FP-A	298	≤ 4x10^{-4}	-	-	Wallington et al.[26]
FP-A	298	< 4x10^{-4}	-	-	Burrows et al.[69]
SC-A	298	≤ 7x10^{-9}	-	-	Daubendiek and Calvert[70]

Recommended Upper Limit: $k_{28} \le 7 \times 10^{-20}$ $cm^3molecule^{-1}s^{-1}$

* Experimental Techniques: A = absorption; CIMS = chemical ionization mass spectroscopy; CK = competitive kinetics; DF = discharge flow; EPR = electron paramagnetic resonance; FP = flash photolysis; FT = flow tube; FTIR = Fourier transform intra-red; GC = gas chromatography; LIF = laser induced fluorescence; LMR = laser magnetic resonance; MS = mass spectroscopy; PLP = pulsed laser photolysis; SC = smog chamber; SP = steady (continuous) photolysis; UVF = ultraviolet flourescence spectroscopy

with laser induced fluorescence detection of NO_3. Inspection of Table VIII reveals that all four studies show that the reaction is extremely slow, with the study by Daubendiek and Calvert (70) recording the slowest upper limit to k_{27}.

Conclusions

The reaction mechanisms of the oxidation of DMS, CH_3SH, DMS and CS_2 which have been discussed in the foregoing sections are summarized in Figure 1. This figure indicates that a substantial amount of information has now been gathered about the primary reactions of the reduced sulfur species with OH and NO_3 radicals. Elementary reactions for which rate coefficients have been obtined are shown by solid lines. Rather less is known about the intermediate pathways. Nevertheless, the combination of elegant experimental studies on some of the elementary reactions under atmospheric conditions, together with a better interpretation of the product studies, has brought together a great deal of circumstantial evidence to connect the primary reactants to the eventual products, indicated by broken lines in Figure 1. Clearly, studies of the elementary reactions of the intermediate radical species (CH_3SO, CH_3SO_2, etc.) and the relatively stable adduct species formed in the reactions of NO_3, although experimentally very demanding, are the next steps required to elucidate completely the oxidation chemistry of these biogenic sulfur species.

Acknowledgments

This work was supported under Grant ATM 87-09802 from the National Science Foundation.

Literature Cited

1. Cullis, C. F.; Hirschler, M. M. Atmos. Environ. 1980, 14, 1263. Toon, O. B.; Kasting, J. F.; Turco, R. P.; Liu, M. S. J. Geophys. Res. D: Atmos. 1987, 92, 943.

2. Lovelock. J. E.; Maggs, R. J.; Rasmussen, R. A. Nature (London), 1972, 37, 452.

3. Andreae, M. O.; Ferek, R. J.; Bermond, F.; Byrd, K. P.; Engstrom, R. T.; Hardin, S.; Houmere, P. D.; LeMarrec, F.; Raemdonck, H.; Chatfield, R. B. J. Geophys Res. D: Atmos. 1985, 90, 12891.

4 Hatakeyama, S.; Okuda, M.; Akimoto, H. Geophys. Res. Lett. 1982, 9, 583.

5. Grosjean, D.; Lewis, R. Geophys. Res. Lett. 1982, 9, 1203.

6. Hatakeyama, S.; Akimoto, H. J. Phys. Chem. 1983, 87, 2387.

7. Niki, H.; Maker, P. D.; Savage, C. M.; Breitenbach, L. P. Int. J. Chem. Kinet. 1983, 15, 647.

8. Grosjean, D. Environ. Sci. Technol. 1984, 18, 460.

9. Hatakeyama, S.; Izumi, K.; Akimoto, H. Atmos. Environ. 1985 19, 135.

10. MacLeod, H.; Poulet, G.; Le Bras, G. J. Chem. Phys. 1983, 80, 287.

11. MacLeod, H.; Jourdain, J. L.; Pulet, G.; Le Bras, G. Atmos. Environ. 1984, 18, 2621.

12. Martin, D.; Jourdain, J. L.; Le Bras, G. Int. J. Chem. Kinet. 1985, 17, 1247.

13. Hsu, Y-C.; Chen, D-S.; Lee, Y-P. Int. J. Chem. Kinet. 1987, 19, 1073.

14. Atkinson, R.; Perry, R. A.; Pitts, J. M., Jr. Chem. Phys. Lett. 1978, 54, 14.

15. Kurylo, M. J. Chem. Phys. Lett. 1978, 58, 233.

16. Wine, P. H.; Kreutter, N. M.; Gump, C. A.; Ravishankara, A. R. J. Phys. Chem. 1981, 85, 2660.

17. Hynes, A. J.; Wine, P. H.; Semmes, D. H. J. Phys. Chem. 1986, 90, 4148.

18. Wallington, T. J.; Atkinson, R.; Tuazon, E. C.; Aschmann, S. M. Int. J. Chem. Kinet. 1986, 18, 837.

19. Cox, R. A.; Sheppard, D. Nature (London), 1980, 284, 330.

20. Barnes, I.; Bastian, V.; Becker, K. H.; Fink. E. H.; Nelsen, W. J. Atmos. Chem. 1986, 4, 445. Barnes. I.; Bastian, V.; Becker, K. H. Int. J. Chem. Kinet. 1988, 20, 415.

21. Atkinson, R.; Pitts, J. N., Jr.; Aschmann, S. M. J. Phys. Chem. 1984, 88, 1584.

22. Atkinson, R. Chem. Rev. 1985, 85, 69.

23. Stockwell, W. R.; Calvert, J. G. J. Geophys. Res. C: Oceans Atmos. 1983, 88, 6673.

24. Dlugokencky, E. J.; Howard, C. J. J. Phys. Chem. 1988, 92, 1188.

25. Tyndall, G. S.; Burrows, J. P.; Schneider, W.; Moortgat, G. Chem. Phys. Lett. 1986 130, 463.

26. Wallington, T. J.; Atkinson, R.; Winer, A. M.; Pitts, J. N., Jr. J. Phys. Chem. 1986, 90, 4640.

27. Wallington, T. J.; Atkinson, R.; Winer, A. M.; Pitts, J. N., Jr. J. Phys. Chem. 1986, 90, 5393.

28. Slagle, I. R.; Baiocchi, F.; Gutman, D. J. Phys. Chem. 1978, 82, 1333.

29. Cvetanovic, R. J.; Singleton, D. L.; Irwin, R. S. J. Am. Chem. Soc. 1981, 103, 3526.

30. Chameides, W. L.; Davis, D. D. J. Geophys. Res. 1980, 85, 7383.

31. Barnes, I.; Becker, K. H.; Carlier, P.; Mouvier, G. Int. J. Chem. Kinet. 1987, 19, 489.

32. Martin, D.; Jourdain, J. L.; Laverdet, G.; Le Bras, G. Int. J. Chem. Kinet. 1987, 19, 503.

33. Atkinson, R.; Perry, R. A.; Pitts, J. N., Jr. J. Chem. Phys. 1977, 66, 1578.

34. Wine, P. H.; Thompson, R. J.; Semmes, D. H. Int. J. Chem. Kinet. 1984, 16, 1623.

35. Hynes, A. J.; Wine, P. H. J. Phys. Chem. 1987, 91, 3672.

36. Lee, H. J.; Tang, I. N. J. Chem. Phys. 1983, 78, 6646.

37. Barnes, I; Bastian, V.; Becker, K. H.; Fink, E. H. Proc. 3rd European Symposium, Varese, Italy, Reidel: New York, 1984; p. 149.

38. Benson, S. W. Chemical Reviews 1978, 78, 23.

39. MacLeod, H.; Aschmann, S. M.; Atkinson, R.; Tuazon, E. C.; Sweetman, J. A.; Winer, A. M.; Pitts, J. N., Jr. J. Geophys. Res. D: Atmos. 1986, 91, 5338.

40. Jones, B. M. R.; Burrows, J. P.; Cox, R. A.; Penkett, S. A. Chem. Phys. Lett. 1982, 88, 372.

41. Jones, B. M. R.; Cox, R. A.; Penkett, S. A. J. Atmos. Chem. 1983, 1, 65.

42. Barnes, I.; Becker, K. H.; Fink, E. H.; Reiner, A.; Zabel, F.; Niki, H. Int. J. Chem. Kinet. 1983, 15, 631.

43. Yang, W. X.; Iyer, R. S.; Rowland, F. S. CACGB Meeting, Peterborough, Ontario, 1987.

44. DeMore, W. B.; Margitan, J. J.; Molina, M. J.; Watson, R. T.; Golden, D. M.; Hampson, R. F.; Kurylo, M. J.; Howard, C. J.; Ravishankara, A. R. Chemical Kinetics and Photochemical Data for use in Stratospheric Modeling, Evaluation No. 7, NASA/JPL publication 85-37, 1985.

45. Perry, R. A.; Atkinson, R.; Pitts, J. N., Jr. J. Chem. Phys. 1977, 64, 3237.

46. Leu, M. T.; Smith, R. H. J. Phys. Chem. 1982, 86, 73.

47. Michael, J. V.; Nava, D. F.; Brobst, W.; Borkowski, R. P.; Stief, L. J. J. Phys. Chem. 1982, 86, 81.

48. Lin, C. L. Int. J. Chem. Kinet. 1982, 14, 593.

49. Black, G.; Jusinski, L. E. J. Chem. Soc., Faraday Trans. 2 1986, 82 2143. Chemical Phys. Lett. 1987, 136, 241; ibid., 139, 431.

50. Black, G.; Jusinski, L. E.; Patrick, R. J. Phys. Chem. 1988, 92, 1134.

51. Black, G.; Jusinski, L. E.; Patrick, R. J. Phys. Chem. submitted.

52. Niki, H.; Maker, P. D.; Savage, C. M.; Breitenbach, L. P. J. Phys. Chem. 1983, 87, 7.

53. Balla, R. J.; Heicklen, J. J. Phys. Chem. 1984, 88, 6314.

54. Balla, R. J.; Nelson, H. H.; McDonald, J. R. Chem. Phys. 1986, 109, 101.

55. Balla, R. J.; Heicklen, J. J. Photochem. 1985, 29, 297.

56. Barnes, I.; Bastian, V.; Becker, K. H., Physico-Chemical Behavior of Atmospheric Pollutants, 4th European Symposium, 1986.

57. Barnes, I.; Bastian, V.; Becker, K. H. Chem. Phys. Lett. 1987, 140, 451.

58. Cox, R. A. J. Photochem. 1974, 3, 291.

59. Castleman, A. W., Jr.; Tang, I. N. J. Photochem. 1977, 6, 349.

60. Davis, D. D.; Ravishankara, A. R.; Fischer, S. Geophys. Res. Letts. 1979, 6, 113.

61. Paraskevopoulos, G.; Singleton, D. L.; Irwin, R. S. Chem. Phys. Lett. 1983, 100, 83.

62. Izumi, K; Mizuochi, M.; Murano, K.; Fukuyama, T. Environ. Sci. Technol. 1984, 18, 116.

63. Wine, P. H., Thompson, R. J.; Ravishankara, A. R.; Semmes, D. H.; Gump, C. A.; Torabi, A.; Nicovich, J. M. J. Phys. Chem. 1984, 80, 2095.

64. Margitan, J. J. J. Phys Chem. 1984, 88, 3314.

65. Martin, D.; Jourdain, J. L.; Le Bras, G. J. Phys. Chem. 1986, 90, 4143.

66. Gleason, J. F.; Sinha, A.; Howard, C. J. J. Phys. Chem. 1987, 91, 719.

67. Bandow, H; Howard, C. J. Atmos. Chem. to be published.

68. Stockwell, W. R.; Calvert, J. G. Atmos. Environ. 1983, 17, 2231.

69. Burrows, J. P.; Tyndall, G. S.; Moortgat, G. K. J. Phys. Chem. 1985, 89, 4848.

70. Daubendiek, R. L.; Calvert, J. G. Environ. Lett. 1975, 8, 103.

RECEIVED October 13, 1988

Chapter 25

OH-Initiated Oxidation
of Biogenic Sulfur Compounds

Kinetics and Mechanisms Under Atmospheric Conditions

A. J. Hynes and P. H. Wine

Molecular Sciences Branch, Georgia Tech Research Institute,
Georgia Institute of Technology, Atlanta, GA 30332

Our current understanding of the rates and mechanisms of the
OH-initiated oxidation of several reduced sulfur compounds is
reviewed with particular emphasis on recent studies in our
laboratory. Important uncertainties are highlighted and
directions for future research are suggested.

Much progress has been made in our understanding of the mechanism of the
oxidation of naturally and anthropogenically produced sulfur compounds (1).
We focus here on our recent studies of the OH-initiated oxidation of reduced
sulfur in the gas phase (2-4) and try to address some questions posed by current
work. Recent reviews (5,6) have summarized progress in laboratory studies,
which can be conveniently divided between direct studies, aimed primarily at
isolating and determining absolute rates of elementary steps in the oxidation
process, and competitive studies aimed at determining relative reaction rates
and also the identities and yields of stable end products. Clearly, the success of
these efforts can be measured by our ability to model the observed
concentration profiles obtained in controlled laboratory studies (7) and the
concentrations and lifetimes observed in the atmosphere (8). In this paper, we
review the current level of understanding of the rates and mechanisms of OH
reactions with several reduced sulfur compounds, and suggest directions for
future research.

OH + CS$_2$ -> Products

Direct studies of reaction (1) which have

$$\text{OH} + \text{CS}_2 \text{ ----> Products} \tag{1}$$

been performed in the absence of oxygen indicate that the reaction proceeds
slowly, if at all (9). In apparent contradiction to this, both competitive rate
studies performed under atmospheric conditions and atmospheric lifetime
measurements of CS$_2$ are consistent with a fast rate for (1) (10-12).
Adduct formation followed by adduct reaction with O$_2$ has been suggested
(10-12) as a mechanism for reaction (1) which would remove the discrepancy
between these results. We have studied reaction (1) (4) using 'Pulsed Laser

0097–6156/89/0393–0424$06.00/0

Photolysis - Pulsed Laser Induced Fluorescence' (PLP-PLIF), a technique which allows us to directly measure OH reaction rates under atmospheric conditions and also permits the sub-microsecond time resolution necessary to observe fast equilibration processes.

In the absence of oxygen, we observed the rapid formation of a CS_2OH adduct and were able to extract elementary rates for its formation and decay (k_{1a}, k_{-1a}), as a function of temperature.

$$OH + CS_2 + M <\text{----}> CS_2OH + M \qquad (1a,-1a)$$

The temperature dependence of the forward rate in 700 Torr N_2 is described by the Arrhenius expression

$$k_{1a} = 6.9 \times 10^{-14} \exp(1150/T) \ cm^3molecule^{-1}s^{-1}.$$

The ratio of the forward and reverse rates gives the equilibrium constant and a Van't Hoff plot of its variation with temperature, shown in Figure 1, gives a heat of reaction $\Delta H = -9.9 \pm 0.8$ kcal/mole. Using known values for the heats of formation of OH and CS_2 (13) leads to a value of 27.5 kcal/mole for the adduct heat of formation.

In the presence of oxygen, the observed reaction rate increases rapidly as a function of oxygen pressure as shown in Figure 2, and a rate of $(1.5 \pm 0.1) \times 10^{-12}cm^3molecule^{-1}s^{-1}$ in 700 Torr of air at 298K is obtained. This is in good agreement with the competitive studies (10-12) which report rates of ~2 x $10^{-12}cm^3molecule^{-1}s^{-1}$ at this pressure and temperature. The temperature dependence of the observed bimolecular rate constant, k_{obs}, defined as the slope of a plot of the pseudo-first order OH decay rate versus the CS_2 concentration, is shown in Figure 3, plotted in Arrhenius form. The rate increases rapidly as the temperature is decreased which is consistent with a mechanism involving reversible adduct formation followed by adduct reaction with oxygen:

$$CS_2OH + O_2 \text{----}> products. \qquad (2)$$

At 251K, the observed bimolecular rate constant reaches its limiting value, $k_{1a} \approx 6 \times 10^{-12}cm^3molecule^{-1}s^{-1}$, at an oxygen partial pressure of ≈100 Torr. At this point, adduct formation becomes the rate limiting step and further increases in oxygen partial pressure produce no increase in the observed rate constant. Applying a steady state analysis to the reaction sequence (1a,-1a,2) gives the expression

$$k_{obs}(T,[O_2]) = \frac{k_{1a}(T)X(T)[O_2]}{1+X(T)[O_2]} \qquad (I)$$

where

$$X(T) = k_2(T)/k_{-1}(T),$$

and allows us to obtain a value of k_2, the adduct + O_2 rate constant (units are $cm^3molecules^{-1}s^{-1}$):

$$k_2 = 1.4 \ ^{+\ 2.5}_{-\ 0.8} \times 10^{-14} \exp[217 \pm 301/T].$$

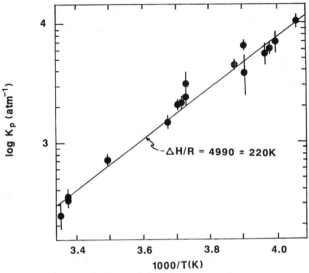

Figure 1. Van't Hoff plot for the equilibrium OH + CS_2 <----> CS_2OH. (Reproduced from Reference 4. Copyright 1988 American Chemical Society.)

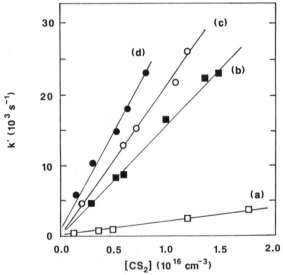

Figure 2. Plots of k' versus $[CS_2]$ for data obtained at 295K and 700 Torr N_2 + O_2. Solid lines are obtained from linear least squares analyses and give the following bimolecular rate constants (units are $10^{-12} cm^3 molecule^{-1} s^{-1}$, errors are 2σ and represent precision only): (a) 0.20 ± 0.01, (b) 1.51 ± 0.09, (c) 2.06 ± 0.10, and (d) 2.63 ± 0.19. For the sake of clarity, data points obtained for $[CS_2]$ = 0 are not shown. (Reproduced from Reference 4. Copyright 1988 American Chemical Society.)

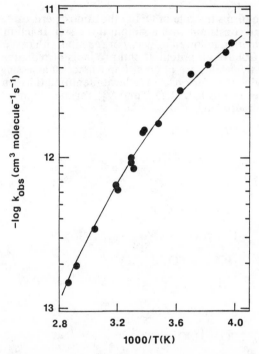

Figure 3. Arrhenius plot for the OH + CS_2 reaction in 680 ± 20 Torr air. The solid line is obtained from Equation II. (Reproduced from Reference 4. Copyright 1988 American Chemical Society.)

A temperature independent rate constant of (2.9 ± 1.1) x $10^{-14}cm^3molecule^{-1}s^{-1}$ is also consistent with our data within the estimated error limits. Fitting (I) to the k_{obs} data shown in Figure 3 gives an expression for the temperature dependence of k_{obs} in one atmosphere of air (units are $cm^3molecule^{-1}s^{-1}$):

$$k_{obs} = \frac{1.25 \times 10^{-16} \exp(4550/T)}{T + 1.81 \times 10^{-3} \exp(3400K)}. \tag{II}$$

This work directly confirms the role of OH in the atmospheric oxidation of CS_2. All of our results are consistent with a simple three step reaction mechanism involving adduct formation followed by decay or reaction with oxygen.

COS, SO_2 and CO are all potential end products of reaction (2), and its mechanism and relative yields are of great importance in assessing the role of reaction (1) in the global sulfur cycle. We have identified thirteen exothermic channels for (2), three of which, (2c), (2g) and (2j), regenerate the OH molecule and would not be measured in our experiments.

$CS_2OH + O_2 \longrightarrow SH + SO_2 + CO$	$-\Delta H = 91$ kcal/mole	(2a)
$SH + SO + CO_2$	86	(2b)
$OH + S_2 + CO_2$	81	(2c)
$H + SO_2 + COS$	79	(2d)
$H + S_2O + CO_2$	79	(2e)
$HOCS + SO_2$		(2f)
$OH + SO + COS$	50	(2g)
$S + HSO + CO_2$	45	(2h)
$SO + HSO + CO$	42	(2i)
$OH + SO_2 + CS$	34	(2j)
$SH + O_2 + COS$	27	(2k)
$S + SO_2 + HCO$	23	(2l)
$S_2 + HO_2 + CO$	20	(2m)

The consistency between our elementary rate constants, $k_{1a}(T)$ and our values for $k_{obs}(T)$, combined with the agreement between our k_{obs} (298K) values and those of Jones, et al. (10,11), leads us to believe that the OH regeneration channels are relatively minor.

The competitive rate studies of Jones, et al. (10,11) and Barnes, et al. (12) monitored both CS_2 decay and the appearance of COS and SO_2, and found the reaction stoichiometry to be

$$CS_2 + 1\frac{1}{2}\ O_2 \longrightarrow COS + SO_2. \tag{1b}$$

Rowland, et al. ([14]) have used radiochemical tracer techniques to study reaction (1) and obtain equal yields of CO and COS. CS_2 oxidation could account for a significant fraction of the atmospheric COS budget ([8]) and definitive measurements of the mechanism of reaction (2), its primary products and the routes to, and yields of the ultimate stable products are needed.

OH + Dimethyl Sulfide

Marine dimethyl sulfide emissions are thought to account for about half of the global flux of biogenic sulfur into the atmosphere ([15]). While there appears to be reasonable agreement on concentration levels, which are in the 100 ppt range in the marine boundary layer, ([16]) estimates of the atmospheric lifetime vary by several orders of magnitude ([17,18]). OH oxidation has been considered to be the major sink for DMS, although NO_3 and possibly IO could play significant roles. There have been numerous studies of reaction (3), ([5])

$$OH + CH_3SCH_3 \dashrightarrow Products. \tag{3}$$

However, agreement between studies is poor and the overall mechanism of oxidation is unclear, although SO_2, methane sulfonic acid (MSA), and dimethyl sulfone ($DMSO_2$) have been identified as stable end products in steady state photolysis experiments ([19-22]).

Studies in our laboratory, utilizing both conventional flash photolysis-resonance fluorescence (FP-RF) ([2,23]) and PLP-PLIF ([2]) indicate that reaction (3) proceeds via a complex mechanism which involves abstraction, reversible addition, and adduct reaction with O_2. In the absence of O_2, the FP-RF studies obtain a 298K rate constant of $4.4 \times 10^{-12} cm^3 molecule^{-1} s^{-1}$. The observed kinetic isotope effect and positive activation energy are consistent with the reaction proceeding via H atom abstraction,

$$OH + CH_3SCH_3 \dashrightarrow CH_3SCH_2 + H_2O. \tag{3a}$$

In the presence of O_2, however, a significant rate enhancement occurs which shows no isotope dependence and increases dramatically as the temperature is reduced, as shown in the Arrhenius plot, Figure 4, for OH + DMS-d_6. As in the case of CS_2, this behavior is consistent with adduct formation followed by adduct reaction with O_2.

$$OH + CH_3SCH_3 + M \xleftrightarrow{\hspace{1cm}} CH_3 - \overset{\overset{\displaystyle OH}{|}}{S} - CH_3 + M \tag{3b,-3b}$$

$$CH_3 - \overset{\overset{\displaystyle OH}{|}}{S} - CH_3 + O_2 \dashrightarrow products. \tag{4}$$

As the temperature is decreased the adduct lifetime increases, resulting in more collisions with O_2 and an increase in the observed reaction rate. Applying a steady state analysis to the reaction sequence (3a, 3b, -3b, 4) gives an expression similar to Equation (I), and assumption of a small negative activation energy for reaction (3b) allows us to fit the data in Figure 4 to the steady state expression and extract values for $k_{3b}(T)$ and $k_4(T)/k_{-3b}(T)$. Although direct observation of equilibration kinetics was not possible because of the fast abstraction channel and short adduct lifetime, observation of a non-linear dependence of k_{obs} on [DMS-d_6] at high DMS-d_6 and low temperature

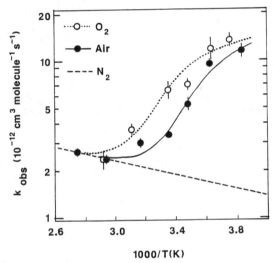

Figure 4. Arrhenius plots for the OH + CD_3SCD_3 reaction in 700 Torr N_2, air, and O_2. (Reproduced from Reference 35. Copyright 1987 American Chemical Society.)

allowed us to obtain an adduct lifetime toward unimolecular decomposition, $1/k_{3b}$, of 400 ns and, hence, a rate for the adduct + O_2 reaction of $(4.2 \pm 2.2) \times 10^{-12} cm^3 molecule^{-1}s^{-1}$ at 261K and 700 Torr N_2 + O_2. Our results indicate that, at 300K, reaction proceeds 25% via addition followed by adduct reaction with O_2, whereas, at 250K, the addition route accounts for 70% of the overall reaction.

In their modeling study of DMS oxidation, Yin, et al. consider both addition and abstraction channels to produce SO_2 and MSA (7). They do not consider other energetically accessible channels such as

$$\begin{array}{cc} OH & O \\ | & | \\ CH_3 - S - CH_3 + O_2 ----> CH_3 - S - CH_3 + HO_2 \end{array} \qquad (4a)$$

$$\begin{array}{cc} OH & O \\ | & | \\ CH_3 - S - CH_3 + O_2 ----> CH_3 - S - CH_3 + OH \\ & | \\ & O \end{array} \qquad (4b)$$

Barnes, et al. (22) have studied reaction (3) in the presence of O_2 using a competitive rate technique which utilizes H_2O_2 as the OH photolytic precursor. This avoids the use of NO_x containing precursors which have been shown to produce unreliable results in studies of OH reactions with organo-sulfur compounds (24,25). They find up to 30% yields of dimethylsulfone, $(CH_3)_2SO_2$, and suggest that reaction (4b) is important. If reaction (4b) is an elementary reaction and OH is regenerated, then our study of reaction (3) would have obtained values of k_{obs} which were too low and, hence, underestimated the importance of the addition channel. One apparent inconsistency between the recent work of Barnes, et al. (23) and our study (2) is that Barnes, et al. observed only DMSO$_2$ as a product of reaction (4); neither DMSO nor MSA was observed. Since we would not have observed an O_2 rate enhancement if the adduct + O_2 reaction proceeded entirely via channel (4b), our results imply that an addition channel other than (4b) must be important. The only addition pathways which seem reasonable are (4a), (4b), and

$$\begin{array}{c} OH \\ | \\ CH_3 - S - CH_3 + O_2 ----> CH_3SOH + CH_3O_2. \end{array} \qquad (4c)$$

The CH_3SOH product of reaction (4c) is expected to react with O_2 to produce MSA. Further studies are needed to establish whether DMSO$_2$ is formed directly via reaction (4b) or via a multistep pathway involving production and subsequent oxidation of DMSO. If DMSO$_2$ is, indeed, formed via reaction (4b), then the source of the oxygen enhancement observed in our experiments needs to be identified.

On the basis of our results, one might expect the end product yields from the OH-initiated oxidation of DMS to be strongly temperature dependent. No temperature dependent laboratory studies have been performed which would allow this hypothesis to be tested. However, atmospheric measurements suggest higher MSA-to-SO$_2$ yield ratios at higher latitudes (i.e. lower temperatures) (17).

At present, modeling simulations of both laboratory experiments (7) and measured atmospheric trace species distributions (8) are unable to reproduce observed concentrations of DMS and its oxidation products. It is clear that direct studies aimed at identifying reactive intermediates, and a better understanding of the chemistry via which these intermediates are converted to end products, is required.

OH + Methyl Mercaptan

While methyl mercaptan (MM) has been observed in the marine atmosphere under favorable conditions at the ppt level (17), its mean concentration appears to be sub ppt and it is not considered to be a significant component of marine sulfur emissions. FP-RF studies (23,29) from our laboratory show that OH reacts rapidly with MM (k_5 = 3.3 x 10^{-11} cm^3molecule^{-1}s^{-1} at 298K),

$$OH + CH_3SH ----> products, \qquad (5)$$

and that k_5 shows very small kinetic isotope effects and a slight negative activation energy; these results are consistent with an addition mechanism. While direct laboratory studies are in reasonable agreement, two competitive studies report considerably faster rate constants in the presence of O_2, suggesting the possibility of reversible adduct formation, followed by adduct reaction with O_2 (26,27). In our PLP-PLIF studies (3) of reaction (5) in one atmosphere of nitrogen, air, and oxygen, shown in Figure 5, we observe rate constants in good agreement with the FP-RF studies with no O_2 rate enhancement. The competitive studies used NO_x containing compounds as photolytic precursors. It has recently been demonstrated that secondary chemistry in these systems produces unreliable rate constants (24,25). The photolysis studies of Hatakeyama and Akimoto (19) appear to indicate that the CH_3SOH adduct has a reasonable lifetime and reacts with NO_x containing compounds or decomposes to CH_3S and H_2O; its reaction with oxygen appears to be slow. Again, direct measurement of CH_3S yields together with a better understanding of CH_3S chemistry are important in understanding MM oxidation.

Reactivity Trends

Comparison of reactions (1), (3), (5) and others which might be expected to behave similarly, i.e.

$$OH + COS ----> products \qquad (6)$$

$$OH + H_2S ----> products, \qquad (7)$$

show no consistent trends. Reaction (6) would appear to be a good candidate for "O_2 rate enhancement." However, a recent study saw no evidence for adduct formation or O_2 rate enhancement (28). Reaction (7) appears to proceed solely via an abstraction mechanism and no "O_2 enhancement" is observed in competitive rate studies (26). Additionally, the reactivity towards O_2 of the adducts formed in reactions (1), (3) and (5) appears to be dramatically different with $(CH_3)_2SOH$ reacting 100 times faster than CS_2OH (2,4), while $CH_3SH(OH)$ appears to react very slowly (19). It is difficult to rationalize the very different behavior of rather similar chemical systems, and future experimental studies might be expected to produce more surprises. A

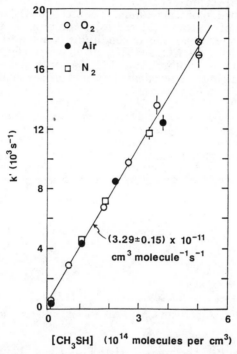

Figure 5. Plot of the pseudo-first order OH decay rate versus [CH$_3$SH] for data obtained in a series of back-to-back experiments at 300K and 700 Torr total pressure but with different buffer gases. The O$_2$ data points labeled with an "x" and a "-" were obtained at particularly low and high laser powers, respectively. (Reproduced from Reference 3. Copyright 1987 American Chemical Society.)

theoretical framework for understanding the observed reactivity trends remains to be developed.

Future Work

We believe that our studies of reactions (1), (3) and (5) have largely resolved the question of the mechanism of initial OH attack and the focus of our attention has shifted to the yields and chemistry of reactive intermediates and stable products. Direct observation of CH_3S and quantitative yield measurements as a function of temperature, pressure, and $[O_2]$ will clarify the detailed mechanisms of reactions (3) and (5). Additionally, the oxidation pathway for CH_3S under atmospheric conditions is not well understood. Reactions of CH_3S with O_2, O_3, NO_2 and HO_2 are all potentially important, but data from these reactions is either limited (30-34) or non-existent and the kinetics and mechanisms of these reactions all require further study.

Based on chemical kinetic evidence, it is now clear that OH reactions with CS_2, CH_3SCH_3, and CH_3SH proceed primarily via formation of relatively short lived addition complexes. However, it should be emphasized that none of these short lived complexes have been observed, and no structural information, either experimental or theoretical, is available for any of them. In the case of the OH...CS_2 adduct, for example, it is not obvious whether OH binds to the carbon atom or to a sulfur atom. A detailed understanding of the reactivity trends discussed above will not be forthcoming until adduct structures are elucidated.

Acknowledgments

This work was supported by the National Science Foundation through grant no. ATM-86-00892.

Literature Cited

1. Plane, J. M. C. In Biogenic Sulfur in the Environment; Saltzman, E.; Cooper, W. J., Eds.; American Chemical Society: Washington, DC, 1988, this volume.

2. Hynes, A. J.; Wine, P. H.; Semmes, D. H. J. Phys. Chem. 1986, 90, 4148.

3. Hynes, A. J.; Wine, P. H. J. Phys. Chem. 1987, 91, 3672.

4. Hynes, A. J.; Wine, P. H.; Nicovich, J. M. J. Phys. Chem. 1988, 92, 3846.

5. Atkinson, R. Chem. Rev. 1985, 85, 69.

6. Heicklen, J. Rev. Chem. Intermed. 1985, 6, 175.

7. Yin, F.; Grosjean, D.; Seinfeld, J. H. J. Geophys. Res. 1986, 91, 14417.

8. Toon, O. B.; Kasting, J. F.; Turco, R. P.; Lui, M.S. J. Geophys. Res. 1987, 91, 943.

9. Bierman, H. W.; Harris, G. W.; Pitts, J. N., Jr. J. Phys. Chem. 1982, 86, 2958, and references cited therein.

10. Jones, B. M. R.; Burrows, J. P.; Cox, R. A.; Penkett, S. A. Chem. Phys. Lett. 1982, 88, 372.

11. Jones, B. M. R.; Cox, R. A.; Penkett, S. A. J. Atmos. Chem. 1983, 1, 65.

12. Barnes, I.; Becker, K. H.; Fink, E. H.; Reiner, A.; Zabel, F.; Niki, H. Int. J. Chem. Kinet. 1983, 15, 631.

13. Benson, S. W. Thermochemical Kinetics; Wiley-Interscience: New York, 1976.

14. Yang, W. X.; Iyer, R. S.; Rowland, F. S. CACGB Meeting; Peterborough: Ontario, 1987.

15. Bates, T. S.; Cline, J. D.; Gammon, R. H.; Kelly-Hansen, S. R. J. Geophys. Res. 1987, 92, 2930.

16. Andreae, M. O.; Ferek, R. J.; Bernard, F.; Byrd, K. P.; Engstrom, R. T.; Hardin, S.; Houmere, P. D.; LeMarres, F.; Raemdanck, H.; Chatfield, R. B. J. Geophys. Res. 1985, 90, 12891.

17. Berresheim, H. J. Geophys. Res. 1987, 92, 13245, and references therein.

18. Nguyen, B. C.; Bergeret, C.; Lambert, G. In Gas Transfer at Water Surfaces; Brutseert, W.; Jurke G. H., Eds.; Reidel: Dordrecht, 1984, p 549.

19. Hatakeyama, S.; Akimoto, H. J. Phys. Chem. 1983, 87, 2387.

20. Hatakeyama, S.; Okuda, S. M.; Akimoto, H. Geophys. Res. Lett. 1982, 9, 583.

21. Grosjean, D. Environ. Sci. Technol. 1984, 18, 460.

22. Barnes, I.; Bastian, V.; Becker, K. H. Int. J. Chem. Kinet. 1988, 20, 415.

23. Wine, P. H.; Kreutter, N. M.; Gump, C. A.; Ravishankara, A. R. J. Phys. Chem. 1981, 85, 2660.

24. Wallington, T. J.; Atkinson, R.; Tuazon, E. C.; Aschmann, S. M. Int. J. Chem. Kinet. 1986, 18, 837.

25. Barnes, I.; Bastian, V.; Becker, K. H.; Fink, E. H.; Nelsen, W. J. Atmos. Chem. 1986. 4, 445.

26. Cox, R. A.; Sheppard, D. Nature 1980, 284, 330.

27. Barnes, I; Bastian, V.; Becker, K. H.; Fink, E. H. Physi-Chemical Behavior of Atmospheric Pollutants; Proceedings of the 3rd European Symposium: Varese, Italy; Reidel: New York, 1984, pp. 149-57.

28. Wahner, A.; Ravishankara, A. R. J. Geophys. Res. 1987, 92, 2189.

29. Wine, P. H.; Thompson, R. J.; Semmes, D. H. Int. J. Chem. Kinet. 1984, 16, 1623.

30. Balla, R. J.; Nelson, H. H.; McDonald, J. R. Chem. Phys. 1986, 109, 101.

31. Black, G.; Jusinski, L. E. J. Chem. Soc. Faraday Trans. II 1986, 82, 2143.

32. Barnes, I.; Bastian, V.; Becker, K. H. Chem. Phys. Lett. 1987, 140, 451.

33. Tyndall, G. S.; Ravishankara, A. R. In Biogenic Sulfur in the Environment; Saltzman, E.; Cooper, W. J., Eds.; American Chemical Society: Washington, DC, 1988, this volume.

34. Hatakeyama, S. In Biogenic Sulfur in the Environment; Saltzman, E.; Cooper, W. J., Eds.; American Chemical Society: Washington, DC, 1988, this volume.

35. Hynes, A. J.; Wine, P. H. In The Chemistry of Acid Rain, Sources and Atmospheric Processes; Johnson, R. W.; Gordon, G. E., Eds.; American Chemical Society Symposium Series 349; Washington, DC, 1987, pp 133-41.

RECEIVED August 30, 1988

Chapter 26

Mechanism of Atmospheric Oxidation of Sulfur Dioxide by Hydroxyl Radicals

Larry G. Anderson, Paul M. Gates, and Charles R. Nold

Department of Chemistry, University of Colorado—Denver, Denver, CO 80204

The $HOSO_2$ radical formed in the reaction of hydroxyl radicals with SO_2 was found to react with both O_2 and NO. The kinetics of these two reactions were studied in a discharge flow-resonance fluorescence system. The rate constant for the $HOSO_2 + O_2$ reaction was found to be about $(4 \pm \frac{4}{3}) \times 10^{-13} \, cm^3/s$, in good agreement with previously reported values. The rate constant for the $HOSO_2 + NO$ reaction was found to be about $(1 \pm \frac{1}{3}) \times 10^{-12} \, cm^3/s$, which is in between the two values previously reported for this rate constant. The $HOSO_2 + NO$ reaction is too slow to be of significance in the atmosphere, when compared to the $HOSO_2 + O_2$ reaction. This latter reaction forms a HO_2 radical, hence the HO reaction with SO_2 does not act as a radical sink under atmospheric conditions.

The oxidation of sulfur dioxide is of importance not only in the polluted atmosphere, but it is also of importance in the natural atmosphere. Toon et al. (1) have recently published a paper on the sulfur cycle in the marine atmosphere. In that work, they conclude that there are no known natural emissions of SO_2 to the marine atmosphere. Yet, based on the measured concentrations of SO_2 and the expected lifetime of SO_2 in the marine atmosphere, there must be a source of SO_2 of about 30 Tg S yr^{-1} in the marine environment (1). The vast majority of this SO_2 is believed to arise from the atmospheric oxidation of dimethyl sulfide, which has an estimated biogenic source strength of 40 ± 20 Tg S yr^{-1} in the marine environment (1). The atmospheric oxidation of other biogenic sulfur containing compounds are also believed to contribute to the SO_2 production. These compounds include carbonyl sulfide, carbon disulfide and hydrogen sulfide. But these compounds are not expected to contribute significantly to the production of SO_2 in the marine environment, since the combined biogenic source strengths for these compounds is believed to be only a few Tg S yr^{-1}. This budget information suggests that most of the biogenic sulfur emissions in the marine atmosphere are oxidized to SO_2. Hence, it is necessary to understand the atmospheric chemistry of SO_2 in order to understand the role of biogenic sulfur in the atmosphere.

In this paper, we will discuss the current understanding of the homogeneous gas phase oxidation of SO_2 by hydroxyl radicals (HO) in the atmosphere. This

0097–6156/89/0393–0437$06.00/0

reaction ultimately leads to the formation of sulfuric acid (H_2SO_4) and is of major importance to acidic deposition processes. The reaction of HO with SO_2 is believed to proceed by the addition of the HO radical to SO_2 (2)

$$HO + SO_2 (+ M) \rightarrow HOSO_2 (+ M) \tag{1}$$

Until recently, our knowledge of the fate of the $HOSO_2$ radical under atmospheric conditions was limited. To avoid the uncertainties in the chemistry of the $HOSO_2$ radical, early atmospheric chemical models for acid deposition (3) have treated the process as

$$HO + SO_2 \rightarrow \rightarrow \rightarrow H_2SO_4 \tag{2}$$

Chemical modeling using this simplified mechanism (2) for H_2SO_4 formation has shown that there are nonlinearities in the changes in H_2SO_4 concentrations calculated as the SO_2 concentration is changed (3-5). When the SO_2 concentration is reduced by 30%, the H_2SO_4 concentration is only reduced by 14%, not the 30% reduction expected if there were a linear relationship (5). This nonlinearity is a major complication in relating SO_2 emissions to H_2SO_4 deposition. Reaction (2) leads to the loss of one HO radical for each SO_2 molecule oxidized, and the formation of one molecule of H_2SO_4. As the SO_2 concentration decreases, this mechanism suggests that the HO radical concentration must increase, thus increasing the efficiency of the SO_2 to H_2SO_4 conversion. Although the amount of H_2SO_4 produced increases, it does not do so in direct proportion to the SO_2 increase. Thus, this mechanism is responsible for the nonlinearity in the relationship between SO_2 concentration changes and H_2SO_4 concentration changes.

In 1983, Stockwell and Calvert (6) presented data which suggested that an alternate mechanism might be more appropriate for the description of the HO-SO_2 reaction in air. In their experiments, they photolyzed HONO/NO_x/CO mixtures in the presence and absence of SO_2. Data were collected on the CO_2 production rate from the photooxidation of CO. This is directly related to the HO radical concentration in the reaction system. The data suggested that there was no significant change in the HO concentration upon the addition of SO_2 to the reaction chamber. They interpreted their data with the following mechanism:

$$HO + SO_2 (+ M) \rightarrow HOSO_2 (+ M) \tag{1}$$

$$HOSO_2 + O2 \rightarrow HO_2 + SO_3 \tag{3}$$

$$SO_3 + H_2O \rightarrow H_2SO_4 \tag{4}$$

In their system, sufficient nitric oxide (NO) was present to react with the HO_2 radical, reforming the HO radical.

$$HO_2 + NO \rightarrow HO + NO_2 \tag{5}$$

These data and the resulting mechanism are consistent with earlier observations of smog chamber processes conducted in the presence and absence of added SO_2, as discussed in references (2,6). In the atmosphere, reaction (4) is expected to be the primary fate of SO_3 due to the presence of large quantities of water and the presumed lack of other important reactions for SO_3. Under atmospheric conditions, this mechanism suggests that HO radicals are effectively converted to HO_2 radicals, if reaction (3) is sufficiently fast and there are no other significant loss processes for the HOSO$_2$ radical. Thus the net reaction under atmospheric conditions becomes

$$HO + SO_2 (+ O_2, H_2O) \rightarrow H_2SO_4 + HO_2 \tag{6}$$

Meagher et al. (7) performed similar experiments in propene/butane/NO$_x$/ H$_2$O mixtures with and without added SO$_2$. These experimental results were compared with model results from the use of reaction (2) or the use of reaction (6) to describe the HO + SO$_2$ reaction. The experimental results compared reasonably well with the model results from the use of reaction (6), and very poorly with those from the use of reaction (2).

Three studies have reported measurements of the rate constant for reaction (3), HOSO$_2$ + O$_2$. These three studies approached the problem differently, and will be discussed individually. In the earliest of these studies, Margitan (8) used a flash photolysis/resonance fluorescence system for the study of the HO radical reaction with SO$_2$. The HO radical decay was followed as a function of time with SO$_2$ present and with and without NO and O$_2$ added. It was observed that the HO radical decay rate was smaller when NO and O$_2$ were added, suggesting that HO radicals were being regenerated when both NO and O$_2$ were present. These experiments demonstrated the necessity for the presence of both NO and O$_2$ for this HO radical regeneration. These experimental results were fit with a simple chemical mechanism and the rate constant for reaction (3), HOSO$_2$ + O$_2$, was determined to be (4 ± 2) x 10^{-13} cm^3/s. It was also found that the HOSO$_2$ radical reacted with NO with a rate constant of about 2.5 x 10^{-12} cm^3/s. The interpretation of the experimental data was complicated by the fact that all of the reactants were present throughout the reaction, hence many reactions occur simultaneously. In addition, the HO radical concentration in this complex reaction system was monitored as a function of reaction time for only about 10 ms. Schmidt et al. (9) have reported very similar experimental results, but have not reported a rate constant for the reaction of HOSO$_2$ with O$_2$.

The second study of the kinetics of the HOSO$_2$ reaction with O$_2$ was reported by Martin et al. (10). They used discharge flow techniques to produce the HO radicals and EPR detection of the HO radicals. Some of their experiments were carried out in a manner analogous to Margitan's. Namely, SO$_2$, O$_2$ and NO were added to the flow system through the same movable injection port. Under these conditions all reactions were occurring simultaneously. They also performed experiments in which they took advantage of one of the characteristics of flow systems, the ability to add reactants at different reaction times. In these experiments, they introduced SO$_2$ and O$_2$ through the same movable port and added NO through a fixed port upstream of the EPR detector. Typical reaction times used in their system were about 20 ms. Chemical modeling was used to estimate the rate constants for various reactions occurring in the system. The HOSO$_2$ + O$_2$ reaction (3) was reported to have a rate constant of (3.5 ± 1) x 10^{-13} cm^3/s, the HOSO$_2$ + NO reaction was ≤ 5 x 10^{-13} cm^3/s.

In the studies discussed above, HO radicals were detected. HO radicals are neither a reactant nor a product of the HOSO$_2$ + O$_2$ reaction (3). Hence, these studies of the kinetics of reaction (3) are indirect. In the work of Gleason et al. (11), the reactant HOSO$_2$ and the product SO$_3$ were monitored. This work at first glance appears to be a direct study of the kinetics of reaction (3), until one realizes that these species were monitored by chemical ionization mass spectrometry. HOSO$_2$ was converted to SO$_3^-$ and SO$_3$ was converted to (Cl·SO$_3$)$^-$ prior to mass spectrometric detection. The detection of HOSO$_2$ and SO$_3$ are based upon a complete understanding of the chemical ionization processes, much in the same way that the previous studies used the understanding of the HO_2 + NO reaction to allow one to measure HO and infer a knowledge of the HO_2 concentration. This chemical ionization technique should not be considered a direct spectroscopic measurement of $HOSO_2$ or SO_3. This study provides the first evidence that SO_3 is formed as a result of reaction (3). From this work, the rate constant for the $HOSO_2$ + O_2 reaction is reported as (4.37 ± 0.66) x 10^{-13} cm^3/s. They did not measure the rate constant for the $HOSO_2$ + NO reaction. In their paper, they also report the results of a study by Bandow and Howard of the $HOSO_2$ + O_2 reaction. In that work, direct spectroscopic measurement of the other product HO_2 was reported using laser magnetic resonance techniques. The rate constant for reaction (3) was reported to be (4.9 ± 2.9) x 10^{-13} cm^3/s.

Several modeling studies have been performed to evaluate the difference between using reaction (6) and reaction (2) to represent the oxidation of SO_2 by HO in atmospheric model calculations. Samson (12) has performed calculations which show a more nearly linear relationship between the change in SO_2 emissions and the change in H_2SO_4 concentrations when using reaction (6) than is found by the use of reaction (2). This is the case in model calculations that at least crudely treat the aqueous phase oxidation of SO_2, as well as the gas phase oxidation. Seigneur et al. (13) have shown that by using reaction (6) to describe the oxidation of SO_2 by HO in an atmospheric chemical model produces a very nearly linear relationship between sulfate concentrations calculated and SO_2 precursor concentrations under clear-sky conditions. Under cloudy conditions, when the aqueous phase processes become more important, the response to the SO_2 precursor concentration changes is no longer linear. Stockwell et al. (14) have shown that by using reaction (6) to describe the HO radical oxidation of SO_2 to H_2SO_4 in the more complete gas phase reaction mechanism of the National Center for Atmospheric Research's Regional Acid Deposition Model (RADM) good linearity is observed between the sulfate produced and the SO_2 precursor concentrations under both rural and urban atmospheric conditions. By using reaction (6) to describe the gas phase oxidation of SO_2 to sulfate, a variety of different models seem to show good linearity between the SO_2 precursor concentrations and the sulfate produced. It is therefore expected that much of the nonlinearity that may actually occur in the atmosphere between SO_2 emissions and sulfate concentrations arises from the aqueous phase or cloud processes for the conversion of SO_2 to sulfate.

The justification for using reaction (6) to describe the gas phase oxidation of SO_2 to sulfate, is based on the belief that the $HOSO_2$ radical reacts quickly with O_2 to form HO_2 and SO_3, and that there is no other significant loss process for the $HOSO_2$ radical. In the current study, a discharge flow-resonance fluorescence system was used to produce and monitor the concentration of HO radicals. The HO radical signal decreases as a function of reaction time upon the addition of SO_2 to the system, reaction (1). The kinetics of this reaction have been studied extensively and have not been reinvestigated in this work. Rather, the subsequent reactions have been investigated in the presence of added NO, or O_2, or NO and O_2 to determine the importance of the proposed mechanism (1), (3) and (5). The rate constants for the $HOSO_2$ reactions with O_2 and NO are reported.

Experimental

A discharge flow-resonance fluorescence system has been constructed for use in these studies. A mixture of H_2 in He was passed through a microwave discharge to produce H atoms. Very low concentrations of H atoms can be generated by passing "pure" He through the discharge. In all of the experiments described in this work, He was used as the diluent and the reactions were studied at a pressure of about 3 torr. A short distance downstream of the H atom source, a mixture of NO_2 in He was added to the flow. The H atoms react quantitatively with NO_2 to produce HO radicals. Further downstream the reactant gases SO_2, O_2 and NO were added to the reaction system. Each of these gases were added to the system through a series of fixed ports, rather than a movable injector. The principal advantage of this technique is that as the position at which reactants are added to the system is changed, one is not changing the surface area to which the radicals are exposed. In flow systems using movable injectors a correction must be applied to the data to account for radical losses on the changing wall surface area. This type of correction is unnecessary for flow systems using fixed ports. In the detection zone of the flow system, the HO resonance fluorescence signal is directly proportional to the HO radical concentration.

Results and Discussion

Experimental Results. Most of the experiments that have been performed in the current studies were carried out with the sequential addition of reactants to the reaction system. In the experiments to be discussed in this work, the SO_2 is introduced into the flow system about 115 ms upstream of the HO detection region. The O_2 was added to the reaction system about 95 ms upstream of the HO detection system. Finally, NO was added to the reaction system at about 15 ms upstream of the detector. Alternatively for studying the $HOSO_2$ + NO reaction, the NO was added to the flow system about 135 ms upstream of the HO detection system, and upstream of the SO_2 inlet. By carrying out the reactions in a sequential manner, the chemistry (or at least the more important reactions) occurring in different regions of the flow system were simplified. Most of the experiments were performed by adding about $0.5 - 2 \times 10^{15}$ molecule/cm^3 of SO_2 to the flow system. In this concentration range, 80% or more of the HO radicals were removed from the reaction system by reaction with SO_2 prior to the HO detection region. Experiments were performed in which the amount of O_2 added to the reaction system varied from $0 - 2.0 \times 10^{15}$ molecule/cm^3. NO was added to the flow system at relatively low concentrations, about 1.0×10^{14} molecule/cm^3. This concentration was sufficiently high to insure virtually complete conversion of HO_2 to HO in the approximately 15 ms reaction time. The NO concentration was sufficiently low and the reaction time was sufficiently short, that there was little effect on the HO radical concentration due to its reaction with NO to form HONO.

Figure 1 shows a plot of a typical set of data. These data were collected with a relatively high initial HO radical concentration of about $5 \times 10^{11}/cm^3$ at the SO_2 inlet port. The SO_2 concentration was 1.1×10^{15} molecule/cm^3 and the initial NO_2 concentration was 3.3×10^{12} molecule/cm^3. Molecular oxygen was added at concentrations between 0 and 2.2×10^{15} molecule/cm^3 and NO was added 15 ms prior to the detection system at a concentration of 1.2×10^{14} molecule/cm^3. At this high concentration of SO_2 and the relatively long reaction time employed, the HO radical concentration remaining at the detector is about $1.1 \times 10^{10}/cm^3$, in the absence of added O_2. As increasing quantities of O_2 are added to the reaction system, the HO radical concentration increases to about $3 \times 10^{10}/cm^3$ when the O_2 concentration reaches about 1×10^{15} molecule/cm^3. These data slightly underestimate the HO radical concentrations in the presence of O_2, since the HO radical fluorescence intensity has not been corrected for the effect of fluorescence quenching by O_2. The O_2 quenching effect could amount to as much as 10% at the highest O_2 concentrations used. Each data point shown in the figures is the weighted

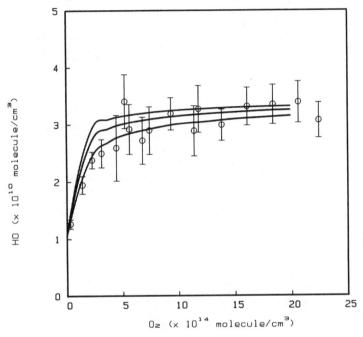

Figure 1. Experimental HO radical concentration data and model results as a function of added O_2 concentration. Initial HO radical concentration - 5 x 10^{11} /cm^3 at the SO_2 port, NO_2 concentration - 3.3 x 10^{12} molecule/cm^3, SO_2 added 115 ms from HO detection zone at concentration - 1.1 x 10^{15} molecule/cm^3, NO added 15 ms from HO detection zone at concentration - 1.2 x 10^{14} molecule/cm^3 and O2 added 95 ms from HO detection zone at concentrations 0 - 2.2 x 10^{15} molecule/cm^3. The curves show the effect of varying the HOSO$_2$ + O_2 rate constant upper - 8.0 x 10^{-13} cm^3/s, middle - 4.0 x 10^{-13} cm^3/s and lower - 2.0 x 10^{-13} cm^3/s.

average of the between two and twenty individual data points collected in a series of similar experiments. Each of the individual data points is the average of ten measurements of the HO radical fluorescence intensity, and hence the HO radical concentration. The individual data points are weighted by the reciprocal of the standard deviations in these data points. The error bars shown in the figures were determined from the standard deviations in the measured HO fluorescence signals and an assessment of the absolute accuracy of the measurement of the HO concentration. From this data, it is clear that the HO radical concentration increases as O_2 is added to the flow system. Similar experiments were performed in the absence of added NO. These experiments show only a very small increase of the HO signal due to the addition of increasing quantities of O_2 to the reaction system. This small effect remains because there is a small quantity of NO present in our reaction system, from the initial production of the HO radical.

Figure 2 shows the results for a similar set of experiments in which the initial HO radical concentration was about 4.5×10^{11} /cm^3 at the SO$_2$ port, the initial NO$_2$ concentration was 3.3×10^{12} molecule/cm^3, the SO$_2$ concentration was 1.0×10^{15} molecule/cm^3, the NO concentration was 1.1×10^{14} molecule/cm^3 and the O$_2$ concentration was between 0 and 2.0×10^{15} molecule/cm^3. This time the NO was added about 135 ms upstream of the HO detection zone. Under these conditions the observed signal will be affected by both the reaction of the HO radical with NO and by the reaction of HOSO$_2$ with NO. The HO radical concentration at the detector is only about 0.5×10^{10}/cm^3 in the absence of added O$_2$, rising to about 1.5×10^{10}/cm^3 when the O$_2$ concentration exceeds about 1×10^{15} molecule/cm^3.

Figure 3 shows a plot of experimental data for a set of experiments in which the initial HO radical concentration was about 3×10^{11} /cm^3 at the SO$_2$ port, the initial NO$_2$ concentration was 3.3×10^{12} molecule/cm^3, the SO$_2$ concentration was 1.0×10^{15} molecule/cm^3, the O$_2$ concentration was 1.1×10^{15} molecule/cm^3 and the NO concentration was varied between 0 and 3.2×10^{14} molecule/cm^3. For this set of experiments the NO was added about 135 ms upstream of the HO detection zone. In these experiments, the HO radical concentration was observed to increase with the smallest quantities of added NO. As the NO concentration was increased further, the HO radical concentration was observed to decrease.

Several other sets of experiments of the types shown in the figures were performed using different initial concentrations of the HO radical and SO$_2$. These figures show an example of the types of experimental results and analysis that have been performed. The conclusions presented are based on the analysis of the results over a broader set of experiments.

Model Results. In order to interpret the results of our experiments, we have undertaken an extensive chemical modeling effort to investigate the chemistry occurring in this reaction system. The reactions used in this model are listed in Table I. This is a more complete mechanism than that used by either Margitan (8) or Martin et al. (10) in their studies of the kinetics of this reaction system. Gleason et al. (11) reported no modeling of their studies. The rate constants for many of the reactions listed in Table I are known (15). The wall loss rate constant for HO radicals used initially in the model was based upon measurements made in our reaction system. The HO$_2$ and HOSO$_2$ wall loss rate constants were determined in our modeling studies. The data of Martin et al. (10) were used to calculate the rate constant for the reaction of HO with SO$_2$ which was appropriate for our experimental conditions. Their rate constant results allowed us to calculate an accurate effective bimolecular rate constant for this reaction in the presence of large quantities of SO$_2$. The initial choice of

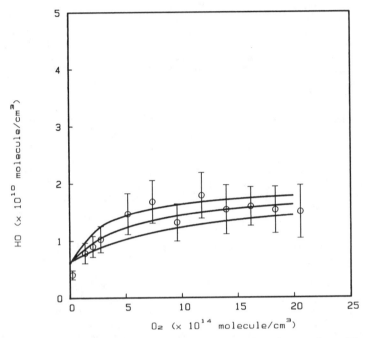

Figure 2.   Experimental HO radical concentration data and model results as a function of added O_2 concentration. Initial HO radical concentration - 4.5 x $10^{11}/cm^3$ at the SO_2 port, NO_2 concentration - 3.3 x 10^{12} molecule/cm^3, SO_2 added 115 ms from HO detection zone at concentration - 1.0 x 10^{15} molecule/cm^3, NO added 135 ms from HO detection zone at concentration - 1.1 x 10^{14} molecule/cm^3 and O_2 added 95 ms from HO detection zone at concentrations 0 - 2.1 x 10^{15} molecule/cm^3. The curves show the effect of varying the $HOSO_2$ + NO rate constant upper - 0.5 x 10^{-12} cm^3/s, middle - 1.0 x 10^{-12} cm^3/s and lower - 2.0 x 10^{-12} cm^3/s.

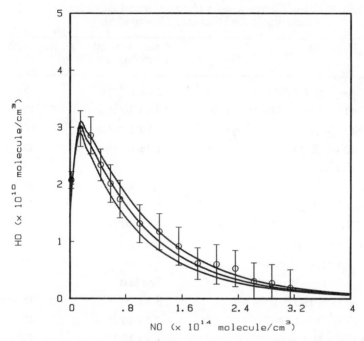

Figure 3.   Experimental HO radical concentration data and model results as a function of added NO concentration. Initial HO radical concentration - 3×10^{11} /cm^3 at the SO$_2$ port, NO$_2$ concentration - 3.3×10^{12} molecule/cm^3, SO$_2$ added 115 ms from HO detection zone at concentration - 1.0×10^{15} molecule/cm^3, O$_2$ added 95 ms from HO detection zone at concentration - 1.1×10^{15} molecule/cm^3 and NO added 135 ms from HO detection zone at concentrations 0 - 3.2×10^{14} molecule/cm^3. The curves show the effect of varying the HOSO$_2$ + NO rate constant upper - 0.5×10^{-12} cm^3/s, middle - 1.0×10^{-12} cm^3/s and lower - 2.0×10^{-12} cm^3/s.

Table I. Chemical Mechanism Used to Describe Chemistry
Occurring in the Discharge Flow System

Reaction	Rate Constant cm^3 molecule^{-1} s^{-1}	Reference
1) $H + NO_2 \rightarrow HO + NO$	1.13×10^{-10}	15
2) $HO + NO (+ M) \rightarrow HONO (+ M)$	6.9×10^{-14}	15
3) $HO + NO_2 (+ M) \rightarrow HONO_2 (+ M)$	2.5×10^{-13}	15
4) $HO + HO \rightarrow H_2O + O$	1.9×10^{-12}	15
5) $O + HO \rightarrow H + O_2$	3.3×10^{-11}	15
6) $O + NO_2 \rightarrow O_2 + NO$	9.3×10^{-12}	15
7) $HO + wall \rightarrow products$	$18 \ s^{-1}$	
8) $HO + SO_2 + M \rightarrow HOSO_2 + M$	1.2×10^{-14}	10
9) $HOSO_2 + O_2 \rightarrow HO_2 + SO_3$	See text	
10) $HOSO_2 + HO \rightarrow products$	5×10^{-11}	10
11) $HOSO_2 + NO \rightarrow products$	See text	
12) $HO_2 + NO \rightarrow HO + NO_2$	8.3×10^{-12}	15
13) $HO_2 + NO_2 \rightarrow HO_2NO_2$	2.0×10^{-14}	15
14) $O + HO_2 \rightarrow HO + O_2$	5.9×10^{-11}	15
15) $HO + HO_2 \rightarrow H_2O + O_2$	7.0×10^{-11}	15
16) $HO_2 + HO_2 \rightarrow H_2O_2 + O_2$	1.7×10^{-12}	15
17) $HO_2 + wall \rightarrow products$	See text	
18) $HOSO_2 + wall \rightarrow products$	See text	

the rate constants for the HOSO$_2$ reactions with HO radicals and NO were taken from Martin et al. (10).

Comparison of Model and Experimental Results. Modeling a wide variety of different experimental conditions while varying the rate constant for the HO + HOSO$_2$ reaction, demonstrated that the results were not significantly affected by the choice of this rate constant up to a value of 1 x 10^{-10} cm^3/s. The experimental results shown in Figure 1 were obtained with NO added to the reaction system for the final 15 ms prior to HO detection. Model results for this type of experiment were found to be largely unaffected by the choice of the rate constant for the HOSO$_2$ + NO reaction.

Figure 1 shows a comparison of the model results with the experimental results. The three curves shown in the plot correspond to three different values of the rate constant for the HOSO$_2$ + O$_2$ reaction; upper - 8 x 10^{-13}, middle - 4 x 10^{-13}, and lower - 2 x 10^{-13} cm^3/s. Similar comparisons between model and experimental results have been made for a wide variety of other experimental conditions. Based upon such comparisons, we have concluded that a rate constant of (4 $^{+4}_{-2}$) x 10^{-13} cm^3/s gives the best match between the experimental and model results, in both an absolute sense and based upon the shape of the O$_2$ titration results. Since there is greater uncertainty in the absolute concentrations of HO radicals than there is in the trend of the HO concentrations with increasing O$_2$, the comparison of the shapes of the experimental and model O$_2$ titration profiles may provide a reliable basis for comparison.

From modeling a variety of different experiments with and without SO$_2$ added to the system, it was found that the HO radical wall loss must increase when SO$_2$ is added to the reaction system. Experimentally we found that an HO wall loss rate constant of 10/s worked quite well in the models of experiments performed in the absence of SO$_2$. When SO$_2$ was added to the reaction system, it was necessary to increase the HO wall loss rate constant used in the model to (18 ± 1)/s to be able to adequately describe the experimental results. The comparison between the experimental and model results shown in Figure 1 are based on the use of a HOSO$_2$ wall loss rate constant of 30/s. Our modeling studies have shown that changing the HOSO$_2$ wall loss rate constant by ± 10/s has little effect on the overall comparisons between the experimental and model results. The HO$_2$ wall loss rate constant used when SO$_2$ was present in the reaction system was 20/s. There was no effect on the modeling results of varying this wall loss rate constant by ± 10/s.

Figure 2 again shows a comparison between experimental and model results. In this experiment, the NO was added to the reaction system about 135 ms upstream of the HO radical detection system. In this case the effect of varying the rate constant for the HOSO$_2$ + NO reaction is being shown. Three different rate constants were used for the HOSO$_2$ + NO reaction; upper - 0.5 x 10^{-12}, middle - 1.0 x 10^{-12}, and lower - 2.0 x 10^{-12} cm^3/s. From several different comparisons between the experimental and model results, we have concluded that the rate constant for the HOSO$_2$ + NO reaction is best described as (1 $^{+1}_{-.5}$) x 10^{-12} cm^3/s. The data shown in Figures 1 and 2 are for essentially the same concentrations of reactants, only the reaction time for the NO in the system is different. The data shown in Figure 2 is more representative of the results that would be obtained if all of the reactants were present throughout the entire reaction period. The model results for these experiments are significantly affected by the choice of rate constants for both the HOSO$_2$ reactions with O$_2$ and NO. By performing experiments such as those shown in Figure 1, one is able to experimentally study a kinetic system which depends significantly on only one unknown rate constant. This demonstrates one of the advantages of flow studies of complex kinetics.

Figure 3 shows the comparison between the experimental and model results for a variation in the rate constant for the $HOSO_2$ + NO reaction between 0.5 and 2.0 x 10^{-12} cm^3/s. These experiments were performed with 1.1 x 10^{15} molecule/cm^3 of O_2 and a variable concentration of NO added to the reaction system 135 ms upstream of the HO detection system. The initial increase in the HO concentration upon addition of NO is due to the introduction of sufficient NO to drive the conversion of HO_2 to HO to completion. The further decay in the HO concentration with added NO is due to the combined effects of the HO reaction with NO and the $HOSO_2$ reaction with NO. Again this data suggests that the rate constant for the $HOSO_2$ + NO reaction is adequately described as $(1 ^{+.5}_{-.5})$ x 10^{-12} cm^3/s. This value of the rate constant is smaller than the 2.5 x 10^{-12} cm^3/s reported by Margitan ([8]), and larger than the ≤ 5 x 10^{-13} cm^3/s reported by Martin et al. ([10]). Our value for the $HOSO_2$ + NO rate constant is only about 2.5 times larger than that for the $HOSO_2$ + O_2 rate constant, hence the reaction of $HOSO_2$ with NO will be of no importance in the atmosphere.

The only other constituent in the atmosphere that is present in sufficient concentration to possibly compete with O_2 is a possible reaction with $HOSO_2$ is water vapor (3.9 x 10^{17} molecule/cm^3 at 50 % R.H.). If the rate constant for an $HOSO_2$ + H_2O reaction were about 5 x 10^{-12} cm^3/s, the reaction with water would be about equally important as the reaction of HOSO2 with O_2. We are currently performing experiments which will allow us to evaluate the possible effects of H_2O on the $HOSO_2$ + O_2 reaction. If the hydration of the $HOSO_2$ radical is as important as Friend et al. ([16]) suggest, the SO_2 oxidation mechanism might be altered by the presence of H_2O. None of the previous kinetics studies have been performed in the presence of H_2O, hence there is no basis for evaluating the potential importance of the $HOSO_2 \cdot OH_2$ complex under atmospheric conditions.

In summary, this work suggests that the rate constant for the $HOSO_2$ + O_2 reaction is $(4 ^{+4}_{-?})$ x 10^{-13} cm^3/s and that the rate constant for the $HOSO_2$ + NO reaction is $(1 ^{+.5}_{-.5})$ x 10^{-12} cm^3/s. This rate constant for the $HOSO_2$ + NO reaction is too slow for this reaction to ever be of any significance in the atmosphere, when compared to the $HOSO_2$ + O_2 reaction.

Acknowledgments

The authors would like to thank Juan L. Bonilla for his assistance with the data acquisition system and software development for these experiments. This research is based upon work supported by the National Science Foundation under Grants ATM-8405394 and ATM-8521192.

Literature Cited

1. Toon, O. B.; Kasting, J. F.; Turco, R. P.; Liu, M. S. J. Geophys. Res. 1987, 92D, 943-63.

2. Calvert, J. G.; Stockwell, W. R. In Acid Precipitation - Vol. 3: SO₂, NO and NO₂ Oxidation Mechanisms: Atmospheric Considerations; Ann Arbor Science Publishers Ann Arbor, MI, 1983, pp 1-62.

3. Rodhe, H.; Crutzen, P.; Vanderpol, A. Tellus 1981, 33, 132-41.

4. Acid Deposition: Atmospheric Processes in Eastern North America, National Academy of Sciences, Washington, D.C., 1983.

5. Samson, P. J.; Small, M. J. In Acid Precipitation - Vol. 9: Modeling of Total Acid Precipitation Impacts; Ann Arbor Science Publishers Ann Arbor, MI, 1984, pp 1-24.

6. Stockwell, W. R.; Calvert, J. G. Atmos. Environ. 1983, 17, 2231-5.

7. Meagher, J. F.; Olszyna, K. J.; Luria, M. Atmos. Environ. 1984, 18, 2095-104.

8. Margitan, J. J. J. Phys. Chem. 1984, 88, 3314-8.

9. Schmidt, V.; Zhu, G. Y.; Becker, K. H.; Fink, E. H. Ber. Bunsen-Ges. Phys. Chem. 1985, 89, 321-2.

10. Martin, D.; Jourdain, J. L.; Le Bras, G. J. Phys. Chem. 1986, 90, 4143-7.

11. Gleason, J. F.; Sinha, A.; Howard, C. J. J. Phys. Chem. 1987, 91, 719-24.

12. Samson, P. J., as quoted in Acid Deposition: Atmospheric Processes in Eastern North America, National Academy of Sciences, Washington, D.C., 1983 and Regional Acid Deposition: Models and Physical Processes, NCAR/TN-214+STR, Boulder, CO, 1983.

13. Seigneur, C.; Saxena, P.; Roth, P. M. Science 1984, 225, 1028-9.

14. Stockwell, W. R.; Milford, J. B.; McRae, G. J.; Middleton, P.; Chang, J. S., accepted for publication in Atmos. Env. 1988.

15. DeMore, W. B.; Margitan, J. J.; Molina, M. J.; Watson, R. T.; Golden, D. M.; Hampson, R. F.; Kurylo, M. J.; Howard, C. J.; Ravishankara, A. R. Chemical Kinetics and Photochemical Data for Use in Stratospheric Modeling. Evaluation Number 7; JPL Report 85-37, Jet Propulsion Laboratory Pasadena, CA, 1985.

16. Friend, J. P.; Barnes R. A.; Vasta, R. M. J. Phys. Chem. 1980, 84, 2423-36.

RECEIVED August 5, 1988

Chapter 27

Atmospheric Reactions of CH_3S Radicals

G. S. Tyndall and A. R. Ravishankara

National Oceanic and Atmospheric Administration, Environmental Research Laboratory, R/E/AL2, Boulder, CO 80303 and CIRES, University of Colorado, Boulder, CO 80309

Rate coefficients have been measured for the reactions of CH_3S radicals with NO_2 and O_2 at 298 K. The rate coefficient for CH_3S + NO_2 is $(6.1 \pm 0.7) \times 10^{-11}$ cm^3 molec^{-1} s^{-1}. NO was found to be the major product of the reaction, consistent with the mechanism $CH_3S + NO_2 ---> CH_3SO + NO$. CH_3SO reacts with NO_2 to give CH_3SO_2 + NO. We were only able to assign an upper limit, $\leq 2.5 \times 10^{-18}$ cm^3 molec^{-1} s^{-1}, to the rate coefficient for the reaction of CH_3S with O_2. Despite this low rate coefficient this reaction could still be important in the marine troposphere. The impact of these rate coefficients on the production of SO_2 and CH_3SO_3H in the atmosphere is discussed.

Tropospheric oxidation of dimethyl sulfide, methyl mercaptan and dimethyl disulfide, the major organo-sulfur compounds released into the atmosphere, is initiated by reaction with OH, or to a lesser extent, NO_3 (1,2). The products of these reactions are not well known, and the details of the subsequent reactions leading to completely or partially oxidized sulfur species are obscure. The knowledge that we do have on this oxidation sequence is primarily derived from indirect studies, most of which used elevated levels (>90 ppb) of NO_x (3-5). For example, all these studies agree that in the oxidation of CH_3SCH_3 (the most abundant of the organic sulfur compounds) SO_2 and CH_3SO_3H (methane sulfonic acid, MSA) are the major observed products. However, considerable discrepancies exist in the yield of SO_2 relative to MSA measured by various laboratories, and the atmospheric observations of SO_2 and MSA (6) do not agree with the product distributions seen in the laboratory.

At present the exact details of the mechanisms leading to each of the products are far from well understood (7,8). However, it seems likely that oxidation of the common naturally-occurring organo-sulfur compounds (dimethyl sulfide, methyl mercaptan and dimethyl disulfide), at least in part, through formation of the methyl thiyl radical, CH_3S (4,9-11). That is the main reason for carrying out direct studies on CH_3S radical reactions where the individual reactions are isolated.

Studies using continuous photolysis to produce CH_3S followed by end product analysis suggest that the reactivity of CH_3S with O_2 is low, with a rate

0097–6156/89/0393–0450$06.00/0

coefficient several orders of magnitude smaller than that with NO or NO_2 (3,4,7). Recent measurements by Balla et al. (12), who followed CH_3S directly in time-resolved experiments, support this conclusion. However, results obtained on the oxidation of CH_3S radicals in the complete absence of NO_x indicate greater than 80% conversion of the sulfur to SO_2 (3,13,14), indicating the possible occurrence of the $CH_3S + O_2$ reaction. Because of these uncertainties in the CH_3S oxidation scheme, we have initiated a systematic study of CH_3S chemistry, using pulsed laser photolysis with pulsed LIF detection. We report here results on the reactions of CH_3S with O_2 (1) and NO_2 (2), which are potential removal mechanisms in the atmosphere.

$$CH_3S + O_2 \ ----> \ products \tag{1}$$

$$CH_3S + NO_2 \ ----> \ products \tag{2}$$

Experimental

All experiments were carried out by pulsed 248 nm laser photolysis of CH_3SSCH_3 (DMDS) in excess bath gas to generate CH_3S, followed by pulsed laser induced fluorescence detection of CH_3S, OH, or NO. CH_3S, OH and NO were detected by exciting the $\tilde{A}(^2A_1) \ <--- \ \tilde{X}(^2E)$ transition at 371 nm, $A(^2\Sigma) \ <--- \ X(^2\Pi)$ transition at 282 nm and $A(^2\Sigma) \ <--- \ X(^2\Pi)$ at 226 nm respectively. Red-shifted fluorescence was isolated by band-pass or cut-off filters and detected by a photomultiplier tube. The pulse height from the PM tube was proportional to the concentration of the CH_3S, OH or NO in the cell and was measured by a gated charge integrator. The delay between the photolysis laser and the probe laser was varied in order to get the kinetic data. The details of such an experiment are described in a previous publication from our laboratory (15). The radiative lifetime and quenching of the $\tilde{A}$ state of CH_3S by various gases have been characterized by Black and Jusinski (16). Our values (17) of these parameters are substantially in accord with those of Black and Jusinski. For the kinetic measurements particular care was taken to optimize the sensitivity for detection of CH_3S radicals so that initial radical densities, and hence secondary reactions, could be kept to a minimum. This was particularly important for the O_2 reaction where an upper limit to the rate coefficient was being sought, and where increased radical concentrations would only lead to a faster removal rate through recombination. The photolysis laser fluence was limited to less than 5 mJ/cm², resulting in CH_3S concentrations less than 10^{12} molec cm⁻³. The detection sensitivities for CH_3S and OH, in 100 torr N_2 and O_2, and NO in 100 torr N_2, were 7×10^{10}, 1×10^9 and 3×10^8 molec cm⁻³ respectively for 100 sec averaging time.

Reaction of CH₃S with O₂

From previous measurements it is known that CH_3S does not react rapidly with O_2. Therefore, large concentrations of O_2 are needed in order to obtain a useful estimate of the upper limit. The decay of CH_3S radicals was studied in back-to-back experiments in up to 300 torr ultra-high purity O_2 and N_2; the decay rates were found to be equal, within measurable limits. At 300 torr the apparent first order loss rates were (48 ± 5) and (50 ± 5) s⁻¹ in N_2 and O_2, respectively (see Figure 1). Some recombination was seen in the first 2-3 ms since high CH_3S concentrations had to be used due to quenching of the fluorescence signal by the bath gas. The fitted lines correspond to data taken at times longer than 3ms. The error limits are 1σ values obtained from the least squares analyses of ln(signal) vs. time data. If the entire loss rate in O_2 is

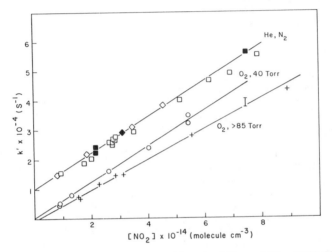

Figure 1. Decay of CH_3S radicals in 300 torr of O_2 and N_2. $[CH_3S]_0 = 8 \times 10^{11}$ molec cm^{-3}. Initial faster decay is due to the CH_3S radical recombination reaction. The fitted decay rates for $t > 3$ ms are (50 ± 5) s^{-1} for O_2 and (48 ± 5) s^{-1} for N_2.

assigned to a chemical reaction we obtain a rate coefficient ≤ 5 x 10^{-18} cm^3 molec^{-1} s^{-1}. In our experiments the major loss process for CH_3S in N_2 should be diffusion out of the detection region since it has less than 0.5 ppm of gases that could possibly react with CH_3S. The diffusion rates of CH_3S in N_2 and O_2 should be nearly identical, and there is less than 0.5 ppm of reactive gases in O_2 also. Therefore the loss observed in O_2 must also be due to diffusion, and it is appropriate to use the difference in loss rate between the two bath gases for the reactive loss of CH_3S. When the uncertainties in the measurements are included, an upper limit for the rate coefficient $k_1 \leq 2.5$ x 10^{-18} cm^3 molec^{-1} s^{-1} is obtained at the 2σ confidence level.

The absence of observable CH_3S loss in our experiments does not prove the absence of this reaction because of two reasons. First, CH_3S and O_2 could form an adduct in a reversible step. This is the case in the reaction of OH with CS_2. In such a case, unless the equilibration process is observed, the loss rate of the reactant would be very close to the loss rate in the absence of a reaction. If such equilibration occurred, the CH_3S + O_2 process could be a significant loss under atmospheric conditions, if the adduct reacts further. Observations made at the shortest times possible (< 1μs) showed no evidence for an initial fast decay in O_2, and we conclude that CH_3S and O_2 do not form an adduct, or that such an adduct must be very weakly bound. Even if the adduct were formed, it can not react rapidly with O_2, since there was no observable CH_3S loss.

The second possibility is that CH_3S could be regenerated due to formation of OH radicals. Balla and Heicklen investigated the dependence of the SO_2 yield on O_2 pressure in the C.W. photolysis of DMDS-O_2 mixtures (13). They postulated a chain mechanism propagated by a reaction between CH_3S and O_2 whose overall effect was to give OH radicals.

$$CH_3S + 2O_2 \text{-----> -----> } SO_2 + HCHO + OH$$

Their analysis yielded a lower limit for this reaction relative to the recombination of CH_3S radicals.

$$CH_3S + CH_3S \text{-----> } CH_3SSCH_3 \tag{3}$$

Balla and Heicklen based their lower limit on a value of $k_3 = 4.1$ x 10^{-14} cm^3 molec^{-1} s^{-1}. This number was apparently wrongly converted from the original work of Graham et al. (18). When the correct value of 4.1 x 10^{-11} cm^3 molec^{-1} s^{-1} obtained by Graham is used, a lower limit of 2.1 x 10^{-16} cm^3 molec^{-1} s^{-1} is obtained for the reaction of CH_3S with O_2 to give OH. We looked for, but did not observe, OH formation following pulsed laser photolysis of DMDS in 100 torr O_2. From the detection limit for OH under these conditions ($\approx 10^9$ molec cm^{-3}) and the known initial concentrations of CH_3S and DMDS we estimate that a reaction channel giving OH must occur with a rate coefficient of 3.5 x 10^{-18} cm^3 molec^{-1} s^{-1} or less. This upper limit is consistent with our measured upper limit for CH_3 S removal.

Reaction of CH₃S with NO₂

The rate coefficient for the reaction of CH_3S with NO_2 was measured in 40 torr to 200 torr of He, N_2, or O_2. In He and N_2 the rate coefficient was found to be (6.10 ± 0.10) x 10^{-11} cm^3 molec^{-1} s^{-1}, independent of the pressure, residence time in the reaction vessel and initial CH_3S concentration. At low pressures of O_2 a similar result was obtained, but as the amount of O_2 was increased the apparent rate coefficient decreased, the value at an O_2 pressure of 100 torr being 25% less than in the absence of O_2. It is probable that this is due to regeneration of CH_3S by OH formation, although the mechanism for OH production is not obvious. Indeed, OH was observed in the reaction of CH_3S with NO_2 in 100 torr O_2: its concentration was not large due to its fast reaction with DMDS ([19]). Figure 2 shows the dependence of the pseudo-first order rate coefficients on the concentration of NO_2 for experiments using He, N_2 and O_2. The He and N_2 data have been offset by 10^4 s^{-1} for clarity. We estimate a maximum uncertainty of 10% related to the measurement of flow rates and NO_2 concentration, leading to a preferred rate coefficient $k_2 = (6.1 \pm 0.7) \times 10^{-11}$ cm^3 $molec^{-1}$ s^{-1}.

Our value for the rate coefficient of the reaction $CH_3S + NO_2$ is about 40% lower than that measured by Balla et al. $(1.1 \pm 0.1) \times 10^{-10}$ cm^3 $molec^{-1}$ s^{-1} using a very similar technique ([12]). This group also found no dependence on pressure, and measured a weak negative temperature dependence, leading them to speculate that the reaction may proceed to give $CH_3SO + NO$ as products. We have measured NO produced in this reaction and found that the NO production had two temporally distinct components. The faster component could be correlated with $CH_3S + NO_2$, while the slower rise is due to the subsequent reaction of CH_3SO with NO_2. We therefore suggest that under these conditions ($\approx 10^{14}$ NO_2 cm^{-3} and ≤ 1 torr O_2) the following reactions occur:

$$CH_3S + NO_2 ----> CH_3SO + NO \qquad (2)$$

$$CH_3SO + NO_2 ----> CH_3SO_2 + NO. \qquad (4)$$

These reactions are analogous to those of HSO, as proposed by Lovejoy et al. ([20]). The NO data were fitted by a non-linear least squares routine to an analytical solution. This yielded values for k_4, the branching ratio of reaction (4) to give NO and the overall yield of NO. The fitted values were not definitive since some of the NO appears to be produced vibrationally excited, and relaxation may not have been complete on the time scale of the experiment. Values of k_4 fell in the range $(8 \pm 4) \times 10^{-12}$ cm^3 $molec^{-1}$ s^{-1}. The overall yield of NO produced was around 1.5 per CH_3S, and we suspect that the yields may actually be close to 1.0 for both reactions, (2) and (4). Further experiments are in progress to elucidate the reaction sequence. More detailed accounts of both the O_2 and NO_2 reactions will be published shortly ([21]).

Discussion

The results of our experiments, along with those of Balla et al. ([12]) confirm the earlier findings of Grosjean ([3]) and Hatakeyama and Akimoto ([4]), that reaction of CH_3S with O_2 is much slower than with NO or NO_2. Grosjean obtained $k_{NO2}/k_{O2} = 2 \times 10^6$, and Hatakeyama $k_{NO}/k_{O2} = 2 \times 10^3$ at 760 torr. These estimates lead to rate coefficients for $CH_3S + O_2$ of 3×10^{-17} and 2×10^{-14} cm^3 $molec^{-1}$ s^{-1} respectively, based on the directly measured rate coefficients for the NO_2 and NO ([12]) reactions. However, the mechanisms used in deriving both of these ratios were incomplete. Our results suggest that CH_3S reacts very slowly with O_2, if it reacts at all. The mechanism proposed by Balla and Heicklen, in which CH_3S is regenerated by OH radicals, would lead to a detectable amount of OH in our system, if the rate coefficient were close to 10^{-16} cm^3 $molec^{-1}$ s^{-1} as their (corrected) result suggests. As stated earlier, no OH was detected and we could place an upper limit for this process which is two orders of magnitude lower than Balla and Heicklen's value.

The efficient conversion of CH_3S to SO_2 in the C.W. photolysis experiments which used very low levels of NO_x points to the possible occurrence of a $CH_3S + O_2$ reaction in the atmosphere. Further experiments are required to assess

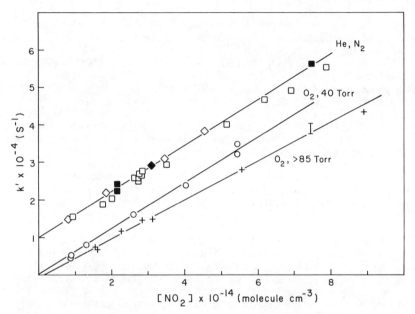

Figure 2. Pseudo first order rate coefficient for $CH_3S + NO_2$ as a function of $[NO_2]$. Data for He and N_2 have been offset by 10^4 s^{-1} for clarity. Symbols used are: □ He, 40 torr; ■ He, 100 torr; ◇ N_2, 40 torr; ◆ N_2, 85 torr; ○ O_2, 40 torr; + O_2, ≥ 85 torr.

this possibility. One explanation is that radical- radical reactions of CH_3S could become important in experiments where SO_2 production is observed under conditions of very low NO_x.

$$CH_3S + HO_2 \longrightarrow CH_3SO + OH$$

$$CH_3S + RO_2 \longrightarrow CH_3SO + RO$$

Akimoto and coworkers detected S_2 following 248 nm photolysis of DMDS, implying that CH_3 radicals are also produced ([22]). Low levels of CH_3O_2 and HO_2 radicals that could be produced in the C.W. experiments would not be removed by reactions with NO_x but rather sustain a chain process leading to SO_2 production.

The upper limit for the $CH_3S + O_2$ reaction rate coefficient determined here is an order of magnitude lower than previous estimates. Even with this lower value we still can not rule out this reaction in the atmosphere. O_2 has a mixing ratio of 0.21 which implies a loss rate for $CH_3S < 12$ s^{-1}. Even though the NO_2 reaction rate coefficient is 6 x 10^{-11} cm^3 $molec^{-1}$ s^{-1}, the reaction with O_2 could be dominant for NO_2 concentrations less than 2 x 10^{11} molec cm^{-3}, or a mixing ratio of ~8 ppb. Since most CH_3S is produced in the marine troposphere, where NO_2 is typically < < 1ppb, the O_2 reaction can not be ruled out. Since the $CH_3S(O_2)$ adduct, if it is formed at all, does not appear to react with O_2, our upper limit is the real upper limit for CH_3S loss in air.

Pertinent to this discussion is the possible occurrence of a reaction between CH_3S and O_3. Several recent studies have found that HS reacts with O_3 with a rate coefficient of approximately 4 x 10^{-12} cm^3 $molec^{-1}$ s^{-1} ([23,24]). In view of the similarity of the rate coefficient for CH_3S with NO_2, 6.1 x 10^{11} cm^3 $molec^{-1}$ s^{-1}, with that for HS with NO_2, 6.7 x 10^{-11} cm^3 $molec^{-1}$ s^{-1} ([25]), it could be expected that CH_3S reacts with O_3 with a rate coefficient in excess of 10^{-12} cm^3 $molec^{-1}$ s^{-1}. Black and Jusinski reported an upper limit for this reaction of 8 x 10^{-14} cm^3 $molec^{-1}$ s^{-1}, but the determination used very high (> 10^{16} molec cm^{-3}) concentrations of O_3 ([26]). Since the potential for regeneration of CH_3S is very high in such a system (cf Friedl et al. ([23]) in the HS + O_3 reaction), a reaction between CH_3S and O_3 should not be excluded.

The rate coefficient for CH_3S with NO_2 reported here is about 40% lower than that reported by Balla et al. While we observed a decrease of up to 30% in the value of the rate coefficient in 100 torr O_2, we believe this to be due to regeneration of CH_3S in our NO_2-rich system. However we feel that the experimental conditions were sufficiently well-controlled that we have eliminated systematic errors.

There have recently been several measurements of the rate coefficient for the reaction of HS with NO_2 ([25] and references therein). Our value for $CH_3S + NO_2$ is very close to the rate coefficient for HS + NO_2 determined by Wang et al. ([25]).

Our results shed some light on the mechanism for the conversion of sulfides to SO_2 and MSA in the atmosphere and in chamber experiments. If the NO_2 (or an O_3) reaction dominates, then CH_3S will be converted to CH_3SO:

$$CH_3S + NO_2 (O_3) \longrightarrow CH_3SO + NO (O_2)$$

CH_3SO, it seems, can further react with NO_2, or, as recently shown by Barnes et al. ([27]), add O_2:

$CH_3SO + NO_2 \text{-----} > CH_3SO_2 + NO$

$CH_3SO + O_2 + M \text{-----} > CH_3S(O)O_2 + M$

Barnes et al. identified a product $CH_3S(O)O_2NO_2$, suggesting that O_2 can effectively compete with NO_2 for CH_3SO at atmospheric pressure.

It seems, therefore, that in systems with high NO_x (i.e., chamber experiments) formation of CH_3SO and CH_3SO_2 may predominate, and this could lead to the elevated levels of MSA detected in these systems. Hatakeyama et al. (28) have recently used isotopic substitution to show that some of the SO_2 formed comes from NO_2 reaction with CH_3S or CH_3SO. Furthermore, at high NO_x the SO_2 yield is independent of O_2 or N_2 (27,28), pointing strongly to the involvement of these radicals in the production of SO_2. One of the outstanding problems remains, therefore, to measure the rate coefficients of reactions of CH_3SO and CH_3SO_2 with, for example, NO_2, O_2, and O_3, and to identify the products. One particularly interesting question is whether the production of CH_3SO radicals in air at low NO_x mixing ratio leads to SO_2 or MSA formation.

Under clean troposphere conditions, however, an oxidation chain initiated by $CH_3S + O_2$ could give different products, and it is essential to understand the mechanism of SO_2 formation at low NO_x. It will be interesting to see whether the same radicals lead to SO_2 formation in the high and low NO_x cases. Since the reaction of CH_3S with O_2 appears to be so slow, it may be necessary to perform C.W. photolysis studies under very carefully controlled conditions in order to determine whether a slow reaction actually occurs, or whether secondary reactions lead to the observed products.

Acknowledgments

This work was supported by NOAA as part of the National Acid Precipitation Assessment Program.

Literature Cited

1. Atkinson, R.; Pitts, J. N., Jr.; Aschmann, S. M. J. Phys. Chem. 1984, 88, 1584.

2. Andreae, M. O.; Ferek, R. J.; Bermond, F.; Byrd, K. P.; Engstrom, R. T.; Hardin, S.; Houmere, P. D.; LeMarrec, F.; Raemdonck, H.; Chatfield, R. B. J. Geophys. Res. 1985, 90, 12891.

3. Grosjean, D. Environ. Sci. Tech. 1984, 18, 460.

4. Hatakeyama, S.; Akimoto, H. J. Phys. Chem. 1983, 87, 2387.

5. Hatakeyama, S.; Izumi, K.; Akimoto, H. Atmos. Environ. 1985, 19, 135.

6. Saltzman, E. S.; Savoie, D. L.; Zika, R. G.; Prospero, J. M. J. Geophys. Res. 1983, 88, 10897.

7. Yin, F.; Grosjean, D.; Seinfeld, J. H. J. Geophys. Res. 1986, 91, 14417.

8. Toon, O. B.; Kasting, J. F.; Turco, R. P.; Liu, M. S. J. Geophys. Res. 1987, 92, 943.

9. Hynes, A. J.; Wine, P. H. J. Phys. Chem. 1987, 91, 3672.

458   BIOGENIC SULFUR IN THE ENVIRONMENT

10. Niki, H.; Maker, P. D.; Savage, C. M.; Breitenbach, L. P. Int. J. Chem. Kin. 1983, 15, 647.
11. MacLeod, H.; Jourdain, J. L.; Poulet, G.; LeBras, G. Atmos. Environ. 1984, 18, 2621.
12. Balla, R. J.; Nelson, H. H.; McDonald, J. R. Chem. Phys. 1986, 109, 101.
13. Balla, R. J.; Heicklen, J. J. Photochem. 1985, 29, 297.
14. Barnes, I.; Bastian, V.; Becker, K. H.; Fink, E. H. In Physico-Chemical Behaviour of Atmospheric Pollutants; Angeletti, G; Restelli, G., Eds.; Fourth European Symposium: Reidel, Dordrecht, 1987; p 327.
15. Wahner, A.; Ravishankara, A. R. J. Geophys. Res. 1987, 92, 2189.
16. Black, G.; Jusinski, L. E. J. Chem. Phys. 1986, 85, 5379.
17. Tyndall, G. S.; Ravishankara, A. R., in preparation.
18. Graham, D. M.; Mieville, R. L.; Pallen, R. H.; Sivertz, C. Can. J. Chem. 1964, 42, 2250.
19. Wine, P. H.; Kreutter, N. M.; Gump, C. A.; Ravishankara, A. R. J. Phys. Chem. 1981, 85, 2660.
20. Lovejoy, E. R.; Wang, N. S.; Howard, C. J. J. Phys. Chem. 1987, 91, 5749.
21. Tyndall, G. S.; Ravishankara, A. R., J. Phys. Chem. 1988, submitted.
22. Suzuki, M.; Inoue, G.; Akimoto, H. J. Chem. Phys. 1984, 81, 5405.
23. Friedl, R. R.; Brune, W. H.; Anderson, J. G. J. Phys. Chem. 1985, 89, 5505.
24. Schönle, G.; Rahman, M. M.; Schindler, R. N. Ber. Bunsenges. Phys. Chem. 1987, 91, 66; Schindler, R. N.; Benter, T. ibid 1988, 92, 588.
25. Wang, N. S.; Lovejoy, E. R.; Howard, C. J. J. Phys. Chem. 1987, 91, 5743.
26. Black, G.; Jusinski, L. E. J. Chem. Soc. Faraday 2 1986, 82, 2143.
27. Barnes, I.; Bastian, V.; Becker, K. H.; Niki, H. Chem. Phys. Lett. 1987, 140, 451.
28. Hatakeyama, S; Akimoto, H., proceedings, this symposium.

RECEIVED September 6, 1988

Chapter 28

Mechanisms for the Reaction of CH$_3$S with NO$_2$

Shiro Hatakeyama

Division of Atmospheric Environment,
The National Institute for Environmental Studies,
P.O. Tsukuba, Ibaraki 305, Japan

Product analyses for the reaction of CH$_3$S with NO$_2$ were carried out in order to elucidate the mechanism for that reaction in air. SO^{18}O was observed by means of FT-IR spectroscopy when NO^{18}O was used as a reactant. This is a clear evidence for the formation of CH$_3$SO and NO as products of the above reaction. Dependence of the yield of SO$_2$ on the initial concentration of O$_2$ and NO$_2$ was observed, which indicates that the secondary reactions of CH$_3$SO with O$_2$ or NO$_2$ are important in the atmospheric oxidation of reduced organic sulfur compounds.

Oxidation of biogenic reduced sulfur compounds such as thiols (RSH), sulfides (RSR), and disulfides (RSSR) is important from a point of view of atmospheric chemistry. Many studies concerning kinetics and mechanisms of those compounds with O(^{3}P), OH, and NO$_3$ are reported. In those studies the reaction of CH$_3$S radical with O$_2$ or NO$_x$ was pointed out to be most significant in the atmosphere. However, the mechanism for the reaction of CH$_3$S with O$_2$ or NO$_2$ is still unclear by now. In our previous study ($\underline{1}$) we reported that CH3S reacts with O$_2$ to give about 90% of SO$_2$ without forming CH$_3$ radicals. The same conclusion was obtained by Balla and Heicklen ($\underline{2}$). In both the studies it was confirmed that CH$_3$S + O$_2$ does not give CH$_3$ radicals.

As for the reaction of CH$_3$S with NO$_2$ little information is available concerning the reaction products and the reaction mechanism. CH$_3$SNO$_2$ was reported as a product of CH$_3$S + NO$_2$ recombination reaction ($\underline{3},\underline{4}$). However, its contribution is reported to be minor ($\underline{3},\underline{5}$). Balla et al. ($\underline{6}$) suggested following reaction pathways in addition to the recombination path.

$$CH_3S + NO_2 \text{ ---> } CH_2S + HONO \tag{1}$$

$$CH_3S + NO_2 \text{ ---> } CH_3SO + NO \tag{2}$$

They also reported ($\underline{6}$) the rate constants for the reactions of CH$_3$S + NO, CH$_3$S + NO$_2$, and CH$_3$S + O$_2$ and estimated the rate-constant ratio to be k(NO)/k(O$_2$)>2 x 10^6 and k(NO$_2$)/k(O$_2$)>5 x 10^6. On the basis of these rate-constant ratio, photooxidation mechanism for reduced sulfur compounds must be re-examined, because most experimental studies used much NO$_2$.

0097–6156/89/0393–0459$06.00/0

In this study we investigated the reactivity of CH_3S radical with O_2 and NO_x utilizing the photolysis of CH_3SNO as a source of CH_3S radical. $NO^{18}O$ was also used to check if the oxygen atom of NO_2 transferred to a product SO_2 molecule.

Experimental

The photochemical reactor used for the photolysis of CH_3SNO was an 11-L cylindrical quartz vessel (120 mm i.d., 1000 mm in length) equipped with multi-reflection mirrors for long-path Fourier transform infrared spectroscopy (Block Engineering Co., JASCO International Inc., FTS-496S). The light source for photolysis was six yellow fluorescent lamps (National FL20S Y-F, $500 < \lambda < 700$ nm, $\lambda_{max} = 600$ nm), which surrounded the reactor coaxially. No photodecomposition of NO_2 took place by use of this light source.

CH_3SNO was prepared according to the method reported by Philippe (4). Namely, CH_3SH and N_2O_3 were mixed at low temperature. $NO^{18}O$ was prepared by mixing pure NO (Matheson) with excess of $^{18}O_2$ (atomic purity > 99.5 %, Nippon Sanso).

Effects of the different concentration of oxygen or NO_2 were checked. Typical reaction conditions were $[CH_3SNO]_0 \sim 8$ mTorr, $[NO]_0 \sim 1$ mTorr, $[NO_2]_0 \sim 1-8$ mTorr, $[O_2]_0 \sim 0-760$ torr under total pressure of 1 atm with N_2. Photolysis was performed for about 2 h and the change in the concentration of reactants and products were monitored by FT-IR (path length: 51.6 m, resolution: 1 cm^{-1}, scan times: 64). All the experiments were carried out at room temperature.

Results and Discussion

At first, photolysis of CH_3SNO in the presence of $NO^{18}O$ was performed in air in order to confirm that the reaction of CH_3S with NO_2 gives SO_2 possibly via $CH_3SO + NO$ formation. Photoirradiation of the mixture ($[CH_3SNO] = 7.5$ mTorr, $[NO_2] = 2.0$ mTorr, $[NO^{18}O] = 8.9$ mTorr under 1 atm of air) gave a mixture spectrum as shown in Figure 1, B. After subtraction of SO_2 a signal at 1341.5 cm^{-1} was clearly seen (Figure 1, C). This signal agrees well with that of $SO^{18}O$ (1341.1 cm^{-1}) reported by Polo and Wilson (8). Since the light emitted by the yellow fluorescent lamps does not photolyze NO_2, the observed $SO^{18}O$ is not a product of the reaction of ^{18}O atom released from $NO^{18}O$. The observed yield ratio of $SO^{18}O/SO_2$ was 0.36 (IR absorptivity of $SO^{18}O$ was assumed to be the same as that of SO_2) and it was consistent with the 1:1 formation of $SO^{18}O$ and SO_2 as expected from the following mechanism when the initial concentration ratio of $NO^{18}O$ to NO_2 was taken into account.

$$CH_3SNO + h\nu \:\text{---}\!\!> CH_3S + NO \qquad (3)$$

$$CH_3S + NO^{18}O \:\text{---}\!\!> CH_3S^{18}O + NO, \; CH_3SO + N^{18}O \qquad (4)$$

$$CH_3S^{18}O + O_2 \:\text{---}\!\!> \text{---}\!\!> SO^{18}O + CH_2O \qquad (5)$$

$$CH_3SO + O_2 \:\text{---}\!\!> \text{---}\!\!> SO_2 + CH_2O \qquad (6)$$

On the basis of these results we can conclude that the reaction of CH_3S with NO_2 gives SO_2 in air. This is the first clear evidence for the formation of SO_2 from $CH_3S + NO_2$ reactions in air. Between two proposed reaction paths, (1) and (2), $CH_3SO + NO$ forming path is the most plausible primary reaction, since substantial amount of $N^{18}O$ was observed as a product and neither CH_2S nor HONO was detected.

Next, the effects of the different concentration of oxygen or NO_2 were checked. Table I shows the yield of major products; formaldehyde, SO_2, and dimethyl disulfide (DMDS). The yield of DMDS decreased as the O_2 concentration increased. SO_2 and formaldehyde were produced in almost the same yield. Thus, these two compounds are produced from the same precursor without forming CH_3 radical. If CH_3 radical is formed in air, CH_3OH must be formed and the yield of CH_2O must be smaller than that of SO_2. The fact that neither CH_3ONO nor CH_3ONO_2 was formed also indicates that CH_3 radical is not formed in this reaction system. Since the production of the compound which is tentatively assigned to CH_3SNO_2 (3,4) was also detected in this reaction system by FT-IR spectroscopy, the recombination of CH_3S with NO_2 should be an additional path for this reaction. However, the sum of the concentration of nitrogenous compounds, i.e., $[CH_3SNO] + [NO_2] + [NO]$, after 2 h of irradiation with about 50% consumption of CH_3SNO during this reaction time is more than 95% of initial $[CH_3SNO]_0 + [NO_2]_0 + [NO]_0$. Thus, the fraction of adduct formation path is minor as pointed out by Niki et al. (3). The increase of the yield of SO_2 with the increase of O_2 also supports the contention that the secondary reaction of CH_3SO radical with O_2 can be the source of SO_2 as depicted by Equations 5 and 6. By analogy with the mechanism for the formation of SO_2 from $CH_2SO_2 + O_2$ (2), the following equation is proposed here to explain the formation of SO_2 from CH_3SO.

$$CH_3SO + 2O_2 \text{ ---> } CH_2O + SO_2 + HO_2. \tag{7}$$

Dependence of the yield of SO_2 on the concentration of NO_2 was also observed (Table I). It suggests that the reaction of CH_3SO with NO_2 can be an additional reaction channel to form SO_2. Recently, Lovejoy et al. (9) reported the reaction of HSO with NO_2 and suggested the mechanism to be the formation of $HSO_2 + NO$. By analogy with this reaction following mechanism is proposed here for the reaction of $CH_3SO + NO_2$.

$$CH_3SO + NO_2 \text{ ---> } CH_3SO_2 + NO \tag{8}$$

$$CH_3SO_2 + O_2 \text{ ---> } CH_2O + SO_2 + OH \text{ (2)}. \tag{9}$$

There is another reaction path for CH_3SO radicals. As shown in Table I, DMDS was produced in a high yield, particularly in the absence of O_2. This is rather a curious thing, because the DMDS is usually thought to be formed by the self reaction of CH_3S radicals ($CH_3S + CH_3S \text{ ---> } CH_3SSCH_3$). However, in the present reaction condition this self reaction cannot compete with the very rapid reaction $CH_3S + NO_2$ (rate constant = 1.1×10^{-10} cm³molecule⁻¹s⁻¹ (6)) in the presence of much NO_2. Thus, there must be another reaction to form DMDS. One possible reaction is

$$CH_3SO + CH_3SNO \text{ ---> } CH_3SSCH_3 + NO_2 \text{ (5)}. \tag{10}$$

From the results of the present study it became clear that the reaction of CH_3S with NO_2 in the presence of O_2 gives SO_2 and CH_2O. It was also suggested that the reactions of CH_3SO and CH_3SO_2 radicals with O_2 play a very important role in the formation of SO_2 from the atmospheric oxidation of reduced sulfur compounds. More detailed studies on the reactions of CH_3S, CH_3SO, and CH_3SO_2 are needed.

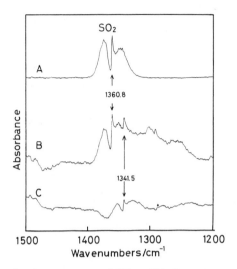

Figure 1.(A) Standard spectrum of SO_2; (B) Spectrum of the products of CH_3SNO photolysis in air with $NO^{18}O$; (C) Spectrum of $SO^{18}O$ obtained by subtraction of SO_2 from (B).

Table I. Initial conditions and product yields for CH_3SNO photolysis at room temperature under 1 atm of total pressure

$[CH_3SNO]_0$	$[O_2]_0$	$[NO_2]_0$	$[NO]_0$	Product Yield (%)		
mTorr	Torr	mTorr	mTorr	DMDS	HCHO	SO_2
9.24	~0	1.55	0.75	71	~0	4
7.16	76	1.06	0.48	52	7	9
9.49	152	1.52	0.97	54	19	19
9.49	380	1.92	1.38	42	18	24
7.60	760	1.86	0.82	20	41	36
8.05	152	1.01	0.65	67	23	20
9.49	152	1.52	0.97	54	19	19
8.76	152	4.73	1.45	18	22	25
7.38	152	6.79	1.94	13	30	37
5.83	152	7.87	4.36	-	45	56

Literature Cited

1. Hatakeyama, S.; Akimoto, H. J. Phys. Chem., 1983, 87, 2387.
2. Balla, R. J.; Heicklen, J. J. Photochem., 1985, 29, 297.
3. Niki, H.; Maker, P. D.; Savage, C. M.; Breitenbach, L. P. J. Phys. Chem., 1983, 87, 7.
4. MacLeod, H.; Aschman, S. M.; Atkinson, R.; Tuazon, E. C.; Sweetman, J. A.; Winer, A. M.; Pitts, J. N., Jr. J. Geophys. Res., 1986, 91, 5338.
5. Balla, R. J.; Heicklen, J. J. Phys. Chem., 1984, 88, 6314.
6. Balla, R. J.; Nelson, H. H.; McDonald, J. R. Chem. Phys., 1986, 109, 101.
7. Philippe, R. J. J. Mol. Spectrosc, 1961, 6, 492.
8. Polo, S. R.; Wilson, M. K. J. Chem. Phys., 1954, 22, 900.
9. Lovejoy, E. R.; Wang, N. S.; Howard, C. J. J. Phys. Chem., 1987, 91, 5749.

RECEIVED July 13, 1988

Chapter 29

Impact of Halogen Oxides on Dimethyl Sulfide Oxidation in the Marine Atmosphere

I. Barnes[1], K. H. Becker[1], D. Martin[1], P. Carlier[2], G. Mouvier[2], J. L. Jourdain[3], G. Laverdet[3], and G. Le Bras[3]

[1]Physikalische Chemie FB 9, G.H.S. Wuppertal, Bergische Universität, 400 Wuppertal, Federal Republic of Germany
[2]Laboratoire de Physico-Chimie de l'Atmosphère, Université de Paris VII, 2 Place Jussieu, 75251 Paris Cédex 05, France
[3]Centre National de la Recherche Scientifique, Centre de Recherche sur la Chimie de la Combustion et des Haute Temperatures, 45071 Orléans Cédex 2, France

Laboratory kinetic measurements have shown that the reactions of DMS with halogen oxide radicals, specially IO, and probably to a much lesser extent ClO and BrO, are potentially important in the DMS oxidation in the marine atmosphere. The reaction of DMS with IO could explain the lower values of the DMS lifetime obtained from different field measurements. These values range from fractions of a hour to approximately 40 hours. The present paper successively reports on these field measurements, the laboratory kinetic data obtained for the reaction of DMS with IO, ClO and BrO, and the possible impact of these reactions in marine air chemistry. The discussion of this impact shows the need for further field and laboratory data to better estimate the role of these reactions in marine air chemistry.

The importance of organosulfur compounds in the atmospheric sulfur cycle became apparent during the seventies when it was realized that a flux of gaseous sulfur compounds from the ocean surface into the atmosphere was needed to balance known sinks. The assumption of the existence of a marine source of reduced sulfur compounds allowed a balancing of the sulfur atmospheric budget and explained the variations of the ratio S(VI)/Na in the atmospheric aerosols in non polluted marine areas. These variations demonstrated that a major fraction of sulfur was not injected into the atmosphere by the bubbling of the sea. It was assumed previously that the biogenic sulfur emitted mostly was hydrogen sulfide. However, in 1977 Lovelock et al. (1) proposed that the key compound was more likely to be dimethylsulfide DMS. Since that time, many studies related to the exchange of trace gases between ocean and atmosphere have confirmed this proposal. Other reduced sulfur species have also been identified: methanethiol, dimethyldisulfide, carbon disulfide, carbonyl sulfide (2-8).

0097–6156/89/0393–0464$06.00/0

The studies on the organosulfur compounds in the atmosphere can be grouped in two main series:
1. geochemical studies which describe the spatio-temporal distribution of these compounds in the atmosphere and which attempt to quantify the air/sea exchange of these substances;
2. physico-chemical studies which establish the mechanisms of chemical transformations of these compounds during their residence in the atmosphere. The present paper deals with this second aspect.

A major difficulty of the physico-chemical studies is to link the observations in the atmosphere and the kinetic data determined in the laboratory (elementary reaction kinetics and atmospheric simulation in smog chambers). The most amenable method of comparing field information with kinetic data is by means of the chemical residence time τ, defined by the relationship:

$$\frac{1}{\tau} = \Sigma_X \ k_{X+A} \ [X]$$

where X is a chemical species which can react with the compound A, and k_{X+A} is the rate constant of the reaction X + A --> products. Together with the k_{X+A} values determined in the laboratory the concentrations of X are needed. [X] has to be measured in the atmosphere which is in many cases very difficult, or information on [X] has to be obtained by model calculations considering all the chemical reactions which control the concentration of X. The studies of reaction products and their yields have also been carried out by laboratory investigations.

The Tropospheric Residence Time of DMS

Three determinations of the DMS residence time in marine atmosphere from field measurements have been reported (4,6-10). These determinations lead to widely different results. The results of these field measurements and their possible uncertainties are briefly described below.

The measurements carried out by the Centre des Faibles Radioactivités (CFR) in France (4,9) are based on the determination of the vertical profile of DMS concentrations in the first few meters above the ocean. These data were obtained during the oceanographic mission OCEAT 1 in the Pacific ocean in November 1982. A sharp decrease of the concentration was observed between the surface and 20 m height from which the residence time could be estimated by making the following assumptions,
- the concentration profile was stationary within the 20 m height during the sampling,
- the DMS oxidation process obeyed a first order rate law,
- the DMS vertical concentration profile was only dependent on the first order oxidation rate constant $\lambda = 1/\tau$ and the vertical eddy diffusion coefficient K_z.
From these assumptions it follows,

$$C(z) = C(o) \ e^{-\left(\sqrt{\frac{\lambda}{K_z}} \ z \right)}$$

where $C(z)$ and $C(o)$ are the concentrations at sea level and a height z, respectively. K_z is assumed to be constant with height and is given by the relation:

$K_z = 0.41 \, u^* \, z$

where the values of u^* and z at an altitude of 10 m are used. The friction velocity u^* is related to the wind velocity u by the expression:

$$u = \frac{u^*}{0.41} \; \text{Log} \; \frac{1000}{z_0} \quad \text{(CGS)}$$

where z_0 is the roughness length given by:

$$z_0 = \frac{u^{*2}}{ag} \quad (a = 0.81 \text{ and } g = 981 \text{ cm s}^{-2})$$

This technique is particularly applicable for the emission study from extended sources which are homogeneous enough on a large scale. The accuracy of these determinations depends strongly on the validity of the K_z estimations. In addition, this model does not take into account the convection which is considered as negligible compared to the turbulence. This approximation seems to be justified for the first few meters above the ocean surface. The lifetimes measured were essentially lower than 9h with values as low as few minutes. The large variability obtained remains to be explained.

The determinations of the Laboratoire de Physico-Chimie de l'Atmosphère (LPCA) in France (10) are based on measurements of horizontal concentrations profiles of organo-sulfur compounds close to a local intensive source (algae fields) such as one finds in the Breton coastal area. These results should be considered as preliminary. Mean concentrations of reduced sulfur compounds measured over periods of 6 hours during a field campaign in September 1983 were collected at three different places: two of them were located on the side of the algae field at altitudes of 5 and 40 m. The third place was situated 4 km inland. The DMS residence time values were obtained assuming the following:
- all the gaseous compounds were equally affected by dispersion and transport - consequently the variations of the concentration ratio of these compounds only depends on physico-chemical processes.
- at a given time, the reduced sulfur compounds were homogeneously emitted from the algae field.
- all depletion rates of the sulfur compounds followed a pseudo first order rate law.
- the dry deposition of gaseous compounds was negligible.
Making some assumptions on the chemical filiation between some organo-sulfur compounds, it was possible to establish the mathematical variation law for the concentration ratio of the various detected species and consequently to deduce the depletion rate constant of these compounds. From the measurements at the "Pointe de Penmarc'h" in September 1983, the DMS lifetime estimations obtained are reported in Table I. This method for determining chemical lifetimes can only be applied for local and intensive sources. The most critical point concerns the chemical relation between the various sulfur compounds which should be verified in order to validate these estimations. However, the other assumptions do not seem to have a significant influence on the lifetime estimation within an order of magnitude.

The third series of measurements have been carried out by Andrea et al. (6,7). These authors have attempted to assess the DMS residence time by means of the model discussed by Chatfield and Crutzen (11). They have determined the standard profile for the daily variation of the DMS concentration in the mixing layer. The best fit between simulation and experimental results was obtained for a mean residence time of 36 h. However, in two recent papers, Andreae et al. (12,13) have concluded from investigations concerning the vertical profiles (from 0 to 6 km) of the DMS concentration in the marine troposphere, that the model of Chatfield and Crutzen, which can be applied only for extremely convective situations, overestimates the "Staubsauger Effect." As a consequence, the residence time of DMS may be shorter (8-10 h) than the previous estimations by these authors. In addition, Bingemer (14) has compared the concentration of DMS measured over the equatorial Atlantic with those calculated from models based on known marine source strengths for DMS and the accepted rate constant for the OH sink reaction. In all cases his measured DMS concentration was significantly under those predicted by the various models and he concluded that an unknown sink for DMS was the most probable cause for the discrepancy.

In conclusion, there are many uncertainties involved in estimating the atmospheric lifetimes of DMS from field measurements. However, even if these uncertainties are taken into account, the field measurements mentioned above (4,9,10,14), show that the DMS atmospheric lifetime is often much shorter than that currently predicted by model observations using the known sink reactions with OH and NO_3. It has been suggested (15) that halogen oxides may be possible, as yet unconsidered, candidates for species which could react rapidly with DMS and consequently explain the short observed DMS lifetimes.

Table I. DMS Lifetime Estimations

Day	09/16/1983				09/17/1983		
Time period (hours)	0 - 6	6 - 12	12 - 18	18 - 24	6 - 12	12 - 18	18 - 24
τ (min.)	< 29	< 34	< 48	< 37	$80 < \tau < 165$	$90 < \tau < 190$	$70 < \tau < 135$

Kinetic Studies of the Reactions of Halogen Oxide Radicals IO, BrO and ClO with DMS

Reaction IO + DMS ------> products (1). The reaction IO + DMS has recently been investigated by two different techniques (16,17). The study of Barnes et al. (16) was carried out in 760 Torr N_2 at 296 K using a simulation chamber technique and FTIR analysis, whereas, Martin et al. (17) used a discharge flow-mass spectrometry system at 1 Torr helium pressure by 298 K.

Both studies showed a fast reaction between IO and DMS. In their study, Barnes et al. (16) obtained reasonable agreement between experimental and calculated concentration-time profiles for the $NO_2/I_2/N_2/DMS$ photolysis system. Within the experimental error limits, they found that for every molecule of NO_2 photolyzed, one molecule of DMS is consumed and one molecule each of NO and DMSO are formed and concluded that the reaction IO + DMS ----> DMSO + I occurs. A best fit between experimental and computed concentration-time profiles including the formation of DMSO was obtained for a value of $k_1 = (3.0 \pm 1.5) \times 10^{-11}$ cm^3 molecule^{-1}s^{-1} for reaction (1).

In their study, Martin et al. ([17]) obtained from 13 experiments the best agreement between experimental and computed values with the rate constant: $k_1 = (1.5 \pm 0.5) \times 10^{-11}$ cm^3 molecule^{-1}s^{-1} for the reaction IO + DMS. Moreover, the mass spectrometric analysis of the products of reaction (1) led to the detection of a peak at mass 78, which corresponds to the parent peak of DMSO. Although a quantitative analysis of DMSO was not possible in the discharge flow system, this observation supports the mechanism proposed by Barnes et al. ([16]).

The two rate constant determinations by very different experimental techniques are in reasonable agreement. This seems to indicate that the rate constant of the IO + DMS reaction cannot exhibit a significant pressure dependence.

Reactions ClO + DMS ----> products (2) and BrO + DMS ----> products (3). Preliminary experiments on the reaction ClO + DMS have been carried out at 760 Torr total pressure in the 420 l glass reaction chamber at Wuppertal. Cl_2/O_3 and $SOCl_2/O_3$ photolysis systems were used to produce ClO radicals through the reaction Cl + O_3 ----> ClO + O_2. Since the reaction Cl + DMS (2^1) was found to be very fast ($k_{298~K} = (2.0 \pm 0.4) \times 10^{-10}$ cm^3 molecule^{-1}s^{-1}) this led to difficulties in interpretating the data and studies on the reaction using steady-state photolysis techniques were consequently abandoned. The reactions ClO + DMS and BrO + DMS are currently being investigated at Wuppertal University using the discharge flow mass spectrometric technique ([18]). Excess DMS was used to ensure pseudo-first order conditions. ClO and BrO were monitored at their parent peaks and were produced by the reactions Cl + O_3 and Br_2 + O, respectively. Cl and O atoms were generated by microwave dissociation of Cl_2 and O_2, respectively. For a DMS concentration ranging from 1.5×10^{14} to 1.5×10^{15} molecules cm^{-3} and for a ClO concentration kept about 5 $\times 10^{12}$ molecules cm^{-3} the following rate constant was obtained at 298 K and 2 Torr for the reaction ClO + DMS ----> products:

$$k_2 = (3.9 \pm 0.5) \times 10^{-14} \text{ cm}^3 \text{ molecule}^{-1}\text{s}^{-1}.$$

However, the intercept of the straightline - d/dt (ln [ClO]) = f ([DMS]) -, which was found to be (80 ± 5)s^{-1} indicates a substantial wall reactivity of ClO in the presence of DMS. Some dependence of ClO wall loss with DMS concentration can exist if the reactor wall is not saturated with DMS in the range of DMS concentration used to derive k_2 from the plots - d ln [ClO] /dt = f([DMS]). In such a case k_2 is an upper limit of the rate constant of the homogeneous reaction. In the absence of any indication of the saturation degree of the reactor wall in this study, k_4 should be considered as an upper limit for the rate constant. Such complications have not been observed for the reaction BrO + DMS ----> products (3). For DMS and BrO concentrations ranging from 1.5×10^{14} to 7.9×10^{14} molecules cm^{-3} and from 1.33×10^{13} to 1.55×10^{13} molecules cm^{-3}, respectively, the following rate constant was determined at 298 K and 1.4 Torr:

$$k_3 = (2.65 \pm 0.65) \times 10^{-13} \text{ cm}^3 \text{ molecule}^{-1}\text{s}^{-1}$$

In both studies, product analysis led to the detection of a peak at mass 78 which corresponds to the parent peak of DMSO. Both reactions appear to produce DMSO via a mechanism similar to the one proposed for the IO + DMS reaction:

$$XO + DMS \text{ -------> } DMSO + X$$

However, quantitative yield determinations of the products have not yet been carried out. Although additional kinetic data are required for the ClO + DMS and BrO + DMS reactions, the trend of rate constants obtained for the XO + DMS reactions (X = I, Br, Cl) is in good agreement with the thermodynamics. Assuming the mechanism discussed above, the reactivity of XO radicals with DMS decreases with increasing X-O bond dissociation energy: E (X-O) = 41.5, 55.1 and 63.4 kcal mole^{-1} for X = I, Br, Cl, respectively, which of course also lowers the reaction enthalpy in the same order for the respective reactions.

Atmospheric Implication of the Reactions of DMS With the Halogen Oxide Radicals IO, BrO and ClO

Role of the reaction of IO with DMS. The IO + DMS reaction may have an impact on both the iodine-ozone chemistry and on the fate of DMS in the atmosphere.

The tropospheric iodine ozone chemistry has been previously discussed by Chameides and Davis ([19]) and Jenkin et al. ([20]). CH_3I, which has been found in the pptv range in the marine troposphere (see for example ([3])), is considered to be the main source of atmospheric iodine. It is photodissociated at a frequency of 3 x 10^{-6} s^{-1} ([19]) (12 hour averages for equinox at 30° latitude) to yield I atoms. I atoms are rapidly converted into IO by reaction with O_3, with typical time constants of a few seconds. IO can be converted back to I atoms by several routes, and lead to catalytical cycles which involve O_3. Among these routes, there are the reactions of IO with NO and the IO photodissociation for which the rate constant and photolysis frequency are 1.67 x 10^{-11} cm^3 molecule^{-1} s^{-1} at 298 K ([21]) and 0.06 s^{-1} ([22]), respectively. This photolysis frequency for IO deduced from very recent measurements of the U.V. - visible absorption cross section of IO ([22]), has been shown by the authors to be more reliable than the value of 0.2 s^{-1} obtained in earlier studies ([21,23]). In their previous modeling of atmospheric iodine chemistry, Chameides and Davis ([19]) had used a value of k_4 = 0.022 s^{-1}, averaged over 12 hours for equinox at 30° latitude. The present value of 0.06 s^{-1} was calculated for solar zenith angles from 0° to 40°. Although the reaction with NO and the photodissociation convert IO into I, these reactions have no net effect on O_3 as the following cycles show:

$$
\begin{array}{ll}
I + O_3 & \text{-------}> IO + O_2 \\
IO + h\upsilon & \text{-------}> I + O \\
& \quad +M \\
O + O_2 & \text{-------}> O_3
\end{array}
\tag{4}
$$

not net change in O_3

$$
\begin{array}{ll}
I + O_3 & \text{-------}> IO + O_2 \\
IO + NO & \text{-------}> I + NO_2 \\
NO_2 + h\upsilon & \text{-------}> NO + O \\
& \quad +M \\
O + O_2 & \text{-------}> O_3
\end{array}
\tag{5}
$$

not net change in O_3

In contrast, the reaction IO + DMS (1) will convert IO into I, as the kinetic studies suggest ([16,17]), consecutively destroying O_3 and reforming IO:

$$I + O_3 \quad \text{-------} > \quad IO + O_2$$
$$IO + DMS \text{-------} > \quad I + DMSO \qquad\qquad (1)$$

net reaction: $DMS + O_3$ -------> $DMSO + O_2$

In this way, the IO initiated oxidation of DMS could catalyze the O_3 depletion.
The importance of the catalytical depletion of O_3 by DMS depends on the fraction of IO which reacts with DMS. This fraction is obtained by comparing the rates of reactions (1), (4) and (5) which convert IO into I. The rates for these reactions are 0.06, 0.06, and 0.2 s^{-1}, respectively, with k_1 = 2 x 10^{-11} cm^3 molecule^{-1} s^{-1}, [IO] = 10^7 cm^{-3}, [DMS] = 3 x 10^9 cm^{-3}, and [NO] = 1.25 x 10^{10} cm^{-3}. The k_1 value is an average of the two independent studies ([16,17]) and the IO concentration appears to be a realistic upper limit as discussed latter in the text. The DMS concentration is a typical measured value ([3-10,12-14]) and the NO concentration is for rural continental areas ([20]). It can be seen that under the above conditions the reaction of IO with DMS could contribute as much as 19% to the conversion of IO into I in the marine atmosphere. However, in remote marine areas the concentration of NO can be much less ([24]) and consequently in such areas the contribution of the IO + DMS reaction will be greater. It has to be stressed that the values taken in this paper for atmospheric trace concentrations are averaged values, with the only objective to qualitatively estimate the potential atmospheric impact of the chemistry presented.

Concerning the role of the reaction IO + DMS on the atmospheric lifetime of DMS, the rate of reaction (1) has to be compared to that of the reaction (6)

$$OH + DMS \text{------} > \text{products} \qquad\qquad (6)$$

which is considered at the present time to be the main diurnal atmospheric reaction of DMS. The rate constant for the reaction of OH with DMS under atmospheric condition is still somewhat uncertain. Two recent studies by Hynes et al. ([25]) applying LIF techniques to measure absolute rate constants and Barnes et al. ([26,27]) who used a competitive technique to measure the relative rate constant under NO_x - free conditions have shown that the rate constant is O_2 dependent. These authors report (6.3 ± 0.3) x 10^{-12} and (8.0 ± 0.5) x 10^{-12} cm^3 molecule^{-1} s^{-1}, respectively, for the reaction at 298 K and 1 atm of air. The higher value is also supported by the work of Wallington et al. ([28]) who report (8.5 ± 0.2) x 10^{-12} cm^3 molecule^{-1} s^{-1} using the reaction $N_2H_4 + O_3$ as OH source for relative rate measurements. Barnes et al. ([26,29]) have shown that in the absence of NO_x the OH reaction with DMS leads to the formation of SO_2 and $DMSO_2$ with yields of 70% and 20%, respectively, however no DMSO has been observed as product of the reaction. The product spectrum of the DMS oxidation by OH seems to depend very much on the abundance of NO_x ([27,29]), that applies also to the formation of methane sulfonic acid.

The rate ratio for reactions (1) and (6) is (k_1 x [IO]/k_6 x [OH]). While typical averaged values of [OH] in the marine atmosphere are 10^6 cm^{-3}, the IO concentrations were originally estimated to be as high as 10^8 - 10^9 cm^{-3} by Chameides and Davis ([19]). Based on the later determination of the IO cross-section by Cox and Cocker ([21]) the original estimates of Chameides and Davis ([19]) had been considered to be too high due to the use of an erroneously low photolysis frequency for IO. Recent model calculations of Brimblecombe et al. ([30]) using this revised photolysis frequency of IO, suggest that the IO concentrations may be of the same order of magnitude as for OH. However, the actual values of IO concentrations are expected to be higher than the calculation of Brimblecombe et al. ([30]), as a result of the new determination of

k_4 by Stickel et al. ([22]). Nevertheless, if the calculation of Brimblecombe et al. ([30]) is considered and if we use the following values: $k_1 = 2 \times 10^{-11}$ at 298 K ([16,17]), $k_6 = 8 \times 10^{-12}$ at 298 K and 1 atm. of air ([26,27]) and $[OH] = [IO] = 10^6$ cm^{-3}, we obtain $(k_1 \times [IO]/k_6 \times [OH]) = 2.5$. Consequently, even with lower values of [IO] than originally calculated and now expected from the recent k_4 data, the reaction IO + DMS may be the main sink of DMS in marine atmosphere. Considering this new sink of DMS, the atmospheric lifetime would be $\tau = 1/(k_1 \times [IO] + k_6 \times [OH]) = 10$ h. This lifetime is significantly shorter that the value of 35 h calculated when OH + DMS is assumed to be the only sink for DMS. Assuming somewhat higher values of [IO] the lifetime comes in the range of the values (0.3 - 3 h) measured in coastal areas of France ([10]). Furthermore, reaction (3) would also be a source of DMSO which has recently been measured at levels of $1-5 \times 10^7$ cm^{-3} in marine atmosphere ([31]). As recently discussed ([32,33]) these levels would be apparently too low to support the assumption that reaction (1) would be the main sink for DMS. However, these low levels of DMSO could also be explained by the existence of unknown sinks of DMSO, rather that assuming that the main oxidation route of DMS does not produce DMSO. A preliminary value of k = $(5.8 \pm 2.3) \times 10^{-11}$ cm^3 molecule^{-1} s^{-1} has been obtained as rate constant for the reaction at OH + DMSO by Barnes et al. ([26]). The product DMSO has been found in the reaction of DMS with IO but not with OH, respectively ([16,26]).

In summary, the IO + DMS reaction appears to be potentially important in iodine-ozone chemistry and in controlling the DMS behavior in the marine atmosphere. However, the role of this reaction remains to be confirmed by direct field measurements of key species such as IO and CH_3I and also by further laboratory measurements on key reactions of the atmospheric iodine chemistry.

Role of the reaction of BrO with DMS. The role of this reaction in the marine atmosphere is expected to be much less important than that of the reaction IO + DMS, since the rate constant is lower by about two orders of magnitude and the BrO atmospheric concentration is not expected to be higher than that for IO. If we take $[BrO] = 6 \times 10^5$ cm^{-3}, obtained from modeling computation by Wofsy et al. ([34]), the rate ratio $(k_3 \times [BrO]/k_6 \times [OH])$ is found to be 2.0×10^{-2}. The reaction BrO + DMS would be consequently negligible compared to the reaction OH + DMS as a diurnal sink for DMS. However, it is not possible completely to exclude the involvement of reaction (3), due to the large uncertainties on the BrO concentrations which depend on both the bromine compounds emissions and their sinks.

The atmospheric bromine chemistry has been reviewed by Wofsy et al. ([34]) and Yung et al. ([35]). CH_3Br is the major bromine species in the troposphere with concentrations around 20 pptv ([36]) in marine air, as for CH_3I. But unlike CH_3I, CH_3Br has a negligible absorption in the solar spectrum ([37]) and does not primarily photodissociate into Br atom, which subsequently would have yielded BrO by reaction with O_3. It is the case also for the other bromomethane, CH_2Br_2 ([37]), and probably also for $CHBr_3$ and $CHClBr_2$, which are produced in the marine environment ([35]). All these bromomethanes react primarily with OH radicals which leads to an atmospheric lifetime of 1 to 2 years, for example for CH_3Br. The primary reactions of OH with bromomethanes produce bromomethyl radicals which are further oxidized into bromomethoxy radicals by reaction with O_2 and NO. For CH_3Br and CH_2Br_2, for example, the following steps occur:

$$\text{CH}_3\text{Br} \xrightarrow{\text{OH}} \text{CH}_2\text{Br} \xrightarrow{\text{O}_2,\text{M}} \text{CH}_2\text{BrO}_2 \xrightarrow{\text{NO}} \text{CH}_2\text{BrO}$$

$$\text{CH}_2\text{Br}_2 \xrightarrow{\text{OH}} \text{CHBr}_2 \xrightarrow{\text{O}_2,\text{M}} \text{CHBr}_2\text{O}_2 \xrightarrow{\text{NO}} \text{CHBr}_2\text{O}$$

The fate of the bromomethoxy radicals is presently not known but from ab-initio rate calculations for the thermal decomposition of the analog chlorine radical (38), we may expect the following pathways:

$$\text{CH}_2\text{BrO} + \text{O}_2 \longrightarrow \text{CHBrO} + \text{HO}_2$$

$$\text{CHBr}_2\text{O} + \text{M} \longrightarrow \text{CHBrO} + \text{Br} + \text{M}$$

The fate of CHBrO produced is also unknown. It is, in particular, not known whether bromine will be released in the form of active bromine atom or as inactive HBr molecules.

There is a need for kinetic data on reactions and on emission strengths of the different bromine source compounds in order to be able to better estimate the BrO concentrations. Such information is needed before further conclusions can be made concerning the influence of the BrO + DMS reaction on chemical processes in the marine atmosphere.

Role of the reaction of ClO with DMS. Considering first the rate constant reported in this paper for the Cl + DMS reaction (2^1), this reaction would be a competitive sink of DMS for a Cl atom concentration of 4×10^4 cm^{-3} considering ($k_2 \times [\text{Cl}]/k_6 \times [\text{OH}]$) = 1. Although there are no field measurements or modelling calculations available for the Cl and ClO concentrations in the lower troposphere, while Cl concentrations in the 10^4 cm^{-3} range do not appear to be unrealistic, ClO concentrations as high as 10^8 cm^{-3} appear unlikely, considering the current atmospheric chemistry of chlorine compounds. The reaction Cl + DMS probably proceeds through an H atom transfer mechanism, as for the dominant path of the OH + DMS reaction under atmospheric conditions. Therefore, DMSO would not be a reaction product in contrast with the reactions of halogen oxide radicals with DMS. The main sources of chlorinated compounds in the marine atmosphere are halomethanes, mainly CH$_3$Cl at 600 pptv levels and HCl at ppbv levels (36,39). These compounds react primarily with OH, with the exception of the fully chlorinated compounds which are unreactive in the troposphere. HCl will release Cl atoms with a time constant of one week. CH$_3$Cl which has a lifetime of about one year, will probably be oxidized to CHClO (38). The fate of CHClO is presently unknown and should be studied in order to establish whether active or inactive chlorine will be released. Concerning the other chloromethanes containing more than one Cl atom, CH$_2$Cl$_2$ and CHCl$_3$, Cl atoms are expected to be formed from the thermal decomposition of the chloromethoxy radicals CHCl$_2$O and CCl$_3$O which are expected to be produced in the OH initiated oxidation of CH$_2$Cl$_2$ and CHCl$_3$ in air (38).

Additional laboratory kinetic data on the oxidation reactions of chlorocarbons are needed together with modeling of the chlorine chemistry in the marine atmosphere in order to assess the role of the reactions of DMS with Cl and ClO.

Conclusion

Recent field measurements have shown that the DMS lifetime in marine coastal areas could be much lower than currently accepted. Laboratory kinetic studies show that the reactions of DMS with IO could explain these field observations. Preliminary results seem to suggest that the reactions of ClO and BrO with DMS are slow. In order to test the hypothesis that XO species have an influence on the DMS lifetime the particular species concentration has to be measured. In the case of IO the reaction can compete with the OH + DMS reaction when the IO concentration exceeds 10^4 cm^{-3}. Field measurements are planned to measure the IO concentration at the coast of Brittany. BrO is probably of negligible importance for the oxidation of DMS. In the case of ClO, although the reaction is slow, the possible impact on the DMS transformation cannot be estimated at present because the concentration of the ClO species could be much higher than that of IO or I since other sources in addition to CH_3Cl might be involved. To confirm this proposal, improved model calculations of halogenated radicals are also needed. Model calculations also require a more complete kinetic data base for atmospheric iodine chemistry and some specific aspects of the bromine and chlorine chemistry as well as results from field measurements on halogenated source species.

The DMS emissions from the ocean represent a significant part of the global sulfur flux to the atmosphere. It might be of particular interest to find new reaction routes which couple biogenic emissions from the sea to the chemistry of tropospheric photooxidants by which important properties of the atmosphere are regulated.

Acknowledgments

This work was financially supported by the European Commission (CEC).

Literature Cited

1. Lovelock, J. E.; Maggs, R. J.; Rasmussen, R. A. Nature 1972, 237, 452-3.

2. Cullis, C. F.; Hirschler, H. H. Atmos. Environ. 1980, 14, 1263-78.

3. Nguyen, B. C.; Bonsang, B.; Gaudry, G. J. Geophys. Res. 1982, 88, 10903-14.

4. Nguyen, B. C.; Bergeret, C.; Lambert, G. In Gas Transfer at Water Surface; Brutsaert, W.; Jurka G. H., Eds.; D. Reidel Publ. Co.: Dordrecht, 1984; pp 539-45.

5. Andreae, M. O.; Raemdonck, H. Science 1983, 221, 744-7.

6. Andreae, M. O.; Ferek, R. J.; Bermond, F.; Byrd, K. P.; Engstrom, R. T.; Hardin, S.; Houmere, P. D.; Le Marrec, F.; Raemdonck, H.; Chatfield, R. B. J. Geophys. Res. 1985, 90, 12891-900.

7. Andreae, M. O. In The Role of Air-Sea Exchange in Geochemical Cycling; Buat-Menard, P., Ed.; D. Reidel Publ. Co.: Dordrecht, 1986; pp 331-62.

8. Moss, M. R. In Sulfur in the Environment, Part 1: The Atmosphere Cycle Nriagu, J. O., Ed.; J. Wiley and Sons: New York, 1978; pp 23-50.

9. Nguyen, B. C.; Belviso, S.; Bonsang, B.; Lambert G. In Physico-Chemical Behaviour of Atmospheric Pollutants; Angeletti, G.; Restilli, G., Eds.; D. Reidel Publ. Co.: Dordrecht, 1987; pp 434-9.

10. Luce, C. Ph.D. Thesis, University Paris VII, 1984.

11. Chatfield, R. B.; Crutzen, P. J. J. Geophys. Res. 1984, 89, 7111-32.

12. Ferek, R. J.; Chatfield, R. B.; Andreae, M. O. Nature 1986, 320, 514-6.

13. Van Valin, C. C.; Berresheim, H.; Andrea, M. O. Geophys. Res. Letters 1987, 14, 715-8.

14. Bingemer, H. Ph.D. Thesis, Frankfurt University, Federal Republic of Germany, 1984.

15. Carlier, P. In Atmospheric Ozone; Zerefos, C. S.;Ghazi, A.; Eds.; D. Reidel Publ. Co.: Dordrecht, 1985; pp 815-9.

16. Barnes, I.; Becker, K. H.; Carlier, P.; Mouvier, G. Int. J. Chem. Kinet. 1987, 19, 489-502.

17. Martin, D.; Jourdain, J. L.; Laverdet, G.; Le Bras, G. Int. J. Chem. Kinet. 1987, 19, 503-12.

18. Barnes, I; Becker, K. H.; Martin, D. Chem. Phys. Letters 1987, 140, 195.

19. Chameides, W. L.; Davis, D. D. J. Geophys. Res. 1980, 85, 7383.

20. Jenkin, M. E.; Cox, R. A.; Candeland, D. E. J. Atmos. Chem. 1985, 2, 359.

21. Cox, R. A.; Coker, F. B. J. Phys. Chem. 1983, 87, 4478.

22. Stickel, R. E.; Hynes, A. J.; Bradshaw, J. D.; Chameides, W. L.; Davis, D. D. J. Phys. Chem. 1988, 92, 1862.

23. Jenkin, M. E.; Cox, R. A. J. Phys. Chem. 1985, 89, 192.

24. Brol, A.; Helas, G.; Rumpel, K. J.; Warneck, P. In Physico-Chemical Behaviour of Atmopsheric Pollutants; Versino, B.; Angelleti, G., Eds.; Riedel, Pub. Comp.: Dordrecht, 1984; pp 390-400.

25. Hynes, A. J.; Wine, P. H.; Semmes, D. H. J. Phys. Chem. 1986, 90, 4148.

26. Barnes, I.; Bastian, V.; Becker, K. H. In Physico-Chemical Behaviour of Atmospheric Pollutants; Angeletti, G.; Restelli, G., Eds.; D. Reidel Publ. Co.: Dordrecht, 1987; pp 327-37.

27. Barnes, I.; Bastian, V.; Becker, K.H. Int. J. Chem. Kinet., in press 1988.

28. Wallington, T. J.; Atkinson, R.; Tuazon, E. C.; Aschmann, S.M. Int. J. Chem. Kinet. 1986, 18, 837.

29. Barnes, I.; Bastian, V.; Becker, K. H.; Niki, H. Chem. Phys. Letters 1987, 140, 451-7.

30. Brimblecombe, P.; Levine, J. S.; Tennille, G. M. AGU Fall Meeting, San Francisco, Dec., 1987.

31. Harvey, G. R.; Lang, R. F. Geophys. Res. Letters. 1986, 13, 49.

32. Brimblecombe, P.; Shooter, D.; Watts, S. 194th A.C.S. National Meeting, Divison of Environmental Chemistry, New Orleans; 30 Aug-4 Sept. 1987, Book of Abstracts, p 96.

33. Charlson, R. J.; Lovelock, J. E.; Andreae, M. O.; and Warren, S. G. Nature 1987, 326, 655-61.

34. Wofsy, S. L.; Mc Elroy, M. B.; Young, Y. L. Geophys. Res. Letters 1975, 2, 215.

35. Yung, Y. L.; Pinto, J. P.; Watson, R. T.; Sander, S. P. J. Atmos. Sciences 1980, 37, 339.

36. Singh, H. B. In Environmental Impact of Natural Emissions Transactions; Aneja, V. P.,Ed.; APCA Specialty Conference: Research Triangle Park, 1984; pp 193-206.

37. Molina, L. T.; Molina, M. J.; Rowland, F. S. J. Phys. Chem. 1982, 86, 2672.

38. Rayez, J. C.; Rayez, M. T.; Halvick, P.; Duguay, B.; Lesclaux, R.; Dannenberg, J. J. Chem. Phys. 1987, 116, 203.

39. Cicerone, R. J. Rev. Geophys. Space Phys. 1981, 19, 123.

RECEIVED September 22, 1988

Chapter 30

Fourier Transform IR Studies of the Reactions of Dimethyl Sulfoxide with OH, NO_3, and Cl Radicals

I. Barnes, V. Bastian, K. H. Becker, and D. Martin

Physikalische Chemie FB 9, G.H.S. Wuppertal, Bergische Universität, 400 Wuppertal, Federal Republic of Germany

Dimethyl sulfoxide (DMSO) has recently been detected in marine air masses. To date nothing is known about the atmospheric fate of DMSO in the gas phase. Reported here are product and kinetic studies on the reactions of OH, NO_3 and Cl radicals with DMSO. The investigations were performed in a 420 l reaction chamber at atmospheric pressure using long path *in situ* Fourier transform (FTIR) absorption spectroscopy for detection of reactants and products.

Using the competitive kinetic technique preliminary rate constants of $(6.2 \pm 2.2) \times 10^{-11}$, $(1.7 \pm 0.3) \times 10^{-13}$ and $(7.4 \pm 1.8) \times 10^{-11}$ cm^3s^{-1} have been obtained for the reaction of OH, NO_3 and Cl with DMSO, respectively. SO_2 and dimethyl sulfone ($DMSO_2$) were major products of the reactions of OH and Cl with DMSO. For NO_3 only $DMSO_2$ was observed as product.

Dimethyl sulfoxide (DMSO) is known to be present in seawater at higher concentrations than dimethyl sulfide (DMS) (1). Because of its low vapor pressure it has largely been neglected in the chemistry of the atmospheric sulfur cycle. However, DMSO has been recently observed in marine rain and marine air masses (2). Further, new laboratory studies (3,4) have shown that the reaction rate of IO radicals with DMS is rather fast and leads quantitatively to the formation of DMSO and I. It has been suggested that the reaction may be a sink for DMS in marine environments (3,4) and could explain the short atmospheric residence times of DMS observed in field measurements over the ocean (5) and in coastal regions (6). In a recent article (7) it has been argued that the reaction of IO with DMS could not be an important sink for DMS because the concentrations of DMSO in the atmosphere and in marine rain are very low compared to those for DMS, MSA (methane sulphonate aerosol), and $NSS-SO_4^{2-}$ (non-sea-salt sulfate). However, in contrast to other reduced organic sulfur compounds in the troposphere nothing is known about the atmospheric fate of DMSO. If a rapid sink reaction exists for DMSO then it would invalidate the argument proposed above.

The major daytime sink for most organic sulfur compounds in the atmosphere is reaction with OH radicals (8,9) and it has been shown that reaction with NO_3 is an important nighttime sink for DMS (10,11). As shown

0097–6156/89/0393–0476$06.00/0

by the recent study on the reaction of IO with DMS our present knowledge of atmospheric chemistry does not allow the exclusion of the possible involvement of other reactions such as those with halogen atoms or halogen oxides in the tropospheric chemistry of sulfur compounds. Interest in the reaction of Cl atoms with organic substances has presently been renewed because of the role that Cl-containing species may play in the recently observed "ozone hole" over the Antarctic (12). Although there is no significant source of chlorine in the troposphere it is known to be emitted from various chemical industries (13). Hov (13) has shown that chlorine is actively involved in the formation of photochemical oxidants in industrial coastal areas. Thus a knowledge of the reactions of Cl with atmospheric sulfur compounds will be of interest for modeling the chemistry of such localized situations.

Presented here are, to our knowledge, the first investigations of the reactions of OH, Cl, and NO_3 with DMSO. A comparison is made with the corresponding reactions of the radicals with DMS and the possible implications for the atmospheric sulfur cycle are briefly discussed.

Experimental

The investigations were carried out at 298 ± 2 K in a 420 l Duran glass reaction chamber surrounded by 24 Philips TLA 40W/05 fluorescence lamps. Details of the experimental set-up can be found elsewhere (14,15) and will not be given here. The only modification to the original set-up was the insertion of a quartz tube through the center of the chamber which contained a germicidal lamp (Philips, TUV 40 W) with maximum output at 254 nm. The investigations on the OH and Cl reactions were carried out at 760 Torr total pressure and those on NO_3 at 500 Torr using either synthetic air or $N_2 + O_2$ mixtures as diluent gas.

OH radicals were generated by the photolysis of either NO_x/hydrocarbon/N_2/O_2 (9,15) or $CH_3ONO/NO/N_2/O_2$ (8,9) reaction mixtures using the fluorescence lamps. In a few experiments the photolysis of H_2O_2 at 254 nm was employed (16). Cl radicals were generated by the 254 nm photolysis of $COCl_2$. The thermal decomposition of N_2O_5 prepared by the reaction of O_3 with NO_2 (10) was used as the source of NO_3 radicals. The concentration-time behavior of the majority of reactants was followed using long-path in situ Fourier Transform infrared (FT-IR)-absorption spectroscopy. The organic reference compounds for the experiments with OH and Cl were analyzed by gas chromatography with flame ionization detection (Hewlett Packard, Model 5710 A) using a 2 m stainless steel column packed with Porasil C.

The competitive kinetic technique was used to determine the rate constants for the reactions of OH, Cl, and NO_3 with DMSO (8). The principals of the technique are the same for all three radicals. In the reaction systems studied, provided that DMSO and the reference hydrocarbon RH are removed solely by reaction with the radical X (X = OH, Cl, or NO_3),

$$DMSO + X \rightarrow \text{products}, k_1 \tag{1}$$

$$RH + X \rightarrow \text{products}, k_2 \tag{2}$$

then the rate constant ratio k_1/k_2 is given by

$$\frac{(d\ln[DMSO]/dt)}{(d\ln[RH]/dt)} = \frac{k_1}{k_2} \tag{3}$$

where (dln[DMSO]/dt) and (dln[RH]/dt) are the first order decays of DMSO and reference compound, respectively, obtained from an analysis of appropiate reactions mixtures. However, under the conditions of the present experiments there is a slow loss of DMSO to the reactor wall which must be corrected for. The wall loss was first order in DMSO and was typically of the order $(1-2) \times 10^{-4}$ s^{-1}. It is represented by the term (dln[DMSO]/dt)w which has to be considered in addition resulting in Equation (4),

$$\frac{(d \ln [DMSO]/dt) - (d \ln [DMSO]/dt)w}{(d \ln [RH]/dt)} = \frac{k_1}{k_2} \tag{4}$$

which was used in the rate constant determinations.

Propene, cis-2-butene, trans-2-butene, or isobutene were used as the reference hydrocarbons in the experiments. The rate constants used for their reactions with OH, Cl, and NO_3, are given in the relevant sections in the text. Reactant concentrations were in the range 5-25 ppm and the concentration of the radical precursors, CH_3ONO, NO_x, H_2O_2, and N_2O_5 were typically 20, 3-6, 10-40, and 15-30 ppm, respectively.

Product analyses were carried out on $DMSO/NO_2$/air, and $DMSO/Cl_2$/air or N_2 photolysis systems and the $DMSO/N_2O_5$/air dark reaction system. Reactants and products were measured using the FT-IR facility and the IR spectra were recorded within the range 400-4000 cm^{-1} with a resolution of 1 cm^{-1}. The corresponding IR absorption spectra were derived from interferograms ratioed against background and converted to absorbance. Typically, between 30 and 60 interferograms were co-added per spectrum and 10 to 15 spectra were recorded over periods up to 30 min. Concentrations of known species were determined by computer-aided subtraction of calibrated reference spectra. DMSO and $DMSO_2$ were calibrated as follows: The reactor was filled with 500 Torr of air. Weighed amounts of the substances were then heated and their vapor was swept into the reactor using a flow of air. At a pressure of 760 Torr a spectrum was immediatly recorded. The whole procedure was performed as quickly as possible, usually within 4-5 min, to minimize wall losses. The calibration for DMSO was reproducible to within $\pm$ 10%, however, the calibration for $DMSO_2$ showed a scatter of $\pm$ 30%.

<u>Results and Discussion</u>

<u>OH + DMSO</u>. Irradiation of CH_3ONO/NO/hydrocarbon/DMSO and NO_2/NO/hydrocarbon/DMSO reaction systems were carried out at 760 Torr total pressure using different mixtures of N_2 + O_2 as diluent gas. The measurements were carried out relative to the reaction of OH with either propene (k = 4.85 x 10^{-12} exp(504/T) cm^3 s^{-1}, (8)), cis-2-butene (k = 1.09 x 10^{-11} exp(488/T) cm^3 s^{-1}, (8)), or isobutene (k = 9.51 x 10^{-12} exp(503/T) cm^3 s^{-1}, (8)) as reference hydrocarbon.

Figure I shows typical data for a NO_2/NO/isobutene/DMSO reaction mixture plotted in the form ln C_t/C_0 versus t. The rate constants for the reaction of OH with DMSO obtained from the analysis of such plots according to Equation 4 are listed in Table I. The values of k_2 measured were independent of the OH source used and an average value of $(6.5 \pm 2.5) \times 10^{-11}$ cm^3 s^{-1} has been obtained as rate constant for the reaction of OH with DMSO in 760 Torr air at 298 K, which is approximately a factor of 10 higher than the corresponding reaction of OH with DMS.

The reaction of OH with DMS is known to proceed primarily by H abstraction (16,18) and H abstraction rate constants are known to scale with

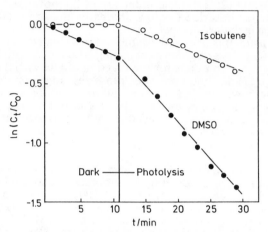

Figure 1. Plots of ln C_t/C_0 against t for DMSO and isobutene obtained from a DMSO/isobutene/NO_2 photolysis system in 760 Torr synthetic air at 298 K.

Table I. Rate constants for the reaction of OH with DMSO obtained in 760 Torr synthetic air at 298 K. For reference rate constants k_2 see text. The errors are 1σ and refer to precision only

Reactant	Ref.	No. Expts.	k_1/k_2	$k_1(10^{-11}\ cm^3\ s^{-1})$
DMSO	propene	3	2.64 ± 0.9	6.9 ± 2.5
DMSO	cis-2-butene	3	1.10 ± 0.4	6.2 ± 2.2
DMSO	isobutene	6	1.24 ± 0.4	6.4 ± 2.0
			average	6.5 ± 2.5

C-H bond energies. Therefore, such a dramatic increase in the reactivity of OH towards DMSO compared to DMS would not be expected if an abstraction mechanism was operative in each case. It is possible that the presence of the O in DMSO or the higher oxidation state of sulfur is increasing the reactivity of the molecule towards an addition type of reaction. The reaction of the addition aduct, OH-DMSO, with O_2 could lead to an enchancement of the effective rate constant. Such a mechanism has been envoked to explain the dependence of the rate constant for the reaction of OH with CS_2 on the partial pressure of O_2 in the reaction system (26,27). The detection of both SO_2 and $DMSO_2$ as reaction products, as described below, indicates that both addition and abstraction reaction pathways are operative.

On the other hand, it is now well established that relative kinetic methods which rely on the photolysis of systems containing NO_x for OH production lead to erroneously high rate constants for the reaction of OH with thiols (9) and organic sulfides (16,17) due to uncontrolled processes. In the present work the O_2 partial pressure was varied between 50 and 760 Torr and the NO_x concentration was varied by a factor of 4. The trend in the measurements was towards higher values of the rate constants with increase in the O_2 partial pressure. However, the observed increase never exceeded the total error limits of the experiments. Changing the NO_x concentration had very little effect upon the measured rate constant.

Recently, Barnes et al. (9,16) have shown that the difficulties associated with the relative method for the determination of rate constants for reactions of OH with organic thiols and sulfides when using NO_x containing precursor for OH generation can be overcome by using the photolysis of H_2O_2 at 254nm as the OH source. A few experiments were performed on DMSO/isobutene/ H_2O_2 mixtures in 760 Torr synthetic air and 760 Torr N_2 at 298 K. The gas phase reaction between DMSO and H_2O_2 in the dark was found to be negligible. However, the rate constants determined for OH + DMSO from the photolysis experiments were always much higher than those obtained using the photolysis of NO_x or CH_3ONO/NO as the OH source and also showed a dependence on the H_2O_2 concentration. The rate constants increased with increasing concentration of H_2O_2. At present this chemical behavior is not understood. One possible explanation is that H_2O_2 is reacting with a OH-DMSO adduct in a similar manner recently reported for the reaction of O2 with a OH-DMS adduct (16,18). This is, however, only speculation and the system needs further investigation to understand the detailed mechanism.

Product analysis on the OH + DMSO reaction were performed using DMSO (20 ppm)/ NO_2 (4 ppm)/ air photolysis system at 298 K. The readily identified products were SO_2, $DMSO_2$, CO, HCHO, CH_3NO_3 and an unstable intermediate compound with IR-absorption peaks at 1766, 1303, and 766 cm^{-1}. This compound has recently been assigned to methylsulfinyl peroxynitrate ($CH_3S(O)OONO_2$) from an investigation on $DMDS/NO_2/air$ photolysis systems (19). Figure 2 shows a difference spectrum for a typical investigation obtained by subtracting a spectrum recorded after 12 min irradiation from a spectrum recorded before irradiation. The absorptions due to HNO_3 and CH_3NO_3 have also been subtracted for clarity. After subtraction of all known products the residual spectrum resembles very closely that of an aerosol consisting of CH_3SO_3H (20).

The yields of SO_2 and $DMSO_2$-were, on a molar basis, 60 ± 10% and approximatly 30%, respectively. Because of a series of difficulties in calibration, wall loss, and aerosol formation it is not possible to indicate whether the observed yield of $DMSO_2$ is being over- or underestimated. As stated above the observed products show that both abstraction and addition reaction pathways are operative,

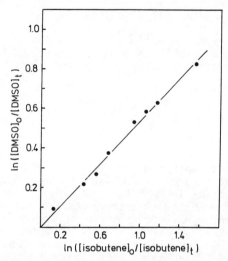

Figure 2. Plot of Equation (3) for results obtained from a DMSO/isobutene/ N_2O_5/air reaction system.

$$OH + CH_3SOCH_3 \quad --> \quad CH_3SOCH_2 + H_2O \tag{5}$$

$$OH + CH_3SOCH_3 + M \quad --> \quad CH_3\underset{|}{S}OCH_3 + M \tag{6}$$
$$\qquad\qquad\qquad\qquad\qquad OH$$

The reactions of CH_3SOCH_2 formed in step (5) are expected to lead to the CH_3SO radical,

$$CH_3SOCH_2 + O_2 \quad --> \quad CH_2SOCH_2O_2 \tag{7}$$

$$CH_3SOCH_2O_2 + NO \quad --> \quad CH_3SOCH_3O + NO_3 \tag{8}$$

$$CH_3SOCH_2O \quad --> \quad CH_3SO + HCHO \tag{9}$$

whose further reactions with O_2 and NO_2 lead to the formation of SO_2 ([19]). $DMSO_2$ is probably being formed in the reaction system by the further reaction of the OH-DMSO adduct formed with O_2, possibily via (10),

$$\underset{\underset{OH}{|}}{CH_3\text{-}SO\text{-}CH_3} + O_2 \quad --> \quad CH_3\text{-}\overset{\overset{O}{\|}}{\underset{\underset{O}{\|}}{S}}\text{-}CH_3 + HO_2 \tag{10}$$

One interesting point to note is that the SO_2 yield of 60% observed for $DMSO/NO_2$/air photolysis systems is considerably higher than the yield of typically 30% reported for DMS/NO_2/air photolysis systems ([20,32-35]) indicating perhaps differences in the respective oxidation mechanisms. It is not possible to say whether the SO_2 is being formed entirely via the H abstraction oxidation pathway of DMSO by OH or if it also being formed partly by the oxidation of the $DMSO_2$ product. Because of difficulties in the calibration of $DMSO_2$ and its possible further oxidation it is not possible to give information on the relative importance of the abstraction and addition reaction pathways. More work is needed on individual reaction steps in order to obtain a better understanding of the reaction mechanisms.

NO$_3$ + DMSO. The rate constant for the reaction of NO_3 with DMSO was measured relative to the reaction of NO_3 with isobutene using DMSO (20 ppm)/isobutene (10-20 ppm)/N_2O_5 (15-30 ppm) reaction mixtures in 500 Torr of synthetic air at 298 K. A value of $(3.2 \pm 0.3) \times 10^{-13}$ cm^3 s^{-1} as determined in this laboratory relative to NO_3 + trans-2-butene ($k = (3.06 \pm 0.30) \times 10^{-13}$ cm^3 s^{-1}, ([21])) was used as rate constant for the reaction of NO_3 + isobutene. The value is in good agreement with the value obtained by Atkinson et al. ([22]) after correction for uncertainties in the N_2O_5 equilibrium constant ([23]) and also with a more recent absolute determination using a discharge flow-mass spectrometer technique by Rahman et al. ([24]).

The experiments were performed by flushing N_2O_5 from an external 20 l bulb into the 420 l reactor containing DMSO and isobutene in 500 Torr of synthetic air. The decays of DMSO and isobutene were monitored for a period of 5 min using FT-IR spectroscopy. Over the time period of the investigation the wall loss of DMSO was small (3%) compared to reaction with NO_3. A typical plot of the results according to Equation (3) is shown in Figure 3. From a total of 6 experiments a rate constant ratio k (NO_3 + DMSO) / k (NO_3 +

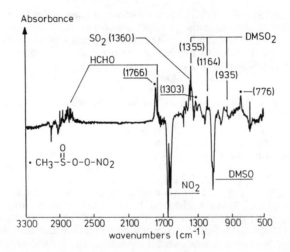

Figure 3. IR spectrum in the range 500 to 3300 cm⁻¹ for a DMSO (20 ppm)/ NO₂ (4 ppm) photolysis mixture in 760 Torr synthetic air. The spectrum is a difference spectrum formed by subtracting a spectrum recorded before irradiation from one recorded after 12 min irradiation. Absorption due to HNO₃ and CH₃NO₃ have been subtracted for clarity.

isobutene) = 0.53 ± 0.10 was obtained. Using the rate constant for NO_3 with isobutene quoted above a value of $(1.7 \pm 0.3) \times 10^{-13}$ cm^3 s^{-1} is obtained as rate constant for the reaction of NO_3 with DMSO. The quoted 1σ error represents experimental precision only.

Product studies on DMSO/N_2O_5 mixtures were performed in 760 Torr of synthetic air. HNO_3 and $DMSO_2$ were the only observed products. HNO_3 can be accounted for entirely by the HNO_3 entering the reactor with the N_2O_5 and also by the decomposition of N_2O_5 at the reactor surface. It could, however, also be formed by the abstraction reaction (11),

$$CH_3SOCH_3 + NO_3 \longrightarrow CH_3SOCH_2 + HNO_3 \tag{11}$$

The further reactions of CH_3SOCH_2 are known to lead to the formation of SO_2. Since SO_2 is not observed in the product spectrum the abstraction reaction is considered to be a minor channel. The results suggest that the major channel is addition of NO_3 to DMSO probably forming an unstable adduct which decomposes to $DMSO_2$ and NO_2,

$$CH_3SOCH_3 + NO_3 \longrightarrow CH_3\overset{\overset{\text{O}}{\|}}{\underset{\underset{\text{ONO}_2}{|}}{S}}CH_3 \longrightarrow CH_3SO_2CH_3 + NO_2 \tag{12}$$

Since an addition reaction pathway appears to play a role in both the mechanisms of the reactions of OH and NO_3 with DMSO it is surprising that the reactivity of NO_3 towards DMSO is less than that towards DMS, whereas, the opposite is true for OH. This could reflect a difference in fate of the NO_3-DMSO and OH-DMSO adducts. The former adduct can probably only decompose via reaction (12), whereas, the OH-DMSO adduct could react with O_2 (reaction (10)) which would probably considerably shorten the lifetime of the adduct and lead to a faster overall rate constant for the reaction. This is only speculation and as mentioned in section 3.1 the possibily that secondary chemistry is effecting the OH rate determination needs to be elimated before meaningful comparisons can be made.

Cl + DMSO. The rate constant for the reaction of Cl with DMSO was measured relative to the reaction of Cl with propene using DMSO (10-20 ppm)/propene (10-20 ppm)/$COCl2$ (20-30 ppm) reaction mixtures in 760 Torr of synthetic air and 760 Torr of N_2 at 298 K. A rate constant of $(24.4 \pm 0.8) \times 10^{-11}$ cm^3 s^{-1} was taken for the reaction of Cl with propene (25).

As for the studies on OH + DMSO a correction had to be made for the wall loss of DMSO which contributed, in this case, approximately 40% to the DMSO decay. After correction for these wall losses rate constants ratios, k(Cl + DMSO)/k(Cl + propene), of 0.3 ± 0.08 and 0.22 ± 0.06 were obtained for the experiments performed in air and N_2, respectively, from a total of five experiments in each diluent gas. Using the rate constant for Cl + propene quoted above values of $(7.4 \pm 1.8) \times 10^{-11}$ cm^3 s^{-1} in air and $(5.4 \pm 1.4) \times 10^{-11}$ cm^3 s^{-1} in N_2 were obtained for the reaction of Cl with DMSO. The errors are 1σ and represent precision only. The rate constant obtained for the reaction of Cl with DMSO in N_2 is approximately 40% lower than the rate constant obtained for the reaction in air. The results suggest a possible O_2 dependence of the reaction rate. The rate constant for the reactions of OH with DMS (16,18) and CS_2 and of Cl with CS_2 (28) are known to be dependent on the oxygen concentration. However, in the present case further kinetic studies are needed to exclude the possibility of the interference of secondary reactions before the observed increase can be attributed more certainly to an "O_2 effect".

The rate constant obtained for Cl + DMSO in air is slightly more than a factor of 2 slower than the value of $(2.0 \pm 0.3) \times 10^{-10}$ cm^3 s^{-1} for the reactiom Cl + DMS (this laboratory, unpublished results). No effect of the O_2 concentration was observed for the reaction of Cl with DMS. However, this rate constant is already close to the collision frequency and a small O_2 effect could remain undetected within the precision of the present experimental method. In air the reaction of Cl with DMSO leads to the formation of SO_2, $DMSO_2$, CO, HCHO, and HCOOH with yields of approximately 42% (S), 14% (S), 15% (C), 18% (C), and 2% (C), respectively. As discussed earlier it is not known whether the yield of DMSO2 is being over- or underestimated. The total sulfur yield was 56% indicating that probably a major sulfur containing product has not been detected. With the inclusion of the contribution from $DMSO_2$ the total carbon yield was 63%. The formation of $DMSO_2$ and SO_2 as products indicates that both, addition (13) and abstraction (14) pathways are operative,

$$Cl + CH_3SOCH_3 \dashrightarrow CH_3SOCH_2 + HCl \qquad (13)$$

$$Cl + CH_3SOCH_3 \dashrightarrow \underset{\overset{|}{Cl}}{CH_3SOCH_3} \qquad (14)$$

The further reactions of CH_3SOCH_2 are expected to yield SO_2. The $DMSO_2$ is possibly being formed by further reaction of the Cl- adduct produced in reaction (14) with O_2, possibly via,

$$\underset{\overset{|}{Cl}}{CH_3SOCH_3} + O_2 \dashrightarrow \underset{\underset{\overset{|}{O}}{Cl\ O}}{\overset{O \atop ||}{CH_3SCH_3}} \dashrightarrow CH_3SO_2CH_3 + ClO \qquad (15)$$

Before the missing products are identified it is not possible to discuss the dominating channel any further.

Conclusions

Rate constants for the reaction of O, OH, Cl, NO_3, and O_3 with DMS and DMSO are listed in Table II. With the exception of the rate constants for the reactions of O, OH, NO_3, and O_3 with DMS which are from references ([16,18, 29-31]), respectively, all other values are either results of this study or of unpublished work from this laboratory. The value quoted for OH + DMS is the average of the two latest studies on this reaction ([16,18]). Apart from the reaction of OH with DMSO which is a factor of 10 faster than the reaction of OH with DMS, DMSO is less reactive towards the other atmospheric species O, Cl, NO_3, and O_3 than DMS. Of the radicals listed in Table II Cl shows the highest reactivity towards both DMS and $DMSO_2$. No field measurements of the tropospheric concentration of Cl have been reported. Its tropospheric concentration is almost certainly lower than the typical atmospheric OH concentration of 1×10^6 molecules cm^{-3}. However, a Cl concentration of 10^4 molecules cm^{-3}, which may occur in industrial regions, would suffice to make the Cl + DMS reaction a significant atmospheric sink for DMS in addition to reaction with OH in such areas. In general the reaction of Cl with DMSO will be of no tropospheric importance.

As pointed out in section 3.1 relative methods which involve the use of NO_x containing precursors for OH production are known to give too high rate constants for the reactions of OH with mercaptans and organic sulfides probably due to secondary chemistry in the reaction systems. The possibility that secondary chemistry is affecting the determination of the rate constants for OH + DMSO in the present study cannot be completely excluded. The quoted value should, therefore, be used with some caution until the reaction rate constant is supported by additional studies applying different techniques. However, if this rate constant is used in conjunction with an average OH concentration of 1 x 10^6 molecules cm^{-3} a value of ~3.9 h is obtained as atmospheric lifetime for DMSO due to reaction with OH. This value is approximately a factor of 10 lower than the corresponding lifetime of DMS. DMSO and $DMSO_2$ are considerably less volatile than DMS and are also water soluble. Dissolution of DMSO and $DMSO_2$ into suspended droplets could also constitute a major sink for DMSO and $DMSO_2$ in the atmosphere. To our knowledge nothing is known about these loss processes for DMSO or $DMSO_2$. A combination of loss through dissolution and reaction with OH could lead to rather short atmospheric lifetimes for DMSO.

Obviously our understanding of the involvement of DMSO and $DMSO_2$ in the atmospheric sulfur cycle is at present very limited. The results from laboratory and field measurements as briefly discussed here imply that DMSO and $DMSO_2$ are important products of the photooxidation of DMS which is the major source of reduced sulfur to the marine atmosphere. Further kinetic studies are needed, however, to elucidate the role of DMSO and $DMSO_2$ in the atmospheric sulfur cycle.

Table II. Comparison of the rate constants (in units of cm^3 s^{-1}) for the reaction of O, OH, Cl, NO_3, and O_3 with DMS and DMSO, respectively, for conditions of 760 Torr total pressure and 298 K

	DMS	DMSO
O	$(3.3 \pm 0.3) \times 10^{-11}$	$(1.2 \pm 0.2) \times 10^{-11}$
OH	7×10^{-12}	$(7.1 \pm 2.5) \times 10^{-11}$
Cl	$(2.0 \pm 0.3) \times 10^{-10}$	$(7.4 \pm 1.8) \times 10^{-11}$
NO_3	$(8.1 \pm 1.3) \times 10^{-13}$	$(1.7 \pm 0.3) \times 10^{-13}$
O_3	$< 8 \times 10^{-19}$	$< 5 \times 10^{-19}$

Acknowledgments

This research was supported by the "Bundesminister fur Forschung und Technologie" and the "Umweltbundesamt".

Literature Cited

1. Andreae, M. O. Limnol. Oceanogr. 1980, 25, 1054.

2. Harvey, G. R.; Lang, R.F. Geophys. Res. Letts. 1986, 13, 49.

3. Barnes, I.; Becker, K. H.; Carlier, P.; Mouvier, G. Int. J. Chem. Kinetics 1987, 19, 489.

4. Martin, D.; Jourdain, J. L.; Laverdet, G.; Le Bras, G. Int. J. Chem. Kinetics 1987, 19, 503.

5. Nguyen, B. C.; Bergeret, C.; Lambert, G. In Gas Transfer of Water Surfaces; Brutsaert, W.; Jirke, G. H., Eds.; Reidel: Dordrecht, 1984; pp 539-45.

6. Carlier, P. In Atmospheric Ozone; Zerefos, C. S.; Ghazi, A. B., Eds.; Reidel: Dordrecht, 1985; pp 815-19.

7. Charlson, R. J.; Lovelock, J. E.; Andreae, M. O.; Warren, S. G. Nature 1987, 326, 655.

8. Atkinson, R. Chem. Rev. 1986, 86, 69.

9. Barnes, I.; Bastian, V.; Becker, K. H.; Fink, E. H.; Nelsen, W. J. Atmos. Chem. 1986, 4, 445.

10. Atkinson, R.; Pitts, J. N. Jr.; Aschmann, S. M. J. Phys. Chem. 1984, 88, 1584.

11. Wallington, T. J.; Atkinson, R.; Winer, A. M.; Pitts, J. N., Jr. J. Phys. Chem. 1986, 90, 4640.

12. McElroy, M. B.; Salawitch, R. J.; Wofsy, S. C.; Logan, J. A. Nature, 1986, 321, 759.

13. Hov, O. Atmos. Environ. 1985, 19, 471.

14. Barnes, I.; Bastian, V.; Becker, K. H.; Fink, E. H.; Klein, Th.; Nelsen, W.; Reimer, A.; Zabel, F. Untersuchung der Reaktionssysteme NOₓ/ClOₓ/ HOₓ unter troposphrischen Bedingungen, BPT Bericht 1/84 ISSN 01761/0777 GSF, München 1984.

15. Barnes, I.; Bastian, V.; Becker, K. H.; Fink, E. H.; Zabel, F. Atmos. Environ. 1982, 16, 545.

16. Barnes, I.; Bastian, V.; Becker, K. H. Int. J. Chem. Kinetics 1988, 20, 415.

17. Wallington, T. J.; Atkinson, R.; Tuazon, E. C.; Aschmann, S. M. Int. J. Chem. Kinetics 1986, 18, 837.

18. Hynes, A. J.; Wine, P. H.; Semmes, D. H. J. Phys. Chem. 1986, 90, 4148.

19. Barnes, I.; Bastian, V.; Becker, K. H.; Niki, H. Chem. Phys. Letts. 1987, 140, 451.

20. Hatakeyama, S.; Izumi, K.; Akimoto, H. Atmos. Environ. 1985, 19, 135.

21. Ravishankara, A. R.; Mauldin, R. L., III. J. Phys. Chem. 1985, 89, 3144.

22. Atkinson, R.; Plum, C. N.; Carter, W. P. L.; Winer, A. M.; Pitts, J.N., Jr.; J. Phys. Chem. 1984, 88, 1210.

23. Atkinson, R.; Aschmann, S. M.; Winer, A. M.; Carter, W. P. L., Environ. Sci. Technol. 1985, 19, 87.

24. Rahman, M. M.; Becker, E.; Schindler, R. N. Ber. Bunsenges. Phys. Chem. 1987, 92, 91.

25. Atkinson, R.; Aschmann, S. M. Int. J. Chem. Kinetics 1985, 17, 33.

26. Barnes, I.; Becker, K. H.; Fink, E. H.; Reimer, A.; Zabel, F.; Niki, H. Int. J. Chem. Kinetics 1983, 15, 631.

27. Jones, B. M. R.; Burrows, J. P.; Cox, R. A.; Penkett, S. A. Chem. Phys. Letts. 1982, 88, 372.

28. Martin, D.; Barnes, I.; Becker, K. H. Chem. Phys. Letts. 1987, 140, 195.

29. Lee, J. H.; Tang, I. N.; Klemm, R. B. J. Chem. Phys. 1980, 72, 1793.

30. Wallington, T. J.; Atkinson, R.; Winer, A. M.; Pitts, J. N., Jr.; J. Phys. Chem. 1986, 90, 5393.

31. Atkinson, R.; Carter, W. P. L. Chem. Rev. 1984, 84, 437.

32. Hatakeyama, S.; Akimoto, H. J. Phys. Chem. 1983, 87, 2387.

33. Grosjean, D.; Lewis, R. Geophys. Res. Letts. 1982, 9, 1203.

34. Grosjean, D.; Environ. Sci. Technol. 1984, 18, 460.

35. Niki, H.; Maker, P. D.; Savage, C. M.; Breitenbach, L. P. Int. J. Chem. Kinetics 1983, 15, 647.

RECEIVED August 11, 1988

Chapter 31

North Sea Dimethyl Sulfide Emissions as a Source of Background Sulfate over Scandinavia

A Model

Ian Fletcher

Central Electricity Generating Board, Central Electricity Research Laboratories, Leatherhead, Surrey, United Kingdom

A model has been developed to study the transport and gas-phase chemical transformation of dimethyl sulphide (DMS) in air which has originated over the United Kingdom and advects over Scandinavia. Emissions of DMS from the North Sea are significant in spring and summer, particularly during periods of phytoplankton bloom. The model predicts that methanesulphonic acid (MSA) could account for 30-50% of the total sulphur acids in Scandinavian air when blooms occur between the beginning of May and early August. If MSA undergoes radical oxidation in solution, North Sea DMS emissions could provide a significant source of "background" non sea-salt sulphate.

Dimethyl sulphide (DMS) is produced in sea-water as a consequence of enzymatic action in phytoplankton and zooplankton grazing. Measurements of DMS indicate an average global sea-to-air flux of 40 Tg of sulphur per year ($\underline{1,3}$). Instantaneous emission rates to the atmosphere, however, can vary considerably, and depend on the nature and concentration of the phytoplankton, the time of year and the prevailing meteorology. For example, high concentrations of Phaeocystis, a prolific producer of DMS, have been found in the North Sea and when it blooms, emissions of DMS may increase by a factor of 40-60 over those in the absence of blooms ($\underline{4}$). The sea-to-air flux is also dependent on the depth at which the phytoplankton is found since DMS is slightly soluble in water and can be oxidized in aqueous solution.

In the atmosphere DMS is oxidised mainly in the gas phase. Oxidation in cloud-water droplets is insignificant as the low solubility of DMS mitigates the effects of its rapid aqueous oxidation by ozone ($\underline{5}$, McElroy, W.J., Central Electricity Research Laboratories, personal communication). Gas-phase oxidation is initiated principally by reaction with OH radicals ($\underline{6}$) and methanesulphonic acid (MSA) is one of the products ($\underline{7}$). MSA has a very low vapour pressure and will be rapidly scavenged by aqueous aerosols and cloud droplets wherein further oxidation to sulphate by OH may occur. Although the kinetics and mechanism of this process have yet to be unambiguously determined, it is possible that emissions of DMS could be both a significant source of "background" sulphur and, upon oxidation, of non sea-salt sulphate.

Long range transport modelling studies of SO_2 in Northern Europe have shown that a "background" source of sulphate is required to explain a poor

0097–6156/89/0393–0489$06.00/0

correlation between SO_2 source strengths and sulphate in precipitation in areas such as Scandinavia which are remote from the sources (8,9). Furthermore, not only does there appear to be a need for a larger "background" component in summer but also that a better fit between measurements and predictions is obtained if the "missing" source could be converted to sulphate without forming SO_2 as an intermediate (8). To date, modelling studies associated with the oxidation of DMS in the atmosphere have concentrated largely on describing individual fluxes and budgets of biogenic sulphur species. Simple box models (10) and one and two-dimensional photochemical models with vertical transport (11,12) have been applied. This paper presents, by means of a simple Lagrangian model, a preliminary assessment of the potential effects of DMS emissions from the North Sea on precipitation sulphate over Scandinavia in air masses originating over the UK.

The Model

The model is similar to that employed in earlier studies of the gas-phase chemistry of a power plant plume from the UK dispersing over the North Sea (13,14). The three part system has been retained to investigate oxidation of DMS in "clean" and "polluted" air from the UK. "Clean" air is defined as that arising from rural emissions alone and corresponds to the first stage in the model whilst "polluted" air also comprises urban and industrial/power station emissions and necessitates running all three stages. Beyond the East coast of England model runs have been extended to encompass the width of the North Sea (ca. 600 km) and a further 400 km of land traversal over Scandinavia.

Kinetic Scheme

The gas-phase chemical kinetic scheme has been discussed previously and comprises 47 species and 102 reactions in the absence of DMS (14). Inorganic chemistry is treated comprehensively whilst generalised species are taken to be representative of each group of hydrocarbons with the exception of those containing one carbon atom, which are considered as specific entities.

There is considerable uncertainty regarding the kinetics and mechanisms of DMS oxidation (15). Potentially important oxidants are OH, NO_3 and IO.

A recent investigation of the kinetics of the OH/DMS reaction (16) has shown that both addition and abstraction occur at the initial step. At 288 K the ratio was determined to be 55:45 in favour of the abstraction route and the combined rate coefficient to be 7.9×10^{-12} cm^3 molecule^{-1} s^{-1}. These values are in substantial disagreement with several earlier experimental studies (7,17-19) which suggest a ratio of 4:1 in favour of addition and a rate coefficient of 9.7×10^{-12} cm^3 molecule^{-1} s^{-1}. The oxidation product of the abstraction process is almost certainly SO_2. MSA is a product of the addition mechanism. Dimethyl-sulphoxide (DMSO) may also be formed. Atmospheric measurements (6,20-23) support the formation of MSA as opposed to DMSO and that MSA production increases with decreasing temperature (10). Oxidation of MSA by OH radicals in cloud-water droplets prior to collection may account for some of the low MSA concentrations observed. Although field measurements favour the data of Hynes et al. (16), unambiguous corroboration is required. As this study is primarily aimed at assessing the potential of North Sea DMS emissions as a "background" source of sulphate, the 4:1 ratio in favour of addition and higher rate coefficient were adopted as standard and DMSO formation neglected to maximise the production of MSA. The effect of a different branching ratio and lower rate coefficient has also been considered.

Subsequent steps in the OH-initiated oxidation of DMS have not been fully established. In the model reactions 1-3 have been considered sufficient to describe the oxidations.

$$OH + CH_3SCH_3 \xrightarrow{2O_2} CH_3O_2 + CH_3SO_3H \text{ (addition)} \tag{1}$$

$$OH + CH_3SCH_3 \xrightarrow{O_2} CH_3SCH_2O_2 + H_2O \text{ (abstraction)} \tag{2}$$

$$CH_3SCH_2O_2 + NO \xrightarrow{2O_2} CH_3O_2 + NO_2 + SO_2 + HCHO \tag{3}$$

The concentration of NO_x over the North Sea is relatively low. Consequently, CH_3S radicals formed in the initial step of reaction 3 will be preferentially removed by reaction with molecular oxygen.

Absolute measurements (24,25) of the rate coefficient for the NO_3/DMS reaction are in reasonably good agreement. A temperature-independent value of 1×10^{-12} cm^3 molecule^{-1} s^{-1} has been adopted in the model. The products of the reaction have yet to be established. On the basis that Tyndall et al. (24) saw no evidence for DMSO production, hydrogen-atom abstraction has been assumed to occur (reaction 4).

$$NO_3 + CH_3SCH_3 \xrightarrow{O_2} CH_3SCH_2O_2 + HNO_3 \tag{4}$$

There is, however, a strong possibility that an adduct is formed between the two reactants.

DMSO is a product of the IO/DMS reaction (26,27). The magnitude of the rate coefficient (26,27) indicates that an IO concentration in excess of 10^5 molecule cm^{-3} would make a significant contribution to DMS removal. IO concentrations as high as 10^8-10^9 molecule cm^{-3} have been predicted (28) but these are subject to considerable uncertainty as a consequence of the chemical scheme employed. Furthermore, high concentrations of IO are only likely near to strong sources of CH_3I. As these tend to be located at coastal sites their contribution to the chemistry of an air parcel advecting at a moderate velocity will only be small. Consequently, IO chemistry has not been included in the current model.

Model Parameters

Night-time emission rates in rural and urban areas are listed in Table I together with initial concentrations and land deposition velocities. The initial concentrations were chosen to reflect unpolluted air arriving at the West Coast of England. Methane is assumed to be present in the atmospheric boundary layer at a constant concentration of 1.6 ppm. Water vapour is also assumed to be invariant in rural and urban air at a concentration of 10^4 ppm. This corresponds to ca. 60% relative humidity at 288 K. The initial concentration and emission over land of DMS have been taken to be zero as have all other species in the chemical scheme which are not listed in Table I. Emissions over land of NO, SO_2, hydrocarbons, CO and H_2 are subject to diurnal variation and this has been treated as before (13,14). Rural emission rates are assumed to prevail throughout the traversal of Scandinavia. All species are assumed to be fully mixed within an atmospheric boundary layer of constant depth, taken to be

Table I: Initial Concentrations, Emission Rates and
Deposition Velocities

Species	Initial Concentration (ppm)	Night-time Rural Emission Rate (ppm m s^{-1})*	Night-time Urban Emission Rate (ppm m s^{-1})*	Land Deposition Velocity (m s^{-1})
NO	5.0×10^{-4}	1.8×10^{-5}	1.1×10^{-3}	6.0×10^{-4}
NO$_2$	5.0×10^{-4}	-	-	7.0×10^{-3}
O$_3$	2.5×10^{-2}	-	-	6.0×10^{-3}
NH$_3$	3.0×10^{-3}	1.0×10^{-4}	1.0×10^{-4}	1.0×10^{-2}
SO$_2$	1.0×10^{-3}	2.0×10^{-5}	5.5×10^{-4}	1.0×10^{-2}
H$_2$SO$_4$	-	-	-	1.0×10^{-3}
CO	0.2	3.0×10^{-4}	1.0×10^{-2}	-
H$_2$	0.4	6.0×10^{-4}	2.0×10^{-2}	1.0×10^{-3}
H$_2$S	2.0×10^{-4}	4.2×10^{-7}	4.2×10^{-7}	2.0×10^{-4}
OCS	4.0×10^{-4}	1.4×10^{-7}	2.8×10^{-6}	1.0×10^{-3}
CS$_2$	2.0×10^{-4}	2.2×10^{-7}	4.7×10^{-6}	1.0×10^{-3}
HCHO	1.0×10^{-4}	4.6×10^{-7}	1.4×10^{-5}	1.0×10^{-3}
RH	1.0×10^{-4}	7.3×10^{-6}	4.5×10^{-4}	1.0×10^{-3}
RCHO	1.0×10^{-4}	1.3×10^{-6}	3.9×10^{-5}	1.0×10^{-3}
CH$_3$OH	-	-	5.8×10^{-8}	1.0×10^{-3}
ROH	1.0×10^{-4}	8.2×10^{-7}	7.2×10^{-5}	1.0×10^{-3}
RCH=CH$_2$	1.0×10^{-4}	3.6×10^{-6}	2.4×10^{-4}	1.0×10^{-3}
PAN	-	-	-	3.0×10^{-3}
HCl	3.0×10^{-4}	-	-	1.0×10^{-2}
ARO$^+$	1.0×10^{-4}	2.0×10^{-6}	1.8×10^{-4}	1.0×10^{-3}

* at 288 K
+ ARO is a generalised aromatic species

1 km in the standard model runs. No exchange with the free troposphere is assumed.

Measurements of SO_2 in Southern Norway during the months of May, June and July (Clark, P.A., Central Electricity Research Laboratories, personal communication) indicate that it is reasonable to consider that on average, polluted air leaving the East Coast of England and Scotland derives from emissions from one large conurbation and one 2000 MW power station. The consequence of above-average levels of pollution are also investigated.

The North Sea is assumed to be a source of DMS only. Measurements of the concentration of DMS in the North Sea (4) suggest a mean value of 10 ng (S) l^{-1} rising to 300 ng (S) l^{-1} during bloom periods. Two annual plankton blooms occur. These are often in early May and towards the end of June (4) but deviations by more than a month are not unknown. A transfer velocity of 3.5×10^{-5} m s^{-1} as used by Andreae and Raemdonck (1) has been adopted in the present study. This is appropriate for a wind speed of 8 m s^{-1}. The mean flux of DMS from the North Sea is therefore in the range 3.5×10^{-13}-1.05×10^{-11} kg (S) m^{-2} s^{-1}. These values compare with a global flux of 3.5×10^{-12} kg (S) m^{-2} s^{-1} ($\approx$40 Tg of sulphur per year from the world's oceans). The flux during bloom periods, however, provides approximately 45% of the total sulphur in air over Southern Norway which has crossed the North Sea in the months of May to July (Clark, P.A., Central Electricity Research Laboratories, personal communication). DMS emissions have been taken to be diurnally invariant as there are insufficient data to suggest otherwise.

Deposition velocities over land of all species other than H_2O, CH_4, DMS, MSA and those listed in Table I have been taken to be 1×10^{-2} m s^{-1} in the case of radicals and water-soluble species and 1×10^{-3} m s^{-1} for the remainder. A value of 1×10^{-3} m s^{-1} has been adopted for DMS. Since MSA is rapidly scavenged by aerosols it is assumed that it behaves as an aerosol and has a deposition velocity equal to that of sulphate. Over land this is 1×10^{-3} m s^{-1}. As previously (14), deposition velocities over the sea under cloudless conditions have been taken to be 10% of the corresponding land values.

Table II summarises the parameters employed in the standard simulations. Cloudless conditions are assumed throughout. The modelling runs performed are listed in Table III. A lower rate coefficient of 4.2×10^{-12} cm^3 molecule^{-1} s^{-1} and a 50:50 abstraction/addition ratio for the OH/DMS reaction were adopted in run 'b'. In run 'e' the triple plume was considered as a single one with corresponding larger initial cross-section.

Results and Discussion

Predicted concentrations of DMS and MSA are given in Table IV. In all cases more DMS is oxidised in "polluted" air than "clean" air. As a percentage of the oxidisable DMS, 10-30% more conversion in the former is predicted by the time the Scandinavian coast is reached. This is principally a consequence of the NO_3/DMS reaction assuming greater importance in "polluted" air which has a larger overnight NO_3 concentration. The effect of the higher NO emissions in "polluted" air diminishes with time as dispersion and entrainment of "clean" air occurs. At the inland site the difference in oxidation of DMS amounts to only a few percent.

The formation of MSA is a function of the OH concentration. Despite mean OH levels over the whole simulation period in "polluted" air exceeding those in "clean" air, predicted MSA concentrations at the inland site do not, in general, reflect these differences. The explanation lies in the amount of DMS oxidation by NO_3 that has occurred over the North Sea. If darkness, and correspondingly increased NO, levels, occur closer to Scandinavia than the UK,

Table II: Model Parameters

Model Date	21 June
Latitude	53.5°N
Temperature	288 K
Wind Speed	8 m s^{-1}
Half-angle for Horizontal Dispersion	5°
Half-angle for Vertical Dispersion	1.8°
Boundary-layer Depth	1 km
Initial Cross-section of Plume	310 m^2
Distance from West Coast to Urban Area	144 km
Length of Urban Area	32 km
Width of Urban Area	20 km
Distance from Urban Area to Power Station	16 km
Distance from Power Station to East Coast	64 km
Width of North Sea	608 km
Length of Traversal over Scandinavia	400 km
Power Station Emissions: NO	500 ppm
SO_2	1500 ppm
HCl	200 ppm
H_2O	4 x 10^4 ppm

Table III: Simulations

a. Standard parameters for North Sea DMS concentrations of 100 and 300 ng (S) 1^{-1} and air masses crossing the East coast of England at 02:15, 08:15, 14:15 and 20:15.

b. Modified OH/DMS kinetics

c. Mixing height 500 m

d. Insolation appropriate to May 1

for DMS concentration of 300 ng (S) 1^{-1} and the 08:15 air mass

e. Emissions from 3 2000 MW power stations

Table IV: Predicted Concentrations (ppb) of DMS and MSA

Time Air Crossed E. Coast	North Sea DMS Concentration (ng (S) l⁻¹)	Scandinavian Coast				400 km Further Downwind			
		"Clean" Air		"Polluted" Air		"Clean" Air		"Polluted" Air	
		DMS	MSA	DMS	MSA	DMS	MSA	DMS	MSA
02:15[a]	100	.11	.054	.071	.065	.022	.093	.014	.084
08:15[a]	100	.11	.03	.069	.03	.017	.1	.012	.082
14:15[a]	100	.091	.036	.067	.04	.01	.076	.008	.072
20:15[a]	100	.088	.055	.07	.069	.02	.062	.013	.068
02:15[a]	300	.35	.16	.26	.19	.086	.31	.061	.29
08:15[a]	300	.41	.09	.27	.092	.062	.34	.044	.29
14:15[a]	300	.33	.13	.22	.14	.047	.27	.031	.25
20:15[a]	300	.27	.18	.21	.21	.092	.2	.061	.22
08:15[b]	300	.47	.03	.31	.03	.19	.14	.14	.12
08:15[c]	300	.85	.19	.56	.17	.094	.69	.068	.57
08:15[d]	300	.42	.069	.26	.062	.091	.29	.062	.23
08:15[e]	300	-	-	.24	.077	-	-	.035	.27

more of the total DMS emitted is oxidised by NO_3. As this is greater in "polluted" air, it is the resulting higher concentration of DMS in "clean" air over Scandinavia which determines the final levels of MSA. Plots of MSA concentration versus time for the 08:15 standard simulations are shown in Figure 1. Initially, OH levels in "clean" air exceed those in "polluted" air in this case. By the end of the first daylight period "polluted" air contains more OH and this is the situation throughout most of the second day. Overnight, MSA production stops whilst more DMS is oxidised by NO_3 in "polluted" air than "clean" air.

OH oxidation of DMS is not a nett sink for OH provided the concentration of NO is not too low. Since DMS levels are < 0.5 ppb any such perturbation of the OH concentration would be expected to be small. It is further reduced by regeneration of OH following the initial OH/DMS reaction. Regardless of whether addition or abstraction occurs one CH_3O_2 radical is produced for each OH radical consumed. OH is then regenerated principally by:

$$CH_3O_2 + NO \overset{O_2}{\text{--}{>}} HO_2 + HCHO + NO_2 \qquad (5)$$

$$HO_2 + NO \text{--}{>} OH + NO_2 \qquad (6)$$

The absence of proportionality between predicted MSA concentrations and the North Sea DMS concentration is attributed to limited oxidant availability. The formation of NO_3 is not sufficiently rapid to compensate for its removal by DMS at the higher concentrations of the latter. Consequently, in such cases relatively greater oxidation by OH takes place.

Emission time is a significant factor determining MSA levels. This is clearly shown in Figure 2, differences being greatest over the North Sea towards the Scandinavian coast. It is predicted that the concentration of MSA between 16:00 and 22:00 could be more than twice that between 04:00 and 10:00 in air off the Scandinavian coast. As DMS is oxidised throughout 24 h it shows less of a dependence on emission time (Table IV).

Reducing the rate coefficient of the OH/DMS reaction and weighting the abstraction and addition mechanisms equally are both significant in determining the decrease in MSA with respect to the standard case. Eventually it is the ratio which principally determines the depletion since only some of the extra DMS in the atmosphere is oxidised by NO_3. Halving the mixing height leads to twice the standard MSA concentration. This reflects the low land deposition velocities of DMS and MSA adopted. Lower insolation depresses photochemical oxidant production and hence MSA formation. The reduction in the latter is > 15% for insolation appropriate to May 1. Increasing NO_x but not hydrocarbons as in the triple plume case has only a small deleterious effect on mean OH concentrations over the simulation period. Enhanced oxidation of DMS by NO_3 is primarily responsible for the lower predicted MSA levels.

The contribution MSA makes to total sulphur acids (MSA plus H_2SO_4 from SO_2 oxidation) is an important quantity. If MSA in aqueous aerosols and cloud droplets is converted to sulphate DMS would be a source of "background" non sea-salt sulphate. The reaction need not be rapid if cloud-water collection times are long. Figure 3 shows the predicted percentage of total sulphur acids as MSA during phytoplankton blooms. Over Scandinavia this percentage reaches 24% in the standard case. About 5% of the predicted sulphate comes from DMS via NO_3 oxidation and OH oxidation via abstraction. This percentage is slightly higher when greater weight is given to the abstraction mechanism and partly accounts for the drop in the fraction of total sulphur acids as MSA by comparison with the standard case. A reduction is also predicted in the triple

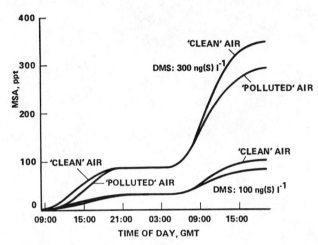

Figure 1. Predicted concentration of MSA in air crossing the East coast of England at 08:15 for standard parameters.

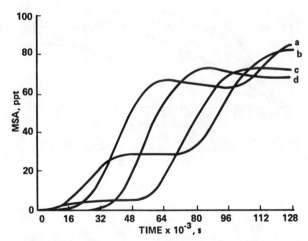

Figure 2. Predicted concentration of MSA in "polluted" air crossing the East coast of England at 02:15(a), 08:15(b), 14:15(c) and 20:15(d) for a North Sea DMS concentration of 100 ng (S) 1^{-1} and standard parameters.

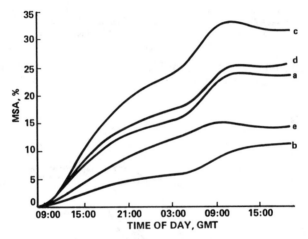

Figure 3. Predicted percentage of sulphur acids as MSA in "polluted" air crossing the East coast of England at 08:15 for a North Sea DMS concentration of 300 ng (S) 1^{-1}.

plume case owing to the enhanced anthropogenic emissions of SO_2. This is an extreme situation and other air masses from the UK should have above average MSA contributions.

Reducing the mixing height to 500 m results in the biggest increase in the MSA fraction with respect to the standard case. Concentrations of both SO_2 and DMS are greater but as the former is deposited more rapidly than the latter, the significance of the OH/DMS reaction is enhanced. The same effect, albeit smaller, appears in the May 1 simulation where lower insolation reduces photochemical oxidant production and results in there being larger amounts of unoxidised SO_2 and DMS.

The corresponding plots to those in Figure 3 for "clean" air are shown in Figure 4 and reveal the same general disposition but the percentage of total sulphur acids as MSA is, however, increased. In the standard simulation values >40% are predicted. A peak percentage of almost 60 is predicted for a mixing height of 500 m. Even in the worst case the MSA contribution exceeds 20%.

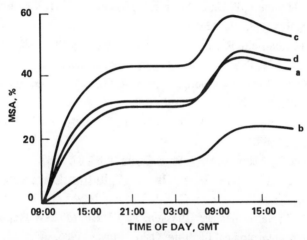

Figure 4. Predicted percentage of sulphur acids as MSA in "clean" air crossing the East coast of England at 08:15 for a North Sea DMS concentration of 300 ng (S) 1^{-1}.

Provided MSA is the major oxidation product of the OH/DMS reaction in the gas phase it is calculated that it could account for 30-50% of the total sulphur acids in Scandinavian air during phytoplankton blooms between the beginning of May and early August. Further, if MSA is converted to sulphate in solution, North Sea DMS emissions, particularly at bloom times, could provide a significant source of "background" non sea-salt sulphate. With the latter proviso, oxidation of DMS by NO_3 and by OH via abstraction are much less efficient means of producing sulphate than is the MSA mechanism. It is clear that a more detailed assessment is necessary. In particular, the kinetics and mechanism of the OH/DMS reaction derived by Hynes et al. (16) will be employed. On the basis of the current study this is expected to reduce the contribution MSA makes to total sulphur acids over Scandinavia by approximately a factor of 2. An air temperature of 288 K is probably not realistic for the North Sea between Scotland and Norway in spring. Adopting a temperature of 283 K for the May 1 simulation would partially counter the effects of the less favourable mechanism. Consequently, North Sea DMS emissions do appear to be a potential source of "background" non sea-salt sulphate.

Acknowledgment

This work was carried out at the Central Electricity Research Laboratories of the Research Division and is published by permission of the Central Electricity Generating Board.

Literature Cited

1. Andreae, M. O.; Raemdonck, H. Science, 1983, 221, 774.

2. Andreae, M. O.; Ferek, R. J.; Bermond, F.; Byrd, K. P.; Engstrom, R. T.; Hardin, S.; Houmere, P. D.; LeMarrec, F.; Raemdonck, H.; Chatfield, R. B. J. Geophys. Res. 1985, 90, 12891.

3. Ferek, R. J.; Chatfield, R. B.; Andreae, M. O. Nature, 1986, 320, 514.

4. Turner, S. M.; Liss, P. S. J. Atmos. Chem. 1985, 2, 223.

5. Zafiriou, O. C.; Joussot-Dubien, J.; Zepp, R. G.; Zika, R. G. Environ. Sci. Technol., 1984, 18, 358A.

6. Saltzman, E. S.; Savoie, D. L.; Zika, R. G.; Prospero, J. M. J. Geophys. Res. 1983, 88, 10897.

7. Hatakeyama, S.; Akimoto, H. J. Phys. Chem. 1983, 87, 2387.

8. Fisher, B. E. A.; Clark, P. A. In Air Pollution Modelling and Its Application IV; de Wispelaere, C., Ed.; Plenum: New York, 1985, p 471.

9. Lehmhaus, J.; Saltbones, J.; Eliassen, A. EMEP/MSC-W Report 1/86; Norwegian Meteorological Institute: Oslo, Norway, 1986.

10. Berresheim, H. J. Geophys. Res. 1987, 92, 13245.

11. Toon, O. B.; Kasting, J. F.; Turco, R. P.; Liu, M. S. J. Geophys. Res. 1987, 92, 943.

12. Chatfield, R. B.; Crutzen, P. J. J. Geophys. Res. 1984, 89, 7111.

13. Cocks, A. T.; Fletcher, I. S.; Kallend, A. S. In Air Pollution Modelling and Its Applicaiton II; de Wispeleaere, C., Ed.; Plenum: New York, 1983, p 137.

14. Cocks, A. T.; Fletcher, I. S. Atmos. Environ. 1988, 22, 663.

15. Yin, F.; Grosjean, D.; Seinfeld, J. H. J. Geophys. Res. 1986, 91, 14417.

16. Hynes, A. J.; Wine, P. H.; Semmes, D. H. J. Phys. Chem. 1986, 90, 4148.

17. Atkinson, R.; Perry, R. A.; Pitts, J. N. Chem. Phys. Lett. 1978, 54, 14.

18. Kurylo, M. J. Chem. Phys. Lett. 1978, 58, 23.

19. Hatakeyama, S.; Izumi, K.; Akimoto, H. Atmos. Environ. 1985, 19, 135.

20. Ayers, G. P.; Ivey, J. P.; Goodman, H. S. J. Atmos. Chem. 1986, 4, 173.

21. Saltzman, E. S.; Savoie, D. L.; Prospero, J. M.; Zika, R. G. J. Atmos. Chem. 1986, 4, 227.

22. Saltzman, E. S.; Savoie, D. L.; Prospero, J. M.; Zika, R. G. Geophys. Res. Lett. 1985, 12, 437.

23. Harvey, G. R.; Lang, R. F. Geophys. Res. Lett. 1986, 13, 49.

24. Tyndall, G. S.; Burrows, J. P.; Schneider, W.; Moortgat, G. K. Chem. Phys. Lett. 1986, 130, 463.

25. Wallington, T. J.; Atkinson, R.; Winer, A. M.; Pitts, J. N. J. Phys. Chem. 1986, 90, 5393.

26. Barnes, I.; Becker, K. H.; Carlier, P.; Mouvier, G. Int. J. Chem. Kinet. 1987, 19, 489.

27. Martin, D.; Jourdain, J. L.; Laverdet, G.; Le Bras, G. Int. J. Chem. Kinet. 1987, 19, 503.

28. Chameides, W. L.; Davis, D. D. J. Geophys. Res. 1980, 85, 7383.

RECEIVED July 20, 1988

THE ATMOSPHERE:
AQUEOUS-PHASE TRANSFORMATIONS

Chapter 32

The Uptake of Gases by Liquid Droplets

Sulfur Dioxide

James A. Gardner[1,3], Lyn R. Watson[1], Yusuf G. Adewuyi[1,4],
Jane M. Van Doren[1], Paul Davidovits[1], Douglas R. Worsnop[2],
Mark S. Zahniser[2], and Charles E. Kolb[2]

[1]Department of Chemistry, Boston College, Chestnut Hill, MA 02167
[2]Aerodyne Research, Inc., 45 Manning Road, Billerica, MA 01821

Heterogeneous reactions involving water droplets in clouds and fogs are important mechanisms for the chemical transformation of atmospheric trace gases. The principal factors affecting the uptake of trace gases by liquid droplets are the mass accommodation coefficient of the trace gas, the gas phase diffusion of the species to the droplet surface and Henry's Law saturation of the liquid. The saturation process in turn involves liquid phase diffusion and chemical reactions within the liquid droplet. The individual processes are discussed quantitatively and are illustrated by the results of experiments which measure the uptake of SO_2 by water droplets.

Heterogeneous reactions involving water droplets in clouds and fogs are increasingly recognized as major mechanisms for the chemical transformation of atmospheric trace gases (1-5). In such heterogeneous reactions, the rate of trace gas uptake is a pivotal factor in understanding the transformation process.

An example of an atmospherically important heterogeneous process is the transformation of SO_2 gas into sulfurous acid and then sulfuric acid as represented in Equations 1a and 1b.

$$SO_2(g) \underset{H_{SO_2}}{\overset{}{\rightleftharpoons}} SO_2(aq) \underset{K_1}{\overset{}{\rightleftharpoons}} HSO_3^- \underset{K_2}{\overset{}{\rightleftharpoons}} SO_3^{2-} \qquad (1a)$$

$$S(IV) \overset{H_2O_2, O_3}{\longrightarrow} S(VI) \qquad (1b)$$

In equilibrium the concentrations of the sulfur (IV) compounds are determined by the Henry's Law Constant H_{SO2} and by the equilibrium constants K_1 and K_2. The conversion of sulfurous acid to sulfuric acid involves further oxidation from state (IV) to state (VI) which requires an oxidizing agent such as O_3 or H_2O_2.

[3]Current address: Air Force Geophysics Laboratory, Hanscom Air Force Base, Bedford, MA 01730
[4]Current address: Institute for Environmental Studies, Environmental Research Laboratory, 1005 W. Western Avenue, Urbana, IL 61801

0097–6156/89/0393–0504$06.00/0
© 1989 American Chemical Society

In order to model the kinetics of such a process one must know the rate of gas uptake for SO_2 gas as well as for the gaseous species involved in subsequent oxidation.

Heterogeneous reactions begin with the gas molecule striking the surface of the droplet and entering into the liquid phase. In this connection the basic parameter determining the transfer of gases into the droplet is the mass accommodation or "sticking" coefficient, which is simply the probability that the molecule which strikes the surface enters into the bulk liquid.

Using simple gas kinetic theory, the flux of gas molecules (J) into a surface is given by

$$ J = \frac{n_g \bar{c} \gamma}{4} \qquad (2) $$

Here n_g is the density of the gas molecules, $\bar{c}$ is the average thermal velocity and γ is the mass accommodation coefficient. This is the maximum flux of gas into a liquid. In many circumstances, however, the actual gas uptake is smaller. It may be limited by several processes, the most important of which are gas phase diffusion and Henry's Law saturation. The treatment of Henry's Law saturation in turn involves liquid phase diffusion and, in some cases, liquid phase chemical reactions.

The measurement of mass accommodation coefficients is a difficult task. In 1975 Sherwood et al. (6) wrote "Not only is there no useful theory to employ in predicting the γ, there is no easy way to experimentally measure it." The experimental problem is due to the difficulty of separating the effects of the various processes on the rate of gas uptake.

Over the past four years we have developed a laboratory method for measuring gas uptake in a way that allows control of the factors affecting gas uptake. Utilizing this method, we have measured mass accommodation coefficients and have studied the effects of the various processes on gas uptake.

In this paper we will describe the experimental apparatus and procedures and then we will discuss the various factors affecting gas uptake, using SO_2 as an example. In our treatment of these phenomena we will focus on the physical nature of the processes. Our derivations will be approximate, lacking mathematical rigor. Our final results, however, are consistent with more exact treatments found in the literature.

Apparatus and Experimental Procedure

A version of our apparatus for the study of gas uptake and the measurement of mass accommodation coefficients is shown in Figure 1. The apparatus consists of three chambers. A highly controlled stream of droplets is produced in the first chamber with a nozzle vibrated by a piezoelectric ceramic oscillator. About 100,000 droplets are produced per second. The radius of the droplets is typically 60 microns and the droplets travel at about 3000 centimeters per second.

The droplets enter the reaction chamber through a 1 mm diameter hole. Here the droplets interact with the reactive trace species which flow through the reaction chamber. The reactive gas can be introduced through one of three inlets along the flow tube. In this way the interaction time can be varied. The droplets pass out of the reaction chamber through a second hole into the third chamber where they may be collected for subsequent chemical analysis. The

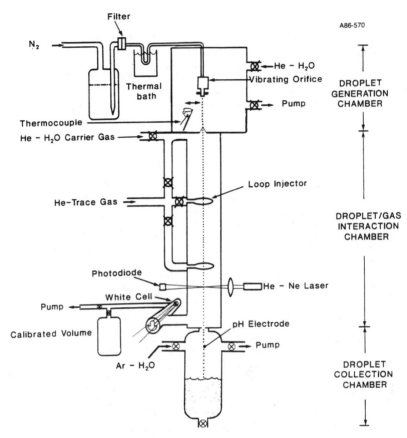

Figure 1a. Schematic of Trace Gas-Droplets Apparatus.

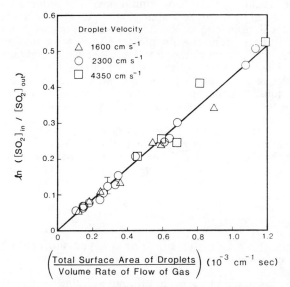

Figure 1b. Insert is a plot of the change in SO_2 concentration as a function of exposed droplet surface area normalized by the volume rate of flow of gas, for different droplet sizes and droplet velocities. (Reproduced with permission from reference 7. Copyright 1987 American Geophysical Union.)

density of the trace gas passing through the reaction chamber is monitored spectroscopically by an infrared tunable diode laser as the droplets are switched on and off.

In the insert of Figure 1 we show a plot of the change in SO_2 density due to uptake by the water droplets as a function of the exposed droplet surface area normalized by the volume rate of gas flow. The mass accommodation coefficient is computed from the slope of the plotted line.

An important aspect of the experimental technique is the careful control of all the conditions within the apparatus. The pressure in the apparatus is kept as low as possible to reduce gas diffusion limitations. The minimum operating pressure is determined by the water vapor partial pressure, which is carefully kept at saturation with respect to the equilibrium vapor pressure of water at the droplet temperature so that the droplets neither grow nor evaporate. Depending on the temperature of the droplets this sets the minimum pressure at 4 to 20 Torr. The transit time of the droplets through the reaction zone is short, on the order of a few milliseconds, in order to avoid saturation of the trace gas in the liquids. Experimental parameters are computer monitored. The details of the technique and of the experimental procedures are discussed in reference (7).

The experiment measures gas uptake into the droplets. The rate of uptake yields the mass accommodation coefficient directly only if the process is governed entirely by the kinetics of Equation 2. Even with the unique design features of this experiment the factors mentioned earlier cannot be completely eliminated and must be taken into account in order to understand the process of gas uptake.

Factors Affecting Gas Uptake

Mass Accommodation Coefficient. For a given molecule the mass accommodation coefficient is a physical constant which depends only on the temperature and on the nature of the liquid surface. The process of the molecule entering the liquid phase might proceed as follows. Since the surface of water is non-rigid it is likely that a molecule which strikes the surface achieves thermal accommodation with near-unit probability. The molecule is bound to the surface in a potential well of depth ΔU_s, where ΔU_s is the binding energy of the molecule to the liquid surface.

On the average, the molecule will remain bound to the surface for a length of time τ_r given by

$$\tau_r = \tau_v \exp^{(\Delta H_s/RT)} \tag{3}$$

Here τ_v is the vibrational period within the potential well (typically on the order of 10^{-12} s) and ΔH_s is the surface enthalpy ($\Delta H_s = \Delta U_s + \Delta[PV]$).

Since the molecule is thermalized it does not possess sufficient energy to penetrate the surface of water. This requires energy on the order of the surface tension of the liquid. Therefore, the entry of the molecule into the liquid phase depends on collective phenomena wherein an opening sufficiently large to accommodate the molecule is created at the surface site of the gas molecule. Designating β as the rate at which sufficiently large holes are formed at the surface per unit area per unit time and A_m as the area of the molecular site, then the mass accommodation coefficient can be written as

$$\gamma = \tau_r \beta A_m \tag{4}$$

It seems reasonable to suggest that ΔU_s is proportional to the solvation energy of the species and β is likely to be inversely proportional to the size of the required accommodation site. Such a model suggests that small molecules with large heats of solvation would have large mass accommodation coefficients and conversely large molecules with small heats of solvation would have small mass accommodation coefficients. Experiments to date support these conclusions.

<u>Gas Phase Diffusion</u>. The gas phase diffusion limitation arises when the diffusional flux of molecules to the surface of the droplet is less than the maximum possible flux of gas across the surface as given by Equation 2. Under these circumstances the gas density near the surface of the droplet (n') is smaller than average volume density (n). The situation can be simply analyzed by writing the rate equation for the total number of molecules N in the neighborhood of the droplet:

$$\frac{dN}{dt} = [\frac{D_g(n-n')}{a}]4\pi a^2 - [\frac{n'\bar{c}\gamma}{4}]4\pi a^2 \tag{5}$$

where a is the droplet radius and D_g is the gas diffusion coefficient ($D_g{\cdot}p = 0.126$ atm cm^2 s^{-1} for SO_2 in air at 298°K) (9). The first term on the right hand side of the equation represents the net arrival rate of molecules at the droplet surface. The second term represents the depletion of molecules due to entry into the liquid droplet. Under steady state conditions $dN/dt = 0$ and the expression for n' is

$$n' = \frac{n}{1 + \frac{\bar{c}a\gamma}{4D_g}} \tag{6}$$

Due to the limitations of gas diffusion, the uptake of the trace gas is reduced by a factor of n'/n. To take this into account, we define a diffusion limited observable mass accommodation coefficient γ_d as

$$\gamma_d = \gamma\frac{n'}{n} = \frac{\gamma}{1 + \frac{\bar{c}a\gamma}{4D_g}} \tag{7}$$

Thus, γ_d is the experimentally measurable diffusion-limited uptake rate, expressed in terms of an effective mass accommodation coefficient.

The rate of gas uptake is dominated by diffusion for γ greater than $4D_g/\bar{c}a$. At atmospheric pressure and 10 μm radius droplets this limitation occurs for $\gamma < 0.02$. The limitations of gas diffusion are discussed in greater detail by Fuchs and Sutugin (8) and by Schwartz (9).

Clearly, an experiment designed to measure γ must be done at a low pressure in order to minimize this diffusion limitation. At the same time, however, the ambient water vapor pressure must be maintained at its equilibrium value which at room temperature is about 20 Torr. In our experiment performed with 60 μm radius droplets, gas uptake is dominated by diffusive transport for γ greater than 0.1. The effect of diffusion on gas uptake is illustrated in Figure 2.

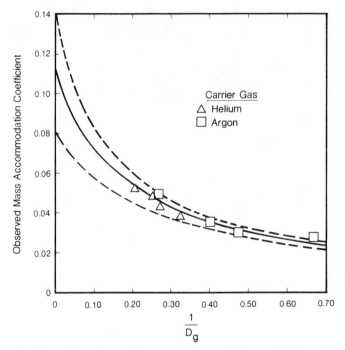

Figure 2. Plot of the observed mass accommodation coefficient of SO_2 on water as a function of the inverse of the calculated diffusion coefficient. Curves are calculated from Equation (7) for a $= 85$ μ and $\gamma = 0.08$, 0.11 and 0.14.

SO$_2$ uptake was measured at total system pressures in the range of 20 to 50 Torr, consisting of 17.5 Torr H$_2$O vapor with the balance either helium or argon. The observed mass accommodation coefficients, γ_d, are plotted in Figure 2 as a function of the inverse of the calculated diffusion coefficient of SO$_2$ in each H$_2$O-He and H$_2$O-Ar mixture. The diffusion coefficients are calculated as a sum of the diffusion coefficients of SO$_2$ in each component. The diffusion coefficients for SO$_2$ in He and in Ar are estimated from the diffusion coefficient of SO$_2$ in H$_2$O ($D_g \cdot p$ = 0.124 (10)) by multiplying this value by the quantity $(m_{He}/m_{H2O})^{1/2}$, and $(m_{Ar}/m_{H2O})^{1/2}$, respectively. The curves in Figure 2 are plots of Equation 7 with three assumed values for γ: 0.08, 0.11 and 0.14. The best fit to the experimental values of γ_d is provided by γ = 0.11. Since gas uptake could be further limited by liquid phase phenomena as discussed in the following section, γ_{SO2} = 0.11 is a lower limit to the true mass accommodation coefficient for SO$_2$ on water.

Henry's Law Equilibrium. The equilibrium relationship between molecules in the gas phase at a pressure p and the molecules dissolved in the liquid is given by Henry's Law.

$$n_l = p \cdot H = n_g RTH \tag{8}$$

Here n_l and n_g are the liquid and gas phase concentrations is moles per liter, and H is the Henry's Law coefficient which for SO$_2$ in water at 25°C is 1.3 M atm^{-1} (11).

As equilibrium is approached, the net uptake of gas decreases and with it the observed mass accommodation coefficient γ_{obs}. Once equilibrium is reached, the rate of absorption of the gas is balanced by the rate of desorption from the liquid and the net uptake of gas is zero. Under dynamic conditions such as found in our experiment, the surface may saturate long before the trace species penetrates into the whole droplet. At that point, the rate of gas uptake is controlled by the rate of liquid phase diffusion of the dissolved molecule away from the surface. The surface density and characteristic time τ_p for the surface to reach saturation can be derived as follows. The net flux J of trace species entering the liquid is

$$J = \frac{n_g \gamma \bar{c}}{4} = \frac{n_l \delta \bar{c}}{4} \tag{9}$$

Here δ is the desorption coefficient which can be related to γ and the equilibrium liquid phase density of the trace species n_l^o as $n_g \gamma = n_l^o \delta$. Using the relationship n_l^o/n_g = RTH, δ can be expressed as $\delta = \gamma/RTH$ and J can be written as

$$J = \frac{\bar{c} \gamma}{4RTH} [n_l^o - n_l] \tag{10}$$

In the absence of a chemical reaction in the liquid phase, the density n_l can be expressed approximately as

$$n_l = \frac{Jt}{\Delta x} \tag{11}$$

Here Δx is the average penetration depth into the droplet by the trace species during time t. This depth is approximately

$$\Delta x = (D_l t)^{1/2} \tag{12}$$

where D_l is the diffusion coefficient of the trace species in the liquid ($D_l = 1.8$ x 10^{-5} cm^2 s^{-1} for SO_2 in water at 25°C) (9). In this derivation it is assumed that Δx is smaller than the radius of the droplet.

Using Equations 10, 11 and 12, we obtain an expression for n_l as

$$n_l = \frac{n_l{}^o}{1 + \dfrac{4RTH\,(D_l/t)^{1/2}}{\bar{c}\,\gamma}} \tag{13}$$

We define the characteristic time, τ_p, for the equilibrium between gas and liquid phases as

$$\tau_p = (4RTH/\bar{c}\gamma)^2 D_l \tag{14}$$

which allows us to rewrite Equation 13 as

$$n_l = \frac{n_l{}^o}{1 + (\tau_p/t)^{1/2}} \tag{15}$$

This expression is developed in a more rigorous treatment by Schwartz and Freiberg (11).

From our measurements of SO_2 uptake, we know that γ is at least equal to our experimental value of $\gamma = 0.11$ shown in Figure 2. This corresponds to $\tau_p = 3$ x 10^{-8} s which is considerably shorter than the transit time of the droplet through the reaction zone. The characteristic time τ_p represents the minimum time for saturation of the surface if there were no chemical removal of the species within the droplet. In fact, as is indicated in Equation 1, sulfur (IV) is dissolved in the liquid also in the form of HSO_3^- and SO_3^{-2}. The capacity of the liquid to hold S(IV) is therefore increased, and saturation is delayed.

The equilibrium concentration of the total amount of S(IV) dissolved in the liquid can be expressed according to Schwartz (9) in terms of an effective Henry's Law coefficient H* as

$$n_{S(IV)_l}/n_{SO_2(g)} = H^*{}_{S(IV)}RT \tag{16}$$

For the SO_2 equilibrium H*$_{S(IV)}$ is given by (7)

$$H^*{}_{S(IV)} = H_{SO_2}\left(1 + \frac{K_1}{[H^+]} + \frac{K_1\,K_2}{[H^+]^2}\right) \tag{17}$$

The equilibrium constants K_1 and K_2 are 1.39 x 10^{-2}M and 6.5 x 10^{-8}M, respectively (8). With these values, H* is 23 at a pH of 3.0 and is 4200 at a pH of 5.2.

Taking into account all the S(IV) species and assuming sufficiently rapid equilibration among them, the liquid phase S(IV) concentration, $n_{S(IV)_l}$, is

$$n_{S(IV)l} = n_{SO_2(g)} RTH^*_{S(IV)} = (H^*_{S(IV)}/H_{SO_2}) n_{SO_2(l)} \tag{18}$$

In this case, Equation 10 becomes

$$J = \frac{\bar{c}\gamma}{4RTH^*_{S(IV)}} [n^o_{S(IV)l} - n_{S(IV)l}] \tag{19}$$

and Equation 13 becomes

$$n_{S(IV)l} = \frac{n^o_{S(IV)l}}{1 + \dfrac{4RTH^*_{S(IV)} (D_l/t)^{1/2}}{\bar{c}\gamma}} \tag{20}$$

Using this equation and substituting τ_p from Equation 14, we obtain

$$n_{S(IV)l} = \frac{n^o_{S(IV)l}}{1 + (H^*_{S(IV)}/H_{SO_2}) (\tau_p/t)^{1/2}} \tag{21}$$

The observed mass accommodation coefficient γ_{obs} is obtained from

$$\frac{n_g \bar{c} \gamma_{obs}}{4} = J \tag{22}$$

Rearranging terms and using Equations 19, 21 and 22, we obtain a time dependent expression for γ_{obs}

$$\gamma_{obs}(t) = \gamma_{obs}(0) \left[\frac{(H^*_{S(IV)}/H_{SO_2}) (\tau_p/t)^{1/2}}{1 + (H^*_{S(IV)}/H_{SO_2}) (\tau_p/t)^{1/2}} \right] \tag{23}$$

If the equilibrium between the S(IV) species were reached instantaneously, $\gamma_{obs}(0)$ would be simply the gas diffusion limited mass accommodation coefficient γ_d given by Equation 7. Equation 21 may be rewritten by substituting Equation 14 for τ_p

$$\gamma_{obs}(t) = \gamma_{obs}(0)/[1 + \frac{\bar{c}\gamma_{obs}(0)}{4RTH^*_{S(IV)}} (t/D_l)^{1/2}] \tag{24}$$

In Figure 3 we plot experimental values of γ_{obs} as a function of the droplet transit time t for two values of $H^*_{S(IV)}$: 23 and 4200. These values were obtained by setting the pH of the droplets surface at 3.0 and 5.2, respectively. This pH includes the effect of the acidification due to the absorbed SO_2 gas. The contact time was varied by the inlet position of the trace gas and by the velocity of the droplets. These data were taken under conditions of water vapor pressure on the order of 20 torr, or as in Figure 2, $1/D_g \sim 0.20$. The solid lines are plots of Equation 23, where $\gamma_{obs}(0)$ is assumed to be 0.059.

A more exact treatment, valid for droplets of arbitrary diameter, has been presented by Danckwerts (13). In the region of applicability, our simplified derivation is in quantitative agreement with that work.

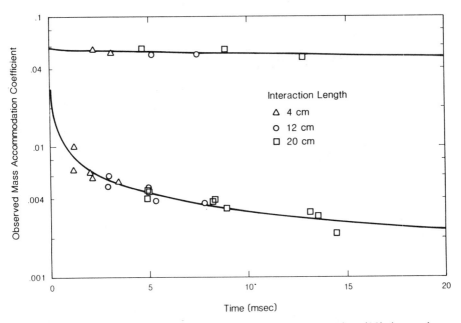

Figure 3. Experimental values and fits calculated by Equation (23) (assuming $\gamma = 0.059$) for γ_{obs} as a function of the droplet transit time for two values of $H^*_{S(IV)}$: Lower curve -- $H^*_{S(IV)} = 23$; upper curve -- $H^*_{S(IV)} = 4200$. These values were obtained by setting the pH of the droplets surface at 3.0 and 5.2, respectively.

Rate of Liquid Phase Reactions. The uptake of gas by a droplet can be affected by the rate of liquid phase chemistry within the droplet. An example is provided by the uptake of SO_2. In the previous section we showed that the uptake of SO_2 by the liquid is enhanced due to the fact that S(IV) is dissolved in the liquid not only as $SO_2(aq)$ but also in the form of HSO_3^- and SO_3^{2-}. However, in the derivation of Equation 23 it was assumed that equilibrium between the S(IV) species is reached instantaneously. This is not the case. The conversion of $SO_2(aq)$ to HSO_3^- occurs at a finite rate k_1 and the corresponding reaction time is $\tau_{rxn} = (k_1)^{-1}$ (11). If τ_{rxn} is longer than τ_p then a significant fraction of $SO_2(aq)$ can evaporate before reacting to form HSO_3^-. In that case, Equation 23 as it stands is not applicable. The finite reaction rate can be taken into account by modifying $\gamma(0)$ in Equation 23.

Early in the absorption process, where the reverse reaction HSO_3^- ---> $SO_2(aq)$ can be neglected, the change in the number density of $SO_2(aq)$ at distance from the surface, x, is given by

$$\frac{\partial n_{SO_2 (l)}(x)}{\partial t} = \frac{D_l \, \partial^2 n_{SO_2 (l)}(x)}{\partial x^2} - \frac{n_{SO_2 (l)}(x)}{\tau_{rxn}} \qquad (25)$$

Both τ_{rxn} and τ_p are short compared to the millisecond scale transit time of droplets through the reaction zone. Therefore, steady state conditions can be applied to Equation 25 and the solution is

$$n_{SO_2 (l)}(x) = n_{SO_2 (l)} (x=0) \, \exp [-x/(t_{rxn}D_l)^{1/2}] \qquad (26)$$

The flux of trace species at the interface on the liquid side of $x = 0$ is

$$J = D_l \frac{\partial n_{SO_2(l)}}{\partial x} = (D_l/\tau_{rxn})^{1/2} n_{SO_2 (l)} (0) \qquad (27)$$

This flux is set equal to the flux as calculated earlier in Equation 10 which yields at the interface the steady state ratio

$$\frac{n^o_{SO_2 (l)}}{n_{SO_2 (l)}(0)} = 1 + (\tau_p/\tau_{rxn})^{1/2} \qquad (28)$$

The approximations used in the derivation make this expression valid for times $t > \tau_p$ but still early in the absorption process. In this early period, $\gamma_{obs} \approx \gamma(t = 0)$.

As before, using the relationship $n_g \, \bar{c} \, \gamma(0) \, t/4 = Jt$ we obtain, consistent with Reference (7) and (9),

$$\gamma_{obs} = \frac{\gamma}{1 + (\tau_{rxn}/\tau_p)^{1/2}}$$

We can compute from this expression a theoretical upper limit for γ_{obs}. The rate for the conversion $SO_2(aq)$ to HSO_3^- was measured by Eigen et al. (13) as 3.6×10^6 s^{-1} which gives a characteristic reaction time $\tau_{rxn} = 2.8 \times 10^{-7}$s.

With the maximum possible value of $\gamma = 1$, we obtain $\tau_D = 3 \times 10^{-10}$s yielding the upper limit for γ_{obs} as 0.03. From Equation 7 and Figure 2 we have $\gamma = 0.11$, correcting γ_{obs} for gas phase diffusion limitations. This value implies that the rate for the conversion of SO_2 to HSO_3 is at least twelve times higher than the one measured by Eigen. However, their results are not necessarily contradicted by this observation. The conversion reaction studied by Eigen occurs in the bulk of the liquid whereas the process observed in our experiment occurs near the interface. It appears that the conversion rate is faster near the surface than in the bulk of the liquid. The conversion rate may be so rapid at the interface that the species enters the liquid as HSO_3^- rather than $SO_2(aq)$. A more complete presentation and analysis of our data on the uptake and conversion of SO_2 by water droplets has been prepared and submitted for publication (15).

Implications for Atmospheric Sulfur Oxidation

The large SO_2 mass accommodation coefficient ($\gamma = 0.11$) indicates that interfacial mass transport will not limit the rate of SO_2 uptake into clean aqueous cloud and fog droplets. Either gas phase diffusion, Henry's law solubility, or aqueous reactivity will control the overall rate of aqueous S(IV) chemistry. This conclusion is demonstrated by modeling studies of SO_2 oxidation in clouds by Chamedies (3) showing that the conversion time of S(IV) to S(IV) is independent of the mass accommodation coefficient for $1 \geq \gamma > 10^{-2}$ Schwartz (16) has also shown that, with γ as large as our measured value, the interfacial mass transport is unlikely to inhibit the oxidation of SO_2 by O_3 or H_2O_2 in cloud droplets for gas concentrations typical of non-urban industrialized regions.

Summary

Heterogeneous reactions involving water droplets in clouds and fogs are important mechanisms for the chemical transformation of atmospheric trace gases. The principal factors affecting the uptake of trace gases by liquid droplets are the mass accommodation coefficient of the trace gas, the gas phase diffusion of the species to the droplet surface and Henry's Law saturation of the liquid. The saturation process in turn involves liquid phase diffusion and chemical reactions within the liquid droplet. The individual processes are discussed quantitatively and are illustrated by the results of experiments which measured the uptake of SO_2 by water droplets.

Acknowledgments

This work has been supported by funds from the Coordinating Research Council (contract number CAPA-21-80(2-84)) and the Electric Power Research Institute (contract number RP 2023-8) to Aerodyne Research, Inc. and by grants from the National Science Foundation (ATM-8400748) and the Environmental Protection Agency (CR 812296-01-0) to Boston College.

Literature Cited

1. Graedel, T. E.; Goldberg, K. I. J. Geophys. Res., 1983, 88, 10865-82.

2. Heikes, B. G.; Thompson, A. M. J. Geophys. Res., 1983 88, 10883-95.

3. Chameides, W. L. J. Geophys. Res. 1984, 89, 4739-55.

4. Schwartz, S. E. J. Geophys. Res. 1984, 89, 11589-98.

5. Jacob, D. J. <u>J. Geophys. Res</u>. 1986, <u>91</u>, 9807-26.

6. Sherwood, T. K.; Pigford, R. L.; White, C. R. <u>Mass Transfer</u>; McGraw-Hill, 1975.

7. Gardner, J. A.; Watson, L. R.; Adewuyi, Y. G.; Davidovits, P.; Zahniser, M. S.; Worsnop, D. R.; Kolb, C. E. <u>J. Geophys. Res</u>. 1987, <u>92</u>, 10887-95.

8. Fuchs, N. A.; Sutugin, A. G. <u>High Dispersed Aerosols</u>; Hidy, G. M.; Brock, J. R., Eds.; Pergamon: Oxford, 1971; pp 1-60.

9. Schwartz, S. E. <u>Chemistry of Multiphase Atmospheric Systems</u>; Jaeschke, W., Ed.; NATO ASI Series: Sprinter-Verlag, Berlin, 1986; Vol. G6, pp 415-71.

10. Kimpton, D. D.; Wall, F. T. <u>J. Phys. Chem</u>. 1952, <u>56</u>, 715-7.

11. Schwartz, S. E.; Freiberg, J. E. <u>Atmos. Environ</u>. 1981, <u>15</u>, 1129-44.

12. Goldberg, R. N.; Parker, V. B. <u>J. Res. N.B.S</u>. 1985, <u>90</u>, 341-58.

13. Danckwerts, P. V. <u>Trans Faraday Soc</u>. 1951, <u>47</u>, 1014-23.

14. Eigen, M.; Kustin, K.; Maass, G. <u>Z. Phys. Chem. N.F</u>. 1961, <u>30</u>, 130-6.

15. Worsnop, D. R.; Zahniser, M. S.; Kolb, C. E.; Gardner, J. A.; Watson, L. R.; Van Doren, J. M.; Jayne, J. T.; Davidovits, P. <u>J. Phys. Chem</u>., in press.

16. Schwartz, S. E. <u>Atmos. Environ</u>., submitted.

RECEIVED December 19, 1988

Chapter 33

Rate of Reaction of Methanesulfonic Acid, Dimethyl Sulfoxide, and Dimethyl Sulfone with Hydroxyl Radical in Aqueous Solution

P. J. Milne, Rod G. Zika, and Eric S. Saltzman

Division of Marine and Atmospheric Chemistry,
Rosenstiel School of Marine and Atmospheric Science,
University of Miami, Miami, FL 33149–1098

Methanesulfonic acid, dimethyl sulfoxide and dimethyl sulfone are potential intermediates in the gas phase oxidation of dimethylsulfide in the atmosphere. We have measured the rate of reaction of MSA with OH in aqueous solution using laser flash photolysis of dilute hydrogen peroxide solutions as a source of hydroxyl radicals, and using competition kinetics with thiocyanate as the reference solute. The rate of the reaction k'(OH + SCN$^-$) was remeasured to be 9.60 ± 1.12 x 10^9 M^{-1} s^{-1}, in reasonable agreement with recent literature determinations. The rates of reaction of the hydroxyl radical with the organosulfur compounds were found to decrease in the order DMSO (k' = 5.4 ± 0.3 x 10^9 M^{-1} s^{-1}) > MSA (k' = 4.7 ± 0.9 x 10^7 M^{-1} s^{-1}) > DMSO$_2$ (k' = 2.7 ± .15 x 10^7 M^{-1} s^{-1}). The implications of the rate constant for the fate of MSA in atmospheric water are discussed.

Organosulfur compounds, of which dimethylsulfide (DMS) appears to be the most abundant, are the principal precursors of sulfur dioxide and non sea-salt sulfate aerosol in non-polluted air, and have a major impact on the global tropospheric sulfur cycle. The primary mechanism for atmospheric oxidation of these compounds is gas phase reaction with OH or possibly NO$_3$· radicals. Until quite recently ([1]) the oxidation product of organosulfur gases in the atmosphere was thought to be primarily sulfur dioxide, which would then be subsequently oxidized, again quantitatively, to sulfate. However, it is now apparent that a number of other intermediate products and chemically reactive transients may also be formed from organosulfur compounds in passing from the (-II) to the (+VI) oxidation state. This work examines the possible further oxidation of one of the more stable oxidation products, methanesulfonic acid, by hydroxyl radicals.

Methanesulfonic acid, although it comprises a relatively small fraction of total non sea-salt aerosol sulfur, has been shown ([2]) to be a ubiquitous component of marine aerosols. Its occurrence and distribution have been suggested as of use as an *in situ* tracer ([3],[4]) for oceanic emissions and subsequent reaction and deposition pathways of organosulfur compounds and dimethyl sulfide in particular.

0097–6156/89/0393–0518$06.00/0
© 1989 American Chemical Society

Specifically we wished to measure the rate of reaction of OH with MSA to enable modelling calculations of the stability of MSA in aerosol droplets. The one reported measurement of this rate (5), using pulse radiolysis techniques, 3.2 x 10^9 M^{-1} s^{-1}, is fast enough to suggest that this reaction pathway could be an important sink for MSA. This is of interest in explaining an apparent discrepancy that exists between laboratory and field studies of the oxidation of dimethyl sulfide. Although a number of laboratory studies (6-9) show that MSA is the major stable product, and SO_2 a minor one, field observation suggest MSA is only a minor (10%) fraction (2) of total non-sea-salt sulfur in marine aerosols. Two possible rationalizations of this are that i) MSA is subject to further reaction in marine aerosols and ii) other reaction pathways of dimethyl sulfide, or perhaps other non-methylated sulfur compounds should be considered.

Dimethyl sulfoxide, (DMSO) and dimethyl sulfone, ($DMSO_2$) have been reported in rain water samples (10). As is the case for MSA, the vapour pressures of these compounds are such that they are much more likely to be partitioned into heterogenous (aqueous) phases than to remain in the gas phase. A second aim of this work was to observe possible reaction pathways of these organosulfur compounds, which are also potentially stable oxidation products of dimethyl sulfide.

Methods and Experimental Approach

The experimental approach used in this study combined time resolved laser flash photolysis with competition kinetics.

Hydroxyl radicals were produced in solution by the laser flash photolysis of dilute solutions (5 - 50mM) of hydrogen peroxide which undergoes the homolytic cleavage

$$H_2O_2 \xrightarrow[h\nu]{} 2OH$$

The excitation source for this step was an excimer laser (EMG 201MSC, Lambda Physik, Acton Ma) operated at the Kr-F (248nm) line. The laser produced a ~25ns width pulse of rectangular cross section, a portion of which was allowed to impinge on a (1cm) quartz flow cell. Test solutions were pumped through the cell by a peristaltic pump connected to an external reservoir, whereby solutions could be exchanged or reagent composition readily adjusted. Absorbance was monitored at right angles to the pump beam with a xenon flash lamp (FX 193 U, EG&G Electro-Optics, Salem, MA) collimated to pass through the irradiated portion of the sample cell and collected with a monochromator (HR-320,Instruments SA, Metuchen, NJ) acting as a spectrograph for an intensified gated diode array (DARSS, TN-6133, Tracor Northern). The diode array was interfaced through an optical multichannel analyzer (Tracor Northern TN-1710) to a laboratory computer (Hewlett Packard HP-85). The timing sequence of the experiment with respect to i) the pump laser flash, ii) the probe beam lamp flash, and iii) the gate pulse on the diode array detector was set with the use of a digitial delay/pulse generator (DG535, Stanford Research Systems, Palo Alto, Ca). In this way, spectra of approximately 300nm width were obtained at 40ns intervals at selected times after t_0, as defined by the laser flash. In the experiments performed here, the region 275-575nm was monitored at intervals up to 2μsec after the photolyzing flash. The experimental system used in the time resolved absorption measurements is outlined in Figure 1.

Analytical grade potassium thiocyanate (Fisher), hydrogen peroxide (Mallinckrodt), dimethyl sulfoxide, (Fluka), dimethyl sulfone (Sigma) were used

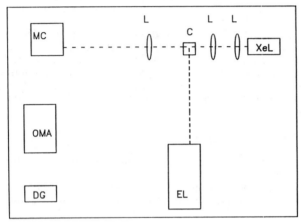

Figure 1. Experimental system used in the time resolved absorption measurements. (EL=excimer laser, KrF, 248nm: DG=delay generator: OMA=optical multichannel analyser: MC=monochromator and gated diode array detector: C=cell: X=xenon flash lamp: L=lenses)

as received. Methanesulfonic acid (Morton Thiokol) was specified as high purity (99.5%), yet two batches had a light brown colouration. LC and NMR examination of neat samples of the methanesulfonic acid did not however reveal any impurity.

Competition Kinetics

Gas phase kinetic studies of the reactions of hydroxyl radical are most conveniently carried out with direct monitoring of the OH radical with time using laser induced fluorescence (11). The low absorption coefficient of the aqueous hydroxyl radical (ϵ_{188nm} ~ 540 M^{-1} cm^{-1}, (12)) precluded the direct measurement of this reactant species by its absorbance. Also, the absence of a readily observable product species for the reaction of OH + MSA at the wavelength range (275-575 nm) easily accessible in our experiments, has lead us to monitor the concentration of OH in solution indirectly by competition kinetics (13), measuring the absorption of the thiocyanate radical anion (ϵ_{480nm} = 7600 M^{-1} cm^{-1} (12)).

Competition kinetics uses the absorption of a transient (or stable) product formed from the reaction of OH with a reference solute. This absorption will be suppressed in the presence of a competitor solute, as OH radicals will react with either the reference or competitor solutes in proportion to the products of the concentration and rate constants of the two competing reactions.

Specifically, for the reactions:

OH + SCN⁻ ---> *absorbing product*

OH + X ---> *non-absorbing product*

it follows that

$$\frac{k(OH+X)}{k(OH+SCN^-)} = \frac{A}{(A_0 - A)} \frac{[X]}{[SCN^-]} \tag{1}$$

Here $k(OH+X)$ is the absolute rate of OH with competitor X, $k(OH+SCN^-)$ the absolute rate of OH with reference solute, SCN⁻, A_0 is the absorbance of the reference product when X is absent, and A is the absorbance in the presence of some concentration of X.

Significant assumptions of this scheme are:
i) the kinetics can be described as simple bimolecular reactions
ii) there are no other absorbing species at the wavelength being measured
iii) there are no secondary reactions, of the reference or the competitor, occurring in the time frame being studied.

Rate of Formation of (SCN)$_2$⁻

Hydroxyl radical oxidation of thiocyanate ion in acid and neutral solution leads to the formation of a transient species which absorbs strongly at a wavelength of 475nm. Figure 2 shows the growth in absorption of this species at intervals from 20ns to 800ns.

Historically, this absorption was first attributed (14) to the SCN radical formed by

OH + SCN⁻ ---> SCN + OH⁻

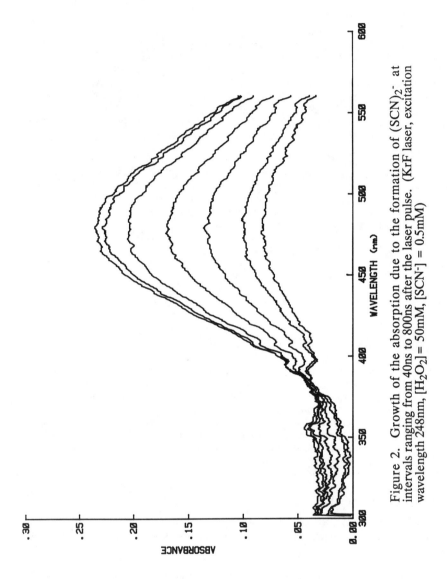

Figure 2. Growth of the absorption due to the formation of $(SCN)_2^-$ at intervals ranging from 40ns to 800ns after the laser pulse. (KrF laser, excitation wavelength 248nm, $[H_2O_2]$ = 50mM, $[SCN^-]$ = 0.5mM)

The data was also shown (15) to be consistent with the transient species being $(SCN)_2^-$, analogous to the dihalide anion radicals, produced by

$$SCN + SCN^- \rightarrow (SCN)_2^-$$

Further investigations (16) have suggested an even more complex description of the mechanism:

$$OH + SCN^- \rightarrow SCNOH^-$$

$$SCNOH^- \rightarrow SCN + OH^-$$

$$SCN + SCN^- \rightarrow (SCN)_2^-$$

$$SCNOH^- + SCN^- \rightarrow (SCN)_2^- + OH^-$$

Correspondingly, some uncertainty has existed as to the value of the rate constant for the reaction of OH + SCN⁻, depending upon the kinetic scheme used to fit the experimental data. Due to the large number of pulse radiolysis competiton kinetic studies performed using thiocyanate as the reference system, the absolute value of this rate is of some concern and has been in fact arbitrarily adjusted (1.1×10^{10} M⁻¹ s⁻¹, (13)) to accord with other reference solutes. Independently, a very similar value ($k' = 1.08 \pm 0.10 \times 10^{10}$ M⁻¹ s⁻¹) was measured (17).

Results and Discussion

Figure 3 shows a plot of the rate of formation of the $(SCN)_2^-$ in our experiments measured at two different thiocyanate concentrations. Linear fits of these and other data enabled evaluation of the slopes of $\ln[(A_m-A)/A_m]$, where A_m is the maximum absorbance. Attempts were made to fit the kinetic data with the reaction scheme suggested by Ellison et al. (17), but the lack of sufficient data points in the time domain, together with the inherently ill-conditioned nature of a triple exponential fit, meant that this was not meaningful for our data set.

Figure 4 shows the observed k' values plotted as a function of the thiocyanate concentration, the slope of which yields a value of $9.60 \pm 1.12 \times 10^9$ M⁻¹ s⁻¹. This value is in fair agreement with an accepted value of 1.1×10^{10} M⁻¹ s⁻¹ (17,13).

Competition experiments in which increasing amounts of the substrate of interest were added to a known concentration of thiocyanate and peroxide before photolysis were carried out for MSA, DMSO, and DMSO₂.

Typical conditions for these experiments were [SCN⁻] = 0.5mM, [H₂O₂] = 50 mM, [MSA] = 0~75mM, [DMSO] = 0~1mM, and [DMSO₂] = 0~150mM. Solutions were not degassed, although an experiment with DMSO which had been sparged and blanketed with nitrogen gave a rate constant similar to that had not. Solutions were also unbuffered, although their pH was monitored (~ pH 2 for experiments with MSA, pH 4-5 otherwise).

In none of the experiments was there indication of absorbance by species other than the thiocyanate radical, which supports the assumption that any reactions between uncomplexed SCN radicals and other species present are unimportant.

Figures 5a),b) show the diminution of the thiocyanate radical absorption at 480 nm (A) as increasing amounts of each added competitor substance were added for the substrates added. Using equation (1), plots of A_0/A vs [s]/[SCN⁻] were made, the slopes of which yielded k(OH + s)/k(OH + SCN), where s is

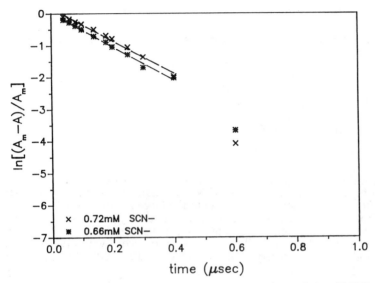

Figure 3. Kinetic analysis of the rate of formation of the $(SCN)_2^-$ at two different thiocyanate concentrations.

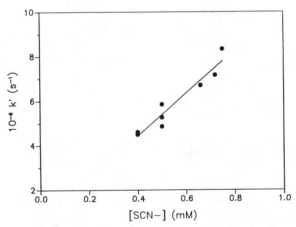

Figure 4. Rate of formation of $(SCN)_2^-$ radical as a function of initial thiocyanate concentration.

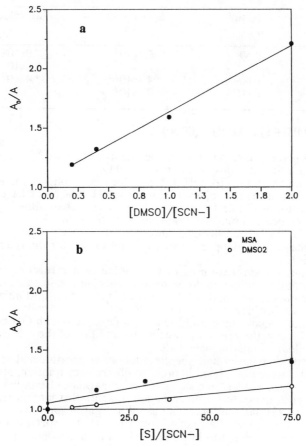

Figure 5. Decreases of the thiocyanate radical anion absorption as a function of added competitor concentration. Absorbances monitored at 480 nm. a) DMSO b) MSA or $DMSO_2$.

the compound being measured. Table I gives values of the determined ratio $k(OH + s)/k(OH + SCN^-)$ for the compounds studied and the corresponding estimates of the absolute rate constants based on the value of $k(OH + SCN^-)$ determined above.

Table I. Determination of $k(OH + s)^a$ from
Competition Kinetic Studies

Compound	Ratio[b]	$k(OH + s)$ $M^{-1} s^{-1}$	Lit. value	
DMSO	0.59	6.4×10^9	7.0×10^9	[18]
			5.8×10^9	[19]
MSA	0.0049	4.7×10^7	1.4×10^9	[5]
DMSO2	0.0026	2.5×10^7		

a) s = substrate
b) Ratio of $k(OH + s)$ to $k(OH + SCN^-)$

There was good agreement between the measured (6.4×10^9) and published ([18,19]) values for the rate constant of OH + DMSO.

The literature value ([5]) for OH + MSA given in Table I is the corrected value for that data, using the adjusted value for $k(OH + SCN)$ of 1.1×10^{10} M^{-1} s^{-1}, which was not available to those authors. Even when allowing for this correction of the original published rate constant, the value determined in this work (4.7×10^7 $M^{-1} s^{-1}$) is substantially lower than that value.

We are unaware of any previously reported values for the reaction rate $k(OH + DMSO_2)$.

One further consideration as to the choice of a reference system in competition kinetic studies should be mentioned. That is, in cases where the competitor substance, here both MSA and $DMSO_2$, react at substantially lower rates than the reference system, the presence of even small amounts of impurities in the system which could react at a faster rate with OH could bias the analysis. Ideally, the reference system rate and that of the substrate being measured should be similar in magnitude.

Experiments were performed wherein solutions of peroxide and MSA alone (no thiocyanate) were photolysed in order to observe any transient absorbance of a product radical, as had been suggested in an earlier pulse radiolysis study ([5]). No such absorbance was seen, although some preliminary measurements with a longer ($\sim \times 3$) path cell have indicated a slight absorbance in the 290-300nm region. The reported extinction coefficient for this species, presumably the $CH_2 \cdot SO_3^-$ radical anion, was low $\epsilon = 850$ M^{-1} cm^{-1}, and would not have been measurable with our experimental system at the concentrations used. Also the pH of our measurements is quite different from that of the reported transient, which was observed at pH ~ 9. This is being investigated further.

Reaction Mechansims

In the absence of detailed product analysis, and of readily observable intermediate species in our experiments, we can merely say that our results are consistent with the conclusions of other more detailed studies ([5,17,20]) of the reaction mechansims of these compounds. Thus it is reasonable that dimethyl sulfoxide, having its S-atom at the centre of a pyramidal structure with a free electron pair pointing to one corner, is readily acessible to electrophilic attack by the OH radical. There is in fact conductometric evidence ([20]) for the existence of an adduct intermediate species, $R_2SO(OH)$, which is formed according to

$$R_2SO + OH ---> R_2SO(OH) ---> R\cdot + RSO_2H$$

At the S-atoms in the other two molecules, adduct formation does not take place, and the OH radical is believed to react by a slower, hydrogen abstraction path from the associated methyl groups

$$CH_3-SO_3H + OH ---> CH_2\cdot-SO_3H + H_2O$$

The expected relative ordering of the reactivity of methyl sulfonic acid and dimethyl sulfone may presumably be accounted for in terms of inductive effects; a more detailed study of the influence of solution pH on the rate of reaction of OH with the methanesulfonate anion might be of use here.

Implications for MSA in Atmospheric Waters

A simple calculation of the lifetime of MSA in cloud water can be made using model estimates of the free radical chemistry of cloud droplets. The OH concentration in cloud water is a complex function of both the gas and aqueous phase chemistry and the dynamics of gas/liquid exchange. A recent model (21) estimated cloudwater OH concentration as $2\text{-}6 \times 10^{-13}$ M for droplets of 5 - 30μm radius. Using the rate constant measured here (4.7×10^7 M^{-1} s^{-1}), this yields a lifetime of 1.2 - 3.5 hours. Considering that the lifetime of a non-raining cloud is on the order of a few minutes to an hour, some fraction of the MSA present could react with OH, presumably being converted to sulfate. While such a process may lower the concentration of MSA in the droplet, it would only have a minor effect on the cloudwater sulfate levels because of the typically low MSA:non-sea-salt sulfate ratio in the aerosol entering the cloud.

Estimating the lifetimes of MSA in the aerosol particles themselves is considerably more difficult due to the problem of estimating an OH concentration in what is essentially an acid brine. A measurement of k(OH + MSA) as a function of ionic strength, possibly using $NaClO_4$ as an ionic strength adjuster, could be carried out using the procedure used in this work.

Acknowledgment

The technical assistance of David Odum and the comments of Dr. J. M. C. Plane were most welcome. This research was supported directly by the National Science Foundation on grant ATM-8709802 and indirectly through the Office of Naval Research.

Literature Cited

1. Cox, R. A.; Sheppard, D. Nature 1980, 284, 330-1.

2. Saltzman, E. S.; Savoie, D. L.; Prospero, J. M.; Zika, R. G. J. Atmospheric Chemistry 1986, 4, 227-40.

3. Saltzman, E. S.; Savoie, D. L.; Zika, R. G.; Prospero, J. M. J. Geophys. Res. 1983, 88, 10897-902.

4. Hatakeyama, S.;Izumi, K.; Akimoto, H. Atmos. Environ. 1985, 19, 135-41.

5. Lind, J.; Eriksen, T. E. Radiochem. Radioanal. Letters 1975, 21, 177-81.

6. Hatakeyama, S.; Okuda, M.; Akimoto, H. Geophys. Res. Lett. 1982, 9, 583-6.

7. Grosjean, D.; Lewis, R. Geophys. Res. Lett. 1982, 9, 1203-6.

8. Niki, H.; Maker, P. D.; Savage, C. M.; Breitenbach, L. P. Int. J. Chem. Kinet. 1983, 15, 647-54.

9. Hatakeyama, S.; Akimoto, H. J. Phys. Chem. 1983, 87, 2387-95.

10. Harvey, G. R.; Lang, R. F. Geophys. Res. Lett. 1986, 13, 49-51.

11. Atkinson, R. J. Chem. Revs. 1987, 85, 69-201.

12. Hug, G. L. Optical Spectra of Nonmetallic Inorganic Transient Species in Aqueous Solutions NSRDS-NBS 69; U.S. Government Printing Office: Washington, DC, 1981.

13. Dorfman, L. M.; Adams, G. E. Reactivity of the Hydroxyl Radical in Aqueous Solutions NSRDS-NBS 46; U.S. Government Printing Office: Washington, DC, 1973.

14. Adams, G. E.; Boag, J. W.; Michael, B. D. Trans. Farad. Soc. 1965, 61, 1674-80.

15. Baxendale, J. H.; Bevan, P. L. T.; Stott, D. A. Trans. Farad. Soc. 1968, 64, 2389-97.

16. Behar, D.; Bevan, P. L. T.; Scholes, G. J. Phys. Chem. 1972, 76, 1537-42.

17. Ellison, D. H.; Salmon, G. A.; Wilkinson, F. Proc. R. Soc. Lond. A. 1972, 328, 23-36.

18. Meissner, G.; Henglein, A.; Beck, G. Z. Naturforsch. 1967, 22b, 13-19.

19. Reuvers, A. P.; Greenstock, C. L.; Borsa, J.; Chapman, J. D. Int. J. Radiat. Biol. 1973, 24, 533-6.

20. Veltwisch, D.; Janata, E.; Asmus, K.-D. J. C. S. Perkin II 1980, 146-53.

21. Jacob, D. J. J. Geophys. Res. 1986, 91, 9807-26.

RECEIVED September 6, 1988

Chapter 34

Oxidation of Biogenic Sulfur Compounds in Aqueous Media

Kinetics and Environmental Implications

Yusuf G. Adewuyi[1]

Institute for Environmental Studies, University of Illinois, Urbana, IL 61801

The liquid phase reaction kinetics and mechanisms of oxidation of biogenic sulfur compounds (H_2S, RSH, CS_2, OCS, CH_3SCH_3, CH_3SSCH_3) by various environmental oxidants (O_2, O_3, H_2O_2, OH, Cl_2) including metal catalysis and photo-oxidation are critically reviewed. Significance of these reactions in aqueous and atmospheric systems; and their practical applications in the detoxification of waste streams are discussed. Suggestions for future studies pertinent to better understanding of the chemical transformations of these compounds in aqueous media are made.

Biogenic sulfur compounds in the environment are believed to impact significantly on acidity of precipitation in remote areas (*1*), formation of sulfur aerosols in marine atmosphere (*2,3*), global climate (*4,5*), atmospheric corrosion (*6*), global sulfur budget; and organic chemistry of seawater and biogeochemical cycling of its elements (*7,8*). In addition, control of emissions of these compounds is a key environmental concern in the retorting of oil shale (*9*) and in kraft pulp and synthetic oil industries due to their toxicity, unpleasant odor and taste even in low levels of concentration. They are also the source of malodorous condition in municipal sewage systems.

Reduced sulfur compounds are ubiquitous in aqueous and atmospheric systems (*10,11*). Natural sources of reduced sulfur species in aqueous environment result from biological reduction of sulfate, anaerobic microbial processes in sewage systems, putrefaction of sulfur-containing amino-acids (*12*), oxidative decomposition of pyrite (*13*), and activities of marine organisms in the upper layers of the ocean (*14,15*). The build-up of sulfides in areas such as the Black Sea is also giving cause for concern (*8*).

[1]Current address: Research Department, Paulsboro Research Laboratory, Mobil Research & Development Corporation, Paulsboro, NJ 08066

0097–6156/89/0393–0529$08.75/0
© 1989 American Chemical Society

The liquid phase reactions of biogenic sulfur compounds with the exception perhaps of H_2S (and its dissociated forms: S^{2-}, HS^-) and the mercaptans, are highly uncertain and unquantified due to lack of adequate liquid phase oxidation kinetics data in the open literature. However, gas phase kinetics have been studied by a number of investigators (16–20). The difficulty in studying the aqueous phase kinetics probably stems from the fact that these compounds have relatively low solubilities in water and that they are fairly resistant to noncatalyzed oxidations. However, as shown in Table I, they have appreciable Henry's law constants. An understanding of the mechanisms and kinetics of chemical transformation of these compounds in aqueous media is essential in determining the fate of these compounds in the environment and in implementing control strategies.

In this paper, an attempt is made to summarize what is known of the general solution chemistry of the compounds dimethyl sulfide (DMS), dimethyl disulfide (DMDS), carbon disulfide (CS_2), carbonyl sulfide (OCS), hydrogen sulfide (H_2S), and thiols (e.g., methyl and ethyl mercaptans) together with such kinetic and mechanistic knowledge as exists. The objective is to bring together data from totally different applications of the subject in an attempt to create a unified picture of the environmental significance of these compounds. The review will concentrate on their oxidation by oxygen, hydrogen peroxide, ozone and free radicals. Photo-oxidation and effect of pH and metal catalysis will also be considered. The oxidation of SO_2 and detailed review of H_2S is deliberately avoided. More complete review of the thermodynamics and kinetics of oxidation of H_2S in aqueous media are given elsewhere (11,21,22, this book). The oxidation of SO_2 by the aforementioned oxidants are fairly understood and has been the subject of many review papers (23–29). Some implications for the fields of application described above are discussed and suggestions for future studies pertinent to better understanding of the subject are made.

Solution Chemistry

Dimethyl Sulfide. DMS has a pungent obnoxious odor and is fairly stable in water. It has a maximum solubility of 17.7g/l in distilled water at room temperatures (30). The oxidation products—dimethyl sulfoxide (DMSO) and dimethyl sulfone ($DMSO_2$)—are relatively odorless, nontoxic and water soluble (31).

Oxidation of DMS by H_2O_2: DMS is oxidized by H_2O_2 in aqueous solutions to DMSO at ordinary temperatures. At higher temperatures (>20°C) and excess H_2O_2 the reaction proceeds to sulfone (32):

$$CH_3SCH_3 + H_2O_2 \longrightarrow (CH_3)_2SO + H_2O \tag{1}$$

$$CH_3SCH_3 + 2H_2O_2 \longrightarrow (CH_3)_2SO_2 + 2H_2O \tag{2}$$

The oxidation to DMSO is first order with respect to both DMS and H_2O_2 and is subject to catalysis by strong acids (Figure 1). As shown in

Table I. Henry's Law Constant (M atm^{-1}) of Reduced Sulfur Compounds in Distilled Water (DW) and Seawater (SW) at Different Temperatures

Compound		0°C	5°C	10°C	15°C	20°C	25°C	30°C	References
DMS	DW	1.613	1.300	1.075	0.840	0.685	0.559	0.463	96
	SW	1.429	1.124	0.901	0.725	0.585	0.478	0.394	
DMDS	DW	3.228	2.466	1.901	1.478	1.159	0.909	0.729	40
	SW	1.448	1.124	0.880	0.695	0.553	0.443	0.358	
CS_2	DW	0.160	-	0.094	-	0.056	0.044	0.032	97
OCS	DW	0.060	0.046	0.037	0.030	0.025	0.021	0.018	5,98
	SW	0.049	0.038	0.030	0.024	0.019	0.015	0.013	
H_2S	DW	0.209	0.177	0.152	0.132	0.116	0.102	0.091	99,100
	SW	0.196	0.168	0.145	0.127	0.111	0.099	0.088	
MeSH	DW	1.116	0.893	0.720	0.585	0.471	0.394	0.326	40
	SW	0.677	0.580	0.499	0.432	0.375	0.328	0.287	
EtSH	DW	0.884	0.693	0.548	0.436	0.350	0.283	0.230	40
	SW	0.675	0.536	0.430	0.347	0.282	0.231	0.190	
SO_2	DW	3.286	2.776	2.280	1.841	1.492	1.245	1.045	22,101,102

NOTE: Values obtained directly from citations or computed from relevant equations.

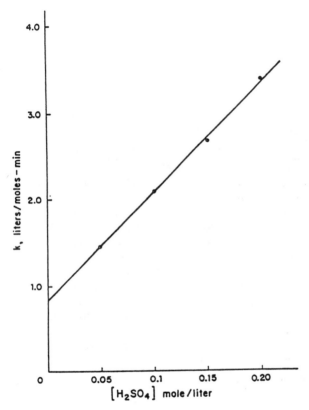

Figure 1. Effect of sulfuric acid on the rate of oxidation of DMS by H_2O_2 at $20°C$. (Reproduced from Reference 33. Copyright 1986 American Chemical Society.)

Figure 2, the rate coefficient is fairly constant between pHs of 2 and 6, increases drastically at pH 1 and below, and decreases substantially at pH 7. The rate law in strongly acidic solution is given by:

$$\frac{-d[DMS]}{dt} = k\,[H_2O_2]\,[DMS] \tag{3}$$

where $k = k'_H + k_H K_H\,[H^+]$, k'_H is the observed rate constant in uncatalyzed neutral solution, the product $k_H K_H$ is the catalytic rate constant for the acid catalyzed reaction, K_H is the equilibrium constant for the reaction of H^+ with H_2O_2, and k_H is the rate constant for the reaction of $HOOH_2^+$ with DMS. Adewuyi and Carmichael (33) found $k(M^{-1}\,S^{-1})$ at $20\,°C$ to be 1.4×10^{-2}, 3.4×10^{-2}, 3.5×10^{-2}, and 8.1×10^{-2} for solutions of pHs 7, 6, 2, and 1, respectively. They also obtained activation energies of 11.14 kcal/mole for reactions in acidic solution (pH < 7) and 3.41 kcal/mole for reactions in neutral medium (pH = 7) and concluded two different reaction mechanisms (as given below) were operative:

Acidic Medium (catalyzed):

$$H^+ + H_2O_2 \rightleftarrows HOOH_2^+ \tag{4}$$

$$HOOH_2^+ + (CH_3)_2S \longrightarrow I \tag{5}$$

where I is the intermediate:

$$CH_3 \qquad O - H$$

$$:S: \to O^+ - H$$

$$CH_3 \qquad H$$

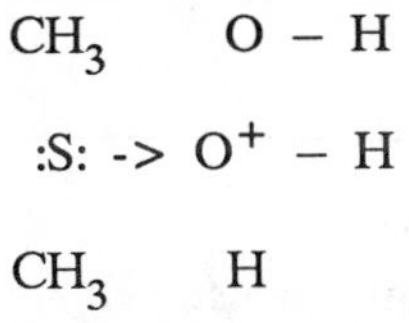

$$I \longrightarrow H_3O^+ + (CH_3)_2SO \tag{6}$$

Neutral Medium (uncatalyzed):

$$HOH + HOOH + (CH_3)_2S \longrightarrow I' \tag{7}$$

where I' is the intermediate:

$$HOH$$
$$|$$
$$|$$
$$H\;\;O\;\;O\;\;H$$
$$\uparrow$$
$$CH_3 \quad S \quad CH_3$$

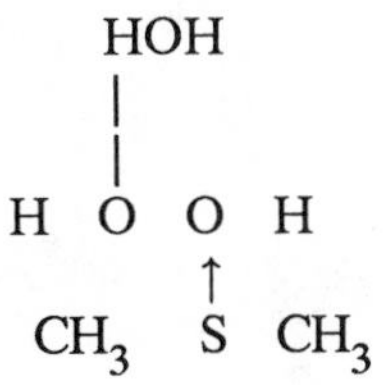

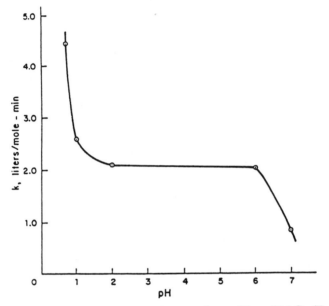

Figure 2. Plots of rate constant against pH at 20 °C. (Reproduced from Reference 33. Copyright 1986 American Chemical Society.)

$$I' \longrightarrow HOH + HOH + (CH_3)_2SO \tag{8}$$

Studies by Brimblecombe et al. (*34*) in seawater and NaCl solutions indicate DMS oxidation by H_2O_2 is first order with respect to DMS and is catalyzed by sea-salt metals. They found rates to be four times faster in seawater than in NaCl solutions. Rates were also found to decrease with increasing pH in NaCl solutions. For example, at $80\,^{\circ}C$ and in 0.7M NaCl solutions (with $[H_2O_2]_0 = 9 \times 10^{-5}$ M), they obtained a first order rate constant of $2.6 \times 10^{-6}\,s^{-1}$ (for pH 7.8) and $6.0 \times 10^{-7}\,s^{-1}$ (for pH 10.2). For the same conditions but at $20\,^{\circ}C$, the value was $2.2 \times 10^{-8}\,s^{-1}$ at pH 7.8. In seawater (pH 8) at $20\,^{\circ}C$ and with $[H_2O_2]_0 = 4.0 \times 10^{-5}$ M, the first order rate constant was found to be $5.7 \times 10^{-6}\,s^{-1}$.

<u>Autoxidation of DMS:</u> DMS is autoxidized by O_2 very slowly in solution at ordinary temperatures but the reaction is catalyzed by metal ions. In a saline solution of pH 8, Brimblecombe and Shooter (University of East Anglia, unpublished data) obtained a first order rate constant of $2.2 \times 10^{-8}\,s^{-1}$ at $20\,^{\circ}C$ and an activation energy of 78 kJ. The rate was also found to decrease with pH (but not linearly dependent on the H^+ ion concentration). In the presence of Cu(II) ions (10^{-4} M) at pH 5.6 and temperatures at $20\,^{\circ}C$, a tenfold increase in first order rate constant was observed ($k_a = 2 \times 10^{-7}\,s^{-1}$). Reactions in NaCl solutions were found to be nearly an order of magnitude slower than in seawater and at least an order of magnitude faster than in distilled water.

According to Correa and Riley (*35*), autoxidation could occur at reasonable rates at high pressures and temperatures in the absence of a catalyst, an initiator, or a photosensitizer and light. The autoxidation was regarded as a radical reaction and occurred in solvents of high polarity. The reaction was found to be first order in sulfide at constant oxygen pressure (72 bar) and linearly dependent on oxygen pressure. Based on the strong dependency on solvent polarity and elevated temperatures ($\geq 105\,^{\circ}C$), they proposed a mechanism involving electron transfer as the initiation step:

$$(CH_3)_2S + O_2 \rightleftarrows (CH_3)_2S^{\cdot+}-OO^- \tag{9}$$

$$(CH_3)_2S^{\cdot+} + O_2 \rightleftarrows (CH_3)_2S^+-OO^- \tag{10}$$

$$(CH_3)_2S^+-OO^- + (CH_3)_2S \longrightarrow 2(CH_3)_2SO \tag{11}$$

At atmospheric pressure, autoxidation of thioethers occurs primarily by abstraction of the α-hydrogen and sulfoxides are produced in low yields by reaction of the unreacted thioethers with intermediate α-hydroperoxides (*36,37*).

<u>Photo-oxidation of DMS:</u> Aqueous DMS is readily photo-oxidized by visible light in the presence of photosensitizers (e.g., methylene blue, anthroquinone) or humic acid and rose bengal which are naturally

occurring in seawater. Brimblecombe and Shooter (38) irradiated a solution of DMS (4.0×10^{-4} M) and methylene blue (5.3×10^{-6}), and observed a rapid oxidation of DMS. Photolysis was also found to be accompanied by a pH drop (from 5.9 to 4.2) and slow production of sulfate anion. With $[DMS]_0 = 6.8 \times 10^{-5}$M in seawater samples containing natural photosensitizers (P_s), they obtained a first order rate constant of 2.4×10^{-5} s^{-1} for photo-oxygenation in sunlight. In a separate experiment, they found DMSO to resist further photo-oxidation in the presence of methylene blue. A likely mechanism for photo-oxidation of substrate at low concentrations will be via singlet oxygen. The mechanism for photo-oxidation of organosulfides by singlet oxygen proceeds via the persulfoxide (39):

$$P_s + h\nu \longrightarrow P_s^* \tag{12}$$

$$P_s^* + O_2 \longrightarrow P_s + O_2^1\Delta_g \tag{13}$$

$$O_2^1\Delta_g + RSR \longrightarrow R_2S^+\text{--}OO^- \tag{14}$$

The investigators also observed photo-oxidation when DMS solution was irradiated with light in the near UV and short visible wavelengths in the absence of photosensitizers. The reaction was sensitive to pH and faster in acidic solutions. A reaction scheme similar to those of reactions 9 through 11 could be assumed:

$$(CH_3)_2S + h\nu \longrightarrow (CH_3)_2S^* \tag{15}$$

$$(CH_3)_2S^* + O_2 \longrightarrow (CH_3)_2S^+\text{--}OO^- \tag{16}$$

$$(CH_3)_2S^+\text{--}OO^- + (CH_3)_2S \longrightarrow 2(CH_3)_2SO \tag{17}$$

Dimethyl Disulfide. DMDS has a nauseating odor and solubility of 3.4g/l in water at $25\,°C$ (40). It is particularly resistant to uncatalyzed oxidation. Its oxidation product—methyl sulfonic acid (CH_3SO_3H)—is odorless, nontoxic, nonvolatile and water soluble.

Oxidation of DMDS by O_2: Oxidation of DMDS in alkaline medium is preceded by a hydrolysis reaction which involves nucleophilic displacement of sulfur from sulfur by hydroxide ion to yield a thiolate ion and a sulfonate ion (41):

$$RSSR + 2OH^- \longrightarrow RS^- + RSO^- + H_2O \tag{18}$$

The hydrolysis step is then followed by oxidation of the mercaptide ion to disulfide and the sulfonate ion to sulfonate. Murray and Rayner (42) obtained the following mechanisms for the reaction of DMDS (1.0g/l) in 0.1N NaOH solution at $100\,°C$:

$$CH_3SSCH_3 + 2NaOH \longrightarrow CH_3SNa + CH_3SONa + H_2O \tag{19}$$

$$4CH_3SNa + 2H_2O + O_2 \longrightarrow 2CH_3SSCH_3 + 4NaOH \tag{20}$$

$$CH_3SONa + O_2 \longrightarrow CH_3SO_3Na \tag{21}$$

$$3CH_3SONa \longrightarrow CH_3SO_3Na + 2CH_3SNa \tag{22}$$

They observed sodium methanesulfonate as the main product of oxidation with trace amount of sulfate ion. Reactions of DMDS with O_2 were found to be negligible in neutral and acidic medium; and copper and iron salts had little catalytic effect on the oxidation rate. The rate of oxygen mass transfer was found to control the overall oxidation rate (i.e., zero order with respect to DMDS) except at very low disulfide concentration when the chemical rate becomes very slow. In the concentration levels typical in black liquor, Murray et al. (*43*) found the oxidation rate strongly dependent on time and temperature, indicating that the chemical reaction rate exerts a strong influence on the overall oxidation rate. Murray and Rayner (*42*) also suggested reaction 22 as rate-determining in the non-oxidative (absence of O_2) alkaline hydrolysis reactions 19 and 22, and reaction 19 or 21 as rate-determining in the chemical reaction between dissolved O_2 and DMDS (reactions 19 through 21).

Oxidation of DMDS by H_2O_2: DMDS is catalytically oxidized by H_2O_2 in acidic aqueous medium. When divalent copper ions (0.001 to 0.1gm per liter of medium) are used as catalyst and the medium is kept at a pH below 6.5, the oxidation could be very efficient at temperatures greater than 40°C. The results of Stas and Biver (*44*) using copper sulfate and ferric sulfate as catalyst at 70°C and with pH kept at 4 (by adding H_2SO_4) are shown in Table II. The medium used for the experiment composed of 250mg of metal ions, 1.5ml DMDS, 10ml of methanol (to enable the DMDS to dissolve) in 5 liters of demineralized solution. The solution thus contained 318mg of DMDS (i.e., 230 mg of sulfides, expressed in equivalents of H_2S per liter). For the same peroxide to sulfide ratio, copper ion is clearly a better catalyst. The investigators obtained good results by using molar ratios of H_2O_2 to sulfide of about 3 to 6. The catalytic effect of acids was also observed by Stoner and Dougherty (*45*) in the oxidation of dithioacetic acid by H_2O_2. The overall reaction (without catalysis and in presence of a base) can be written as follows :

$$RSSR + 5H_2O_2 \longrightarrow 2RSO_3H + 4H_2O \tag{23}$$

$$RSSR + 5H_2O_2 + 2OH^- \longrightarrow 2RSO_3^- + 6H_2O \tag{24}$$

Disulfides are generally oxidized by H_2O_2 to produce unstable sulfenic acid (RSOH) which then disproportionates (by unknown path) to sulfonic acid (*46*):

$$RSSR + H_2O_2 + H^+ \longrightarrow RS^+ + [RSOH] + H_2O \tag{25}$$

Table II. Examples of the Oxidation of Dimethyl Disulfide by Hydrogen Peroxide

Example	Catalyst	H_2O_2/Sulfide(in terms of H_2S) moles/mole	Time mins.	Residual DMDS mg/l	Degree of destruction of DMDS(%)
1	$Fe_2(SO_4)_3$	1.44	30	86	73
2	$CuSO_4$	1.44	30	46	86
3	$Fe_2(SO_4)_3$	2.87	60	48	85
4	$CuSO_4$	2.87	30	14	96
5	$CuSO_4$	2.87	60	9	97
6	$CuSO_4$	4.31	30	4	99
7	$CuSO_4$	4.31	60	1	99.7

SOURCE: Stas, G.; Biver, C. U.S. Patent No. 45–955–77, June 17, 1987.

Oxidation of DMDS by O_3: When an aqueous solution of DMDS is ozonized, the major product formed is methanesulfonic acid with trace amount of methyl methanethiolsulfonate ($CH_3SO_2SCH_3$). The reaction is complex mechanistically but a simple sequence can be envisaged as:

$$CH_3SSCH_3 + O_3 \longrightarrow CH_3SO \cdot SCH_3 \xrightarrow{O_3} CH_3SO_3H \qquad (26)$$

According to Barnard (47), initial attack of O_3 on the DMDS appears to be the rate-determining step of the overall reaction. Thiosulfinates formed as a result of this attack are generally orders of magnitude more reactive than disulfides in reactions leading to cleavage of S–S bond (48). In an experiment with $[DMDS]_0 = 1 \times 10^{-4}$ M, $[O_3]_0 = 6 \times 10^{-4}$ M, Erickson et al. (49) observed the reaction to be complete in about one half minute and a final pH of the reaction mixture to be 5.3. The overall reaction for the oxidation of DMDS by O_3 is:

$$CH_3SSCH_3 + 2O_3 + 2H_2O \longrightarrow 2RSO_3H + O_2 + 2H^+ \qquad (27)$$

The sulfonic acid formed is a strong acid ($pK_a = -6.0$) and ionizes completely in water:

$$CH_3SO_3H + H_2O \rightleftarrows CH_3SO_3^- + H_3O^+ \qquad (28)$$

Oxidation of DMDS by OH Radical: Dimethyl disulfide is oxidized by hydroxyl radicals (50–53) according to the reaction:

$$CH_3SSCH_3 + OH \cdot \longrightarrow CH_3SSCH_3^+ + OH^- \qquad (29)$$

Mockel et al. (51) determined the bimolecular rate constant for reaction 29 to be 1.5×10^{10} M^{-1} s^{-1} in solution of low DMDS concentrations (10^{-5} M) and observed the positive ion ($RSSR^+$) to decay by a second order process with $2k = 4.2 \pm 0.5 \times 10^9$ M^{-1} s^{-1} in neutral and slightly acidic solutions. From the results of the investigators, the decay was mixed (first and second) order with a half life of 50 microseconds at pH 8.05. However, the rate of decay increased upon addition of more OH^- ions and then became pseudo-first order at pH above 9.5. The second order rate decay can be ascribed to disproportionation reaction of the type:

$$2CH_3SSCH_3^+ + H_2O \longrightarrow CH_3SSCH_3 + CH_3SO \cdot SCH_3 \qquad (30)$$

(with the thiosulfinate eventually yielding methyl sulfonic acid). In basic solution, the decay could be attributed to neutralization reaction:

$$RSSR \cdot^+ + OH^- \longrightarrow RSS(OH)R \cdot \longrightarrow products \qquad (31)$$

and dissociative process:

$$RSS(OH)R \cdot + OH^- \longrightarrow RS^- + RS(OH)_2 \cdot \qquad (32)$$

(where $RS(OH)_2^{.}$ is hydrated $RSO^{.}$. The bimolecular rate constant k_{32} ($R = CH_3$) is $1.5 \pm 0.1 \times 10^8$ M^{-1} s^{-1} (50). In slightly acidic solution, the decay was found to increase (i.e., half-life decreases) with increase in ionic strength of the solution indicating a bimolecular reaction of two species with a charge of the same sign. In basic solution, the rate of decay was pseudo-first-order and the half-life was dependent on the OH^- concentration. Increase in ionic strength decreases the decay rate (i.e., half-life increased) as expected for reaction of OH^- with a positively charged species.

Possible reaction mechanisms postulated by Mockel et al. (51) for the oxidation of disulfides by OH radicals are as follows:

$$\text{Addition:} \quad RSSR + OH^{.} \longrightarrow RSSOHR^{.} \quad (33)$$

$$\text{Dissociative Capture:} \quad RSSR + OH^{.} \longrightarrow RSSOH + R^{.} \quad (34)$$

$$RSSR + OH^{.} \longrightarrow RSOH + RS^{.} \quad (35)$$

$$\text{Hydrogen Abstraction:} \quad RSSR + OH^{.} \longrightarrow RSSR(-H)^{.} + H_2O \quad (36)$$

The formation of thiols is also possible in acid medium via the following pathway:

$$RSSOHR^{.} + H^+_{(aq)} \longrightarrow RSH + RSOH^+ \quad (37)$$

$$RSOH^+ \longrightarrow RSO^{.} + H^+_{(aq)} \quad (38)$$

Bonifacic et al. (50) found $RS\overset{v}{O}D$ to be formed to a considerable extent in acid solutions of pH less than 6.

Mercaptans. Thiols possess obnoxious odor and corrosive properties. The solubilities of MeSH, EtSH and propylthiol at 25°C are 39, 12, and 3.8 g/liter respectively in water (40). Their oxidation products are mainly disulfides and sulfonic acids.

Oxidation of Mercaptans by O_2: Mercaptans are autoxidized in the presence of O_2 in alkaline medium. In general, the oxidation is slow in the absence of catalyst because of unfavorable spin state symmetries that result from differences in the electronic configuration of the reactants (54). However, the reaction proceeds rapidly in the presence of traces of metal ions or transition metal phthalocyanines (55–58). The catalyst tends to alter the electronic structure of either the reductant and/or O_2 so as to surmount the activation energy barrier imposed on the reaction by spin-state symmetry restriction. The coupled oxidation system in the presence of catalyst can be represented by:

$$2RSH + 2OH^- \longrightarrow 2RS^- + 2H_2O \quad (39)$$

$$2RS^- + 2H_2O + O_2 \xrightarrow{\text{catalyst}} RSSR + H_2O_2 + 2OH^- \quad (40)$$

The thiolate anion (RS^-) is the species susceptible to reaction with oxygen.

Cullis et al. (*58*), studied the oxidation of EtSH in alkaline solution (2M NaOH) at 30°C (and 700mm O_2 pressure) and found the overall stoichiometry to be:

$$4C_2H_5SH + O_2 \longrightarrow 2(C_2H_5)_2S_2 + 2H_2O \tag{41}$$

Rates for the uncatalyzed reactions were found to consist of two stages. The initial rate was first order with respect to both EtSH and O_2 and became zero order with respect to EtSH after 10-30% of the reaction:

$$\frac{-d[EtSH]}{dt} = 4.86 \times 10^{-2}[EtSH][O_2]M^{-1}\,s^{-1} \tag{42}$$

$$\frac{-d[EtSH]}{dt} = 2.01 \times 10^{-2}[O_2]\,s^{-1} \tag{43}$$

The activation energy of the reaction and pre-exponential factor were determined to be 16.5 kcal/mole and $10^{10}\,s^{-1}$ respectively. Reaction mechanisms involving the formation and dimerization of thiyl radicals (RS·) were utilized to explain the kinetic data. The active species in such mechanisms is the thiolate anion which is formed in a reversible step with the base catalyst (*59*). The resulting anion supposedly reacts with O_2 via an electron transfer mechanism to give peroxide ion:

$$RSH + B \rightleftharpoons RS^- + BH^+ \tag{44}$$

$$RS^- + O_2 \longrightarrow RS\cdot + O_2^- \quad \text{rate–determining} \tag{45}$$

$$RS^- + O_2^- \longrightarrow RS\cdot + O_2^{2-} \tag{46}$$

$$2RS\cdot \longrightarrow RSSR \tag{47}$$

$$O_2^{2-} + H_2O \longrightarrow 2OH^- + 1/2O_2 \tag{48}$$

Direct oxidation of the thiol by the peroxide formed is also possible via:

$$RSH + \cdot O_2^- \longrightarrow RS\cdot + HO_2^- \tag{49}$$

$$RSH + HO_2^- \longrightarrow RS\cdot + HO^- + \cdot OH \tag{50}$$

Cullis et al. (*58*) also found the reactions to be catalyzed by metal ions with concentration $\leq 10^{-3}M$ and Cu, Co and Ni were the most effective. They postulated the following reaction mechanism in the presence of metals ions:

$$2M^{n+} + O_2 \longrightarrow 2M^{(n+1)+} + O_2^{2-} \tag{51}$$

$$2RS^- + 2M^{(n+1)+} \longrightarrow 2M^{n+} + 2RS\cdot \tag{52}$$

$$2RS\cdot \longrightarrow RSSR \tag{53}$$

$$O_2^{2-} + H_2O \longrightarrow OH^- + 1/2O_2 \tag{54}$$

The formation of peroxide is also possible via reaction:

$$O_2^{2-} + 2H_2O \longrightarrow H_2O_2 + 2OH^- \tag{55}$$

The kinetics of the reaction were found to be dependent on the nature of the metal, its surrounding ligands and the homogeneity of the system. Cobalt phthalocyanine was observed to show high catalytic activity. This was attributed to the formation, in the complex, of six-membered rings which enhance the ability of the cation to donate an electron to oxygen and stabilize each oxidation state of the cation. Phthalocyanines are macrocyclic tetrapyrrole compounds that readily form square-plane coordination complexes in which the metal atom is bonded to four pyrrole nitrogen atoms. The rates of reactions were found to be limited by oxygen diffusion at high concentrations (10^{-3} M) of catalyst for Cu and Ni. The final rate constants (at 30°C) obtained in the presence of Cu (10^{-5}) and Co (10^{-3}) were 23.3×10^{-2} s^{-1} and 14.7×10^{-2} s^{-1} respectively. The corresponding activation energies (between 20 and 40°C) were 4.3 and 7.5 kcal/mole for Cu and Co respectively. The final O_2 uptake rate (mole l^{-1}s^{-1}) in the presence of 10^{-5}M Cu (as CuSO$_4$·5H$_2$O), 10^{-3}M Mn (as MnSO$_4$·4H$_2$O) and 10^{-3}M Fe (as FeSO$_4$·7H$_2$O) were 26.8×10^{-6}, 3.2×10^{-6}, and 3.2×10^{-6} respectively. In the autoxidation of 2-mercaptoethanol catalyzed by Cobalt(II)-4,4′,4″,4‴-tetrasulfophthalocyanine, Leung and Hoffmann (60) suggested the principal reactive species in the catalytic cycle is RS$-$Co$''$TSP-RS-Co$'''$TSP-O$_2^-$.

The oxidation of CH$_3$SH in alkaline medium (pH $\geq$ 12), according to Bentvelzen et al. (61), follow a two-term sequence:

$$2CH_3SH + 1/2O_2 \longrightarrow CH_3SSCH_3 + H_2O \tag{56}$$

$$CH_3S^- + H_2O \rightleftarrows CH_3SH + OH^- \tag{57}$$

They obtained the following rate expression for the oxidation:

$$-\frac{d[MeS^-]}{dt} = k_1' \, [MeS^-]^{0.75} \, [P_{O_2}]^{0.66} \tag{58}$$

where $k_1' = kH$ and H is the Henry's law constant. The values for k_1 were found to be 0.012, 0.018, 0.025 and 0.04 (mole/l)$^{0.25}$/(psig)$^{0.66}$ min; at 10, 15, 20, and 25°C respectively. The activation energy of the reaction was calculated to be 17 kcal/mole. Fractional order indicates that the reaction is more complex than represented by equation 56. Rates were found to be independent of pH above 12 while they decreased with decreasing pH from 12 to 8. Disulfide formation was also found to be directly proportional to CH$_3$S$^-$ concentration. These observations suggest the rate is controlled by methyl mercaptide ion concentration. Methyl mercaptan is a weak acid with a pK$_a$ of 9.9 at 25°C (62). Ionization should be essentially complete above pH 12 while the mercaptide ion would be expected to be hydrolyzed to a greater extent as the pH is lowered from 12 to 8.

Harkness and Murray (*63*) studied the reaction at 30°C in strongly alkaline solutions (1M NaOH and 0.1M CH_3SH) with metal salts as catalysts; and found the reaction to be first order with respect to both CH_3SH and dissolved oxygen. They observed both disulfide and H_2O_2 as reaction products and found the amount of total O_2 consumed to be greater than the stoichiometric requirements. Quantitative analysis of CH_3SH consumed, and H_2O_2 and CH_3SSCH_3 formed, indicated the excess O_2 was used in the formation of H_2O_2. They also found the reaction to be catalyzed by metal ions with the order of decreasing catalytic activity among metal salts as: $CoCl_2$ > $CuSO_4$ > $Ni(OOCH)_2$ > $Fe(SO_4)$ > $ZnSO_4$ > $AgNO_3$. Giles et al. (*55*) obtained similar results discussed above in the oxidation of n-propylthiol and concluded the overall coupling reaction proceeded in two stages:

$$2RSH + O_2 \xrightarrow{Cu/base} RSSR + H_2O_2 \tag{59}$$

$$2RSH + H_2O_2 \longrightarrow RSSR + 2H_2O \tag{60}$$

Formation of H_2O_2 was found to be pronounced at lower reaction temperatures, higher oxygen pressure, higher solution pH, higher catalyst concentration and lower substrate concentration. With $[RSH]_o = 0.07M$, $[Cu]_T = 10^{-5}M$, $[NaOH] = [NaCl] = 0.1M$ and oxygen pressure 84 cm Hg, they obtained an activation energy of 9.5 ± 1.2 kcal/mole for the reaction. The reaction was found to be insensitive to free-radical scavengers (e.g., sulfite, hydroquinone), eliminating the possibility of radical chain mechanism. The reaction was also found to show pronounced sensitivity to stirring just as for the case of EtSH (*58*). However, controlled experiment indicated these effects to be due to the tendency of copper-thiolate and disulfide to cluster and reside at the water-disulfide interface and not due to O_2 mass transfer. The reaction had first order dependency on O_2 pressure and copper concentrations, and the order switched from a pseudo-zeroth order to a first and, then, second order as the reaction approached completion.

Based upon the above Michaelis–Menten type kinetic behavior, the visual observation of thiolate complex in a copper(I)-thiolate solution, and compatibility of "soft" thiolate and disulfide ligands with soft copper(I) (*64–65*), Lim et al. (*66*) postulated the following mechanism for the copper-catalyzed, aerobic coupling of alkaline thiols:

$$Cu^+(RSSR)_2^+ + RS^- \rightleftharpoons (RSSR)Cu^+(RS^-) + RSSR \tag{61}$$

$$(RSSR)Cu^+(RS^-) + RS^- \rightleftharpoons (RSSR)Cu^+(RS^-)_2^- \tag{62}$$

$$(RSSR)Cu^+(RS^-)_2^- + O_2 \xrightarrow{H_2O} Cu^+(RSSR)_2^+ + HO_2^- \quad \text{rate–determining} \tag{63}$$

$$RS^- + H_2O_2 \longrightarrow RSOH + OH^- \quad \text{rate-determining} \tag{64}$$

$$RS^- + RSOH \longrightarrow RSSR + OH^- \tag{65}$$

where the stability constants for complex formation and the rate constant for reaction 63 are given by:

$$K_{61} = \exp(-11,000K/T + 33.6) \quad (K_{61} \text{ is dimensionless}) \tag{66}$$
$$K_{62} = \exp(4270K/T - 10.5) \quad (K_{62} \text{ is in L/mole}) \tag{67}$$
$$k_{63} = \exp(30.0 - 5260K/T) \quad (k_{63} \text{ is in L/mole-min}) \tag{68}$$

and T is in degree Kelvin (K). The first rate-determining step (with an activation energy of 10.4 kcal/mole) involves copper mediated transfer of an electron from each thiolate ion to oxygen molecule which is then reduced to peroxide as the thiolate ions are oxidized pairwise first to mercapto radicals and then immediately to disulfide. Results of O_2 uptake measurement also suggested slow conversion of soluble disulfide to sulfonic acids over the course of several days (67).

Oxidation of Mercaptans by H_2O_2: Mercaptans undergo rapid oxidative coupling reaction in alkaline medium in the presence of H_2O_2 without the necessity of metal catalysis. The overall reaction in the presence of a base can be written as:

$$RSH + 3H_2O_2 + OH^- \longrightarrow RSO_3^- + 4H_2O \tag{69}$$

Giles et al. (55), in the oxidation of n-propyl mercaptan (C_3H_7SH), in solutions containing NaOH (> 0.2M), observed a first order dependence on the concentrations of the thiolate ion (RS^-) and the nondissociated H_2O_2 and obtained an activation energy of 12.3 kcal/mole for the reaction. They also observed the rate of reaction to decrease with increase in basicity at pH above the pK_a value of the thiol (i.e., 10.7). The reaction was also found to be insensitive to free-radical scavengers (e.g., sulfite, hydroquinone). The rate data suggested a nucleophilic displacement of the thiolate ion on H_2O_2 with the formation of transient sulfonic acid (RSOH) as the rate-determining step as in reactions 64 and 65.

The following rate expression was consistent with the observed data:

$$-\frac{[RS^-]}{dt} = 2k_{64} [RS^-] [H_2O_2] \tag{70}$$

where $k_{64} = \exp(26.8 - 6190K/T)$ and k_{64} is in liter/mole-min.

Leung and Hoffman (68), in the oxidation of 2-mercaptoethanol (HOC_2H_4SH) by H_2O_2 at pH 10.4 also observed a first order dependence with respect to both reactants and obtained mechanisms similar to equations 64 and 65 above. They observed slow rate of oxidation above and below the pH range of 9 to 12. and concluded that $HOC_2H_4S^-$ was the reactive mercaptan species for the rate-determining steps involving H_2O_2 and HO_2^-. Typical values for the reaction of thiolate ion with H_2O_2 and HO_2^- were 12.64 and 0.93 $M^{-1}s^{-1}$ respectively at 20 °C. They attributed the rapid decrease of oxidation rate above pH 10.5 to the fact that H_2O_2 was more reactive than HO_2^-. The difference in the rate may also be rationalized by a charge consideration. Reaction between RS^- and HO_2^- is hindered by a like-charge repulsion.

Oxidation of Mercaptans by O_3: Oxidation of methyl mercaptan in aqueous solution by O_3 produces methane sulfonic acid as the major product. The reaction mechanisms are complex with the formation of dimethyl disulfide (CH_3SSCH_3), methyl methane thiolsulfonate ($CH_3SO_2SCH_3$), and methyl methanethiolsulfinate (CH_3SOSCH_3) as minor products. Continued ozonation could result in slow formation of sulfuric acid :

$$CH_3SH + O_3 \longrightarrow CH_3SO_3H + \text{minor products} \qquad (71)$$

$$CH_3SO_3H + O_3 \xrightarrow{\text{slow}} H_2SO_4 \qquad (72)$$

Results by Erickson et al. (*49*) indicate the reaction is very fast and requires less than 2 moles of O_3 (about 1.74) per mole of thiol. With $[CH_3SH]_0 = 7 \times 10^{-5}$ M and $[O_3]_0 = 3 \times 10^{-4}$ M, they discovered the reaction was over in about 0.1 seconds. O_3 was found to react with CH_3SO_3H to produce sulfuric acid much faster in basic solutions than in acidic solutions. However, the oxidation of sulfonic acid to sulfuric acid is very slow for all practical purposes.

Kirchner and Litzenburger (*69*) obtained the following rate expression for the oxidation of ethyl mercaptan in the pH range of 6 to 3.4:

$$-\frac{dc}{dt} = k_{oz} \cdot r \cdot C_{oz} C_{EM} \qquad (73)$$

The stoichiometric factor (r) (i.e., the ratio of the change in moles of ozone to that of mercaptan) and the bimolecular rate constant (k_{oz}) were determined experimentally to be 2 and $3.0 \pm 0.8 \times 10^8$ M^{-1} s^{-1}, respectively. They also observed the value of k_{oz} to be independent of pH in the range of 0.85 to 4.7.

Carbon Disulfide. CS_2 is a poisonous and volatile liquid with pungent smell. It has a solubility of 2.2 g/liter at $22°C$ (*70*) and hydrolyzes very slowly in water to form OCS and H_2S (*71*).

Oxidation of CS_2 by H_2O_2: CS_2 is fairly stable in acidic solutions but is quantitively oxidized to sulfuric acid by H_2O_2 in alkaline solutions (*72,73*). The overall stoichiometry for the reaction can be written as:

$$CS_2 + 8H_2O_2 + OH^- \longrightarrow HCO_3^- + 2HSO_4^- + 2H^+ + 6H_2O \qquad (74)$$

Hovenkamp (*74*) studied the hydrolysis kinetics of CS_2 in strong alkaline solutions (NaOH > 0.1M) and found the rate-determining step to be first order with respect to both CS_2 and OH^- ion concentrations. In the oxidation of CS_2 by H_2O_2 to sulfate in alkaline medium (pH $\leq$ 11),

Adewuyi and Carmichael (73) observed the reaction of CS_2 with OH^- to form a dithiocarbonate complex as the rate-determining step for the overall reaction 74 above. The formation of sulfate was found to be dependent on the rate of hydrolysis and to be preceded by long induction periods. As shown in Figure 3, the formation rate of sulfate was also found to be pH dependent and to increase exponentially with time.

Based upon the aforementioned observations and the catalytic effects of OH^- ion, Adewuyi and Carmichael (73) proposed the following mechanism for the oxidation reaction:

$$CS_2 + OH^- \longrightarrow CS_2OH^- \quad \text{rate-determining} \tag{75}$$

$$CS_2OH^- + OH^- \longrightarrow CSO_2H^- + HS^- \tag{76}$$

$$CSO_2H^- + OH^- \longrightarrow CO_3H^- + HS^- \tag{77}$$

$$HS^- + H_2O_2 \longrightarrow HSOH + OH^- \tag{78}$$

$$HSOH + H_2O_2 \longrightarrow S(OH)_2 + H_2O \tag{79}$$

$$S(OH)_2 + H_2O_2 \longrightarrow SO_2 \cdot H_2O + H_2O \tag{80}$$

$$SO_2 \cdot H_2O \rightleftharpoons HSO_3^- + H^+ \tag{81}$$

$$HSO_3^- + H_2O_2 \longrightarrow HSO_4^- + H_2O \tag{82}$$

Figure 4 illustrates the linear dependence of pseudo-first order rate constants for both the formation and decomposition of the dithiocarbonate complex on OH^- ion concentrations. The decomposition of the complex was found to be an order of magnitude faster than the formation. For instance, with $[CS_2]_0 = 1.372 \times 10^{-2}$ M and $[H_2O_2]_0 = 7.41 \times 10^{-2}$, k_{75} and k_{76} were determined to be 2.28×10^{-4} and 2.63×10^{-3} respectively at 20°C and pH 9. The activation energies for the formation and decomposition of the dithiocarbonate complex were also found to be 11.96 and 12.34 kcal/mole respectively. In equation 78, HS^- acts as a nucleophile in an attack on H_2O_2, which then undergoes a heterolytic breakdown with hydroxide as a leaving group. The rate expression for sulfate production as a result of CS_2 oxidation in alkaline medium is :

$$\frac{d[S(VI)]}{dt} = 4 k' [CS_2]_0 e^{-2k't} \tag{83}$$

where k' is the pseudo-first order rate constant for alkaline hydrolysis of CS_2 ($k' = k_{75} [OH^-]$). With pH $\leq$ 7.41, the hydrolysis step is very slow and the predominant product of oxidation is colloidal sulfur (73). The reaction sequence by which sulfur is produced from HS^- is given in equations 93 through 98.

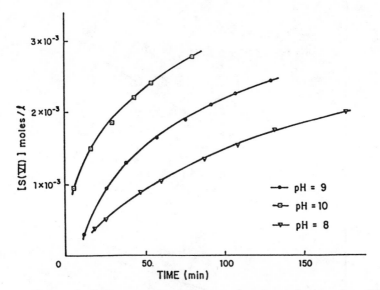

Figure 3. Sulfate production as a function of time for different pH values and $[CS_2]_o = 2.228(10^{-3})$ M, $[H_2O_2]_o = 7.4(10^{-2})$ M, T = 20°C. (Reproduced from Reference 33. Copyright 1986 American Chemical Society.)

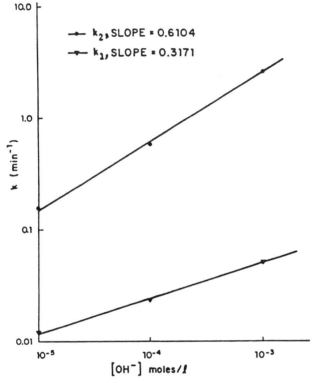

Figure 4. Effect of $[OH^-]$ on reaction rates for $[CS_2]_o$ = $4.573(10^{-4})$ M, $[H_2O_2]_o$ = $4.94(10^{-2})$ M. (Reproduced from Reference 33. Copyright 1986 American Chemical Society.)

Carbonyl Sulfide (OCS). OCS is odorless and tasteless but viciously poisonous and very corrosive in humid conditions. It hydrolyzes slowly in water to form H_2S and CO_2 with pseudo-first order rate constants of about 2.2 $\times$ 10^{-5} s^{-1} at 25°C and pH < 6 (75,76).

Oxidation of OCS by H_2O_2: OCS is very stable in acidic solutions but is easily oxidized by H_2O_2 to sulfate in alkaline medium. The removal of OCS from mixture of gases by alkaline liquid absorbents and oxidation of subsequent solutions to sulfate is an important industrial practice. For instance, OCS can easily be absorbed into ammonia solution and the resulting ammonia thiocarbamate oxidized to sulfate by H_2O_2 (77):

$$OCS + 2NH_3 \longrightarrow NH_2COSNH_4 \qquad (84)$$

$$NH_2COSNH_4 + 2NH_3 + 4H_2O_2 \longrightarrow (NH_4)_2SO_4 + (NH_4)_2CO_3 + 2H_2O \quad (85)$$

Philpp and Dautzenberg (78) observed that the rate-determining step in the alkaline hydrolysis of OCS was the formation of monothiocarbonate (MTC) intermediates according to a bimolecular course:

$$OCS + OH^- \longrightarrow CSO_2H^- \qquad (86)$$

They also observed a linear increase of the hydrolysis constant as a function of OH^- concentration and a bimolecular rate coefficient of 4.8 $\times$ 10^{10} $e^{-6643/T}$ M^{-1} s^{-1} for the reaction with pH $\geq$ 10. Activation energies of 13.2 and 23.1 kcal/mole were obtained for the formation and faster decomposition reactions of MTC respectively. Recent studies by Elliot et al. (76) give values of 3.8 (8°C), 8.0 (22°C) and 13.4 (29°C) for the bimolecular rate coefficient (M^{-1} s^{-1}) in the pH range of 9.1 to 9.4.

The decomposition of MTC to sulfide and subsequent oxidation to sulfate should follow the same steps as in the case of CS_2 (reactions 77 to 82) and thus the overall stoichiometry can be written as:

$$OCS + OH^- + 4H_2O_2 \longrightarrow HCO_3^- + HSO_4^- + H^+ + 3H_2O \qquad (87)$$

Given the data above, the overall kinetics should be governed by the rate of hydrolysis of OCS. The rate of hydrolysis of OCS by OH^- is faster than for CS_2 but slower than reactions 77 through 82 (73,79). Hence, in analogy to CS_2 oxidation, sulfate production rate may be assumed to be given by:

$$\frac{d[S(VI)]}{dt} = k'' [OCS]_o \, e^{-k''t} \qquad (88)$$

where k'' is the pseudo-first order rate constant, $k'' = k_{86} [OH^-]$.

Hydrogen Sulfide. H_2S is malodorous, toxic and corrosive especially at high humidities. It abounds in anoxic marine environments and is the major cause of corrosion in concrete sewer lines. Its oxidation products are mainly sulfur and sulfate.

Oxidation of H_2S by O_3: Hydrogen sulfide (H_2S/HS^-) is rapidly oxidized in aqueous solution by O_3 even in its protonated form (80,81). The rate expression can be written as:

$$\frac{-[H_2S]_T}{dt} = k_T[H_2S]_T[O_3] \qquad (89)$$

where $k_T = k_0\alpha_{H_2S} + k_1\alpha_{HS^-} + k_2\alpha_{S^=}$ and α is the mole fraction of the species in solution. Hoigné et al. (82) obtained a pH independent rate constant (k_T) of 3×10^4 (Ms)$^-$ for reactions at pH < 2 and temperatures $22 \pm 2°C$. For pH > 2, they observed that the apparent rate constant increased roughly by one order of magnitude per pH unit except at pH of about 3. The rate constants at pH 5, 6, and 8 were 6×10^7, 6×10^8 and 3×10^9 (Ms)$^{-1}$ respectively. The rate constant at pH 3 was found to be about 1×10^6 M^{-1}s^{-1}. They attributed the deviation from linearity at pH 3 to changes in $O_3/\Delta[H_2S]_T$ stoichiometry resulting from the faster oxidation rate of an intermediate presumably HSO_3^-. The overall sequence for the reaction is given by:

$$H_2S + O_3 \longrightarrow H_2SO_3 \xrightarrow{O_3} H_2SO_4 \qquad (90)$$

Oxidation of H_2S by H_2O_2: H_2O_2 oxidizes H_2S to sulfur in acidic and neutral conditions and to sulfate in alkaline conditions (83). The stoichiometry in acidic solution can be represented by:

$$H_2S + H_2O_2 \longrightarrow 2H_2O + S \qquad (91)$$

Hoffmann (84) studied the kinetics of oxidation of H_2S and HS^- to sulfur and sulfate by H_2O_2 in the pH range of 8 to 3 and obtained a two term rate law as follows:

$$\frac{-d[H_2S]}{dt} = k_1 [H_2S][H_2O_2] + k_2[HS^-][H_2O_2] \qquad (92)$$

The rate constants k_1 and k_2 were found to be 0.5 M^{-1} min^{-1} and 29.0 M^{-1} min^{-1} respectively at pH 5.05 and 25°C. The rate law and other data suggest a nucleophilic displacement by the bisulfide ion (HS^-) on H_2O_2 as the rate-determining step with subsequent formation of polysulfide as intermediates. The rate of the reaction was found to decrease as HS^- ion in solution decreases and hence the optimal pH for oxidation was determined to be 7. They postulated the following mechanism for the second term in the rate expression:

$$H_2S \rightleftharpoons H^+ + HS^- \qquad (93)$$

$$HS^- + H_2O_2 \longrightarrow HSOH + OH^- \quad \text{rate-determining} \qquad (94)$$

$$HS^- + HSOH \longrightarrow HS_2^- + H_2O \qquad (95)$$

$$HS_n^- + HSOH \longrightarrow HS_{n+1}^- + H_2O \qquad (96)$$

$$HS_9^- \longrightarrow S_8 + HS^- \qquad (97)$$

The formation of higher polysulfides by the reaction of the intermediate HSOH continues in subsequent steps until HS_9^- is formed. The final step involves the formation of c-S_8 (the only stable form of sulfur at STP) by an intermolecular displacement of HS_9^-. Considering the complexity of the mechanism, the more appropriate overall stoichiometry is:

$$H_2S + H_2O_2 \longrightarrow 1/x \, S_x + H_2O \qquad (98)$$

Addition of excess H_2O_2 was found to increase the rate but resulted in a higher percentage of sulfate formed. For instance, at pH 6.7, 41% total recoverable sulfur appeared in the form of sulfate. However, in basic sol-solution, sulfate was the only product formed. The mechanism by which sulfate is formed follows the same steps as in reactions 78 through 82.

Satterfield et al. (85) studied the rate of the reaction at pH of 4.0, 1.5, and 1.2. They found a first-order dependence on H_2O_2 and that the the rection rate was a strong function of the acidity. The value of the rate constant was also found to vary inversely to about 0.4 to 0.5 power of the H^+ ion concentration. Recent studies by Millero et al (86), give overall rate constant ($k_o = k_1\alpha_{H_2S} + k_2\alpha_{HS^-}$) which agree closely with Hoffmann's results in the pH range of 5 to 8, but deviate significantly at lower pH values. They also determined the effect of temperature (5 to 40 °C) at pH 8 and found the activation energy to be 19.4 ± 1.3 kJ/mole irrespective of the ionic strength of the solution. At pH 8 and 25 °C, k_o was found to be 100 M^{-1} min^{-1} for reaction in seawater and 56 M^{-1} min^{-1} for reactions in water and 6m NaCl solution.

Metal-Catalyzed Autoxidation of H_2S: The accelerating effects of metal ions on the oxgenation of sulfide was studied in detail by Chen and Morris (87). As little as 10^{-5} M Ni^{2+} was found to increase the rate of oxygenation of 10^{-2} M sulfide at pH 8.65 about a 10-fold; and 10^{-4} M of Ni^{2+} or Co^{2+} reduced the oxidation time from days to several minutes. In general, the trend of catalysis was found to be $Ni^{2+} > Co^{2+} > Mn^{2+} > Cu^{2+} > Fe^{2+}$ with Ca^{2+} and Mg^{2+} having lesser but significant effects. The observed kinetics of the metal catalysis was explained either in terms of two chain initiating reactions involving free radicals (equations 101 and 102) or formation of metal sulfide complexes which lower the activation energy of autoxidation and result in rapid precipitation of sulfur (equations 99 and 100):

$$M(HS^-)^{(n-1)+} + O_2 \longrightarrow M^{n+} + S + HO_2^- \qquad (99)$$

$$HS^- + HO_2^- \longrightarrow 2OH^- + S_{(s)} \qquad (100)$$

$$M^{n+} + O_2 + H^+ \longrightarrow M^{(n+1)+} + HO_2 \tag{101}$$

$$M^{(n+1)+} + HS^- \longrightarrow M^{n+} + HS \cdot \tag{102}$$

Hoffmann and Lim (88) studied the catalytic effects of trace amount of Co(II), Ni(II) and Cu(II) 4,4',4'',4'''-tetrasulfophthalocyanine and found the experimental rate law

$$\frac{-d[HS^-]}{dt} = k_{obsd} \, [\text{Catalyst}] \, [HS^-] \tag{103}$$

to apply in all cases. At pH range 8.3 to 10, a mechanism involving the formation of a tertiary activated complex in which O_2 and HS^- are reversibly bound to the metal complex, was found to be consistent with kinetic behavior:

$$CoTSP^{2-} + O_2 \rightleftarrows CoTSP\text{-}O_2^{2-} \tag{104}$$

$$CoTSP\text{-}O_2^{2-} + HS^- \rightleftarrows HS\text{-}CoTSP\text{-}O_2^{3-} \tag{105}$$

$$HS\text{-}CoTSP\text{-}O_2^{3-} \longrightarrow CoTSP^{2-} + HSO_2^- \tag{106}$$

$$HSO_2^- + O_2 \longrightarrow HSO_4^- \tag{107}$$

The rate constant, k_{obsd} was found to be $1.0 \pm 0.2 \times 10^5 \ M^{-2} \ min^{-1}$ and the apparent order of catalytic activity was CoTSP > NiTSP > CuTSP over a wide range of pH. The catalytic turnover numbers for the reactions were generally found to be greater than 100,000 (23).

Oxidation of H₂S by Chlorine: Chlorination is one of the most common methods of hydrogen sulfide control. The oxidation of dilute sulfide solutions by chlorine or hypochlorite involves a primary oxidation of sulfide to sulfur and a simultaneous secondary oxidation of a portion of the sulfur to sulfate. The overall reactions are as follows:

$$H_2S + Cl_2 \longrightarrow 2HCl + S \tag{108}$$

$$H_2S + 4Cl_2 + 4H_2O \longrightarrow H_2SO_4 + 8HCl \tag{109}$$

$$H_2S + OCl^- \longrightarrow S + HCl + OH^- \tag{110}$$

$$H_2S + 4OCl^- \longrightarrow H_2SO_4 + 4Cl^- \tag{111}$$

Black et al. (89) found the overall oxidation to go almost completely to sulfate within the first minute of the reaction at pH of slightly less than 5 but the extent of reaction decreased with increasing pH to minimum at about pH 9. Choppin et al. (90), in their studies with hypochlorite, found a distinct minimum in the production of sulfate in the region of about pH 10, with increased sulfate formation at values above and below this point. In addition, they found high sulfide concentrations increased the formation of sulfur and high hypochlorite concentration increased sulfate formation.

Discussion

Atmospheric Considerations. It is probably premature to assess the role of liquid phase oxidation of reduced sulfur compounds in atmospheric chemistry with the limited kinetic data available in the open literature. However, it is appropriate to discuss certain conclusions obvious from the information presented above.

Homogeneous oxidation of these compounds by H_2O_2 may not be an important source of acidity in cloudwater and atmospheric water droplets. Oxidation of CS_2 and OCS is preceded by the rate-controlling hydrolysis step which is significant only at high pH (≥ 8). Belviso et al. (*91*) recently suggested the production of OCS within raindrops by the reaction of strong acids (e.g., dilute H_2SO_4) and scavenged thiocyanate according to the hydrolysis reaction:

$$SCN^- + 2H^+ + H_2O \longrightarrow OCS + NH_4^+ \tag{112}$$

However, most cloudwater and rain droplets have pH of about 6 and below and both CS_2 and OCS are very stable in acids solutions. The oxidation of H_2S by H_2O_2 is preceded by a rapid pre-equilibrium to give HS^- but product of oxidation at pH ≤ 7 is mostly sulfur and not sulfate. On the time scale of atmospheric processes, the rates of oxidation of DMS by H_2O_2 are too slow to compete with photo-oxidation and other heterogeneous pathways (*18,92,93*). It is unlikely, for example, that the aqueous oxidation of DMS by H_2O_2 can account for the presence of DMSO in rain and snow observed by Andreae (*94*). Finally, given H_2O_2 levels in rainwater of 0.1 to 7×10^{-5} M and in clouds of 10^{-6} to 10^{-5} M and the relatively low solubility of these reduced sulfur compounds it seems unlikely that reactions of H_2O_2 could be fast enough to be significant even in the presence of metal catalysis. However, the oxidation of DMS by H_2O_2 in collected rainwater samples may be appreciable on a time scale of hours to a few days.

Oxidation of H_2S, RSH and RSSR by O_3 in aqueous solutions is fast and results in the formation of strong acids (H_2SO_4, CH_3SO_3H), but it is difficult to judge the importance of these rates in the atmosphere for several reasons. O_3 has limited solubility in water with Henry's law constant at $25\,°C$ of about 0.01 M atm^{-1}. With $[O_3]$ roughly 3 to 6 $\times$ 10^{-10} M^{-1} in cloudwater and aqueous-phase rate constant limited by molecular diffusion to be at most 10^{10}M^{-1}s^{-1}, reaction with these compounds could never be sufficiently fast. The reason aqueous-phase oxidation of SO_2 can be significant is because of greater SO_2 solubility and its dissociation constants values which give S(IV) relatively large "effective" Henry's law constant. Moreover, it is yet to be shown that thiols and disulfides are present in significant amount in the upper atmosphere. Only in the vicinity of kraft paper mills have mercaptans and sulfides been monitored in appreciable concentrations (95). In regions where O_3 and these compounds co-exist, some reaction takes place but probably in the gas phase for the most part. However, in plumes from pulp mills, where suspended water droplets are in high concentration, some aqueous phase reactions may occur if O_3 is present.

The reaction of OH radical with DMDS is fast and OH is moderately soluble in water (Henry's law constant is 25 M atm^{-1} at 25 °C. But again, it depends on the concentration levels of DMDS in clouds and rainwater. In addition, the low solubility of DMDS in water makes heterogeneous oxidation in water droplets unlikely.

Environmental Pollution Control. The importance of chemical oxidation of reduced sulfur compounds lies in its application to treatment of wastestreams from municipal sewage systems, acid mine drainage and industrial plants such as tanneries, paper and pulp mills, oil refineries and textile mills.

Ozonation of wastewaters containing these compounds results in the formation of CH_3SO_3H and H_2SO_4 which could be discharged into effluent after neutralisation. Scrubbing of gaseous effluent with ozonized water should present no major technological difficulties since the reaction is fast and CH_3SO_3H is nonvolatile and water soluble. These are all positive attributes that could be taken advantage of in designing pollution equipments. In the oxidation of EtSH with O_3, Kirchner and Litzenburger (69) found the reaction or reaction rate of the mercaptan with O_3 dissolved in water to be the rate-determining step. In selecting a suitable type of scrubber, this means that the absorption rate may depend on liquid hold-up in the absorber.

H_2O_2 oxidizes SO_2 to sulfate, H_2S to sulfate and sulfur, RSH and RSSR to sulfonic acid and sulfate and RSR to sulfoxides and sulfones. The products of oxidation are all odorless. Hence, H_2O_2 may provide an economic effective means for odor and wastewater quality control in kraft mills. For the case of RSSR which are resistant to complete oxidation, catalytic oxidation by a peroxide in acidic medium can be employed. The fact that H_2O_2 is a liquid completely miscible with water and does not give solubility (or mass transfer) problems under any conditions, makes it an attractive choice for pollution control.

Aerobic coupling of thiols has added advantages. For instance, when alkaline propylthiol is oxidatively coupled at room temperatures, the aqueous solubility drops from 9.0 to 0.032 g/l (55). This large solubility reduction provides simple and attractive means of removing mercapto compounds from aqueous solutions. This also provides a means of recovery of biochemical products from dilute solutions. For instance, substances like cysteine which contains sulfhydryl groups are susceptible to oxidative coupling attack. The disulfides which result from oxidation contain weak S-S bonds and can easily be broken in the presence of hydrogen donors. This permits the original mercapto compounds to be easily regenerated. As illustrated in Figure 5, aerobic coupling of thiols in alkaline medium could

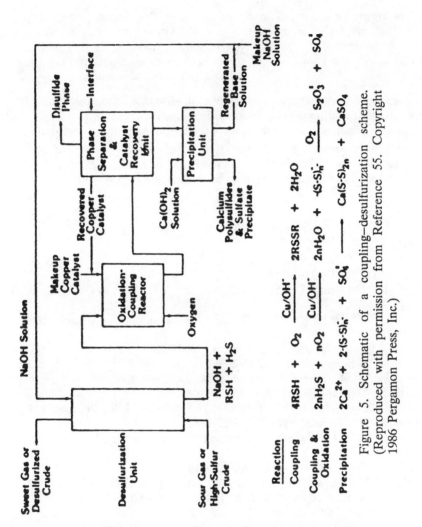

Figure 5. Schematic of a coupling–desulfurization scheme. (Reproduced with permission from Reference 55. Copyright 1986 Pergamon Press, Inc.)

also be employed in the desulfurization of high-sulfur crude and sour gas, and treatment of toxic wastes in the petroleum refining industries.

Summary

It is clear from the information presented above that liquid phase oxidation of reduced sulfur compounds play an important role in natural processes in water and in wastewater treatments. However, further work is needed to clarify the role of these compounds in cloud chemistry and precipitation acidity. Areas of further investigations should include:

1. Photochemistry of these compounds in solution and the catalytic effects of transition metals on oxidation rates.
2. Reaction kinetics and mechanisms of oxidations of these compounds by O_3 and OH radicals.

It is hoped that this presentation has stimulated thinking beyond the cases presented and should be regarded as a basis for future work.

Literature Cited

1. Galloway, J.N.; Liken, G.E.; Keene, W.C.; Miller, J.M. *J. Geophy. Res.* **1982**, *87*, 8771–8786.
2. Saltzmann, E.S.; Savoie, D.L.; Zika, R.G.; Prospero, J.M. *J. Geophy. Res.* **1983**, *88*, 10897–10902.
3. Yin, F.; Grosjean, D.; Seinfeld, J.H. *J. Geophy. Res.* **1986**, *91*, 14417–14438.
4. Turco, R.P.; Whitten, R-C; Toen, O.B.; Hamill, P. *Nature* **1980**, 238–286.
5. John, J.E. *Geophy. Res. Lett.* **1981**, *8*, 938–940.
6. Graedel,D. *Science* **1981**, *212*, 663–665.
7. Andreae, M.O. *Limnol. Oceanogr.* **1980**, *25*, 1054–1063.
8. Dyrssen, D. *Mar. Chem.* **1985**, *15*, 285–293.
9. Sklarew, D.S.; Hayes, D.J.; Petersen, M.R.; Olsen, K.B.; Pearson, C.D. *Environ. Sci. Technol.* **1984**, *18*, 592–600.
10. Niragu,J.O. *Sulfur in the Environment*; Wiley-Interscience: New York 1978.
11. Millero, F. *Mar. Chem.* **1987**, *18*, 121–147.
12. Elliot, L.F.; Travis, T.A. *Soil Sci. Soc. Amer. Proc.* **1973**, *37*, 700–702.
13. Stedman, D.H. *Geophy. Res. Lett.* **1984**, *1*, 858–860.
14. Challenger, F. *Adv. Enzymol.* **1951**, *8*, 1269–81.
15. Cline, J.D.; Bates, T.S.; *Geophy. Res. Lett.* **1983**, *10*, 949–52.
16. Atkinson, R.; Perry R.A., Pitts J.N., Jr. *Chem. Phys. Lett.* **1978**, *54*, 14–18.
17. Logan, J.A.; McElroy, M.B.; Wofsy, S.C. Prather, M.J. *Nature.* **1979**, *281*, 185–188.
18. Hatakeyama, S.; Okuda M.; Akimoto, H. *Geophy. Res. Lett.* **1982**, *9*, 583–586.
19. Nguyen, B.C.; Bonsang, B.; Gaudry A. *J. Geophys. Res.* **1983**, *88*, 10903–14.

20. Wallington, T.J.; Atkinson, R.; Winer, A.M. Pitts, Jr. J.N. *J. Phy. Chem.* **1986**, *90*, 5393–96.
21. 21. Kuhn, A.T.; Kelsall, G.H.; Chana, M.S. *J. Chem. Tech. Biotechnol.* **1983**, *33A*, 406–414.
22. Adewuyi, Y.G. PhD. Thesis, University of Iowa, Iowa City, 1985.
23. Hoffmann, M.R. *Environ. Sci. Technol.* **1980**, *14*, 1061–1066.
24. Linek, V.; Vacek V. *Chem. Eng. Sci.* **1981**, *36*, 1747–1768.
25. Braga, T.C.; Connick, R.E. In *Flue Gas Desulfurization*; Hudson, J.L.; Rochelle, G.T., Eds.; ACS Symposium Series No. 188; American Chemical Society: Washington, DC, 1982; pp. 153–171.
26. Hoffmann, M.R.; Jacob, D.J. In *SO_2, NO and NO_2 Oxidation Mechanisms: Atmospheric Considerations*; Calvert J.G., Ed.; Butterworth: Boston, 1984; p. 101.
27. Hoffmann, M.R. *Atmos. Environ.* **1986**, *20*, 1145–1154.
28. Martin, L.R. In *SO_2, NO and NO_2 Oxidation Mechanisms: Atmospheric Considerations*; Calvert J.G., Ed.; Butterworth: Boston, 1984, p. 63.
29. Hoffmann, M.R.; Boyce, S.D. In *Trace Atmospheric Constituents: Properties, Transformations, and Sulfates*; Schwartz S., Ed.; Wiley: New York, 1983, Adv. Environ. Sci. Technol. 12, p. 147.
30. Koptyaev, V.G. *Hyg. Sanit.* **1967**, *32*, 315–319.
31. Martin, D.; Hauthal, H.G. *Dimethyl Sulphoxide*; Halsted Press: New York, 1975, p. 50.
32. Gutsche, C.D.; Pasto, D.J. *Fundamentals of Organic Chemistry*, 2nd ed.; Prentice-Hall: New York, 1975, p. 450.
33. Adewuyi, Y.G.; Carmichael, G.R. *Environ. Sci. Technol.* **1986**, *20*, 1017–1022.
34. Brimblecombe, P.; Watts, S.; Shooter, D. Proc. 194th ACS Mtg, 1987, p. 96.
35. Correa, P.E.; Riley, D.P. *J. Org. Chem.* **1985**, *50*, 1787–8.
36. Bateman, L.; Cuneen, J.I. *J. Chem. Soc.* **1955**, 1596–1603.
37. Bateman, L.; Cuneen, J.I., Ford J.J. *Chem. Soc.* **1957**, 1539–44.
38. Brimblecombe, P.; Shooter, D. *Mar. Chem.* **1986**, *19*, 343–353.
39. Kacher, M. L.; Foote, C.S. *Photochem. Photobiol.* **1979**, *29*, 765–69.
40. Przyjazny, A.; Janicki, W.; Chrzanowski, W.; Staszewski, R.J. *J. Chromatgr.* **1983**, *280*, 249–260.
41. Danehy, J.P. *Intl. J. Sulfur Chem.* **1971**, *12*, PartB, 103–114.
42. Murray, F.E.; Rayner, H.B. *Pulp Paper Can.* **1968**, *69*, 64–67.
43. Murray, F.E.; Tench, L.G.; Asfar, J.J.; Burkell, J.E. AICHE Sym. Ser. 1973, 69, 102–105.
44. Stas, G.; Biver, C. U.S. Patent 4595 577, 1986.
45. Stoner, G.G.; Dougherty, G. *J. Am. Chem. Soc.* **1941**, *63*, 1291–92.
46. Pryor, W.A. *Mechanisms of Sulfur Reactions*; McGraw Hill: 1962, p. 57.
47. Barnard, D. *J. Chem. Soc.* **1957**, Pt4, 4547–4555.

48. Kice, J. L. In *Advances in Phys. Org. Chem.* **1980**, *17*, 65–181.
49. Erikson, R.E.; Yates L. M., Bufalini, J.J. 1976, EPA-600/3-76-089.
50. Bonifacic,M.; Schafer, K.; Mockel, H.; Asmus, K.-D. *J. Phys. Chem.* **1975**, *79*, 1496–1502.
51. Mockel H.; Bonifacic, M.; Asmus K.-D. *J. Phys. Chem.* **1974**, *78*, 282–4.
52. Bonifacic, M.; Asmus K.-D. *J. Phys. Chem.* **1976**, *80*, 2426–30.
53. Bonifacic, M.; Asmus, K.-D. *J. Chem. Soc. Perkin Trans. II*, **1986**, 1805–1809.
54. Leung, P.K.; Hoffmann, M.R.; Proc 194th ACS Mtg, 1987, p 612.
55. Giles, D.W.; Cha, J.A., Lim, P.K. *Chem. Eng. Sci.* **1986**, *41*, 3129–40.
56. Schutten, J. H.; Zwart, J. *J. Molec. Catal.* **1979**, *5*, 109–123.
57. Schutten, J.H.; Beelen, T.P.M. *J. Molec. Catal.* **1981**, *10*, 85–97.
58. Cullis, C.F.; Hopton, J.D., Trimm., D.L. *J. Appl. Chem.* **1968**, *18*, 330–344.
59. Oswald, A.A.; Wallace, T.J. In *The Chemistry of Organic Sulfur Compounds*; Kharasch, N.; Meyers, C.Y., Eds.; Pergamon Press: 1966, vol.2 p. 205.
60. Leung, P.K.; Hoffmann, M.R. *Environ. Sci. Technol.* **1988**, *22*, 275–282.
61. Bentvelzen, J.M.; McKean, W.T.; Gratzi, S.Y., Lin S.Y., Tucker, W.P. *Tappi* **1975**, *58*, 102–105.
62. Kreevoy, M.M.; Eichinger,B.E.; Stary, F.E.; Katz E.A., Sellstedt, J. H. *J. Org. Chem.* **1964**, *29*, 1641–44.
63. Harkness, A.C.; Murray, F.E. *Atmos. Environ.* **1970**, *4*, 417–424.
64. Pearson, R.G. Hard and Soft Acids and Bases, Dowden, Hutchison, and Ross, Stroudsburg, 1973, p. 20.
65. Pearson, R. G. *Chemistry in Britain* **1967**, *3*, 103–107.
66. Lim, P.K.; Giles, D.W.; Cha, J.A. *Chem. Eng. Sci.* **1986**, *41*, 3141–3153.
67. Capozzi, G.; Modena G. In *The Chemistry of the thiol group*; Patai S., Ed.; Wiley-interscience: New York, 1974, Pt 2, p. 785.
68. Leung, R.S.K., Hoffmann M.R., *J. Phys. Chem.* **1985**, *89*, 5268–71.
69. Kirchner, K.; Litzenburger, W. *Chem. Eng. Sci.* **1982**, *37*, 948–950.
70. Vinogradov, P.B. *Hyg. Sanit.(USSR)* **1966**, *31*, 13–17.
71. Peyton, T.D.; Steelé, R.V., Mobey, W.R. Stanford Research Institute, Report 68-01-2940, 1976.
72. Hanley, A.V., Czech, F.W. In *Analytical Chemistry of Sulfur and Its Compounds*; Karchmer, J.H., Ed.; Wiley: New York, 1970, Vol. 1 Chapt. 8
73. Adewuyi, Y.G.; Carmichael, G.R. *Environ. Sci. Technol.* **1987**, *21*, 170–177.
74. Hovenkamp, S.G. *J. Polym. Sci. Pt C* **1963**, No. 2, 341–355.
75. Thompson, H.W.; Kearton, C.F., Lamb, S.A. *J. Chem. Soc.* **1935**, 1033–37

76. Elliot, S.; Lu. E.; Rowland, F.S. *Geophy. Res. Lett.* **1987**, *14*, 131–134
77. Ferm, R.J. *Chem. Rev.* **1957**, *57*, 621–637.
78. Philipp,B.; Dautzenberg, H. *Z. Physik. Chem.* **1965**, *229*, 210–224.
79. Hoffmann, M.R.; Edwards, J.O. *J. Phys. Chem.* **1975**, *79*, 2096–98.
80. Risenfeld, E.H., Egidius; T.F. *Z. Anorg Chem.* **1914**, *85*, 217–246.
81. Penkett S.A. *Nature Phys. Sci.* **1972**, *240*, 105–106.
82. Hoigné, J.; Bader H., Haag, W.R., Staehelin, L. *Water Res.* **1985**, *19*, 993–1005.
83. Kibbel, W.H., Raleigh C. W., Shepherd, J.A. *27th Annual Ind. Wastes Conf.* **1972**, *Pt2*, p. 824.
84. Hoffmann, M.R. *Environ. Sci. Technol.* **1977**, *13*, 1406–14
85. Satterfield, C.N.; Reid, R.C.; Briggs, D.R. *J. Am. Chem. Soc.* **1954**, *76*, 3922–23.
86. Millero, F.J.; LaFerriere, A.; Fernandez M.; Hubinger,S. *Proc. 194th ACS Mtg* **1987**, p. 103.
87. Chen, K.Y.; Morris J.C. *J. Sanit. Eng. Div., Proc. ASCE* **1972**, *98*, 215–227.
88. Hoffmann, M.R.; Lim, B.C. *Environ. Sci. Technol.* **1979**, *13*, 1406–1414.
89. Black, A.P.; Goodson, J.B. *J. Amer. Works Assoc.* **1952**, *44*, 309–316.
90. Choppin, A.R., Faulkenberry, L.C. *J. Amer. Chem. Soc.* **1937**, *59*, 2203–2207.
91. Belviso, S.; Mihalopoulos, N.; Nguyen, B.C. *Atmos. Environ.* **1987**, *21*, 1363–1367.
92. Sze, N.D., Ko, M.K.W. *Atmos. Environ.* **1980**, *14*, 1223–1234
93. Grosjean, D. *Environ. Sci. Technol.* **1984**, *18*, 460–468.
94. Andreae, M.O. *Limnol. Oceanogr.* **1980**, *25*, 1054–1063.
95. Slatt,B.J.; Natusch, D.F.S.; Prospero, J.M.; Savoie, D.L. *Atmos. Envron.* **1978**, *12*, 981–991.
96. Dacey J.W.H.; Wakehma, S.G.; Howes, B.L. *Geophys. Res. Lett.* **1984**, *11*, 991–994.
97. Seidell, A. *Solubility of Organic Compounds*; Van Nostrand: New York, 3rd, Ed. Vol. 2, p 442.
98. Rasmussen, R.A.; Hoyt, S.D.; Khalil, M.A.K. *2nd Symp. Amer. Meteor. Soc.,* **1982**, p. 265.
99. Douabul, A.A., Riley J.P. *Deep Sea Research* **1979**, *26A*, 259–268.
100. Gerrard, W. *Gas Solubilities, Widespread Applications*; Pergamon Press: New York, 1980, p. 161.
101. Adewuyi, Y.G. M.S. Thesis, University of Iowa, Iowa City, 1980.
102. Adewuyi, Y.G.; Carmichael, G.R. *Atmos. Environ.* **1982**, *16*, 719-729.

RECEIVED October 26, 1988

Author Index

Affiliation Index

Subject Index

Production by Joyce A. Jones
Indexing by Deborah H. Steiner

Elements typeset by Hot Type Ltd., Washington, DC
Printed and bound by Maple Press, York, PA